Knowledge Management in Emerging Economies:
Social, Organizational and Cultural Implementation

Minwir Al-Shammari
University of Bahrain, Bahrain

Information Science REFERENCE

INFORMATION SCIENCE REFERENCE
Hershey · New York

Director of Editorial Content: Kristin Klinger
Director of Book Publications: Julia Mosemann
Acquisitions Editor: Lindsay Johnston
Development Editor: Dave DeRicco
Publishing Assistant: Casey Conapitski
Typesetter: Casey Conapitski
Production Editor: Jamie Snavely
Cover Design: Lisa Tosheff

Published in the United States of America by
Information Science Reference (an imprint of IGI Global)
701 E. Chocolate Avenue
Hershey PA 17033
Tel: 717-533-8845
Fax: 717-533-8661
E-mail: cust@igi-global.com
Web site: http://www.igi-global.com

Library of Congress Cataloging-in-Publication Data

Knowledge management in emerging economies : social, organizational and cultural implementation / Minwir Al-Shammari, editor. p. cm.
Includes bibliographical references and index. Summary: "This book focuses on knowledge management theoretical models and empirical research findings for developing economies to create an understanding of the social, organizational, and cultural implementation aspects of knowledge management in the context of developing economies"--Provided by publisher. ISBN 978-1-61692-886-5 (hbk.) -- ISBN 978-1-61692-888-9 (ebook) 1. Knowledge management--Developing countries. I. Al-Shammari, Minwir, 1962- HD30.2.K637244 2011
658.4'038--dc22

British Cataloguing in Publication Data
A Cataloguing in Publication record for this book is available from the British Library.

Table of Contents

Section 3
Knowledge Management Practices

Section 4
Knowledge Management Issues

Detailed Table of Contents

This chapter explains the use of web mining for gathering strategic competitive intelligence. It first discusses some concepts relating to web mining. This is followed by a discussion of an exploratory study on how web mining for strategic intelligence is actually used by competitive intelligence professionals in some South African organizations. The findings are discussed under the three major themes, namely general intelligence practices; Web-based intelligence; and skills development and education. The bulk of the chapter discusses a practical methodology for using web mining to gather strategic business intelligence. Some practical comments about the impact of particular circumstances of emerging African countries on this methodology are also given.

In the Czech Republic, Knowledge Management-related problems occur at all knowledge stages with difficulties more predominant at an organisational level. In principal, they originate from the lack of an utilisable and detailed KM implementation methodology; varying perceptions of KM coupled with the notion, that KM equates and is limited to information technologies and the like. As a reaction to these problems, the Faculty of Informatics and Management at the University of Hradec Králové (FIM UHK) developed a new KM implementation methodology. It aims to provide interested parties with a set of ground rules distributed among a number of phases. Further development is underway to elaborate the method by integrating project management tools, benchmarking exercises and critical success factors.

This study reports the findings from an interpretive exploration of tacit knowledge construction and application among the Malaysian academicians. In the knowledge era, higher learning institutions are increasingly exposed to marketplace pressures similar to other business. Therefore, it is pertinent for higher learning institutions to leverage on their knowledge for sustainability. However, efforts to leverage on tacit knowledge faced difficulties due to lack of understanding of tacit knowledge within its contextual boundary. The ambiguities surrounding tacit knowledge require deeper understanding of its construction and application to manage it as an important resource for academic institutions. The study discovered that tacit knowledge is constructed through implicit learning from the accumulated past experiences. Tacit knowledge is applied by the academicians as their adaptation strategies in appropriate job situation to fulfil the academic excellence norm.

Chapter 6
Supply Chain Knowledge Integration in Emerging Economies 104
Ryan Atkins, Queen's University Belfast, Northern Ireland

Firms in developed economies have been purchasing from firms in emerging economies for years, but they typically purchase low value, low margin items. Opportunities have arisen for emerging economy manufacturers to supply higher value products and services to firms in developed economies. In today's global, knowledge-based economy, suppliers must have cutting edge knowledge, and they must constantly upgrade their knowledge to maintain a competitive advantage. When supplying high-value products, complex knowledge based interactions between the buyer and supplier is necessary. A conceptual model is developed in this chapter, proposing that if suppliers in emerging economies are committed to long term relationships with developed economy customers, they can increase the level of knowledge integration in the relationship, and in turn, improve performance. The primary contribution of this chapter is to show that firms in emerging economies can achieve sales growth by becoming critical links in today's global, knowledge based supply chains.

Chapter 7
A Social and Technical Investigation of Knowledge Utilization from a Repository Knowledge Management System 122
Kamla Ali Al-Busaidi, Sultan Qaboos University, Oman

The need for developing countries to empower themselves through knowledge management (KM) cannot be underestimated. Knowledge utilization is the process where the benefits of knowledge management systems (KMS) are actually realized. This study investigated the effects of knowledge quality, system level, system form, service quality, management support, knowledge trust and rewards policy on knowledge utilization behavior, and benefits of knowledge utilization from a repository KMS in an Omani organization. Findings suggested that determinants of knowledge utilization were not related to system technical characteristics or information technology service quality; they were related to knowledge quality, management support, knowledge trust, and rewards policy. Results also suggested that knowledge utilization results in individual benefits. Findings implied to practitioners that the development of high-level knowledge management systems does not guarantee their success. Organizations need to establish a knowledge-oriented culture and develop standards that ensure high knowledge quality to promote KMS and knowledge utilization behavior.

Section 2
Knowledge Sharing and Transfer

This chapter examines how knowledge management (KM) develops within the Malaysian housebuilding industry. The objectives of this chapter are twofold. First, it explains how innovation is able to encourage or influence housing developers in generating the new Build Then Sell (BTS) system. Second, it identifies the characteristics of successful adopters so that other developers can imitate them and funnel resources to those areas that successful adopters emphasise. Focus group and semi-structured interviews were conducted to obtain in-depth responses from one house buyers' organisation and successful BTS adopters. The effort to implement BTS is seen as being the result of a strategic investment in new knowledge by developers responding to the pressure they have experienced on the part of the house buyers and the government. Discussion and the sharing of both explicit and tacit knowledge between the main stakeholders have generated the finer details of the BTS. A strong learning culture and the development of KM tendencies also lead to successful BTS implementation.

The need for the adoption of knowledge management practice, particularly knowledge sharing is pressing in any modern organisation today in order to ensure their survival in this highly competitive world. It is evidence that knowledge sharing has given untold advantages to organisations such as the ability to make good decisions and formulate new strategies. This in turn enables organisations to remain competitive in their respective industry. Knowledge bears no value unless it is shared and well disseminated, particularly among workers within the organisation. Despite its advantages, knowledge sharing seems to be taken lightly by many organisations in Malaysia and only a few studies have been carried out. Therefore, the authors take an initiative to identify factors affecting knowledge sharing in Malaysian public sector. An extensive literature review has been conducted in this area in order to determine such factors. Synthesizing from the literatures, this paper proposes a theoretical framework that takes into consideration the individual, organizational and technological dimensions that might affect knowledge sharing practise in Malaysian public sector. It is not the intention of this paper to include the findings of this study and therefore the current study only develops a framework based on literature review. The findings of the study will be reported upon completion of the entire study.

high-growth companies. Technical, organizational and behavioral practices, successes and problems of knowledge management in Mozzartbet are covered, and future plans of top managers in that area are revealed. In this case study we described a relaxed, academic-like organizational culture, organizational structure, strategy of development, industry and regional environment, and other internal and external factors important for knowledge management. Afterwards we presented conclusions about knowledge management in Mozzartbet and future directions, according to contingency approach

In offshore business process outsourcing (BPO) arrangements between US client firms and Indian vendor firms, a "culture gap" has been reported between the organizations. This gap originates from (1) macro level differences between the two firms in their operating procedures and (2) micro level differences in the cultural proclivity of the individual workers. This gap often hinders the much needed bi-directional knowledge sharing in the outsourcing organization. To close this gap and improve knowledge sharing, effective organizational practices are needed. This chapter studies a case where a set of technical support processes were outsourced by an US based multinational client to an Indian vendor. Data was collected using interviews of client-side, process and business managers to understand what challenges were faced in the organization after BPO and what practices facilitated cross cultural knowledge sharing. A set of organizational enablers and collaboration tools were reported by the interviewees to be successful in bridging the gaps and enhancing bi-directional knowledge sharing. This case demonstrates the benefits of three organizational practices – (1) bridging, (2) bonding and (3) linking, that build social capital in the form of relational norms and connections among workers. These relational practices, together with social capital, complement the use of collaboration tools and technologies to improve knowledge sharing between the client and vendor teams.

Within small-size firms, it is too difficult to understand how the process of knowledge transfer happens. It is through a close observation of their mind-set and ethos that it is possible to construct a conceptual map of their transfer of knowledge processes. The relations among footwear manufacturers enabled the generation and sharing of knowledge. This chapter seeks to introduce the behavior of knowledge management and its source by studying the social behavior of micro-entrepreneurs from Peru and Brazil that create relationships, practice communities and other forms of knowledge sharing. Best knowledge-based practices emerged from the adaptation and imitation of others. Three ways were recognized: distance-to-market; distance-to-consumer, and inner circle. Three ways of social memory creation within the core of the small-firms were observed: memory of product, market report and social memory.

Section 4
Knowledge Management Issues

Organizational knowledge and human capital are increasingly regarded as key levers of competitive advantage in today's global, dynamic and complex business environment. People are the ultimate knowledge creators and bearers; although organizations may have memory systems, they do not think by themselves. Human resource management (HRM) has a strategic role facilitating knowledge management (KM) initiatives, specifically in activities such as: recruitment, training and compensation. Although many studies raise the question of how HRM shapes KM, limited research in Latin America has explored this matter. This chapter aims to explore how HRM policies and practices have been designed in three organizations located in Mexico, in order to support KM initiatives. The findings suggest that the strategic role of HRM in supporting KM initiatives must be expanded. HRM policies, practices, and compensation systems, such as incentives and bonus packages to motivate employees to create and share knowledge, need to be redefined. Recruitment, selection and, training and development must be tailored to obtain a successful implementation of the KM program.

A challenge for developing countries is to enter the global economy. This is impeded by weak infrastructure, poor educational and other formidable barriers as these countries embark on essential projects to improve living standards and ultimately create some sort of technological base. This paper describes perceptions of knowledge-based technology management in Afghanistan, one of the poorest and least developed countries in the world. Afghanistan's politics and history have always confronted managers with unique demands. The paper examines the perceptions of Afghan managers regarding technology management by studying the relationship between the importance of different factors in managing new technology, and the extent to which a manager can control them. An importance-control grid framework is used to isolate individual parameters and to assess these in relation to a manager's environment. The research highlights imbalances between importance and control, and suggests reasons there for. Some broader implications for managers are also discussed.

Preface

OVERVIEW

Business organizations of the new economy operate in a highly competitive and turbulent business environment. Organizations are faced with fierce competitive pressures arising from technological advancements, globalization of markets, deregulation, liberalization, privatization, as well as rapid political and governmental changes that place huge demand on firms to remain flexible, responsive, and innovative (Drucker, 1995; Teece et al., 1997).

The resource-based view of the firm recognizes the importance of organizational resources and capabilities as a principal source of creating and sustaining competitive advantage in market competition. According to this approach, resources are the main source of an organization's capabilities, whereas capabilities are the key source of its competitive advantage (Grant, 1991).

The dramatic shifts in the knowledge economy and global competition makes the most valuable resource that creates sustainable competitive advantage for a business organization is what it knows not what it owns. Knowledge is evolving in the new economy as a distinctive core competency and as a main source of wealth for many countries as well as companies. Important knowledge can represent knowledge on products, customers, markets, projects, as well as processes. Managing knowledge, as a strategic corporate resource, can help in the creation of sustainable competitive advantage for today's competitive and globalized business world.

KNOWLEDGE AS A STRATGEIC CORPORATE RESOURCE

The beauty of using knowledge as a base of sustainable competitive advantage is that it is a non-depleting resource. Unlike other business resources that diminish once shared, knowledge development follows the law of increasing returns - the more knowledge is used, the more value it creates. Furthermore, the more knowledge is shared, the more new knowledge is generated. Knowledge sharing, therefore, is becoming a successful way to increase the value of 'intellectual assets' in improving knowledge-intensive processes and adding value to customers and profitability to the business. There are several key attributes of knowledge, which must be factored into KM practices (Kluge, et al, 2001; and Davenport and Prusak, 2000):

- Subjectivity: context and individual background shape the interpretation of knowledge.
- Transferability: knowledge can be extracted and transferred to other contexts.

- Embeddedness: knowledge is often in static and buried form that makes it difficult to extract.
- Self-reinforcement: knowledge is the only unlimited resource, the one asset that its marginal utility increases and does not decrease once used or shared.
- Perishability: knowledge can become outdated.
- Serendipity (spontaneity): knowledge can develop unexpectedly in a spontaneous or incidental process (e.g., water cooler knowledge exchanges).
- Velocity: speed with which knowledge moves through an organization (e.g., computers and networks excel at enhancing the velocity of knowledge).
- Viscosity: richness or stickiness of detailed or subtle knowledge transferred (e.g., apprenticeship or mentoring relationship).

KNOWLEDGE MANAGEMENT IN EMERGING ECONOMIES

As knowledge is one of the principal organizational resources, effective usage and management of knowledge resources is a prerequisite of gaining sustainable distinctive advantage. Knowledge Management (KM) value chain starts with generation and ends with application. It involves strategic planning, capturing or generation of knowledge, sharing it among partners or suppliers, and using it for various business applications.

Many international development agencies have put forward contributions for knowledge-based sustainable national development in different regions of the world. The World Bank Institute (Aubert and Reiffers, 2003) offered its conceptual model that is composed of the following five 'pillars':

- An economic and institutional model that provides incentives for the efficient creation, dissemination, and use of knowledge to promote growth and increase welfare;
- An educated and skilled population that can create and use knowledge;
- An innovation system of firms, research centers, universities, consultants, and other organizations that can tap into the growing stock of global knowledge, adapt it to local needs, and transform it into products valued by markets;
- A dynamic information infrastructure that can facilitate the effective communication, dissemination, and processing of information.
- An intangible component that makes a society function efficiently and move forward, such as the capacity to formulate a vision, the level of trust and self-confidence, and the appropriateness of guiding values.

Although KM ideas have been available to managers for developed countries for decades, they have only recently been integrated into the mainstream of corporate planning and decision making process in emerging economies that not only encompass a large proportion of the earth's population, but they are a major and growing source of economic output. As many emerging economies are shifting from information-based to knowledge-based economies, they need to be aware that successful implementation of business concepts and practices in industrial Western economies may not necessarily replicate itself with success in emerging economies, if implementation was not carefully customized to fit the available organizational infrastructure and to the specific society and culture.

Knowledge management orientations and practices may differ from one socio-cultural context to another. Lessem and Neubauer (1993) argued that the Anglo-Saxon orientation to knowledge creation can be categorized as pragmatic, the French-Nordic as rational, Japanese as holistic, Latin as humanistic. The work of Zhu (2004) questioned the claim that KM is a universal concept and, instead, proposed KM in cross-cultural contexts that clearly reveal the unique associated problematic of different KM styles: the American, the Japanese, the European, and the Chinese (Figure 1). Al-Shammari (2007) added a new profile of KM for the Arab region (Figure 2).

Zhu (2004) presented evidence that, despite growing overlaps, heterogeneity among KM styles is likely to continue due to differences in histories, national cultures, and institutional forces. As a way out, he called for an interactionist, rather than a universal, strategy of constructing, connecting, and sharing knowledge across-cultural contexts, wherein cultural differences and diversity are important sources of competence rather than obstacles to overcome. The KAFRA model for KM (Okunoye and Bertaux, 2006) also called for an increased sensitivity to global diversity that could serve as a link between organizations and their environments in the creation of KM. The model ensures that KM is approached with consideration to the environment in which organizations operate. In the Tunis KM forum, papers have addressed the theory and practice of KM, with a distinctive focus on two complementary perspectives (Francophone and Anglo-Saxon), but were able to integrate the discussion within the local environmental settings and needs of specific organizations (Passerini, et al., 2007).

Figure 1. Connecting and contrasting cross-cultural KM styles

Dimension	American	Japanese	European	Chinese
Motto	I succeed, therefore, I am	I love, therefore, I am	I practice, therefore, I am	I learn, therefore, I am
Mentality	Performanism	Groupism	De(Constructivism)	Pragmatism
Ideal-type	Knowledge as resource	Knowledge as relationships	Knowledge as power	Knowledge as virtue
Embodiment	Knowledge base	Knowledge company	Knowledge agent	Knowledge life
Mechanism	Knowledge economy	Knowledge culture	Knowledge discourse	Knowledge governance
Aim	Near-term gains	Long-term advantage	Legitimacy	Kingliness-sageliness
Focus	Explicit-objectified knowledge	Tacit-subjective knowledge	Situated-constructed knowledge	Useful-workable knowledge
Strategy	Leverage	Creation	Politicization	Integration
Process	Re-using	Converting	Enacting	Contextualizing
Means	Rationality Technology Markets	Vision/emotion Trust/care Socialization	Identity/meaning Participation Negotiation	Material-cognitive Psycho-cognitive Socio-political
Metaphor	Picking low hanging fruit	Nurturing an originating ba	Stories in the market	The Master is free from four negatives
Exemplar	Zack (1999). Managing codified knowledge	Nonaka & Takeuchi (1995). The knowledge creating company	Balckler & McDonald (2000). Power, mastery and organizational learning	Zhu (2001). What is to be managed: knowledge, knowing or knower, and does it matter?

Source: Zhu, Z. (2004). Knowledge management: towards a universal concept or cross-cultural contexts? Knowledge Management Research & Practice, (2), 67-79.

Figure 2. A profile of Arab KM syle

KM Dimension	Arab KM Style	Description
Motto	I codify, therefore, I am	The KM culture that prevails is 'data processing'-centric, wherein 'codification' is emphasized and 'personalization' is de-emphasized.
Mentality	Functionalism/Incrementalism	Knowledge development is not looked at from a holistic process-perspective; rather, development of KM projects on a gradual and localized exploitation basis prevails. Silo knowledge mentality among employees in different departments is enhanced by functional structures. Silo mentality hinders knowledge sharing among employees.
Ideal-type	Knowledge as forgotten resource	Knowledge is rarely considered as a valuable resource to companies. Competition is traditionally product-centric, but is showing gradual concern with achievement of customer-centric competitive advantage. Personal knowledge is not usually shared.
Embodiment	Knowledge system	Emphasis on formal KM systems to capture/create structured knowledge, which in turn is stored, shared, and disseminated.
Mechanism	Knowledge outsourcing	Knowledge is created through internal and external sources. External sourcing dominates the knowledge scene and refers to out-of-the-house development or acquisition of ready-made KM systems. Sometimes, it follows a vendor-push mode where KM systems are pushed by vendors rather than pulled by companies.
Aim	Diffusion of environmental pressures	A critical response activity to environmental pressures such as competitors, liberalization of markets, advancement of ICT, and growing power of customers. Sometimes, KM projects end-up as vendor-push ICT projects introduced merely to follow the latest 'fashion' of the market. Vendors play a significant role in selling ideas of new ICT-based solutions that may end-up with little or no alignment with business strategies.
Focus	Explicit-structured knowledge	Explicit- structured knowledge is emphasized, whereas human and tacit-subjective knowledge is de-emphasized.
Strategy	Ad hoc	KM strategies are chaotic and still in their infancy. Lack of clear identification of business requirements, and more focus on computerized systems. A formal KM strategy hardly exists at a corporate level, although senior managers constantly stress the need to create, share and distribute knowledge.
Process	Buying	Vendors sell KM solutions; formulate KM strategies; identify key business deliverables and present recommendations of process and structural changes to best exploit knowledge.
Means	ICTs People Vendors	ICTs People Vendors
Metaphor	Rich in data, yet poor in knowledge	Plenty of data exist about customers, for example, but rarely have these data been converted into knowledge and used to create sustainable competitive advantage.
Exemplar	Al-Shammari (2005). Implementing a knowledge-enabled CRM strategy in a large company	

Source: Al-Shammari, M. (2008). Toward a knowledge management strategic framework in the Arab region, International Journal of Knowledge Management, 2008, 4(3), 44-63.

TARGET AUDIENCE

This book will be useful for professionals, researchers, executives, academics, policy makers, government executives, and corporate executives concerned with the management of knowledge, both at the corporate and national levels, in developing economies.

ORGANIZATION OF THE BOOK

This book is divided into four sections, each is dedicated to an area of knowledge manageemnt research in emerging economies. The following paragrpahs describe the importance of each section.

Section 1 adresses KM generation, integration, and utilization. Lynnda Wagner and Jean-Paul Van Belle discuss an exploratory study on how web mining for strategic intelligence is actually used by competitive intelligence professionals in South African organizations and discuss a practical methodology for using web mining to gather strategic business intelligence. Richard Brunet-Thornton and Vladimír Bureš explain how the Faculty of Informatics and Management at the University of Hradec Králové in the Czech Republic developed a new KM implementation methodology. Rosana Silveira Reis Ylenia Curzi analyze knowledge integration in the creative process of six product development projects, carried out by globally distributed teams belonging to a Swedish company working with France and Brazil; and an Italian company working with Tunisia. Ian Hipkin addresses the interaction between technology transfers, production competence and knowledge generation in enhancing performance in four case studies from the manufacturing sector, two from South Africa and two from the United Kingdom. Haryani Haron and Rose Alinda Alias report the findings from an interpretive exploration of tacit knowledge construction and application among the Malaysian academicians as an adaptation strategy to fulfil the academic excellence norm. Ryan Atkins develops a conceptual model, proposing that if suppliers in emerging economies are committed to long term relationships with developed economy customers by becoming critical links in today's global knowledge based supply chains, they can increase the level of knowledge integration in the relationship, and in turn, improve sales growth. Kamla Al-Busaidi investigate the effects of knowledge quality, system level, system form, service quality, management support, knowledge trust and rewards policy on knowledge utilization behavior, and benefits of knowledge utilization from a repository KM systems in an Omani organization.

Section 2 discusses knowledge sharing and transfer. Nor'Aini Yusof and Mohd Wira Mohd Shafiei examine how innovation encourages or influences housing developers in generating the new Build Then Sell (BTS) system and identify the characteristics of successful adopters so that other developers can imitate. Zawiyah Mohd Yusof and Mohd Bakhari Ismail propose a theoretical framework that takes into consideration the individual, organizational and technological dimensions that might affect knowledge sharing practise in Malaysian public sector. Abdul Kadir Othman Hazman Shah Abdullah offer several testable propositions to explain the influence of emotional intelligence on tacit knowledge sharing in service businesses. Khairul Shafee B Kalid explores the usage of knowledge-embedded stories in a Malaysian institute of higher learning and provides insights on the culture of storytelling as a method of knowledge transfer. Kalsom Salleh, Syed Omar Sharifuddin Syed Ikhsan, and Syed Noh Syed Ahmad present a knowledge sharing model that interconnects KM enablers and knowledge sharing process for KM implementation success in the public sector accounting organization in Malaysia.

Section 3 looks into KM practices. Tatiana de Almedia Furquim and Sueli do Amaral highlight the importance of understanding the practices of KM and describes knowledge acquisition, protection, transfer, and application practices in the context of a Brazilian software organization. Mladen Čudanov and Kathrin Kirchner discuss technical, organizational and behavioral practices, successes and problems of KM in a high-growth company in Serbia. Biswadip Ghosh studies cross-cultural KM practices in a case where a set of technical support processes were outsourced by an US based multinational client to an Indian vendor. Jose Manuel Cárdenas Medina and Mauro de Mesquita Spinola study the behavior of KM and the social behavior of micro-entrepreneurs from Peru and Brazil that create relationships, practice communities and other forms of knowledge sharing.

Section 4 discusses KM issues. Laura Zapata-Cantú, Jacobo Ramírez, and José Luis Pineda discuss how human resource management policies and practices have been designed in three organizations located in Mexico, in order to support KM initiatives. Ian Hipkin describes perceptions of Afghan managers regarding knowledge-based technology management by studying the relationship between the importance of different factors in managing new technology, and the extent to which a manager can control them. Juha Kettunen and Manodip Ray Chaudhuri demonstrate how knowledge management can be used as a general framework which integrates the approaches of strategic and change management in India. Neeta Baporkiar explains different perspectives on KM in relation to entrepreneurship through two cases from India.

CONCLUSION

There is greater realization in world economies of opportunities in the creation, acquisition, distribution, and use of organizational knowledge. The importance of knowledge to business firms has created tough 'rivalries' among competitors to utilize knowledge as a major weapon in competition. Knowledge is newly evolving as a main non-replenishing and difficult-to-imitate source of distinctive core competencies in emerging markets that could help them in the creation of sustainable competitive advantage. Once knowledge embedded in organization's core business processes and strategic decision-making processes in emerging markets, it may help to create a profound impact on organizational performance, productivity, and possibly competitive advantage. Success of KM in emerging economies shouldn't be confined to the success in technical implementation of ICT-based systems, but should be expanded to the adoption side relative to its local context. Whereas successful implementation of a KM system relates to technical success in the development of a KM solution, successful implementation of a KM strategy relates to the wider business-context success, from an internal perspective (e.g., operational efficiency) as well as an external perspective (e.g., customer satisfaction).

Emerging economies in particular, many of which are known for their lower input costs of scarce resources, may find utilization of knowledge, as a non-depleting resource, the key for achieving sustainable competitive. While the literature of KM initiatives conducted in Western countries is abundant, it is scanty in developing countries. This book seeks to cover this gap in the KM literature by focusing on KM theoretical models and empirical research findings for developing economies. Successful implementation of business concepts and practices (i.e. KM) in Western environmental contexts may not necessarily replicate with success in a non-Western context, if implementation was not carefully customized to fit the available organizational infrastructure and to the specific society and culture. This book specifically seeks to understand the social, organizational, and cultural implementation aspects of KM in the context of developing economies, and to discuss issues, challenges, and trends surrounding this implementation. This book covers a wide variety of KM applications and models in government, non-government, and private sector entities.

Minwir Al-Shammari
University of Bahrain, Bahrain

REFERENCES

Al-Shammari, M. Toward a Knowledge Management Strategic Framework in the Arab Region, *International Journal of Knowledge Management*, 2008, 4(3), 44-63.

Aubert, J.E., & Reiffers, J.L. (2003). *Knowledge economies in the Middle East and North America: toward new development strategies*. World Bank Institute Series, Washington D.C.: The World Bank Press,

Davenport, T.H., & Prusak, L. (2000). *Working Knowledge: how organizations manage what they know*. Boston: MA: Harvard Business School Press.

Drucker, P. (1995). *Managing in a time of great change*. New York: Truman Tally.

Grant, R. M. (1991). The resource-based theory of competitive advantage: implications for strategy formulation. *California Management Review*, (33), 113-135.

Kluge J., Wolfram S, & Licht T. (2001). *Knowledge Unplugged: The McKinsey & Company global survey on knowledge management*, Palgrave Macmillan.

Lessem and Neubauer (1993). *European Management Systems: Towards Unity Out of Cultural Diversity*, McGraw-Hill.

Okunoye, A. & Bertaux, N., (2006). KAFRA: A Context-aware framework of knowledge management in global diversity. *International Journal of Knowledge Management*, 2(2), 26-45.

Passerini, K., Jennex , M.E., & Khaldoun, B.T., (2007). Perspectives from North Africa: the Tunis-2006 knowledge management forum. *International Journal of Knowledge Management*, 3(1), 100-107.

Teece, D., Pisano, G., & Shuen, A. (1997). Dynamic capabilities and strategic management. *Strategic Management Journal*, 18(7), 509-533.

Zhu, Z. (2004). Knowledge management: towards a universal concept or cross-cultural contexts? *Knowledge Management Research & Practice*, (2), 67-79.

Section 1
Knowledge Generation, Integration, and Utilization

Chapter 1
Web Mining for Strategic Competitive Intelligence:
South African Experiences and a Practical Methodology

Lynnda Wagner
University of Cape Town, South Africa

Jean-Paul Van Belle
University of Cape Town, South Africa

EXECUTIVE SUMMARY

This chapter explains the use of web mining for gathering strategic competitive intelligence. It first discusses some concepts relating to web mining. This is followed by the findings of an exploratory study on how web mining for strategic intelligence is actually used by competitive intelligence professionals in some South African organizations. The findings are discussed under the three major themes, namely general intelligence practices; Web-based intelligence; and skills development and education. The bulk of the chapter proposes a practical methodology for using web mining to gather strategic business intelligence. Some practical comments about the impact of particular circumstances of emerging African countries on this methodology are also given.

INTRODUCTION

In today's rapidly changing global business environment, it is important for companies to conduct systematic local and global environmental scans to equip executive management with the necessary knowledge to plan, lead, organize, and control their business. Ideally, organizations employ mechanisms for efficient strategic intelligence collection to empower management to forecast and influence the companies' future direction. This is especially important for smaller businesses in emerging countries, which increasingly have to compete globally but are faced with infrastructural limitations such as high bandwidth costs and more difficult access to information about global technology and market trends (Viviers, Saayman, Muller & Calof, 2002).

Although the Web can be an excellent source of current, insightful, and easily accessible in-

DOI: 10.4018/978-1-61692-886-5.ch001

formation, the useful, relevant and high-quality information is often hard to find systematically in the ocean of lower quality information. Mining the Web for nuggets of intelligence may give an organization competitive advantage, and should form an integral part of the strategic planning process. Companies now have the opportunity to devise a web mining system to rapidly assess its external environment. This will enable companies to be aware of trends and/or changes outside of their operations that can negatively or positively impact their business. This empowers businesses to become proactive towards emerging market issues. Also, it allows management to detect opportunities and threats, and evaluate how to adapt their operations accordingly. Perhaps most of all, it facilitates decision-making with regards to the future direction of the company. Overall, the information gained can be used to formulate corporate strategies.

Not surprisingly, the Internet has become one of the top sources of intelligence for intelligence gathering professionals. For example, the Internet Intelligence Index (I3 available from www.fuld.com), is an index of about 600 intelligence-related websites divided into three areas: 1) general business Internet resources; 2) industry-specific Internet resources; and 3) international Internet resources (Kennedy, 1998).

Unfortunately, the amount of information on the Web is staggering and growing fast. As such, it is not practical to cover the full extent of the Web manually. Although search engines are powerful tools, they are still incapable of providing tailored information, which means that users must still sift through copious search results before finding something within the right scope. Technologies are being developed to enable online data analysis with the ultimate goal being to gain insight from online data. Thus, Web mining has emerged as an active area of research and development.

Web mining provides extremely useful and automated tools and techniques to gather the information required for competitive intelligence, business intelligence and for the strategic information processes which can be readily incorporated as part of any knowledge management process. Thus any knowledge management strategy should carefully consider competitive analysis and web mining as an integral part of the knowledge management process (Liebowitz, 2006). Many commercially available tools already exist (see Table 1) and most offer easy integration points into larger knowledge management systems.

The objective of this chapter is to suggest a practical way to implement a process for supplying strategic managers with effective strategic competitive intelligence through web content mining. The suggested process (or 'methodology') was extracted from the literature but moderated against real world needs and possibilities through an exploratory survey on how Web data is collected and assimilated into intelligence by South African business intelligence professionals. The process, aimed to be fully automatable, consists of a number of steps as described in detail below. Some suggestions for supporting the process using various tools are given. Finally, some specific considerations for web content mining in the context of South Africa as an emerging market are given.

COMPETITIVE AND STRATEGIC INTELLIGENCE

Currently, there is a lot of interest in and around *competitive intelligence* (CI), which usually forms part of the business function known as business intelligence (Liebowitz, 2006). CI is here understood to be "public and private information regarding competitors' present and future activities as well as the behaviors of suppliers, customers, technologies, acquisitions, markets and the general industry environment" (Velder, Vaneck, Guynes & Capel, 1999, p.109). *Business intelligence* (BI) supports the information needs of an organization's internal operations (day-to-day processes, research and development (R&D), material and labor efficiencies) and external circumstances (customer needs, threats from competition, supplier reliability, domestic and international regulation, technology). Although BI covers a larger area than CI, a strong focus of BI is on gathering competitive information and thus the terms BI and CI are sometimes used interchangeably in practice (Thuraisingham, 2003).

Strategic intelligence involves the collection of information about the external factors that have the potential to impact the business' mission and strategic direction. Since companies operate in a dynamic, sometimes volatile, business environment, executives rely far more on external than internal information. Therefore, the focus of this research is on strategic intelligence which addresses an organization's external environment.

Strategic analysis deals with the evaluation of strategic intelligence to form an overall picture of the factors influencing the organization's environment. The information is collected, disseminated, digested and incorporated into policy decisions. Environmental scanning enables a business to determine trends in the macro-environment that are important to the company, detect potential threats and opportunities or changes for the business as a result of those trends, be future-thinking and alert management to converging trends. Generally, when organizations conduct an environmental scan, they look at two facets of the environment: the societal environment and the competitive, or industry, environment. The (elements within the) societal environment of interest are those macro-level forces that have an *indirect* impact on the company whereas the competitive environment has a *direct* impact on the organization. The societal environment deals with broader forces that do not necessarily impact upon an organization's operations but greatly influence its future direction. It includes economic trends, developments in technology, social pressures and changes in the government or political arena. The framework under the acronym of PEST – political, economic,

Table 1. Mapping of some commercially available web mining tools on the proposed process steps

Tool	Taxonomy	Metasearcher	Web Crawling	Web Clustering	Information Extraction	Text Miner
METIS	✓		✓			
Temis Luxid	✓	·	·	·	✓	✓
KartOO	·	✓	·	✓	·	·
Autonomy	·	✓	✓	✓	·	✓
Clarabridge	·	·	✓	·	✓	✓
Convera NewOwl	·	·	✓	✓	✓	✓
ClearForest	·	·	·	·	✓	✓
InXight	·	·	·	✓	✓	✓
NStein	·	·	·	✓	·	✓
WordMap	✓	✓	·	·	·	✓
Copernic	·	✓	✓	·	·	·
SAS TextMiner	·	·	✓	✓	✓	✓
RapidMiner (open source)	·	·	·	·	✓	✓
Copernic	·	✓	✓	✓	·	·

social and technical – provides a good structure for a societal environmental scan.

Much of the information required for scanning the political and legal environment is available on the Internet. Government websites host a plethora of information on macro-level environmental forces such as public policy, business regulation, product safety, government contracts, census data about demographics and so forth. The technological environment can be assessed through an assortment of online publications. Business wires and press releases can be good sources of intelligence. Most companies have websites which can provide information about their products and services, mission and vision, prices, distribution, management, financial position, etc. Private companies without a Web presence can be researched through government sites that provide online access to public record information. Subscriber-based online forums such as Listservs and Mailing Lists offer yet more resources. One can learn about an industry and its key players through these roundtables. Similarly, newsgroups can draw interest groups within a niche market (Westphal, 2008).

From the preceding discussion, it is clear that CI forms a subset of the intelligence gathered in both the BI and strategic intelligence activities where BI emphasizes a mainly internal, short to medium term, operational or tactical focus and strategic intelligence takes a longer term, strategic view with stronger emphasis on external information (Liebowitz, 2006).

WEB CONTENT MINING

Web mining can be classified into three areas: content mining, structure mining and usage mining. Web content mining relates to discovering meaningful information from the contents of web documents. Web structure mining deals with the categorization of web pages based on the link structures. And web usage mining draws conclusions about how users interact with Web sites (Kosala & Blockhell, 2000). This paper focuses on the use of web content mining in gathering strategic intelligence. Because the use of web structure and web usage mining has different applications and purposes, the term "web mining" in the context of CI and strategic intelligence is generally understood to refer to web *content* mining in the literature and by the CI professionals, thus this usage has also been adopted in this chapter.

Generally, Internet search engines are inadequate for web content mining. Common problems include: low precision (due to irrelevant search results), low recall (missing results), inability to create new knowledge from Internet data and lack of personalization of the information (Kosala & Blockhell, 2000). The low recall is directly linked to the inability of many search engines to index the hidden web. The hidden Web refers to a significant portion of the web not indexed by engine spiders – typically pages that are dynamically created from databases (Adams, 2001). Even the use of meta-search engines – which combine and process the results of many different individual search engines – does not remedy most of the above problems. The next few paragraphs explain some of the key concepts relating to web content mining.

Knowledge discovery in databases (KDD) is the method of gaining insight from data that is stored in an organization's databases. Web mining is essentially the application of KDD processes to the Internet. Therefore, Web mining is the automated process of finding, analyzing, retrieving and storing meaningful, applicable information from the Internet. Web content mining activities include information retrieval (IR), categorization and clustering of Web documents and information extraction (IE) from Web pages (Chen, 2003).

Web crawlers, also referred to as spiders, agents, bots, ants, etc., are programs that retrieve information from Web resources. They are widely used by search engines to crawl through the Web and index Web pages. They can also be used to personalize searching e.g. to collect data for Web mining (Hemenway & Calishain, 2004). During the crawling process, there are two ways that Web data can be analyzed: content-based and link-based (Chen, Chau & Zeng, 2002). In the content-based approach, the body text of the web page is scrutinized to determine whether it is relevant to the search criteria. Link analysis uses the hyperlinks and the anchor text describing those links to evaluate page content. Link analysis is widely used and is the basis for the most effective page ranking algorithm (Silverstri, Perego & Orlando, 2004).

Since search results usually return thousands of pages, a clustering process is normally used to organize and group the retrieved information. Unfortunately, effective clustering relies upon the parameters and metrics used to evaluate the similarity among Web pages (Silverstri, Perego & Orlando, 2004). Some of these parameters to be determined include: identification of relevant attributes; weighting; method selection and similarity measure; limitations on computational/memory resources; speed and reliability of retrieved results; ability to make changes in the database and selection of ranking algorithms (Kobayashi & Takeda, 2000).

Finally, information extraction is the process of pulling out data from retrieved documents. Its main objective is to extract data and transform free text into structured data, usually in XML format, to be stored in a database. Encoding Web data into database entries will allow better retrieval, organization and analysis of Web data (Adams, 2001; Flesca, Manco, Masciari *et al.*, 2004). The computational linguistics community applied the same concept of discovering trends and patterns through statistical computation to real text data mining (Hearst, 1999). By using text category assignments to find new patterns or trends and the discovery of new themes within those text collections, real trends may be discernable. Grimes (2004) describes text mining as a 'cousin' of data mining and as it uses statistical analysis to extract concepts, detect relationships, and classify unstructured documents into categories.

Web Intelligence (WI) is a recent discipline that explores the fundamental roles and practical impacts of Artificial Intelligence and advanced Information Technology on the next generation of Web-empowered products, systems, services, and activities" (Zhong, Liu & Yao, 2003, p.1). WI-related topics include: Web agents; Web mining and farming; Web information retrieval; Web knowledge management; the infrastructure for Web intelligent systems; and social network intelligence. Some of the techniques proposed in the literature include MDR - Mining Data Records in Web pages (Liu, Chin & Ng, 2003), superpage classification and pruning (Ester, Kriegel & Schubert, 2002), and a number of approaches related to the semantic Web, including IBM's WebFountain project (Gruhl, Chavet, Gibson *et al*, 2004).

The challenge to build a system for scanning hordes of information on the Web for strategic information calls for a practical web mining methodology. However, such methodology should, if not be grounded in, at least take into account current practices, needs and practical possibilities in the real world. Thus the authors first embarked on a small-scale exploratory empirical study on Web mining as reported in the following section. Although the proposed methodology is not derived systematically from the literature survey, the latter provides basis for the methodology. It is also our belief that it gives a concrete context and background to the discussion.

WEB-MINING IN SOUTH AFRICA: RESULTS FROM AN EXPLORATORY SURVEY

This section reports some crucial findings from a small-scale exploratory survey conducted among CI practitioners in South Africa. This empirical study on the use of Web mining for strategic information in South Africa was exploratory and qualitative in nature. Although the literature suggests how Web (content) mining can be used *in principle* to collect intelligence, it is important to know how the Web is *actually* used for intelligence purposes in South Africa.

Research Methodology

Research Design

Semi-structured telephone interviews were chosen to facilitate open-ended discussions through organized questioning. A completely structured approach can cause rich data to be overlooked. Open-ended questions that probe, clarify and draw out details can get a participant to expound upon his/her experience. The approach is used in order to gain an understanding of intelligence gathering practices in South African companies.

A semi-structured interview questionnaire was used as a guideline and consisted of a combination of structured, Likert scale and open-ended questions. A few of the questions were derived from a study on CI practices in South Africa (Viviers, Saayman, Muller & Calof, 2002) which, itself, was based on a Canadian survey. The following are some sample questions included in the semi-structured interview questionnaire:

- How does your company gather intelligence? (open)
- Do your SI efforts focus on the following external forces? (list)
- Which are the *key* areas for your company and what sources do you use? (open)
- Is the Internet an intelligence source for your company (Yes/No). If so, what types of websites are used (give examples)? (open)
- The Internet is a good source for intelligence? (Lickert scale) Why? (open)
- Does your company have a process for searching the Internet? (Y/N) If so, what is the process? (open)
- How is the Internet intelligence evaluated and tested for validity? (open)
- Are [any of the following] information management techniques used? (list). If so, what technique is used and how?
- What types of tools would be helpful for gathering intelligence from the Internet? (open)
- Do you think your competitors use the Internet for intelligence? (Y/N)

Interviewees were inter-dispersed throughout South Africa. Therefore, it was more practical to conduct telephone interviews since they are less expensive and less time-consuming than face-to-face interviews. Plus, telephone interviews are generally conducive to studies with a small set of questions and a short timeframe.

Sampling Frame and Data Collection

Given the fact that CI in South Africa is still relatively new and the limited research resources, only a small number of companies could be studied. For the telephone interviews, the target audience was South African companies that have a CI discipline. The rationale behind this was to find individuals who already have some foundational knowledge of the intelligence process. Since associations already exist around the field of CI, it was sensible to target members of CI circles or participants in CI workshops. Therefore, the sample frame included members of the South African chapter of the Society of Competitive Intelligence Professionals (SCIP) and attendees of a workshop on Competitive Intelligence for Exporters.

It is difficult to ascertain the representativeness of the sample for South African organizations as a whole. However, for exploratory research purposes a judgmental sampling technique is considered appropriate.

An e-mail was first sent to each targeted interviewee introducing the scope of this study and inviting that person to participate in a telephone interview. At the end, there were 13 interview participants out of a total of 36 targeted. The interviews ranged from 15 minutes to 1 ½ hours in duration.

Five of the interview participants were CI consultants who compile intelligence for clients as a core business. It was very constructive to speak to these field experts. Eight of the interviewees, including the five mentioned, were full-time intelligence employees. The participants operated in a multitude of industries ranging from the wine industry to human capital management. Although a majority of the interviewees worked for a medium-sized company, small and large companies were also represented. On the whole, although small, the sampling frame yielded very high quality interviewees from a broad base of industries.

Thematic Analysis

A standard approach for analyzing qualitative data obtained from interviews is a thematic analysis. Accordingly, a thematic analysis was used to extract meaning from the semi-structured interviews. In effect, the thematic process described by Ganesh *et al.* (2003) was employed to garner themes from the interview data. As such, the basic structure of the questionnaire was used to draw up categories for the collected data. All interviews were transcribed and coded by one of the authors. The raw data was summarized via paraphrases and assigned to the categories. For each participant's data, major and minor themes were noted. Concepts with dominating patterns were distinguished as the major themes. Thereafter, themes from the sub-samples were compared across the entire sample to identify the common themes. The identified themes were coded. The thematic codes were then assigned across the sub-samples to identify all manifest themes. Finally, the data was sorted by category. Figure 1 below shows an extract.

The major themes can be classified into three major areas: general intelligence practices; Web-based intelligence; and skills development and education. In addition, two significant but minor themes revolved around ethics and counter-intelligence.

General Intelligence Practices

Intelligence Process

All of the participants have some mechanism for collecting intelligence. Seven out of the 13 interviewees have a systematized intelligence

Figure 1. Extract from Thematic Analysis categorization stage.

	A	B	C	D	E	F	G
1	Category	Major theme	Minor theme	Note	Comments	Theme	Interviewee
161	Skills	education	skilled search		A good librarian will find the most amazing information because he/she knows how the books/sources are organized and indexed.	limtations are due to operators	W
162	Skills	education	skilled search		Internet is only as good as you use it.	limtations are due to operators	M
163	Skills developme	awareness	CI knowledge		AI workshops – self and director of business	general	W
164	Skills developme	awareness	CI knowledge		course from knowledge management.com	general	S
165	Skills developme	awareness	CI knowledge		workshops at Potch Univ	general	d
166	skills developme	awareness	CI knowledge		workshops at Potch Univ	general	G
167	Skills developme	awareness	CI knowledge		workshops on CI	general	V
168	Skills developme	awareness	CI knowledge		none, attendance of workshops	general	Lu
169	Skills developme	education possessed	CI knowledge		Attend courses regularly from local consultants and CI conferenc	focused	Br
170	Skills developme	education possessed	CI knowledge		good internal training and development	focused	W
171	Skills developme	education possessed	CI knowledge		courses i.e. Jonathan Calof on internet search and CI	focused	Mi
172	Skills developme	education possessed	CI knowledge		They've developed courses for internal training. They also had Jonathan Calof conduct training.	focused	Sr

process as a whole. These individuals follow the SCIP model of planning, collection, analysis and communication of intelligence, although this reflects biasness in the sampling frame. One CI consultant described the process used for all their clients as, "The firm's competitive environment is systematically collected, analyzed and disseminated to policymakers who can act on it." Although the techniques may vary, the process remains the same. One of the CI consultants explained that the methodologies used depend on the clients' needs but they are conducted in the "standard SCIP way".

The remaining six participants practice intelligence in an ad hoc way. Although an intelligence policy is non-existent, they do make an effort to gather intelligence as it is perceived to be important in their businesses.

Intelligence Sources

Secondary sources provide a majority of the information and can be in hardcopy or electronic formats. These sources include industry/market authorities such as trade magazines, industry reports, publications by professional associations, research reports and so on. Fee-based commercial databases such as LexisNexis, Factiva and Inetbridge are also used to search for industry- or company-specific information. Government publications and databases were also mentioned as being used for the collection of intelligence. Information is being gathered from government agencies such DTI, Registrar of Companies, ITC, Justice Bureau, CSS, WTO, UN, etc.

Sources are generally documented especially by those with a structured intelligence process. Either a library of hardcopies is kept or electronic copies are stored in a PC. Some of the interviewees even have a more sophisticated system such as an in-house database and an indexing system.

Data Analysis Techniques

Among interviewees that follow a structured intelligence cycle, a qualitative data analysis is generally conducted. Various qualitative analysis models are applied depending on customer needs. Techniques that were mentioned include industry analysis, market analysis, Porter's Five Forces model, competitor analysis, risk analysis, customer profiling, management profiling, patent analysis, experience curve analysis, growth strategy analysis and stakeholder analysis.

Qualitative data analysis is conducted manually. Statistical or data mining software can be used to conduct quantitative data analysis as an adjunct to the qualitative analysis. However, qualitative data analysis is the primary method used for analyzing data. This is performed manually since computer programs are not sophisticated enough to carry out this task.

One of the consultants elaborates: "Computers still cannot answer the question 'so what?' That is an intellectual exercise. Computers cannot think and AI is too expensive. IT can be used for collection, collation and organizing data. Analysis is putting the puzzles together. A computer cannot put all the pieces of multiple puzzles together."

Another consultant remarked that available BI programs are not well-suited for analyzing external data: "BI tools are not really great for external information but they are good for internal information. For external data, you need to apply your mind to it and it cannot be done with a machine. BI tools cannot come up with conclusions."

Web-Based Intelligence

Web Use

The Web is used by all of the interviewees in the intelligence gathering process. They all use search engines to retrieve data from the Web. While some used basic search functions by entering rudimentary queries and clicking on the resultant links, others used advanced search terminology to achieve more precise results. Using semantically fine-tuned search queries was a sub-theme of advanced searches. One consultant pointed out that business librarians are sometimes consulted for industry-specific lexicons which are then used in query semantics. Although the same types of search engines were used, these advanced searches yield much more relevant search results. Another CI consultant stated that their aim is to retrieve no more than 20 documents at a time through focused queries.

With exception to one interviewee, the Web is checked regularly, often on a daily basis. This is in order to stay current on events. For example, news alerts are used to ensure that the intelligence worker is notified of up-to-date information. Another reason for daily usage is to conduct searches on different facets of the external environment. A third reason is that intelligence professionals conduct regular searches for different clients.

Opinion of the Web as a Source

An overwhelming majority of the interviewees were of the opinion that the Web is a good source of global information. Some of the interviewees use the Web quite extensively. One person claims that 90% of their company intelligence is obtained from the Web and likened the Web to a big brother who "is always there when you need him and who knows everything." A consultant's statement about the Web as a source of information was, "It's an effective resource. There's added value from the Internet." One analyst used to believe that market research always required hands-on experience. Now, he has realized the amount and type of information that can be found on the Web. His comment about the Web was, "It is an invaluable source with an awful lot of information!" Others maintain that the Web is a good source but the information available is limited. One interviewee said, "It's excellent for first-level information but poor for finer details." He cites that company websites, as an example, only provide general information.

A sub-theme that emerged is that the Internet is underutilized in South Africa. This point was conveyed through the following comment: "South Africa lacks information on the Internet. [...] We don't use the Internet to the full extent. There is very limited or older information on the Web compared to other international players [i.e. other countries]. The Internet should be leveraged better."

Consequently, the practitioners use multiple sources to supplement Web information. For example, proprietary databases are consulted for details on South African companies.

Web Data Validation

Since data quality is important in strategic intelligence, the need for Web data validation emerged. One consultant stressed that, even though Web data is usually found to be valid, it still needs to be tested. This was reiterated by one senior analyst who said, "It [the Web] is a good source but you must validate the information."

Information retrieved from the Web is typically validated through cross-referencing against reliable sources. Reliable sources include specialized databases, personal contacts and other niche publications. Another identified technique for testing data validity was a critical evaluation of the document source or author. The reputation of the source publishing the information is evaluated. The author's reputation can be another factor. One consultant explained that if the author is known from past experience or historical data, then the author's interest in the subject matter is evaluated. The consultant explained that more confidence is held in an article that covers an author's personal interests rather than his/her company's interests.

Systems and Tools

Various systems are used in the intelligence process. Some of the interviewees have an in-house system for storing and organizing extracted information. An analyst shared that his company has a full-fledged in-house system to support the intelligence process. A few have an actual data warehouse with years of compiled history. The data warehouse is then mined and used in conjunction with current information. While others have basic electronic files of reference lists of URLs that have been organized into different industry categories.

Some software packages are used to organize information. Two interviewees spoke positively of Brimstone. One consultant found Brimstone to be the "best product for CI". He explained that Brimstone has two components: a CRM component and a capability of supporting intelligence. Brimstone allows a user to search the Web using search engines, extract the Web pages and organize the documents. However, there is no package that can effectively analyze data. For this reason, data analysis is conducted manually.

Skills Development and Education

Advanced Web searching is more effective and efficient. One interviewee shared that the team of analysts at his company make a point of retrieving no more than 20 Web pages. This requires highly fine-tuned search queries. Another individual asserts that an experienced analyst does not need exhaustive Web data since he/she will know how to retrieve just the right amount of information to get to an answer. Obtaining precise search results is predicated upon skilled searching.

Proficient searching stems from the searcher's knowledge of the appropriate terminology to use, sites to explore and Web search tools. Some of the companies have addressed the importance of skills development by developing their own training program or sponsoring in-house training by intelligence experts. Some of the interviewees recognize the value of skilled searching and have expressed a desire to learn better searching techniques and about search tools that are available.

Education empowers intelligence workers with knowledge. Those who have a structured intelligence approach regularly attend conferences, workshops and courses on CI, knowledge management, Internet searching, etc. One consultant is doing a PhD on the skills required for effective CI.

Knowledge also delivers better intelligence products. Intelligence gatherers should be mindful of the intelligence process because it is how data is collated and presented that transforms it into intelligence. This was supported by a consultant who said that, "Internet information is not intelligence. It's what is done with the information that makes it intelligence."

Minor Themes

Two of the minor themes which were uncovered in the thematic analysis are also worth mentioning.

Need for Ethical Intelligence Practices

A few participants have highlighted the need for an ethical means of collecting data. Investigations must be done in a legal and proper manner. There is a lawful and ethical way of deploying an intelligence strategy. The Society for Competitive Intelligence Professionals has a code of ethics that can be used in formulating an intelligence policy. One of the consultants emphasized the need for an awareness of ethical intelligence practices. He said that he first educates his clients on what CI is and is not. He states that awareness must be created because some people have the wrong idea about intelligence. He explained that intelligence must be collected in a legal manner. For example, privacy rights must not be infringed. He further clarified that CI is not industrial espionage. Publicly available information can be used effectively for intelligence.

Counterintelligence

Another minor theme of interest that emerged had to do with counterintelligence. Counterintelligence is essentially defensive intelligence. This includes counter measures taken when intelligence surveillance is detected. For example, one of the interviewees expressed a need to discern between *real* intelligence from measures taken by the competitor to mislead him. A proactive counterintelligence approach can be taken as well. One consultant mentioned that he and his intelligence team ensure that they cover their virtual track as a counterintelligence tactic. This makes sure that their intelligence gathering trail cannot be traced back to them.

A PRACTICAL WEB MINING PROCESS

The above study on intelligence collection in South Africa revealed that Web mining can indeed be used to assist business decision-making. This is advocated further by the fact that professionals who operate an intelligence consulting company use the Web to retrieve strategic information for their clients. This section proposes some practical steps which a smaller, global-thinking organization which is operating in an emerging economy context, can undertake to create their own web mining process or methodology. These steps – loosely organized in a sequential methodology – can and indeed should be customized and/or supplemented by other activities relevant to the experience of the organization. It also draws from and maps directly onto the initial phases of the proprietary SCIP Intelligence Cycle which is used by most of the Intelligence Officers which were surveyed.

Building a Company-Specific Taxonomy

As a starting point for Web mining, search queries must be fine-tuned in order to help alleviate low recall precision. A key strategy for improving precision in IR is to create a company-specific taxonomy. This should include all possible terminology and relevant concepts used in the industry, sector, niche market and company. Not only is it industry-specific, it is also company-specific.

There are online information services that provide industry-specific taxonomies. These fee-based services can be consulted when constructing the company-specific taxonomy. An example of this is the Taxonomy Warehouse (http://www.taxonomywarehouse.com/). It can be beneficial to use an already-compiled domain-specific taxonomy to save time and effort. However, company-specific terms, such as the list of actual competitors, major customers and suppliers, etc., must still be added.

In addition, inputs from the PEST (political, economic, sociological and technological) and Porter's Five Forces model (new entrants, buyers, suppliers, rivalry and substitutes) should also be included in the company-specific taxonomy. A list of terms within each contributory factor must be compiled. For example, when considering the technological environment within a PEST analysis, a list of specific technologies, related components or tools, companies heavily involved in R&D within the technology sector, etc., must be generated first. This is done for each of the four PEST components as well as Porter's five competitive factors. Relevant website addresses should also be included. The accumulated lexicon will be the basis of the Web search parameters for each aspect of the environmental analysis.

Meta-Searching

Meta-search engines combine the results from multiple search engines. This allows for greater coverage of the Web since limitations of one search engine may be accommodated by another search engine. Furthermore, meta-search engines incorporate algorithms to eliminate duplicate Web pages. It makes economical sense to exploit this technology as a part of a Web mining system. If the meta-search engine allows users to specify their own search engines, it is important to include South Africa-specific engines such as Ananzi, Clipps, Everything SA, Grapple or ZuluAlpha.

Terms from the company-specific taxonomy will be used in the meta-search engine query fields. The meta-searcher will retrieve URLs that relate to the query terms. This can range from customers to niche-market product categories. This is solely dependent on what terms are queried. When researching a particular area of interest, the terms from the related section of the PEST and/or Porter's Competitive Forces taxonomy in addition to the general company-specific terms will be used. It is important to set the meta-searcher to retrieve the maximum number of pages from each search engine in order to have a large base of Web data to be processed further.

Another search strategy is to use reverse link analysis. This type of search is possible in most meta-search engines, as well as standalone search engines, by typing "link:" followed by the URL. A reverse link analysis retrieves the URLs of pages with an in-link, or hyperlink, to a specific website (i.e. competitor or customer site). This type of search can reveal some interesting information. For example, a potential customer's website which cites a competitor's website may disclose some form of integrated supply chain arrangement. An alternative use of reverse link analysis is merely to ensure inclusion in public Web directories, or other listings.

Meta-search results require a lot of fine-tuning in order to establish relevancy within an intelligence strategy. According to Grimes (2004, p.24), "Enterprise search tools must negotiate access to content stores, apply techniques that classify and organize results, and embed the search function behind the scenes in operational applications." Essentially, the meta-search process merely provides the raw content; mining techniques need to be applied to organize and process the data in order to transform it into intelligence.

Crawling Web Pages

Running personal spiders on a client machine allows more CPU time and memory to be allocated to the search process and provides more options (Chen *et al*, 2002). Most search engines re-crawl based on an average of once a month and, therefore, possibly retain 'dead pages' at any given time. An intelligence analyst needs more up-to-date data. For this reason, a personal crawler that can refresh data without having to rely on search engine updates is more appropriate.

Personalized Web crawling can also customize search further and improve Web coverage. Since meta-search engines automatically rank pages based on the PageRank algorithm, certain Web pages may be omitted because they were not within the top ranked pages. A company's ranking order will not always be identical to the search engines' since the relevancy of Web pages can be subjective. In order to circumvent the possibility of omissions, a personalized Web crawling process is ideal.

The Web crawler is fed with all the URLs obtained from the meta-search results. Likewise, the website addresses listed in the company-specific taxonomy (i.e. customer sites, competitor sites) are also fed into the crawler. The user must designate how many levels of linked Web pages the crawler must investigate. This will be dependent on how comprehensive the company-specific taxonomy is and how deep the desired search must be. Cache memory, processing speed and storage capacity are other determinants. The user must use discretion in deciding the level of scope the width and depth of the crawl. If the objective is to obtain an overview of the whole competitive environment, the scope is wide but may only be two levels of link analysis depth. On the other hand, in-depth competitor profiling would require a very narrow search scope and very deep link analysis.

Many commercial spidering and crawling tools are available. In addition, it is relatively easily to build customized tools for specific needs or purposes (Hemenway *et al*, 2004).

Searching Online Proprietary Databases

It is necessary to supplement Web data with data from other credible sources such as commercial, governmental or private databases. This will augment Web data in case limited information is available on the Web. However, another reason for doing this is to validate Web data by cross-referencing it against reliable sources. Therefore, data extracted from proprietary databases must be clustered and summarized separately from the bulk of the retrieved data. This will likely result in smaller data sets used merely to test the reliability of the retrieved data. These standalone data sets will be compared to the results of Web clustering. Once data validation is satisfied, these data sets can also be added to the extracted data.

Some typical examples of proprietary databases with South African information include the following:

- LexisNexis (www.lexisnexis.co.za): Info on company, directors, auditors, history; property and deeds.
- I-Net Bridge (www.inet.co.za): Economic & financial market data; corporate market intelligence.
- ShareDate (www.profile.co.za): Financial, accounting and related data on listed companies
- ListSA (www.listsa.co.za): Contact lists of key managers (listed and non-listed businesses)
- McGregorBFA (www.mcgregorbfa.com): Statutory, financial, news and market surveys.
- TBO Reports (www.tboreports.com): Financials, annual reports, organisation charts, news (and news maps) for public companies in sub-Saharan Africa.
- ICT-E database of BEE firms (www.dti.gov.za): Profiles of black empowerment businesses.
- CIPRO (www.cipro.co.za): Statutory information for companies and CCs; registered intellectual property (patents, trademarks).

Searching the 'Hidden Web'

Larger companies have their own searchable websites. These sites make up the 'hidden Web'. Since conventional search engines do not index dynamic Web pages, Web crawlers do not have access to the database from which the dynamic pages are created. Since an off-the-shelf solution is not known at this time, the only option is to perform a manual search within the websites or use custom spider to automate some of these (Hemenway *et al*, 2004). Again, relevant terms from the company-specific taxonomy can be used as the search criteria.

Clustering Web Pages

Once the Web crawler has retrieved potentially relevant pages, clustering will be conducted to organize the pages into categories. Since it is probable to be bombarded with excessive data, it is important to perform clustering before actual extraction in order to avoid downloading irrelevant data. This is especially important in the context of economies where bandwidth is relatively expensive (such as South Africa). Consequently, user interaction is important at this stage of the Web mining process. The user must select which clusters of data will go to the next stage of data processing.

It is advisable to use a real-time, automatic clustering technique. This would save the time and effort required to manually construct a categorical hierarchy. Several researchers recommend that automation is used to cope with Web information overload (Adams, 2001; Chen, 2003; Song, Song & Hu, 2003). In addition, ad hoc clustering would allow for groupings to emerge organically from retrieved data. Allowing information to surface is usually inherent in knowledge discovery.

Using a clustering tool which takes stop-words into account is important. Elimination of stop-words can reduce the level of noise in data. Precision and efficiency can be significantly improved when noise is filtered out from the clustering process. Furthermore, comparisons between text documents can be conducted more accurately when non-discriminative words are purged.

The resultant catalog of clusters is often referred to as a taxonomy. This is not to be confused with the original company-specific taxonomy although the cluster-inspired taxonomy is likely to inform and refine the company-specific taxonomy which can then be used in future searches. This cycle will repeat as and when new clusters are generated.

Extracting Web Data

In the previous stage, clustering was used to help filter out key concepts from the hordes of retrieved data. In the next phase, IE aims to identify and remove pertinent data from the selected clusters. As recommended by Feldman, Aumann, Liberzon *et al.* (2001) this is the final stage of preprocessing unstructured data prior to text mining. Again, the objective of IE is to transform unstructured data into structured database entries. From there, text mining techniques are used to analyze the text (McCallum, 2005).

An IE system needs to know how to locate information for extraction. A wrapper induction system that is trained to learn the extraction rules is ideal. The user then provides the system with sample documents from which extraction rules are to be generated automatically. From the training documents, the system is able to decipher how to locate the start and finish of data to be removed. A Web induction system that can formulate extraction rules from a small set of examples is ideal.

The wrapper also needs to be designed to convert data into XML format. This is the transformation of unstructured data into structured data. Once the XML format has been created, the XML document is then stored in a database.

Mining Text for Information

The final stage of the proposed Web mining system is text mining. This process uses linguistic analysis, statistical and computational techniques, to identify entities, entity relationships and interrelated concepts (Fan, Wallace, Rich & Zhang, 2005). This is the point where data analysis leads to knowledge discovery, or intelligence.

Once data compiled from multiple information sources is aggregated, text mining can reveal unexpected associations or confirm hunches about environmental trends. Associations between organizations and/or concepts may also emerge. The user should be equipped with as much of a synopsis as possible. A summarization and/or visual representation of the text mining results is valuable. A system which allows for the user to interact with the results would be a bonus.

It is important to note that text mining software is not meant to conduct a thorough analysis since text mining technologies are not well-developed enough to do that. Again, summarization and visual representation can be helpful in presenting patterns and trends. However, the user must have some analytical skills and be familiar enough with the data set to really understand what the result means (Robb, 2004).

Repeating the Cycle

The above process is likely to be repeated at regular or ad-hoc intervals. In this case, a baseline historical database of original web information is kept and special rules can be introduced to weigh incremental and new information against static baseline and historical information. Indeed later cycles are believed to benefit greatly from historically collected information (as incorporated in the company specific taxonomy) and differential data items will typically highlight significant evolutions and trends.

Also, any of the phases can feedback into earlier stages for future iterations e.g. the company-specific taxonomy will be updated and refined with results from the cluster analysis.

Commercial Web Mining Tools

It is important to consider the use of commercially available Web mining tools. An illustrative selection of some tools which were available in 2009 is given in Table 1. This serves to demonstrate that most software packages include multiple components. Note that this market space is quite dynamic and the market offerings as well as tool functionality subject to change. For a more recent list, refer to http://wissensexploration.de or similar portals. When one is building an integrated or enterprise wide knowledge management system, one should consider integrating multiple Web mining components in order to achieve a complete solution.

Integration of Web Mining Components

The proposed Web mining architecture consists of the following components: 1) company-specific taxonomy; 2) personalized Web crawling with meta-searching; 3) manual search and extraction of data in proprietary databases, including the Hidden Web; 4) Web IE and 5) text mining. No web mining tool appears to be available that addresses all of these components. Many tools cover one or two functions very well, e.g. the are many dedicated text mining tools, web crawlers, automated ontology and taxonomy builders and meta-search tools.

In practice, additional components will be included in the final web mining architecture. Typical value-adding components are visualization modules, RSS feeds and news mash-ups. Typically, organizations should use their custom taxonomy to subscribe to RSS feeds from various online financial, business, marketing/PR and industry publications. Examples of general online South African publications are the Financial Mail, Business Day, Business Report, Finance24.co.za, ITWeb, Biz-Community, PR Newswire.

Although proposing a detailed technical architecture is beyond the scope of this chapter, a key consideration is the ability to integrate all of the components into one Web mining system. As shown, while many of the leading software packages incorporate multiple components into their suites, an all-encompassing program is yet unavailable. Therefore, the issue of integration must be addressed.

A services-based architecture is likely to facilitate the integration of standalone or disparate components. Wrappers can be written to translate data into XML format and their functionality exposed through web services. Thereafter, XML schema transformations can be used to facilitate the flow of XML data from one component to the other and to massage the data into the required form.

SPECIAL CONSIDERATIONS FOR WEB MINING IN SOUTH AFRICA AS AN EMERGING ECONOMY

The discussion above touched on a number of aspects which make web mining in South Africa, as an emerging economy, different from web mining in developed countries.

A critical issue, and subject of considerable debate, is the very *high cost of internet bandwidth*, not only in South Africa, but for most of sub-Saharan Africa. Bandwidth is up to 100 times more expensive in South Africa than in developed countries. This is due to a combination of the long-time monopoly of the fixed line telecommunications provider Telkom and the lack of capacity on the SAT-3 fibre-optic cable link linking South Africa with the Northern hemisphere. The monopoly is slowly being eroded with rights granted to a competing fixed line infrastructure communications provider Neotel. The available international bandwidth on undersea cables will explode for most of Africa due to the completion of three or four additional high-capacity cables, the first of which came online in late 2009. In the meanwhile, a number of technical solutions exist to reduce the bandwidth requirements. Since especially steps 3 and 6 have high bandwidth requirements and involve high peak computing resources, cloud computing services can provide an economic and scalable alternative. For example, the flexible "pay-as-you-go" computing capacity of the Elastic Computing Cloud (EC2) from Amazon Web Services (AWS) could provide an effective solution in a bandwidth or compute-constrained environment (Buyya, Yeo & Venugopal, 2008). An additional advantage is the availability of the Alexis search engine. An alternative solution is to do the meta-searching, crawling, information extraction and clustering for international information on servers located overseas and just send the results – which is a greatly reduced dataset – down to South Africa.

A direct consequence of the expense of Internet access and bandwidth is the much *lower penetration of the Internet* and thus a smaller amount of available Internet information. Internet adoption and Internet presence by business community in South Africa lagged that of developing countries. However, at least for the corporate market, a lot of the backlogs were caught up in recent years, although the general population still does not enjoy the same degree of Internet participation as Europeans, North-Americans and a number of Asian countries.

Another potential issue is the *cost of software*. Although South Africa does not have the forex constraints faced by some other African countries, software is still comparatively expensive, especially if priced at the same (or even higher) levels than in the US or Europe. Emerging economies generally have lower levels of purchasing power and many pay premiums on software imports due to foreign exchange regulations. However, the cost of a CI solution can be substantially mitigated through the use of open source alternatives. It is relatively easy to obtain free ontology editors and many open source web crawlers exist, e.g. Heritrix or Nutch. A very popular open source information extraction and text mining tool is RapidMiner (formerly known as YALE) and it features amongst the most widely adopted CI tools (KDNuggets, 2009).

Also a critical issue is the relatively lack of depth in the *skills* available in South Africa. Understanding the discipline of CI and the correct use of CI tools is a fairly specialized field and the number of practitioners is relatively small. The premium commanded by relatively rare skills is very high and makes the relative cost of CI higher than in developed countries. This is again not particular to South Africa but also the case for most other African countries.

A final constraint is the relatively *small size of the economy* which is characterized by a relatively small number of competitors in most industries, and generally a less transparent and less efficient exchange of corporate and news information. This reduces the depth and breadth of many proprietary databases when compared to overseas counterparts. Nevertheless, South Africa finds itself in a much better situation than most other African emerging economies as evidenced by a number of relatively sophisticated information providers such as LexisNexis, I-Net Bridge, McGregorBFA, TBO Reports, CIPRO and others listed above.

CONCLUSION AND IMPLICATIONS

This chapter started with an exploratory survey on some of the competitive intelligence (CI) practices in South Africa. It revealed that Web mining is indeed used to assist decision-making, especially for businesses operating in the context of an emerging economy yet having to compete in an increasingly global arena. Moderating the theoretical literature on CI with the practical issues faced by CI professionals in an emerging economy, a practical web (content) mining process and architecture was proposed.

In general, the Internet is relatively underutilized in South Africa, as in many other emerging economies, limiting data extraction to free Web data will negatively impact the quality of the Web mining system. It is therefore recommended that data from proprietary databases is incorporated, especially for areas or industries when there is limited information on the public Web.

Other special considerations in the context of South Africa were the relatively shallow skills pool for which no quick fix solution exist; the high cost of bandwidth which can be partially addressed by technical solutions such as overseas servers or cloud computing; and the relatively high cost of software which makes it more important to include open source tools in the organizational CI tool evaluation and decision making process.

It must be noted that the proposed Web mining process only addresses part of the strategic CI process, namely the phases between planning and processing. Further analysis must still be conducted on the data emanating from the web (content) mining process. Results of the analysis must be assimilated into a report. The report must then be distributed to the relevant parties to be used, or considered, in strategy formulation. Feedback from strategic managers can lead to the next cycle. In addition, ongoing evaluation of the specific processes should be conducted. The proposed Web mining architecture can also deal with the evaluation of data validity. This is done by cross-referencing the retrieved data with information from authoritative sources. Nevertheless, an ongoing appraisal of the strategic intelligence system, as a whole, is a good discipline.

One of the main implications of the research is the opportunity which presents itself for small and medium-sized organizations in emerging markets to compete in this global market. Given the amount of strategic information that can be collected from the Web and the availability of tools that facilitate the Web mining process, small to medium-sized organizations can leverage Web data to gain a competitive local and global advantage. By using a Web mining system to retrieve and process Web data, companies can conduct their own environmental analysis without having to spend a lot of money on consultants. The cost of especially commissioned market analysis or ad-hoc consultation can be circumvented.

Finally, it must be conceded that the proposed Web mining process is mainly conceptual at this point. Testing the Web mining architecture can only be done through the real-world implementations. Future research will hopefully refine the technical architecture significantly. Additionally, the actual process steps may need to be updated as the CI software market becomes more sophisticated and the evolving web creates new CI possibilities, e.g. the impact of the semantic web, web services and Web 2.0 tools beyond RSS. In addition, the careful documentation of a few case studies to illustrate the problems and benefits experienced by a number of organizations using the proposed web content mining process would be extremely beneficial to the CI community.

REFERENCES

Adams, K. C. (2001). The Web as a database: new extraction technologies & content management. *Online*, *25*(2), 27–32.

Buyya, R., Yeo, C. S., & Venugopal, S. (2008). Market-Oriented Cloud Computing: Vision, Hype, and Reality for Delivering IT Services as Computing Utilities. *Proc. of the 10th IEEE International Conference on High Performance Computing and Communications*, China, 2008.

Calof, J. L., & Viviers, W. (2001). Adding competitive intelligence to South Africa's knowledge management mix. *Africa Insight*, *31*(2), 61–67.

Chen, H. (2003). Introduction to the JASIST special topic section on Web retrieval and mining: A Machine Learning Perspective. *Journal of the American Society for Information Science and Technology*, *54*(7), 621–624. doi:10.1002/asi.10254

Chen, H., Chau, M., & Zeng, D. (2002). *CI Spider: a tool for competitive intelligence on the Web*. [Electronic] Available: http://ai.bpa.arizona.edu/~mchau/papers/CISpider.pdf

Du Toit, A. S. (2003). Competitive intelligence in the knowledge economy: what is in it for South African manufacturing enterprises? *International Journal of Information Management*, *23*(2), 111–120. doi:10.1016/S0268-4012(02)00103-2

Ester, M., Kriegel, H., & Schubert, M. (2002). Web site mining: a new way to spot competitors, customers and suppliers in the world wide Web. *Proceedings of SIGKDD*, Edmonton, 249-258.

Fan, W., Wallace, L., Rich, S., & Zhang, Z. (2005). Tapping into the power of text mining. *Communications of the ACM*, *49*(2), 76–82.

Feldman, R., Aumann, Y., Liberzon, Y., Ankori, K., Schler, J., & Rosenfeld, B. (2001). Information retrieval and text mining: a domain independent environment for creating information extraction modules.*Proceedings of the CIKM'01*, Atlanta, 586-588.

Flesca, S., Manco, G., Masciari, E., Rende, E., & Tagarelli, A. (2004). Web wrapper induction: a brief survey. *AI Communications*, *17*(2), 57–61.

Ganesh, U., Miree, C. E., & Prescott, J. (2003). Competitive intelligence field research: moving the field forward by setting a research agenda. *Journal of Competitive Intelligence and Management*, *1*(1), 1–12.

Grimes, S. (2004). Consumer and enterprise search: not an exact match. *Intelligent Enterprise*, *7*(9), 22–31.

Gruhl, D., Chavet, L., Gibson, D., Meyer, J., Pattanayak, P., Tomkins, A., & Zien, J. (2004). How to build a WebFountain: an architecture for very large-scale text analytics. *IBM Systems Journal*, *43*(1), 64–77. doi:10.1147/sj.431.0064

Hearst, M. A. (1999). Untangling text data mining. *Proceedings of ACL '99: 37th annual meeting of the association for computational linguistics*, University of Maryland, June 20-26, pp. 1-8.

Hemenway, K., & Calishain, T. (2004). *Spidering Hacks: 100 Industrial-Strength Tips & Tools*. Sebastopol, CA: O'Reilly.

KDNuggets. (2009). Top Data Mining Tools Used. KDNuggets, 2009(10), item 1. [Electronic] Available: http://www.kdnuggets.com/news/2009/n10/1i.html.

Kennedy, S. D. (1998). Competitive intelligence on the Web 101. *Information Today*, *15*(9), 40–41.

Kobayashi, M., & Takeda, K. (2000). Information retrieval on the Web. *ACM Computing Surveys*, *32*(2), 144–173. doi:10.1145/358923.358934

Kosala, R., & Blockheel, H. (2000). Web mining research: a survey. *Proceedings of SIGKDD*, *2*(1), 1–15. doi:10.1145/360402.360406

Liebowitz, J. (2006) *Strategic Intelligence: Business Intelligence, Competitive Intelligence, and Knowledge Management.* Boca Raton FL: Auerbach Publications (Taylor & Francis)

Liu, B., Chin, C. W., & Ng, H. T. (2003). Mining topic-specific concepts and definitions on the Web. *Proceedings of WWW 2003*, May 20-24, Budapest, 251-260.

McCallum, A. (2005). Information Extraction: Distilling Structured Data from unstructured Text. *ACM Queue; Tomorrow's Computing Today*, *3*(9), 48–57. doi:10.1145/1105664.1105679

Robb, D. (2004). Taming text. *Computerworld, 38*(25), 40–41.

Silvestri, F., Perego, R., & Orlando, S. (2004). Assigning document identifiers to enhance compressibility of Web search engines indexes. *Proceedings of SAC '04*, Nicosia, 600-605.

Song, M., Song, I.-Y., & Hu, X. (2003). KPSpotter: a flexible information gain-based keyphrase extraction system. *Proceedings of WIDM'03*, New Orleans, pp. 50-53.

Thuraisingham, B. M. (2003). *Web Data Mining and Applications in Business Intelligence and Counter-Terrorism*. Boca Raton, FL: CRC Press. doi:10.1201/9780203499511

Velder, R. G., Vaneck, M. T., Guynes, C. S., & Cappel, J. J. (1999). CEO and CIO perspectives on competitive intelligence. *Communications of the ACM, 42*(8), 109–116.

Viviers, W., Saayman, A., Muller, M.-L., & Calof, J. (2002). Competitive intelligence practices: A South African study. *South African Journal of Business Management, 33*(3), 27–37.

Westphal, C. (2008). *Data Mining for Intelligence, Fraud and Criminal Detection: Advanced Analytics and Information Sharing Technologies*. Boca Raton, FL: CRC Press. doi:10.1201/9781420067248

Zhong, N., Liu, J., & Yao, Y. (2003). *Web intelligence*. Berlin: Springer.

Chapter 2
Meeting Czech Knowledge Management Challenges Head-On:
KM-Be.At-It

Richard Brunet-Thornton
IMCA/GARC, UK; University of Economics-Prague, Czech Republic

Vladimír Bureš
University of Hradec Králové, Czech Republic; City University of Seattle, Slovakia

EXECUTIVE SUMMARY

In the Czech Republic, Knowledge Management-related problems occur at all knowledge stages with difficulties more predominant at an organisational level. In principal, they originate from the lack of an utilisable and detailed KM implementation methodology; varying perceptions of KM coupled with the notion, that KM equates and is limited to information technologies and the like. As a reaction to these problems, the Faculty of Informatics and Management at the University of Hradec Králové (FIM UHK) developed a new KM implementation methodology. It aims to provide interested parties with a set of ground rules distributed among a number of phases. Further development is underway to elaborate the method by integrating project management tools, benchmarking exercises and critical success factors. Despite this endeavour, there is an on-going need for further research given the reality that small and medium sized enterprises constitute the majority of organisations in the Czech Republic.

INTRODUCTION

Knowledge Management (KM) as a managerial discipline emerged over twenty years ago. Globally, both practitioners and academicians have adopted it and the Czech Republic is not an exception. Its acceptance has been complicated as it holds a wide range of descriptions and interpretations that are reflected in the various existing KM narratives. Therefore, several analyses on KM perceptions such as those formulated by Bureš (2005), and Wallace (2007) are available. In general, before analysing the situation in the Czech Republic, one considers KM as defined by the expert Karl Wiig "KM is the systemic, explicit, and deliberate building, renewal, and application of knowledge

DOI: 10.4018/978-1-61692-886-5.ch002

to maximize an enterprise's knowledge-related effectiveness and returns from its knowledge assets (Beckman, 1999). It is common to experience problems connected with KM at the conceptual and operational levels. The Czech Republic as a nascent market-economy and transitional society in Central Europe is not an exception. Moreover, there are certain specifics related to the Czech cultural and commercial environment. It is possible to identify several problems. In addition, there is a general lack of information on the state of KM in the Czech lands due to infrequent surveys. Moreover, even current KM resources lack a connotation of the cross-cultural differences and are not readily available in the Czech language. This chapter describes the situation of KM in the Czech Republic and provides an insight to the long-term effort in this field based on the essential features of a particular methodology. Equally, it exposes the measures undertaken to ameliorate research.

From both a practical as well as a research perspective, one raises the question as to the role of cross-cultural differences and the unique character of the Czech Republic. Some KM critics may cite the Czech failures as additional fodder to the statement, "...like many a business concept, KM has evolved from a hot buzzword to a phrase that now evokes more scepticism than enthusiasm" (Frappaolo, 2006). Have our Czech colleagues widely abandoned a notion that they perceive as inept and an imported fad consciously or do cultural constraints and framework render the question moot.

Overall, the discussion that follows addresses three primary issues relative to the Czech KM experience: KM specifics from a purely Czech perspective, the KM methodology developed by the University of Hradec Králové - a work in progress introduced as KM-Be.At-It and lastly, Cross-Cultural Management issues associated with the further development of KM in the Czech Republic.

KNOWLEDGE-RELATED ISSUES IN THE CZECH REPUBLIC

Bureš and Čech (2008) identify four knowledge levels depicted in Figure 1. It describes the overall environment wherein knowledge-related activities arise. Each level differs from the other in many aspects. An increase of the resolution of any one level permits the discovery of another. Exemplars of this refinement are particular economic sectors (tourism, engineering, telecommunications, and education), or clusters that have different knowledge needs, utilisable tools, techniques and procedures. They are frequently uncovered between the national and the organisational level demonstrating that the demarcation between levels is ambiguous.

Supranational Level

This level is both the broadest and the most general. It operates by means of concepts of the Knowledge Economy (KE) and Knowledge Society (KS) in which tangible knowledge does not play a role. Many forces such as, international affairs; cultural and technological developments; political circumstance; cross-border amalgamates and institutions (the World Bank, the OECD, the United Nations) shape this level.

Within a Czech context, this level distinguishes itself particularly through numerous strategic documents issued by the European Union (EU). Recently the European Council published the "Lisbon Strategy." The report emphasises KE as a strategic and fundamental objective. The document states that the EU desires to become a world-leading economy that is both competitive and knowledge-based. It states that "the Union has today set itself a new strategic goal for the next decade: to become the most competitive and dynamic knowledge-based economy in the world capable of sustainable economic growth with more and better jobs and greater social cohesion" (European Council, 2000). Although revised and objectives modified several times, this strategy influences particular member states and their partners. In this vein, KE and knowledge-based competition are no longer theoretical concepts limited to academicians or articles published in scholarly journals (Drucker, 1993; Nonaka and Takeuchi, 1995). Subsequently, the position of the Czech government reflects this EU initiative.

Equally, the many scientific and research projects either fully or partially funded by the EU form part of this level. The majority of these projects focus on different aspects of KM. Other supranational institutions that influence and guide both the Knowledge Economy and the Knowledge Society also manipulate the Czech Republic. Examples are the OECD and UNESCO. The OECD outlines its orientation in its various reports particularly one released in 2004 and the earlier report entitled, "The Knowledge-Based Economy" (1996). In a similar fashion, UNESCO issued its "Towards Knowledge Societies" in 2005. Given the titles of these publications, their point of reference is perceptible and the content and recommendations consistent with other supranational institutions influence the awareness of the KE and KS within Czech society.

National Level

With globalisation, there exists a noticeable increase in the flow of goods and services, and the migration of people, ideas, and wealth on an international basis. The advantages both tangible and intangible of this movement are numerous and significant. Companies of a national character must compete with international enterprises in their domestic market. In the Czech environment, former privately or state owned corporations are now controlled by management councils located outside national borders. Companies must remain vital in the current highly competitive and trans-border environments. The lack of critical knowledge on an appropriate mix of personnel, technology, intellectual assets, and physical assets may lead to the demise of indigenous businesses and significantly influence organisational and national culture.

At the national level, KE and KS exhibit a Czech dimension reflecting the historical, cultural, economic, legislative, and geopolitical specifics. KE and KS are associated with authentic conditions and the needs of a particular organisation in the country. On the national stage, the Czech business environment exhibits the appropriate conditions and the opportunities to regard knowledge

Figure 1. Knowledge levels (adopted from (Bureš, 2009))

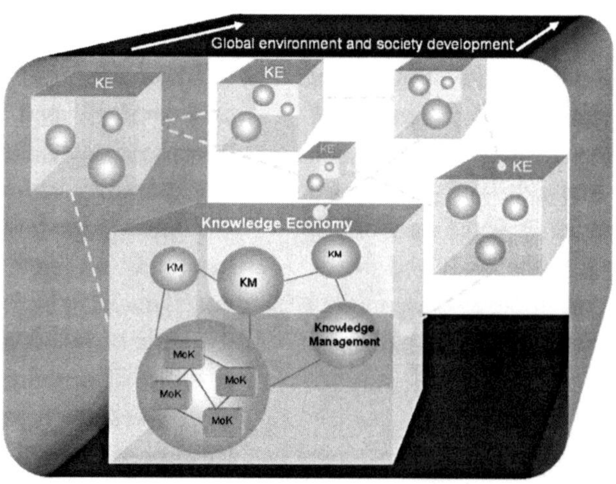

and knowledge assets as important resources for the same reasons that many foreign enterprises located outside the Czech borders have done so successfully. This recognition of knowledge goes far beyond business. Numerous government position papers and declarations such as, the Strategy for Economic Growth of the Czech Republic; the Strategy for Human Resource Development for the Czech Republic and the Strategy of the Government of the Czech Republic within the EU framework identify this position.

With the publication of the Strategy for Economic Growth of the Czech Republic, many discussions ensued producing a considerable number of journal and newspaper articles as well as reports in the popular media. Interestingly the first chapter is entitled "Czech Republic – Knowledge and Technological Centre of Europe with a growing standard of living and high employment" (Jahn, 2005). It further suggests to "...support the creation KM centres and its instruction at universities and public research institutions" (Jahn, 2005).

Organisational Level

Genuine KM transpires at this level. KM encompasses a knowledge-based and knowledge-orientated organisational management irrespective of organisational mandate or nature. Therefore, KM may be present in business organisations, educational institutions, or even, civil administration. Many Czech institutions including commercial interests realise that traditional assets are no longer the sole resources managed during the transition to KE and KS. In doing so, highlight the prominence to knowledge resources and knowledge processes.

Many problems exist at this stratum within the Czech Republic. The transition to KE and KS is in parallel with other major and historical changes. The Czech economy continues in a transitory state of flux vis à vis its historical past of sixty years of oppressive regimes. The reason is that the Czech Republic as a country with an evolutionary financial system encounters many problems originating from its cultural heritage of the socialist era. Cultural aspects, including social and individual barriers, contribute significantly (Bureš, 2003). The multiple perceptions of KM create additional barriers. These substitute KM with off-the-shelf solutions of software-based applications. Although necessary in the overall scheme as implementation or deployment tools, KM in both theory and practice is more concerned with the human dimension and therefore, constitutes a process. Lastly, there is a lack of KM implementation methodology available in and specific to the Czech Republic. Case studies of successful KM implementation tend to deal mostly with larger concerns outside the Czech borders involving large and impressive organisations like Chevron, British Airways, or BP Amoco (Ahmed, Lim and Loh, 2002). The majority of Czech companies neither project the magnitude nor posses the power of the latter. Subsequently, available models are not applicable to the Czech environment contributing to the lack of successful KM implementation. For these reasons, there exists a compelling need for further research and model-development to address the distinct Czech reality. Without the appropriate "Home-Made" tools, Czech organisations revert to services provided by consultancy companies that due to cost often render KM out of reach. In other instances, local enterprises often seek foreign-developed methodologies. In both cases, methods and procedures created in different settings that ignore the cross-cultural connotations confront the Czech organisation. Moreover, they lack sufficient detail and are seldom available in the Czech language.

Management of Knowledge (MoK) Level

MoK represents the lowest but also the most basic of levels wherein attention to knowledge is practiced. Incorporated at this fundamental level are methods, techniques, and principles of research work within fields such as information science, knowledge engineering, knowledge-based systems, pedagogy, psychology, sociology, or other cognitive sciences. MoK focuses, for example, on database data mining (Witten and Frank, 2005); knowledge systems design and implementation (Mikulecký, Ponce and Hynek, 1997); push technologies (Mikulecký, 2001); knowledge representation (Olševičová, Ponce and Haviger, 2006); cognitive maps (Gavalec and Mls, 2005); knowledge ontology (Ponce, 2007); recommender systems (Čech and Olševičová, 2009); or computer supported decision making (Tučník and Mohelská, 2009). A relationship between managers and MoK exists. KM achieves its goals by utilising MoK outputs within the scope of its own activities. Managers are not usually cognisant of the basic principles of the MoK products. The situation is comparable to other work activities in which the operations behind the result are transparent. For instance, how many managers know the workings associated with the generation of pseudo-random numbers in the popular MS Excel application? Even with daily job specific tasks such as, financial reports and financial indicators one would be surprised at the number of managers who are not cognisant of the calculations and/or formulae require to prepare these compilations. Obviously, MoK products or services require a degree of awareness and intent. Although many tools and applications are "easy-to-use" and quickly implemented such as, wikis, discussion forums, videoconferencing, and document management systems, other applications that entail expert systems or multi-agent technologies require awareness of knowledge acquisition, or knowledge storage.

Relationship among and between Single Levels

Each level generates one coherent system that comprises its own considerable relationships. However, as MoK entails the basis for all knowledge related activities, if it is not operational or functions improperly, activities at preceding levels are at risk to be incomplete and compromised. This is of considerable concern as subsequent products function within the organisational level. Despite this strong dependency on technology, the MoK level does not support a technological approach to KM nor does it stress its technological origins. It does nonetheless; accentuate the fact that modern technologies catalyse change. Briefly, the organisational level constitutes the basis for the KE at the national and supranational level.

The higher the level, the lower is the significance of specific knowledge. While MoK operates with real knowledge - no matter the acquisition, progression, distribution or exploitation of the knowledge - and develops tools and procedures at the organisational level it is not feasible to manage knowledge. Rather one creates the appropriate conditions for successful knowledge manipulation. Consequently, managers concentrate on schematic (purpose, strategy, culture, and structure), content (participants, artefacts), and knowledge resources (Holsapple and Joshi, 2001). At the national and supranational levels, the objective of all activities remains the creation of an environment as an economical, political, legislative framework from which lower levels can operate successfully.

Once KM embeds into the context of a greater vision substantial relationships at any single level or among particular levels is not only preserved, but also emphasised.

Institutions that deal with supranational and national affairs that include institutions related to the EU public administration or specific Czech national authorities present another interconnection of single parts. KM at the organisational level and products of the MoK level can be utilised by these organisations for improved performance.

Within these levels, individuals who focus on knowledge are more often politicians, managers, information technology experts, and academicians. There are various setbacks at these levels from the Czech perspective. They demand a systemic view wherein KM is entrenched into the context of a larger custom. In this way, substantial relations at a single level and among particular levels are both preserved, and emphasised.

METHODOLOGY OF KM IMPLEMENTATION

There exist many methodologies and approaches to KM implementation that include the Standardized KM Implementation (Coviello, 2001), as well as the On-To-Knowledge (Sure, 2004). In general, methodology is an explicit way of structuring and thinking. According to Jayaratna (1994) it contains a model that reflects selected views on reality that originate from a set of philosophical paradigms. Methodology dictates which steps are to be conducted, why and in what sequence. From the systems engineering perspective, a methodology provides the conceptual structure and by which a task or series of tasks are to be performed whereas, a project provides the infrastructure from which the methodology is acted upon and finally implemented.

At the Faculty of Informatics and Management (University of Hradec Králové), a new methodology, KM-Be.At-It, was developed. It aims to overcome some of the difficulties, barriers, and other problems encountered. The name represents one of two views. The colloquial KM-Be.At-It promotes the action of doing as in Knowledge Management "be at it" thereby, inspiring enterprises and practitioners -to be at it- with interest and zeal. On the other hand, the more insightful significance KM-Be.At-It compels the audience to master KM as in to beat KM-related issues as well as to beat a purely IT-based approach, grounded in a complex systemic.

Using Tiwana's KM Toolkit (2000), Wiig's building blocks (1999) and Kim's P2-KSP (2003) among others as theoretical foundation the creative process of design and development comprised several stages. The central consideration omnipresent throughout this process is the applicability to the Czech reality. The initial focus of the preliminary phase is the definition of KM subsequent to lengthy analyses of the abundant sources. An in-depth review of the strengths and weaknesses of existing methodologies as well as the definition of the basic attributes for the new adaptation followed this preliminary examination. Figure 2 schematically depicts an overview of the content associated with the KM-Be.At-It methodology.

KM-Be.At-It consists of several phases. Each phase comprises an objective, purpose and content, basic prerequisites relative to the initiation, completion criteria, resource documentation, critical success factors, as well as the effort and the relationship of these actions to each other. Every phase consists of several activities and tasks. Since KM-Be.At-It works with a deep-level resolution, and includes a specification for each activity including the specific goals, inputs, outputs and examples of utilisable methods, techniques and tools. Single phases, graphically depicted in particular figures, provide a view of the sequence of activities. Dashed lines indicate the possibility of several iterations before the activity that follows commences. The width of the rectangle refers to the relative suggested time interval for performing each activity. For this reason, this waterfall approach originates from the systems engineering standpoint (Hitchins, 2007).

The formation of a project team in the initial phase has, as principal objective, the acquirement of support from higher management and/or the owners of the organisation. The project team is responsible for the whole process of KM implementation and is foremost in the acquisition of management sponsorship. This phase consists of:

(1) Without the patronage and support of higher echelons of management and in the case of the Czech small and medium sized enterprises, the owner, or owners, all further

Figure 2. KM implementation process using the KM-Be.At-It methodology

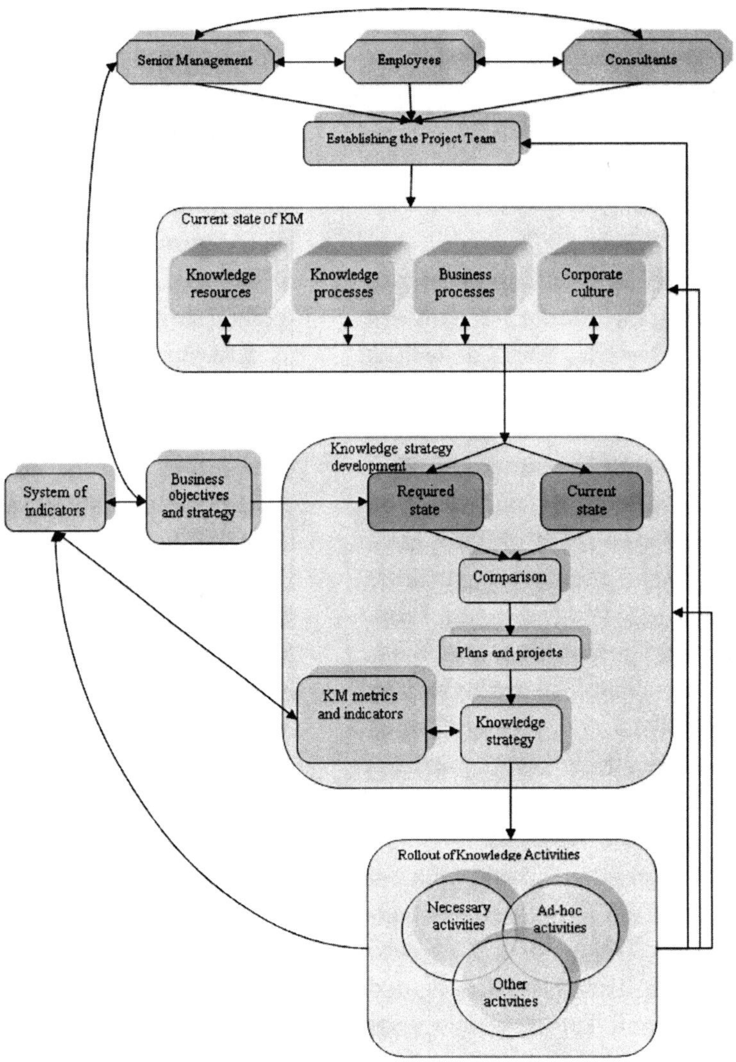

steps are futile. Management must not only be aware of the KM process and benefits but they must sanction the associated costs. This initial step may prove to be the most arduous.

(2) The evaluation of the possibilities and capabilities inherently available within the organisation required to undertake a KM project.

(3) The "GO / NO-GO" decision process during which all parties agree to pursue or to table the notion of KM implementation. For many Czech enterprises, the decision is a crucial one requiring serious review. The KM-Be.At-It documentation suite contains a number of templates with one entitled 'Business Case' that is especially useful at this decisive juncture.

(4) Once approved, establish a team as KM gatekeepers whose major functions include educating the organisation, developing the organisational strategy, and implementing

the latter. The team embraces a cross-section of the organisation thus avoiding groupthink.

(5) From within the team, assign a group lead or team manager. Following this nomination, clearly define and document roles and responsibilities. Determine and establish a method of financial tracking as part of the project management procedures. Templates and procedural guides constitute part of the methodology documentation set.

(6) As not all team members are required for the entire duration of the endeavour, it is imperative to define assignment completion dates.

Whereas the first phase concentrates on setting ground rules and identifying the key players, phase two identifies the current state of KM within the organisation prioritising the areas requiring improvement. In essence, this phase consists of eight steps:

(1) Survey of existing knowledge resources and the compilation of a comprehensive inventory;
(2) Detailed description of the nature and components of each identified knowledge resource;
(3) Identification, definition and cataloguing of existing knowledge processes including ownership;
(4) Evaluation of the existing state of knowledge processes within the organisation;
(5) Identification, definition and cataloguing of organisational processes;
(6) Investigation, documenting and defining the current state and type of organisational culture;
(7) Compilation and coordination of the results;
(8) Publication of results including in-depth analyses and recommendations to move forward.

The third phase is extremely crucial to the successful KM implementation and for the healthy survival of KM within the organisation. The objective of this phase is to finalise the knowledge strategy that supports the business strategy, knowledge-related activities, and ultimately, the KM goals that in turn affect overall business objectives. Not to minimise the importance of this step, the seven major activities (with others incorporated depending on local circumstances and initiatives) include:

(1) Determine and define realistic KM goals;
(2) Compare the existing with the proposed and identify gaps;

Figure 3. Activities in the phase of project team formation

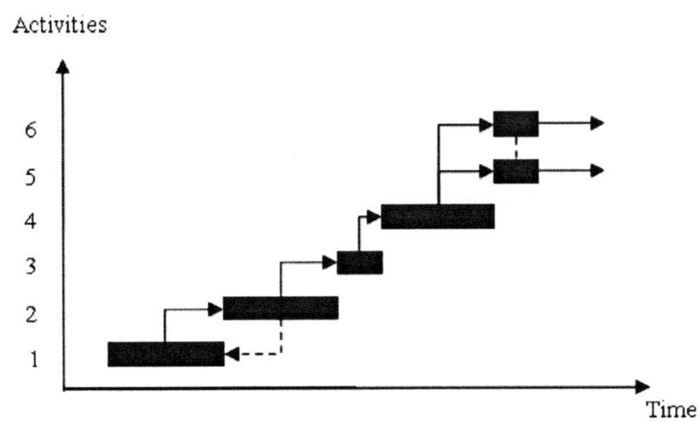

Figure 4. Activities in the phase of current state analysis

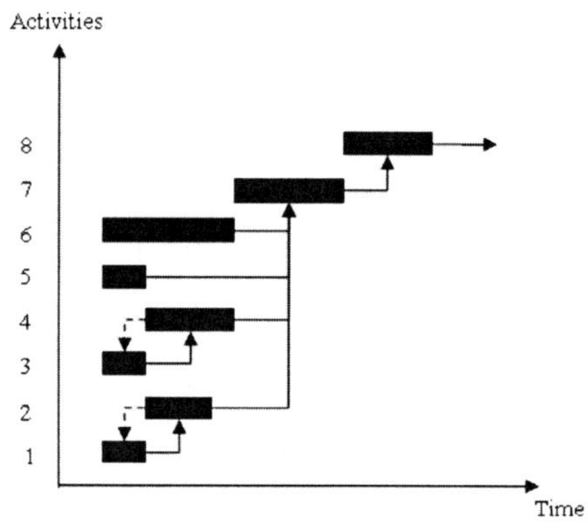

(3) Draft the list of activities required to address the gaps;
(4) Prioritise activities and the creation of a schedule;
(5) Elaborate an action plan considering tactics and policies relative to the overall project;
(6) Finalisation of the knowledge strategy and subsequently, the business plan and strategy;
(7) Identify KM metrics follow-up plans and benchmarking practices.

The last phase in the process accomplishes the actual realisation of KM activities, if not in its final state, at least in a preliminary or trial mode. The KM-Be.At.-It methodology promotes a flexible but well documented approach as outlined in the supplied documentation. Although structured the process permits additional local innovation to accommodate the Czech management actuality.

During this end phase of the process the completion of various activities, sub-projects, and action plans continues. Typically these are associated with the actual deployment and may include among others, educational and motivational programmes for increased employee knowledge sharing and involvement; the naming of an individual as the "knowledge prime" who continues the function as knowledge group leader in the future. Based on the identified KM strategy, the various elements of the KM programme are rolled-out in varying states of finalisation that can include a Community of Practice, intranet sites, on-going training and discussion groups to tackle learning and sharing barriers. The management or the ownership team promotes on-going efforts across the organisation as part of their sponsorship role.

Figure 7 depicts the activities in each phase of the whole methodology illustrating the iterations and loops within each.

Figure 5. Activities in the phase knowledge strategy development

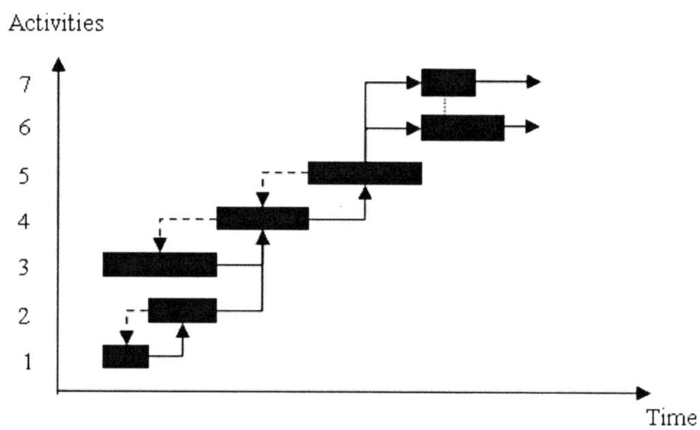

Figure 6. Realisation of particular knowledge activities

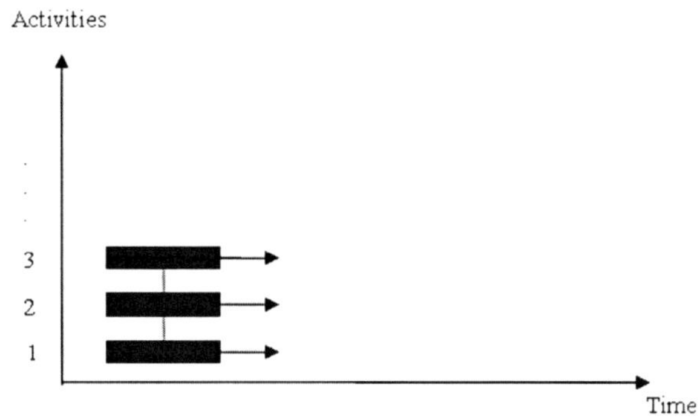

Figure 7. Phases and activities of KM-Be.At-It

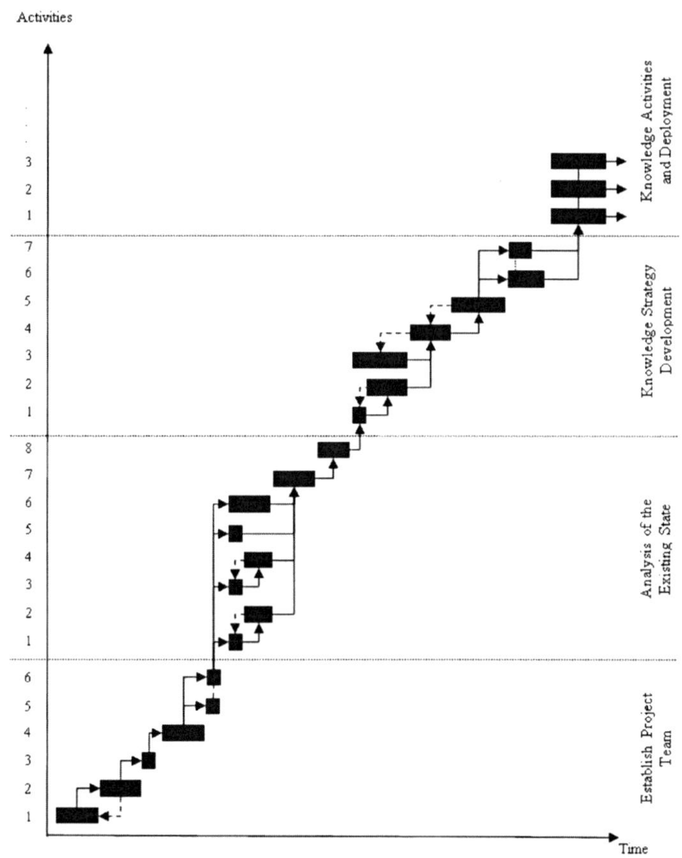

Continuous Knowledge Management

Although an important movement forward the KM-Be.At-It methodology is only the initial step on the long journey to KM (Figure 8). Once instituted no matter how meagre the attempt, it is necessary to recycle from the beginning at regular intervals. As knowledge evolves and reinvents

Figure 8. Continuous KM

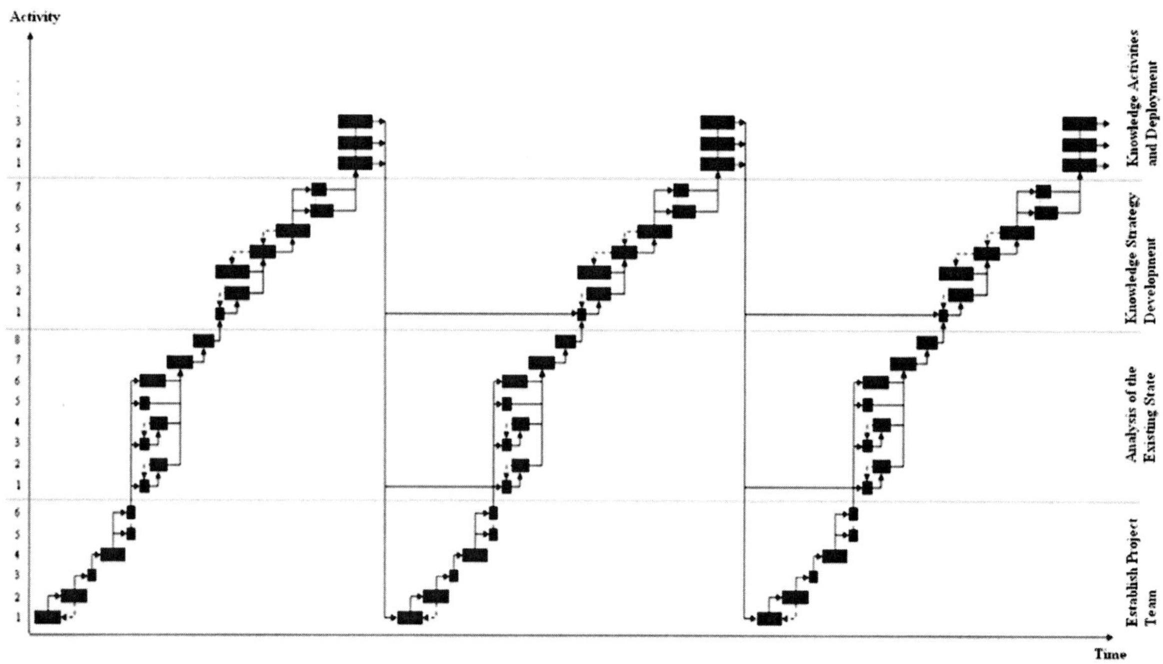

itself, so must KM. Knowledge resources perform extensively throughout the project phases and in order to maintain this knowledge-centric exercise the momentum continues. If not, all is spent in vain in fulfilment of yet another management fad rather than an enriching experience leading to shared knowledge and tangible benefits.

The aim of the first cycle is to set things in motion. Continuous application of the methodology leads to the desired changes within the organisation.

Benefits of KM-Be.At-It

KM-Be.At-It comprises numerous benefits and advantages while respecting the recommended requirements, characteristics and criteria of methodologies, or KM methodologies in particular (Jayaratna, 1994; Hoog et al., 1999; McElroy, 2004). Furthermore, it is in accordance with the KM frameworks prescribed by Wiig, Hoog, and Spek, (1997) and respects the viewpoints outlined by Beckman (1999). To avoid the Czech perception of KM being synonymous with turnkey technological products, KM-Be.At-It assumes the utilisation of information, communication, and knowledge technologies but does not emphasise these solutions within the process. Rather it stresses a "green" usage of what is already at hand. This approach also reinforces the position that KM need not be costly while it respects the economic as well as the collective business objectives. This theme of changing what is currently available equally applies to the organisational culture and its impact on the success of a KM initiative. This avoids the expensive expenditures often associated with IT-based solutions.

As defined in its literature KM-Be.At-It sustains a high degree of neutrality in that it is an approach developed by a group of academicians and practitioners. Unlike some consultancy firms, the only profitability concerns are those of the implementers. The goal of the methodology is specifically to meet the requirements of the Czech audience and its objectives do not reflect one specific branch within industry. The steps and phases associated with deployment, documented in the Czech language, include a high degree of flexibility permitting customisation to meet individual institutional differences. Moreover, it incorporates both a "top-down" and "bottom-up" approach. The "top-down" is extremely beneficial in the initial phase seeks approval and consensus. The "bottom-up" is essential during the phases that follow and the on-going recycling.

Critical Success Factors

As with any process of magnitude, KM-Be.At-It relies on a set of predetermined factors critical to its success and acceptance by its future adherents. The methodology's toolkit discusses the following in more depth and cites recommendations for overcoming obstacles and the on-going task of educating on the benefits of KM in practice. The methodology stresses two groups of critical success factors, the human, and the strategic. Within these groups, the major success criteria consider the higher management/ownership team; employees; KM project team; business strategy, and knowledge strategy.

Higher Management/Ownership Team

This category is crucial not only at the onset of the initiative but throughout the process. As the KM endeavour is not static but fluid so must be the sponsorship by members of the firms' leadership team. Three of the more common roadblocks encountered are: (1) the under- or overestimation of KM and its deployment; (2) lack of significant involvement from the firms' leadership and finally, (3) an over burdened level of requirements and responsibilities either assigned or expected from the KM project team. Management theory teaches that practitioners often "oversell" or "undersell" a project to gain favour. This happens when delivering what the receiving party expects to hear despite tangible indications to the contrary. As with all similar cases, this somewhat dangerous approach creates either the underestimation or overestimation of KM and its potential. In the first case in point, the leadership team believes that KM is simply another management vogue and therefore, pays lip service to its implementation. On the other hand, when overestimated, KM becomes the sole method of survival expecting customers to start immediately knocking at the door. In all cases, equilibrium is the objective and reality primary on the agenda. To avoid such cases of extremes, additional emphasis is imperative on the fact-finding abilities of the project team during scrutiny of the existing knowledge state as well as in the preparation of the business case. Lip service alone is insufficient and will ultimately result in the premature cessation of any KM venture especially during the recycling or continuous KM periods. What is more disconcerting occurs when management believes that KM is restricted to experts and not exclusive to an internal organisational mandate. Early in the KM-Be.At-It methodology, the process defines the relationship as one of strategic management in which all members of the organisational culture become and are on-going participants. Organisational executives are made cognisant of their role, encourage KM and recognise the benefits of this common effort.

Employees

One of the most salient results of KM implementation is that it affects the way the enterprise conducts business and the way that the members work and operate on a daily basis. Literally, it influences the entire organisation but most importantly the base, the employee-knowledge workers. There are two main considerations in this area: the lack of sufficient and continuous training and the unwillingness to support change at the grassroots. KM has been, especially in the Czech Republic, limited to management, academia, or the politicos. Even within these circles the actual mechanics supportive of this process remain at best, vague; hence, the reliance on oftentimes foreign consultancy firms. From the onset and throughout the KM process, employees must be aware of its nature, the purpose, and the beneficial results. This is especially important in cases where non-management members are unionised. To avoid viewing KM as a threat, the organisation defines and communicates the communal advantages. Training, information, and KM progress reports are de rigueur.

Change within management and an organisation is a constant dilemma worthy of its own discipline recognised as change management. No matter the size of the establishment, it is inherent of human nature to resist change as it involves the unknown. Without the proper KM induction, training and subsequent follow-up employees resist KM until they realise the favourable components. The project-process team, although hyphenated represent the same group of people, are those most heavily monitored as to the success factors and the many obstacles to which they are subject. There are a myriad of impediments and many vary on the size and the nature of the organisation and other unique conditions subject to circumstance. Due to the significance and predominance of this group, individual headings treat the factors worthy of concern.

The Composite Nature of the Team

Given that the methodology concentrates on addressing the needs of the Czech small to medium range of enterprises, one assumes that budgetary restrictions exclude outside sources and that the effort is internal. The KM-Be.At-It package includes ample background information on all aspects of KM using a number of pedagogical methods including the popular "cookbook" technique. If necessary, an instructor or advisor is available for classroom instruction. The induction sessions (in the Czech language) discuss in lay terms and illustrate with examples the why, where, when and how. The training is self-paced and contains the necessary reviews. Also vital to the success of the project team is its composition in that it represents a segment of all members. During this learning cycle, there is no longer an employer-employee bias as learning holds no status and knowledge is ignorant of social rank. The exclusive delegation of KM-related activities to a low level of the organisational hierarchy risks allocating the responsibility to a group who may not possess the necessary experience, power, and exposure to succeed.

The investigation and planning cycles of the KM process require adequate concentration and studious review. This cycle is central to the success of the project as it determines the costs and benefits both tangible and intangible attributable to the deployment. The attempt is considerable and conducted with the highest degree of professionalism and exactitude. The development of forecasts and projections is an exercise of planning and organisation. This exercise is not superficial or frivolous but one of analysis of reality both current and expected. Failing to do so create the over- and underestimating scenarios discussed in previous paragraphs that affect the project negatively and lead to disappointment and suspicion. As with any plan, identify risks from the project inauguration.

An abundance of methods, tools, and techniques facilitate the implementation process. Although the project team members are encouraged to investigate and evaluate how the aforementioned may assist the project to progress, avoid over dependence on any one of these. Avoid any approach that boasts of an overall solution no matter how attractive due to its popularity, modernity, or cost-savings. The implementation process is about the coordination of numerous approaches and is not about the perfect application of only one of them.

Business Strategy

The business strategy forms the apex for all activities within the organisation. In essence, it is the raison d'être of the enterprise. KM focuses on the support of the business strategy. Thus, any flaws in the latter mirror in the KM approach. If the business strategy is non-existent, is antiquated, or unrealistic, the KM process suffers and may actually contribute to the demise of an enterprise. A short-term strategic orientation may overly concentrate on one goal such as the reduction of production costs while ignoring equally pressing issues. Similarly, a long-term orientation in the same case creates rigidity. Both long- and short-term strategies can be successful if they are well balanced, flexible and reviewed on a regular basis.

If the knowledge strategy is a result of an outdated business strategy, the lack of congruency between the strategies and the current circumstances and goals jeopardises the potential for a successful realisation. The development of the business and knowledge strategy in parallel is the framework for an identified period. The organisation continues to review and update both when necessary.

Knowledge Strategy

The knowledge strategy renders sense and meaning to all KM related activities. The knowledge strategy establishes the direction for all predetermined activities. Any failures, unseen risks left undetected, or omissions within the knowledge strategy impinge on the successful achievement of the KM undertaking in proportion of their gravity. The content of the knowledge strategy is constantly scrutinised and re-evaluated to avoid these possible repercussions.

Comparable with the business strategy, it owns its own lifecycle that demands monitoring. Although its existence derives from the business strategy, its survival depends largely upon the flexibility of the latter. Monitoring of the knowledge strategy confirms its compatibility with the business goals and aspirations. The on-going alliance of both strategies assures the realisation of objectives and the narrowing of the gap between the existing and the sought conditions.

In conclusion, the global market environment influences transition economies in many possible ways. It demands changes that organisations have to produce. Therefore, some enterprises seek knowledge as a unique resource. KM ensures a higher level of competitiveness. Nevertheless, this higher level materialises only through proper KM deployment. Critical success factors during implementation require supervision, and control and specifically related to higher management, employees, the project team, and of equal significance the business strategy and the knowledge strategy.

THE CULTURAL PERSPECTIVE

Despite the rapid growth of KM over the past decades, organisations that implement an IT-based solution allocate meagre consideration for this effort. Associated with one of the major principles of KM, knowledge sharing is often implemented by enterprises who use sophisticated web based products to enable the sharing under the assumption that all employees at all levels of the enterprise will avail themselves of these IT services. Management invests money, time and technology so that corporate members consult the intranet sites for vital information, save crucial documents such as best practices, communicate through forum and consult their contents for sales leads. What is commonplace however, although the site is user friendly, many pages remain blank, "under construction" or require updating.

In this instance, KM has hit a "cultural wall," wherein the human-machine interface has not been exploited to its fullest. The intended audience lacks sufficient initiation whether through training or informative sessions on the utilities and benefits of the intranet. KM necessitates changes in corporate culture but also in the fashion in which technologies modify work procedures. While IT enables searching, storing, manipulating, and sharing of considerable volumes of knowledge, the human factor remains the foremost concern. The essence that knowledge depends on individuals is understandable.

These same individuals, the primary force in an exercise of knowledge sharing, must enjoy the ability to interpret the knowledge within context. The work techniques used by employees ultimately determine the degree of efficiency of the technologies deployed to convey and share knowledge. Again, this demonstrates that in the KM process, IT plays a substantial role but not a central one. At first glance, it appears that the easiest method to ameliorate knowledge sharing is to subject corporate culture to a stringent overhaul. Prior to embarking on this journey one must realise that this change requires time, energy and financial resources. The corporation alters culture to accommodate KM or KM adjusts to accommodate the existing corporate culture. In both situations, common cultural barriers subsist.

Recognition of knowledge in both a subjective and an objective nature is necessary. In its objective form, it is explicit, implicit, and tacit whereas the subjective components constitute social practices, working groups and the like. Both share equal importance. Institutional communication flows from individuals to groups and vice versa. Social practices, the elements of corporate culture, therefore affect the knowledge and consequently, the act of sharing. Knowledge in this instance constitutes the accumulation of activities within this context but depends on its own creation by the social group. Once created this collective knowledge and not individual knowledge becomes the basic asset of the community, institution, or enterprise. Ideally, this environment produces members who actively and willingly share knowledge with repositories, reports, and verbal exchanges as an integral part of their daily routines. This utopia in turn results in higher productivity, effectiveness, and innovation. Yet prior to reaching this ideal state or any varying grade possible within realistic terms, issues of corporate culture, individual and social barriers demand review.

Individual barriers: Largely, individual barriers are related to the employees' sense of worth, value and sense of security within the organisation. The axiom of knowledge is power compels some members to fear a loss of power whether real or imagined. To the employee the exclusivity of influence once reduced entails the loss of face and potentially, income. Just as there is fear through the revelation that by sharing this knowledge has intrinsic value, the reverse is also valid in that the knowledge holds little worth. In both instances, embarrassment descends onto the worker. Junior workers due to their lack of experience, seniority, and possibly status at the workplace may feel that there is little merit in sharing their knowledge. They are usually on the receiving end of the knowledge flow; their inexperience acts as a hindrance and possible humiliation. Sharing knowledge and subsequent experience appears for some as being outside the normal job functions and is therefore, additional work. This creates the expectation of supplementary rewards-either tangible or intangible. Finally, there is a difference between awareness and knowledge. Some are unable to describe the corporate mission statement but are very knowledgeable on corporate processes.

Social barriers: Many of the social barriers compose daily corporate life and are not obvious without conscientious reflection. One frequent and common example is the use of language. Acronyms, abbreviations, technical terms, and reacted jargon differ from business unit to unit that lead to obstruction within internal communication and the distinct possibility for knowledge sharing. In addition, if the sender is not able to estimate the state and size of the receiver's knowledge base, knowledge sharing does not have to be fecund even when the communication process occurs. Rigid bureaucracy and the tendency to adhere to strict administrative guidelines may lead to incompatible paradigms of personal versus corporate interests. It may also breed an atmosphere in which members avoid risk and innovation as not to deviant from the status quo. Higher management whose attitude is patronising may underestimate the talents and abilities of their workforce. Under these circumstances management, if they do opt for a KM solution, will pursue one heavily dependent on IT technologies to store past information under the assumption that future needs remain the same. This approach renders the knowledge worker as a passive receiver only and management as the knowledge providers. By doing so, management gathers the content of the organisational memory as the final product in design time and then disseminates its content. This style operates contrary to the notion that the knowledge worker creates the new knowledge at the workplace. The knowledge worker as expert solves new and ambiguous problems daily. Workers not management create the organisation's knowledge base as side effects of the tasks and challenges encountered. Emotions also affect the willingness of employees to share knowledge with others. It is difficult to compel two employees to share knowledge with each other when the relationship is unpleasant or difficult. Similarly, pseudo-innovators are those who exaggerate their opinions as "necessary" although unwanted or unsolicited improvements to the ideas of others to emphasise their importance and indispensability.

Motivational Conflict

One of the main reasons of insufficient knowledge sharing is motivational conflict at an individual level. This clash is often a frustrating experience as it frequently involves postponing the necessary or covetable reaction or decision. Decision-making is repeatedly awkward due to the reciprocally incompatible tendencies. K. Lewin's model (2008) defines conflict as the "opposition of approximately equally strong field forces." The definition encompasses five basic conceptual components: appetence-appetence; aversion-aversion; appetence-aversion; "I do not want to, but I have to" and "I want to, but I can not."

Basic to the discussion are two assumptions:

1. Organisations employ a wide-range of incentives. These incentives should ensure knowledge sharing among individuals, individuals and groups and groups within the establishment thereby overcoming existing barriers. This incentive system including financial and intangible rewards, is perceived positively by employees and thus motivating;
2. The majority of individuals perceive knowledge sharing as negative, unpleasant, or undesirable due to the five conceptual components cited above.

The *appetence-appetence* conflict arises when the individual has to choose between two positive alternatives. The solution lies in the sudden higher attraction of one alternative over the other. If this type of conflict does occur, it is a positive dilemma in that the employee has to select one incentive or reward from the system.

The opposite is true with the *aversion-aversion* conflict whereupon an individual has to select from two negative alternatives. The individual usually postpones the final decision as long as possible. With time, the negativity of one emerges as the same for the other alternative. If one grows less unpleasant, the individual selects it. The individual may abandon the idea, but in the end still has to decide.

An *appetence-aversion* conflict arises when the object, the knowledge sharing, is both inviting and forbidding. This conflict type is the most frequent. Incentives exist, but it is necessary to share knowledge to attain them. For some employees this entails a questioning of their positions, and values. From a psychological perspective, the intensity of aversion grows in proportion to this awareness. The reciprocal is also valid. Once the consciousness draws on the negative feelings, the barriers increase. N. E. Miller (1950) in research based on this conflict type postulated principles of ambivalent behaviour:

1. the tendency to approach to the goal is stronger when the subject is nearer to this goal;
2. the tendency to avoid a negative alternative is stronger when the subject is closer to this alternative;
3. the tendency to avoid a negative alternative increases upon approaching this alternative when compared to the tendency to approach the positive alternative;
4. appetence and aversion gradients depend on the intensity of encouragement; the tendencies are more intensive with stimulation that is more concentrated.

In the third point, the individual realises that the dissonance appears with the effort to reappraise the decision. Immediately after the decision to share knowledge, the member accepts both the positive (rewards) and the negative attributes (potential loss of ownership). At the same time, the individual acknowledges the refusal of the positive attributes of not sharing knowledge (increased self-confidence, power, promotion). This reassessment occurs due to the reduction in conflict and consists mainly of overestimating the positive qualities of knowledge sharing and in underestimating the negative. Knowledge non-sharing invokes the reverse process. By pursuing these processes, the employee assures himself/herself of the appropriate choice. The role of the organisation in this progression is one of guidance and enlightenment.

The conflict, "*I do not want to, but I have to*," has typically negative motivational goals for the individual swayed by external values. The individual feels obliged to achieve the goal. The conflict is common in circumstances in which higher management introduces KM as part of an overall company-wide endeavour without having properly introduced the programme or obtained feedback from the employee team.

The "*I want to but I can not*" conflict is the inverse to the previous however the mechanism is the same. This condition is not as common as the previous in that one would find it difficult to envision an organisation directly preventing the sharing of knowledge among its employee base.

Individual and social barriers often prevent effective knowledge sharing. It is therefore necessary to identify and eliminate the maximal number of these barriers. Some of them are possible to remove completely but others remain. The remaining barriers are necessary to minimise in relation to the given conditions. One of these remaining barriers is motivational conflict arising from knowledge management processes particularly, knowledge sharing. This barrier is deeply rooted in the social system and often supported by incentives. In combating individual and social barriers management call upon tools and techniques such as story-telling, expressive communication, trust, people management, teamwork, and balanced usage of codification and personalisation approaches, design and workplace organisation. In addition, the use of social information systems such as Babble (Thomas, Kellogg and Erikson, 2001), or KM realisation models such as SER (Fisher and Osvald, 2001) assist management in removing these obstacles.

The Cross-Cultural View

An area deficient in research attention is the cross-cultural relationship to Knowledge Management. Hofstede (1980, 2001) states that managers cannot assume that organisational cultures exist independently of the national cultures in which they operate. Furthermore, different national cultures emphasise distinct values by means of diverse languages. These additional factors influence the ability to implement extensive initiatives successfully. In an increasingly international business environment, organisations need to comprehend if and how cross-cultural differences influence knowledge management and KM-methodologies.

As noted throughout this chapter, the predominant challenge facing enterprises aspiring to gain a competitive advantage through knowledge management is the creation of a culture and environment in which knowledge sharing thrives. Without the appropriate culture, knowledge transfer is difficult, and does not occur (Davenport and Prusak, 1998). Hofstede argues that organisational culture resides within the national culture and influences the why the nation and its adherents conduct business. Furthermore, studies investigating the effectiveness of knowledge transfer between international partners discover that cultural differences affect information flow, knowledge management processes, and knowledge transfer (Lyles and Salk, 1996). Simonin (1999) concludes that the cultural distance between firms increases the difficulty of performing these processes successfully. Previous research suggests that national culture and cross-cultural distance may influence knowledge processes.

Hofstede defines national culture as "the collective programming of the mind which distinguishes the members of one human group from another" (1980, p. 25) and collective programming as "a broad tendency to prefer certain states of affairs over others" (1980, p. 19). Members of a culture have similar sets of preferences on they view the world. Hofstede proposes four factors that became the dimensions of the taxonomy: Individualism/Collectivism; Uncertainty Avoidance; Power Distance, and Masculinity/Femininity. In 2001, Hofstede developed a fifth dimension, Long Term Orientation. The differences in these dimensions direct the knowledge processes within any organisation. For instance, cultures that measure high on Individualism value the accomplishments of the individual, whereas Collectivistic cultures place more value on group or team accomplishments. One possible implication is that Individualistic cultures demonstrate more difficulty in knowledge sharing as knowledge is a source of power and a tool for success. Similarly, cultures that are high in Masculinity (competitiveness) have more difficulty in knowledge transfer if the competitiveness is between individuals and not limited to organisations.

Power Distance and Long Term Orientation alter knowledge processes (or the adoption of knowledge processes). Countries that have a high Power Distance possess hierarchical structures, therefore knowledge is more likely to flow from the top down, or is more limited than knowledge flows within a culture that has less Power Distance. If a culture has a Long Term Orientation, then the members of that culture work for the long-term goals and benefits that accompany knowledge management. Since benefits of knowledge management are not realised in the very short term, a Long Term Orientation promotes and values knowledge management processes.

In summary, national cultures and knowledge transfer intersect in the following ways:

a) Cultures that rank high on Individualism have more difficulty in knowledge transfer than cultures that value Collectivism.
b) Cultures dominant in the Power Distance dimension possess a top-down flow of knowledge than cultures that are lower on the Power Distance scale.
c) Cultures that are high in the Masculinity aspect conduct less knowledge transfer between organisational members if competitiveness is individually based. There may be no difference if competitiveness is organisationally centred.
d) Knowledge transfer between heterogeneously composed cultural groups is more difficult- requiring more time and effort- than a knowledge transfer within homogenously composed cultural groups.

The aspect of cross-cultural management and the influence on knowledge transfer and KM in general merits a full discussion in its own right. Despite this brief discussion, there are two characteristics worthy of note. KM is predominantly a process of the Anglo-American management realm. As identified previously, the concentration within the Czech reality is specific to organisations that do not resemble the typical case studies. There are many examples in the current literature. Three of these consulted deal with cultural dimensions in global virtual teams (Usoro, 2005); cultural impacts within selected European Union countries (Gulev, 2005) and Deutsche Bank's human resources database (Heier and Borgman, 2002). Within these case studies, each author relies heavily upon Hofstede's (1980) cultural dimensions to substantiate the dissimilarities between the American KM model and those studied. Furthermore, the conclusions attribute any perceived organisational problems to the cultural differences. This premise consequently alleviates the burden from the authors to conduct any further research as what may have been the true origin of the difficulties such as, the lack of training or language.

Heier and Borgman study the impact on national culture on the development, rollout, and usage of KM-based projects. Once identified and categorised the attributes of the research are validated against a conceptual framework based entirely on Hofstede's cultural dimensions. In a similar vein, Gulev seeks to develop two criteria from which to evaluate the cultural impacts on knowledge flows and knowledge sharing in four European countries: Germany, Austria, Denmark, and Slovenia. Data collection is part of a larger project using questionnaires and follow-up interviews via telephone or face-to-face. The authors use two methods of empirical data analysis, Spearman rank correlation and the Standard linear correlation tests. Results are categorised into logical groupings sharing the same attributes. These two examples represent only a meagre sampling of the available research. The methods deployed by the authors are however, consistent with other studies consulted in that they rely principally on the use of surveys, questionnaires, and face-to-face interviews to secure their data.

RESEARCH CHALLENGES

Knowledge Intensity

Although introduced to the public, the KM-Be. At-It methodology did not produce the forecasted results and recognition. Managers- Czechs are not an exception-require motivation to deploy and to appreciate the anticipated changes. Therefore, it is necessary to establish a concept of organisational knowledge intensity and to confirm the positive influence of a KM programme within such

It is understandable that the knowledge intensity of specific nations, industries, or organisations is of crucial importance to the whole knowledge hierarchy. While measuring the benefits of KM implementation or the intellectual capital development in organisations is quite common today, overall organisational knowledge intensity quantification including both static (knowledge resources) and dynamic (knowledge processes) aspects of KM is rare. Autio, Sapienza and Almeida (2000) define knowledge intensity as the extent to which a firm depends on its knowledge as a source of competitive advantage. Davenport and Smith (2000) assert that knowledge intensive companies will allocate more resources to knowledge management. However, the criteria enabling the classification of organisations according to their knowledge intensity lacks description. Likewise, knowledge intensity may not be associated with KM programmes whatsoever. Every organisation possesses knowledge resources and realises knowledge processes, but few recognise such consciously.

In general, the concept of knowledge intensity is widely known. It consists of minimally four factors: engagement of knowledge and understanding required to perform assignments; expertise necessary for the competent handling of work-related unpredictability; severity of consequences of potential work inaccuracy, and swiftness of action. However, there lacks a generally accepted methodology used to measure organisational knowledge intensity from a managerial perspective. This observation enables managers to view their organisations from a different standpoint. Firestone (1999) considers the organisational information and knowledge infrastructure to consist of two models. The first represents the organisational and human components and their continual interaction. Design does not dictate this model, but it emerges from the dynamics of the enterprise process and its activities. Thus, it is natural. The second model is a conceptually distinct and integrated system that arises from the former. It includes computers, software, or electronic components. This model is artificial and developed by design. The resulting information-knowledge processing infrastructure is often a blend of the two. Since organisations are complex systems, humans are not able to design fully the infrastructure leaving one natural phenomenon to propagate an unexpected and emergent behaviour (Čech, 2005). Monitoring of the organisational knowledge intensity assists managers in evaluating how the representations are interrelated, developed and their weak and strong points.

The Project

Knowledge management aims to intensify the comprehension of knowledge processes and the utilisation of knowledge resources. Nevertheless, these are present at some level in every organisation regardless how explicit the nature of their development. Existing explicit knowledge management initiatives are probably more intensive than those without similar functions. Therefore, the proposed project investigates organisational knowledge intensity models, representation, monitoring, and usage. The research proposal concentrates on several goals:

1. systemic exploration and acquisition of deep knowledge on the research performed at the organisational level specifically in the area of knowledge management, knowledge intensity, and methodologies used to evaluate overall organisational results;
2. based on results of the above, the creation of state-of-the-art tools to compile existing attitudes and measurements of both the organisational overall performance and knowledge management. The process incorporates the description of mutual relationships and the existing strengths and weaknesses inherent with these approaches;
3. development and design of a complex methodology for the quantification and monitoring of organisational knowledge intensity and its experimental verification on an international basis, utilising results acquired from Czech, Slovak, and Canadian companies.

Simultaneously, the project seeks answers to the following questions:

- What is a definition of organisational knowledge intensity and what are the possibilities of its quantification?
- In what manner do particular factors such as cross-cultural management, organisational processes, customer base, information and communication technologies, or knowledge technologies contribute to organisational knowledge intensity?
- What is the perception of organisational knowledge intensity (described or modelled)?
- For organisational knowledge intensity quantification which methods, techniques, and tools are used? How does one design and implement such within a managerial environment?
- How does one benchmark the knowledge intensity of organisations in the knowledge economy?

The Research Perspective

From the research perspective, the entire knowledge hierarchy involves many scientific disciplines such as economics, international trade theory, management science, political science but also informatics, sociology and psychology. One current problem is that these disciplines usually analyse at a single strata without deep consideration or linkage to other levels. The more one escalates in the hierarchy, the more evident are the research problems related to economics or management whereas, in the opposite direction, research problems related to informatics or psychology are evident. Consider this perspective when conducting research at the organisational level and organisational knowledge intensity.

Therefore, the application of a systems approach and thinking is necessary while studying organisational knowledge intensity in the context of the complete hierarchy. Moreover, a systems approach is applied when contemplating organisational knowledge intensity and using it as a practical indicator. Unfortunately, the majority of managers are not prepared as training concentrates on a mechanism rather than a systems approach.

Increased knowledge and insight originate from an interdisciplinary and multidisciplinary research. The proposal emphasises the multidisciplinary to explore organisational knowledge intensity. Knowledge economy development (Chen, 2005), knowledge intensive industries, cross-cultural management, knowledge management, modern managerial measurement systems (Kaplan, 1996), organisational culture, human resources, and knowledge technology applications (Mikulecký, Ponce and Hynek, 1997) are only some of the facets considered. Throughout the proposed study, attention focuses on the possibilities of the further development of organisational knowledge intensity, the performance of particular organisations in knowledge management activities and on the general indicators of their performance.

The proposed research concentrates on five major industry segments - public administration, banking, ICT and telecommunications, academia, and tourism services) and uses the following methods:

- Survey Research: Self-administered questionnaires comprise the nature of the survey research. The participant completes the self-administered questionnaire online. The web site contains both the survey instructions and questions. The goal is to acquire data relative to methodological modification and enrichment.
- Qualitative Field Research: Establishing a focus group from selected organisations used for final methodology verification. In addition, in depth face-to-face interviews conducted with senior management. The forecast is to interview two from each industry segment in the Czech Republic, Slovakia, and Canada.

The Significance to Practice

The methodology design ensures practical value enabling organisations to quantify their knowledge intensity and benchmark their performance from a knowledge viewpoint with others at an international level. The practical implication evolves from the composition of the research team. Moreover, the benefits of the project include not only the methodology but also amelioration to the existing body of knowledge. The theoretical results enhance managerial familiarity of the relationship between KM and organisational knowledge intensity. Organisational knowledge intensity relates to KM, wherein it can serve as a tool for the justification and substantiation of related initiatives. Therefore, project output supports and facilitates managerial decision-making. KM-related decisions reinforced by managers' knowledge and awareness of processes, and resources are the new process. Equipped with this information management takes efficient and more effective decisions as to resource allocation, financial budgeting, and in other day-to-day tasks. Incorporating cultural aspects within the total problem ensures an undisputable relationship between the proposed project and practice.

Knowledge oriented activities are viewed at several levels of the knowledge hierarchy. At the supranational level, management interpret objectives as guidelines for national and organisational development. The Czech Republic is not immune as Czech enterprises strive to improve their performance and competitiveness. Unfortunately, the contemporary vision excludes systematic KM as a viable option. Several reasons previously cited contribute to this condition. Previous efforts demonstrate that there is no "turnkey" solution available to meet the special circumstances of the majority of companies in the Czech Republic.

To support further development on KM the proposed project focuses on the organisational knowledge intensity. Its ultimate result – a methodology for monitoring organisational knowledge intensity –provides managers with an appropriate tool for the justification of a KM programme and monitoring of its development. Once familiar with this tool, managers who lack system-thinking ability should avoid the seemingly straightforward solutions that are obviously insufficient to implement KM.

CONCLUSION

2009 marks the twentieth anniversary of the Velvet Revolution. During this brief span of time, the Czech Republic experienced the fall of communism, the re-emergence of a market economy and the adhesion to NATO and the European Union. Within one generation, the nation transformed itself from a Soviet satellite to a democracy that held the Presidency of the European Union for the first half of year 2009. The "return to Europe" is an exceptional adventure for the Czechs manifest by many successes despite its share of *Euro-sceptics* dispersed throughout all levels of society, the intelligentsia, and government. As with other areas, the business environment expands and equals that of other developed economies. Nevertheless, the Czech environment possesses several unequivocal characteristics that influence every national activity including KM implementation. These characteristics are illustrated with the help of knowledge levels that include the supranational level, national level, organisational level, and management of knowledge level. The described specifics of the Czech Republic comprise the lack of an utilisable and detailled KM implementation methodology that reflects Czech needs as well as the many diverse perceptions of KM coupled with the notion that KM equates and is limited to information and/or knowledge technologies. Any individual, who has not had exposure to other societies and cultures, may tend to assume that managerial standards practiced in the home country are the same throughout the world. The notion of globalisation reinforces this tendency. Moreover, unless assigned to an international project many practitioners lack contact with colleagues outside their own jurisdictions. The same is true about Knowledge Management. KM thrives in the United Kingdom, the United States, and Australia; therefore, there exist no impediments to a successful deployment in the Czech Republic. The authors demonstrate in this chapter that such is not the case. Preliminary research indicates that unique historical and cultural circumstances hinder the winning conditions to prevail in the Czech Lands. The initiation of certain innovations such as the KM-Be.At-It methodology assist in determining the root causes to the KM barriers. This methodology represents a general framework consisting of several stages that need to be performed in order to introduce effectively KM into an organisation. To do so, particular organisations have to develop their own instance from this framework customised to suit their own specifics and needs, i.e. create their own KM project with the help of the general methodology.

The exercise is ongoing. The need for further research combined with an intense proliferation of the methodology aimed at the small to medium size category of Czech businesses accounts for the majority of the planned activities. Nonetheless, detailled investigation remains the sole method by which a permanent solution takes form. Consequently, the brief description of research activities conducted in the Czech Republic outlined in this chapter emphasise topics related to KM that require exploration while suggesting questions that should be answered.

ACKNOWLEDGMENT

This paper was created with the support of the GAČR project SMEW - Smart Environments at Workplaces, project num. 403/10/1310.

REFERENCES

Ahmed, K. A. (2002). *Learning Through Knowledge Management*. Oxford, UK: Butterworth-Heinemann.

Autio, E. (2000). Effects of age at entry, knowledge intensity, and imitability on international growth. *Academy of Management Journal*, (Oct): 2000.

Beckman, T. J. (1999). The Current State of Knowledge Management. In Liebowitz, J. (Ed.), *Knowledge Management Handbook*. Boca Raton, FL: CRC Press.

Bureš, V. (2003). Cultural Barriers in Knowledge Sharing. *E+M Economics and Management (6)* special issue, 57-62.

Bureš, V. (2005). Methodology of Knowledge Management Implementation. *3rd International Conference on Multimedia and Information & Communication Technologies*. Cáceres, Spain, Vol. 3, pp. 963-968.

Bureš, V. (2009). Conceptual Perspective of Knowledge Management. *E+M Economics and Management (12)*2, 84-96.

Bureš, V., & Čech, P. (2008). Knowledge Management: the Central European Experience. *Effective Executive*, *6*(XI), 47–53.

Čech, P. (2005). *The Potential of Knowledge in Organizations*. Ph.D. Thesis, Hradec Králové: University of Hradec Králové.

Čech, P., & Olševičová, K. (2009). Proposal of Intelligent Assistant for Graduation Thesis Preparation. *Intelligent environments: workshops proceedings of the 5th international conference*, Barcelona, Spain, pp. 321-327.

Chen, D. H. C., & Dahlman, C. J. (2005). The Knowledge Economy, the KAM Methodology and World Bank Operations. Washington DC: World Bank. Retrieved, March 4, 2009, from http://siteresources.worldbank.org.

Coviello, A. (2001). *Standardised KM Implementation Approach. IST Project No 2000-26393, Deliverable D 3.1*. European KM Forum.

Davenport, T., & Prusak, L. (1998). *Working Knowledge: How Organizations Manage What They Know*. Boston: Harvard Business School Press.

Davenport, T. H., & Smith, D. E. (2000). Managing Knowledge in Professional Service Firms. In *The Knowledge Management Yearbook 2000-2001*. Oxford, UK: Butterworth-Heinemann.

Drucker, P. F. (1993). *The post capitalist society*. Oxford, UK: Butterworth–Heinemann.

European Council. (2000). *Presidency Conclusions - Lisbon European Council*. Lisbon. Retrieved December 20, 2008, from http://europa.eu.int.

Firestone, J. (1999). The Artificial Knowledge Manager Standard: A "Strawman". *Working Paper No. One*. KM ANSI/ISO Standards Committee Meeting. Executive Information Systems, Inc. Retrieved March 12, 2009, from http://www.dkms.com/papers/akmsstrawbak.pdf.

Fisher, G., & Ostwald, J. (2001). Knowledge management: problems, promises, realities and challenges. *IEEE Intelligent Systems*, *1*(16), 60–72. doi:10.1109/5254.912386

Frappaolo, C. (2006). *Knowledge Management*. London: Capstone Publishing.

Gavalec, M., & Mls, K. (2005). Individual cognitive maps and evaluation of their dynamic changes. *International conference Kognice 2005,* Hradec Králové, Czech Republic, pp.24-29.

Gulev, R. (2005). Economic Culture Impacts on Knowledge Flows and Knowledge Sharing within Selected EU Countries. *Journal of Knowledge Management Practice*. Retrieved September 20, 2009, from http://www.tlainc.com/articl102.htm.

Heier, H., & Borgman, H. P. (2002). Knowledge Management Systems Spanning Cultures: The Case of Deutsche Bank's HRBASE. *Symposium conducted at the meeting of ECIS 2002 in Gdansk*, Poland.

Hitchins, D. (2007). *Systems Engineering: A 21st Century Systems Methodology*. Chichester, UK: John Wiley and Sons.

Hofstede, G. (1980). *Culture's Consequences: International Differences in Work-Related Values*. London: Sage Publications.

Hofstede, G. (2001). *Culture's Consequences* (2nd ed.). London: Sage Publications.

Holsapple, C. W., & Joshi, K. D. (2001). Organizational Knowledge Resources. *Decision Support Systems*, *1*(31), 39–54. doi:10.1016/S0167-9236(00)00118-4

Hoog, R. (1999). Investigating a Theoretical Framework for Knowledge Management: A Gaming Approach. In Liebowitz, J. (Ed.), *Knowledge Management Handbook*. Boca Raton, FL: CRC Press.

Jahn, M., et al. 2005. *Strategy of economic growth of Czech Republic*. Prague: Czech Republic. Retrieved May 12, 2009, from http://www.hospodarskastrategie.org.

Jayaratna, N. (1994). *Understanding and Evaluating Methodologies – NIMSAD A Systemic Framework*. London: McGraw-Hill.

Kaplan, R. S., & Norton, D. P. (1996). *Balanced Scorecard: Translating Strategy into Action*. Boston: Harvard Business School Press.

Kim, Y. G., Yu, S. H., & Lee, J. H. (2003). Knowledge strategy planning: methodology and case. *Expert Systems with Applications*, *3*(24), 295–307. doi:10.1016/S0957-4174(02)00158-6

Lewin, K. (2008). *A Dynamic Theory Of Personality: Selected Papers*. Lewin Press.

Lyles, M. A., & Salk, J. E. (1996). 'Knowledge Acquisition from Foreign Parents in International Joint Ventures: An Empirical Examination in the Hungarian Context. *Journal of International Business Studies*, *27*(5), 877–903. doi:10.1057/palgrave.jibs.8490155

McElroy, M. (2004). How to Operationalize KM, KM World. Santa Clara, California. Retrieved May 12, 2009, from http://www.macroinnovation.com.

Mikulecký, P. (2001). *Information Filtration in Knowledge Management Support* (pp. 190–193). Bratislava, Slovak Republic: Quantitative Methods in Economy and Business.

Mikulecký, P., Ponce, D., & Hynek, J. (1997) Knowledge-Based Projects in Industrial Applications, *Proceedings of 4th International Seminar on Mathematical Simulation in the Glass Melting*, Horní Bečva, Czech Republic, pp. 191-201.

Miller, N. E., & Dollard, J. (1950). *Personality and Psychotherapy*. New York: McGraw-Hill.

Nonaka, I., & Takeuchi, H. (1995). *The Knowledge Creating Company*. New York: Oxford University Press.

OECD. (1996). *The Knowledge-Based Economy*. Paris. Retrieved December 27, 2008, from http://www.oecd.org.

Olševičová, K., Ponce, D., & Haviger, J. (2006). Procedural knowledge Representation in Learning Management System. *5th European Conference on e-Learning*, (pp.262-272) Reading, UK.

Ponce, D. (2007). Model of slow process as ontological design pattern. *International conference Znalosti 2007*, (pp.348-351). Ostrava, Czech Republic.

Simonin, B. (1999). Transfer of marketing know-how in international strategic alliances: An empirical investigation of the role and antecedents of knowledge ambiguity. *Journal of International Business Studies*, *30*(3), 463–490. doi:10.1057/palgrave.jibs.8490079

Sure, Y., Staab, S., & Studer, R. (2004). On-To-Knowledge Methodology. In Staab, S., & Studer, R. (Eds.), *Handbook on Ontologies*. Berlin: Springer-Verlag.

Thomas, J. C., Kellogg, W. A., & Erickson, T. (2001). The knowledge management puzzle: human and social factors in knowledge management. *IBM Systems Journal*, *4*(40), 863–884. doi:10.1147/sj.404.0863

Tiwana, A. (2000). *The Knowledge Management Toolkit, Practical Techniques for Building a Knowledge Management System*. Upper Saddle River, NJ: Prentice Hall.

Tučník, P., & Mohelská, H. (2009). Multicriterial Decision-Making Can Drive Strategy, International conference Hradec Economic Days 2009, Hradec Králové, Czech Republic, pp. 80-83.

Usoro, A. (2005). Cultural Dimensions of Knowledge-Sharing in Global Virtual teams: An Exploratory Study. *Computing and Information Systems Journal*, *3*(9), 85–91.

Wallace, D. P. (2007). *Knowledge Management: Historical and Cross-Disciplinary Themes*. Westport, CT: Libraries Unlimited.

Wiig, K. (1999). Introducing Knowledge Management into the Enterprise. In *J. Liebowitz (1999). Knowledge Management Handbook*. Boca Raton, FL: CRC Press.

Wiig, K., de Hoog, R., & van der Spek, R. (1997). Supporting Knowledge Management: A Selection of Methods and Techniques. *Expert Systems with Applications*, *1*(13), 15–27. doi:10.1016/S0957-4174(97)00019-5

Witten, I. H., & Frank, E. (2005). *Data Mining: Practical Machine Learning Tools and Techniques*. San Francisco: Elsevier.

ADDITIONAL READING

Apostolou, D., & Mentzas, G. (1999). Managing Corporate Knowledge: A Comparative Analysis of Experiences in Consulting Firms. *Knowledge and Process Management*, (6): 129–138. doi:10.1002/(SICI)1099-1441(199909)6:3<129::AID-KPM64>3.0.CO;2-3

Autio, E. (2000). Effects of age at entry, knowledge intensity, and imitability on international growth. *Academy of Management Journal*, (43): 909–924. doi:10.2307/1556419

Beckman, T. J. (1997). A Methodology for Knowledge Management, International Association of Science and Technology for Development. *AI and Soft Computing Conference*, Banff.

Boland, R., & Tensaki, R. (1995). Perspective making perspective taking. *Organization Science*, *4*(6), 350–372. doi:10.1287/orsc.6.4.350

Bureš, V. (2006). Knowledge Management and its Implementation. *Proceedings of the 2nd International Conference on Web Information Systems and Technologies (WEBIST 2006)*. Setubal, Portugal: INSTICC, pp. 115-118.

Choi, B., & Lee, H. (2003). An empirical investigation of KM styles and their effect on corporate performance. *Information & Management*, *5*(40), 403–417. doi:10.1016/S0378-7206(02)00060-5

Disterer, G. (2001). Individual and social barriers to knowledge transfer. *Proceedings of the 34th Hawaii international conference on system sciences*.

Hempel, M. D. (2001). Building a Knowledge Sharing Culture to Promote Knowledge Creation. *Elluminati, White paper*. Retrieved September 12, 2008, from http://www.elluminati.com.

Lefley, F. (2004). Manufacturing investments in the Czech Republic: An international comparison. *International Journal of Production Economics*, (88): 1–14. doi:10.1016/S0925-5273(03)00129-4

Liebowitz, J., & Megbolugbe, I. (2003). A set of frameworks to aid to project managers in conceptualizing and implementing knowledge management initiatives. *International Journal of Project Management*, (21): 189–198. doi:10.1016/S0263-7863(02)00093-5

Macintosh, A. (1996). *Position Paper on Knowledge Asset Management*. Scotland: Artificial Intelligence Applications Institute, University of Edinburgh.

McDermott, R., & O'Dell, C. (2001). Overcoming cultural barriers to sharing knowledge. *Journal of Knowledge Management*, *1*(5), 76–85. doi:10.1108/13673270110384428

Münich, D. (2006). Recent Publication Productivity of Czech Economists. *Czech Journal of Economics and Finance*, *11-12*(56), 522–533.

Petrash, G. (1996). Managing Knowledge Assets for Value. *Knowledge-Based Leadership Conference*. Linkage Inc., Boston.

Quinn, J. B., Anderson, P., & Finkelstein, S. (1996). Managing professional intellect: making the most of the best. *Harvard Business Review*, *2*(74), 71–80.

Sunassee, N., & Sewry, D. (2002). A Theoretical Framework for Knowledge Management Implementation. *Proceedings of SAICSIT, 2002*, 235–245.

Sveiby, K. (2000). *Frequently Asked Questions*. Retrieved, January 3, 2009, from http://www.sveiby.com.

UNESCO. (2005). Towards Knowledge Societies. UNESCO Publishing. Retrieved January 2, 2009, from http://www.unesco.org.

van den Berg, P. T., & Wilderom, C. P. M. (2004). Defining, Measuring, and Comparing Organisational Cultures. *Applied Psychology: An International Review*, *53*(4), 570–582. doi:10.1111/j.1464-0597.2004.00189.x

van der Spek, R., & Spijkervet, A. (1997). *Knowledge Management: Dealing Intelligently with Knowledge, Knowledge Management and Its Integrative Elements*. Boca Raton: CRC Press.

van Schaik, T. (2002). Social Capital in the European Values Study Surveys. *Symposium conducted at the meeting of the OECD-ONS Conference on Social Capital Measurement,* London, September 25-27.

Wickramasinghe, N. (2002). Practising what we preach: are knowledge management systems in practice really knowledge management systems? *Proceedings of the 35th Hawaii international conference on system sciences*.

Wiig, K. (1997). Knowledge Management: Where did it come from and where will it go? *Expert Systems with Applications*, *1*(13), 1–14. doi:10.1016/S0957-4174(97)00018-3

Wilson, T.D. (2002). The nonsense of 'knowledge management'. *Information Research, 8*(1).

KEY TERMS AND DEFINITIONS

Knowledge Management: Represents a managerial approach of organisational leadership, the creation of an organisational environment and the achievement of organisational objectives based on the alignment of knowledge and processes through the usage of knowledge resources and appropriate methods, techniques and tools.

Management of Knowledge: Represents a set of disciplines such as psychology, informatics, pedagogy, and cognitive science, together with particular tools, techniques, and methods for work with knowledge. Outcomes from the management of knowledge are used at the organisational level for implementation and realisation of knowledge management.

Knowledge Management Processes: (Also knowledge processes) identify a number of processes considered the foremost essentials in the evolution of knowledge in an organisation. Particular knowledge processes follow the general knowledge life cycle. Therefore, these include knowledge creation, purchase, development, modification, storage, sharing, distribution, usage, application, renewal, and replacement.

Methodology: Represents an explicit and structural way of thinking. Methodology contains a model that reflects selected views on reality and which derive from a certain set of philosophical paradigm. Methodology determines which steps are performed, how these are performed, but above all, it explains why everything is done in the specific order. In other words, methodology is a set of recommended phases, approaches, principles, rules, documents, methods, techniques, and tools, that answer the questions who, when, what and why.

Knowledge Hierarchy: A model of the global knowledge society consisting of several knowledge levels. These are supranational level, national level, organisational level, and management of knowledge level.

Organisational Infrastructure: New units and roles created to establish, coordinate, and manage the technology and tools, and to facilitate the capture, development, and distribution of knowledge. These units generally direct the function of ensuring the usage and institutionalisation of common approaches.

Technological Infrastructure: ICT aims at the development, identification, and application of appropriate technological approaches for the management of ICT-supported learning that equips employees with the knowledge required to perform successfully their engagements. ICT augments and interconnects resources enabling the distribution of information, consistent with the organisational requirements for team-based management and responsiveness to change.

Power Distance Dimension: Reflects the degree to which the culture believes that institutional and organisational power should be distributed unequally and the decisions of the power holders should be challenged or accepted.

Uncertainty Avoidance Dimension: The extent to which the culture feels threatened by ambiguous, uncertain situations and attempts to avoid them by establishing more structure.

Chapter 3
Knowledge Integration in the Creative Process of Globally Distributed Teams

Rosana Silveira Reis
University of Bologna, Italy

Ylenia Curzi
University of Modena and Reggio Emilia, Italy

EXECUTIVE SUMMARY

The aim of this chapter is to analyse knowledge integration in the creative process of globally distributed teams as they develop new products. Mainstream researches and studies focus on knowledge integration in terms of transference of knowledge; they highlight its relevance with respect to innovation and creativity, and investigate the conditions that assure or inhibit it. The creative process in globally distributed teams is fairly unexplored by academic literature. With only a few exceptions, the literature focuses on virtual teams: i.e., distributed teams where factors such as culture, time zone and language are irrelevant to the development of the activities carried out by team members. The authors concentrate their efforts in looking for how knowledge integration happens in the creative process in globally distributed teams. For this reason, they shall rely on a research method founded on the notions of adequate causation and objective possibility. On this basis, they have compared two empirical cases in order to answer their research question. The authors have thus analysed six global product development projects carried out by globally distributed teams belonging to a Swedish company working with teams in France and Brazil; and an Italian company working with teams in Tunisia. The data has been gathered through participant observations, semi-structured interviews and document analysis from 2007 to 2009. This contribution is grounded in the analysis of the existing literature and in the data collected on the field.

DOI: 10.4018/978-1-61692-886-5.ch003

INTRODUCTION

"How and why does knowledge integration occur in the creative process of GDT as they develop new products"?

Using a qualitative approach, this chapter aims to investigate knowledge integration in the creative process of globally distributed teams (GDT) involved in new product development projects. The question being posed has it focus on "how" and "why" phenomena occur. "How" and "why" are explanatory questions, yet they are more concerned with the conditions occurring within a timeframe, rather than with their frequency and incidence. For this reason, recurring to the notions of adequate causation and objective possibility (Weber, 1904, 1906) seems to be more appropriate than recurring to the notions of probable or necessary causation.

The spreading of GDT can be related to a series of circumstances, amongst which there is the interest of bringing together often fragmented knowledge and information from branches of multinational organizations scattered across the globe. This working configuration has proliferated because of globalization, which entices organizations to recognise dispersed competencies.

When we interact with people we have an opportunity to learn, to influence, and to make things happen (Gurteen, 1998). "Team creative process" has been defined by Gilson & Shalley (2004) as members working together in such a manner that they link ideas from multiple sources, delve into unknown areas to find better or unique approaches to a problem, or seek out novel ways of performing a task. According to Baba, Gluesing, Ratner & Wagner (2004) team members of different cultures can bring together divergent bodies of knowledge whose integration yields new organizational capabilities - but only after they recognize both the existence and the validity of their differences.

On the basis of the comparison between two empirical cases, our proposal is to assess whether some conditions can be regarded as adequate conditions that allow the occurrence of knowledge integration in the creative process of GDT involved in new products development projects. We will therefore rely upon the conditional scheme of explanation put forward by Weber (1904, 1906).

BACKGROUND

In this section we first present a review of the literature, and then clarify our perspective on knowledge integration and our position with respect to the creative process in GDT.

The new wave of digital technologies has given organizations an enormous opportunity to bring together their distributed workforce and develop the ability to work together in spite of being bodily apart (Prasad & Akhilesh, 2002). Moreover, as the business environments are no longer confined to physical borders, geographically dispersed teams overcome the limitations of time, space, and the organizational affiliation of traditional local teams (Piccoli, Powell & Ives, 2004). In the literature it has been argued that GDT can be defined as a work team made up of culturally different members based in two or more nations who share collective responsibilities (Baba et al., 2004), who are brought together by information and communication technologies to jointly accomplish one or more organizational tasks and to reach a common goal in the organization (Paulus, 2000; Cohen & Bailey, 1997).

According to the existing literature, the concept of creativity is strictly connected to that of innovation. Creativity can be defined as "the production of novel and useful ideas in any domain" (Amabile, 1996). Innovation is generally defined as a phenomenon associated with purposeful change (Drucker, 2007), reflecting the capacity to imagine what does not exist, and is a process with different stages that extends from an idea, or a new combination of existing ideas (Hargadon & Bechky, 2006; Schumpeter, 1942), to its implementation. Innovation can thus be seen as the successful implementation of creative ideas within an organization. As a result, creativity is the starting point of innovation and innovation is the implementation of creative results. Creativity is therefore a part of the innovation process (Alves, Marques & Saur, 2007).

Focusing on the development of a new product it has been suggested that creativity can be seen as the driving force in the development of a new product, and the locus for potential innovation (Leenders, Van Engelen, & Kratzer, 2003). The creative process, in turn, can be defined as the set of activities occurring when a person or a team is creating (Nemiro, 2004), i.e., the set of activities leading to a generation of ideas that are both novel and appropriate, which subsequently lead, through innovation, to a new product or process or a new way of performing a task in organizations.

Yet compared to other fields of inquiry, such as those concerning innovation, the literature on creativity may be viewed as a relatively new scientific effort. Out of the three possible focal points - individual, team or organization - the majority of investigation has focused on the individuals' creative process (Bergh & Stuhlfaut, 2006). At the organisational level, the literature looks for the factors which can improve or obstruct creativity; whereas the studies on team creativity inquire about conditions which can improve or inhibit creativity within specific teams (Paulus, 2000).

Nevertheless, creativity in GDT is a fairly unexplored topic: almost all of the available studies focus on the creative process in co-located teams, and only a few studies focus on the creative process in distributed teams, i.e., virtual teams made up of members based in the same country. Amabile & Gryskiewicz (1987), Ensor, Cottam & Band (2001) and Baer & Oldham (2006) identified the factors which interfere in co-located teams' creative processes. Some factors identified as influencing the creative process can be summarized as follows: cultural diversity and interpersonal relationships mediated by computer.

It is argued that cultural diversity can affect the team creative process both positively and negatively. The stream of research focusing on the negative impact of cultural diversity draws attention to the obstacles created by communication, work styles, and decision-making patterns, which in turn can cause the participants to experience culture shock (Gassmann, 2001). In addition, Mostert (2007) points out that cultural diversity might decrease the level of creativity in an idea-generating team, and that it does not represent the key for the successful generation of new ideas. The author proposes that it is instead the diversity of minds that prevails.

It has also been argued that interpersonal relationships mediated by computer negatively affect the team creative process due to the fact that the creative process relies upon trust among team members (Mostert 2007), which in turn needs "touch" to be developed (Handy, 1995).

Another important aspect to be addressed is knowledge integration or combination. With regard to this topic, it is worthy of being underlined that mainstream literature refers to knowledge integration in terms of knowledge transfer, highlights the relevance of knowledge transfer with respect to innovation and creativity and investigates the conditions that impact on knowledge transfer. Argote & Ingram (2000) define knowledge transfer as "the process through which one unit (e.g., the team) is affected by the experience of another" (p. 151). The importance of knowledge and knowledge transfer is widely acknowledged in the literature on both creativity and innovation.

In the literature on creativity, it has been pointed out that one of the main features which is important for the creative process is the members' knowledge of their own work (Gilson & Shalley, 2004). Teams made up of members from different professional backgrounds, knowledge, skills and abilities will be more creative than those teams in which the members of which have similar professional backgrounds, knowledge, skills and abilities. This is due to the fact in the first case the teamwork benefits from the useful and different perspectives brought to a given issue or problem by the team members (Paulus, 2000). Gilson & Shalley (2004) emphasize that a moderate amount of tenure is desirable for engaging in creative efforts. Allen's (1977) research demonstrates that innovators get creative ideas from a variety of sources, including searches of knowledge bases with unknown sources. When we add Information Technology to creativity and knowledge, we get a potent combination which can promote collaborations across divergent disciplines and perspectives (Kao, 1996).

In the literature on innovation it has been pointed out that "a central activity for companies aiming at repeatedly generating new products and processes, as well as improving already existing ones, is to create and exploit knowledge within and between organizations in a fruitful manner" (Magnusson, 2004, p. 261).

The transfer of knowledge can bring huge benefits to an organization (O'Dell & Grayson, 1999). But this requires breaking down hierarchies in the organization, because organizations that maintain hierarchical levels and silos will not encourage knowledge transfer (Nonaka, 1994). Another relevant aspect that can impact on knowledge transfer is the willingness of groups or individuals to work with others and share knowledge to their mutual benefit (Goh, 2002).

Knowledge transfer provides a larger shared base of comprehensible knowledge, effectively improving the ability of a social system to interpret diverse information stimuli collectively (Alavi & Tiwana, 2002). Product development is often mentioned as one way to integrate dispersed knowledge of different natures in an innovative manner (Grant, 1996), and to generate new and effective knowledge to improve operational competences. On the product development level, individuals working in teams share their own knowledge in order to solve problems and find innovative solutions (Revilla & Cury, 2008).

According to Alavi & Tiwana (2001), one of the preconditions for knowledge integration in virtual teams is to know who has the relevant knowledge and expertise, as well as where the relevant knowledge and expertise are located and where they are needed. Moreover, teamwork requires an emerging process of rich exchanges and joint problem-solving to integrate and apply knowledge and expertise to the task at hand in a coordinated manner. This process in turn requires a supportive environment in terms of easy, frequent, content- and context-rich interpersonal interactions.

Knowledge Integration

With regard to what has been highlighted above, it is worth stressing one basic presupposition used by mainstream literature on knowledge transfer. This concept expresses the idea that knowledge is detachable from the course of an action that agent develops over time: knowledge is therefore regarded as being transferable to other agents and actions. Mainstream literature accordingly emphasises information and communication technologies and the willingness of individuals, groups and organizations to share their own knowledge as conditions that positively affect knowledge transfer.

In this contribution we take on a different point of view, as we presuppose that knowledge is not detachable from the course of the action an agent develops over time. Therefore, knowledge is not transferable to other agents or actions. Accordingly, when different domains of knowledge are required to accomplish an action, and when these domains of knowledge are possessed by different agents, we assume that the only way to accomplish the action is to assign its execution to these different agents so that they act simultaneously.

In light of this assumption, we shall hereby use the term 'knowledge integration' or 'knowledge combination', instead of 'knowledge transfer', when referring to the situation in which a task to be accomplished requires different domains of knowledge possessed by different agents, and its accomplishment is assigned to various agents to act simultaneously.

Creative Process in GDT

Studies dedicated to the development of the creative process in teams reveal some inconsistencies. There are studies which argue that "in a team more ideas of higher quality are produced", and that "creative teams seem to be more efficient" (Paulus, Larey & Dzindolet, 2001), but we also can find authors such as Simonton (2004), who defends the idea that although many individuals may contribute towards a final product, their efforts are not submerged in the collective creation because they maintain a certain level of identifiable individuality.

In the same line, Ekval (1997, p.195), emphasises that the "creative product originates from an individual's performance, even though other individuals may have provided clues and stimulations". In the light of this perspective it is utterly important for this chapter to highlight the differences between the "individual mind" and the ideas arising from the group synergy. In this regard we agree with Kurtzberg & Amabile (2001), who argue that a group of people may produce something that one single person would not have been able to do alone.

The individuality of the individual plays a key role in the execution of the team's activities, but it is the dynamic within the team that strengthens the individual's creative process. Within GDT there is the dispersion of team members in time, the diversity of expertise and/or culture, the absence of a work history among the team members, and linguistic misunderstandings. These aspects could constrain the development of shared understanding among team members. Nevertheless, we assume that these elements are the same factors that contribute to the richness of the creative process where there is effective communication among team members (Reis & Chevalier, 2009).

RESEARCH METHODOLOGY

The aim of this contribution is to put forward possible answers for *"how and why does knowledge integration occur in the creative process of GDT as they develop new products"?* Our analysis is concerned with 'how' and 'why' this empirical social phenomenon occurs, rather than with its frequency and impact. For this reason we have used a research method based on the notions of adequate causation and objective possibility (Weber, 1904, 1906).

In this section we first discuss the method applied, and in order to contextualize the field of investigation, we briefly outline data collection, the development of ERP at Delta, and the development of vehicles at Alpha.

Discussion of the Method

Weber (1904) specifies that the researcher should rely on the concepts of adequate causation and objective possibility when the aim of his research is to account for the individuality of an empirical social phenomenon, and the research presupposition is that this individuality is unlimited. This presupposition means that the aspects which make an empirical social phenomenon individual and different are endless. It also signifies that the number and kind of causes which have brought about any single human event are endless. According to Weber (1904) yet, there is no innate characteristic in human events which allows the researcher to identify a single element to be studied.

As a result, the researcher has to refer to his personal values in order to account for the individuality of an empirical social phenomenon. The reference to his personal values allows the researcher to single out the aspect which he deems worthy of being known, and the possible causes to which it can be causally attributed. In other words, the reference to the researcher's personal values permits him to define the domain within which he can account for the individuality of an empirical social phenomenon, thereby expressing a subjective point of view from which the research starts.

This in turn implies two relevant consequences. Firstly: with regard to any particular social phenomenon, it is possible to give different accounts by virtue of the diversity of various particular, specific and unilateral points of view from which a research can start. Secondly: these accounts are always partial, because they highlight only a finite complex of antecedents and not the totality of causes which brings about the totality of individual qualities of the phenomenon. Thus we can say that any partial account, based on a particular point of view, will highlight a finite complex of conditions (besides others which are equally important, but neglected by the biased account) which concur to the occurrence of a certain phenomenon regarded in its full and endless individuality. "Concur" in this sense signifies that the complex of conditions, which has been isolated from the overall set of possible antecedents, cannot bring about the occurrence of the individual phenomenon but is *adequate* to make it objectively *possible*.

Weber (1906) specifies that any partial account that is formulated on the basis of a subjective point of view must be subjected to a counterfactual validation. This implies that any account of an individual empirical social phenomenon lies in the assessment of the *adequacy* of the conditions for the *objective possibility* of a particular phenomenon (or course of an event). The basic operation needed for this assessment is the construction of a hypothetical process to which the actual process (that is, the course of the event that has become thus and not otherwise) must be compared (Weber, 1906).

The hypothetical process is constructed by means of a conceptual elaboration that involves a series of abstractions. The first and crucial abstraction is accomplished by firstly imagining one or several objective causal components of the actual course of the event as absent or altered in a certain direction, and then by asking oneself if whether under the changed conditions, the same (or what other) consequence was to be expected, according to general rules of experience, with regard to the aspect of the course of the event in which we are interested.

Drawing upon general rules of experience thus allows the researcher to judge the causal significance of one or more of the conditions under consideration. More specifically, it makes it possible to distinguish between causally adequate conditions and causally accidental conditions.

Weber (1906) clarifies that a condition is a causally accidental condition (i.e. a condition without causal significance) if, according to general rules of experience, we expect the same consequence to occur in the specific incident that we are examining regardless of whether or not that is absent or altered. Conversely, a condition is regarded as a causally adequate condition where, according to general rules of experience, we expect different consequence to occur if that condition had been absent or altered.

Weber (1922) clarifies that general rules of experience, which are required to formulate a judgment of objective possibility, can be provided by a comparison between empirical cases which are different in the condition under consideration, but homogeneous in other aspects. A comparison between these kinds of empirical cases which also shows that the differences in the condition under consideration are associated with differences in the aspect of the course of the event under investigation

Using this latter clarification provided by Weber as a basis, we sought to answer our research question by comparing two empirical cases where the teams are similar with respect to the following aspects:

a) Global new product development projects characterized by a similar degree of complexity and requiring the integration and combination of different and complementary domains of knowledge;

b) A globally and geographically dispersed workplace;
c) Relationships and communications mediated by computer;
d) Cultural diversity (different ways of thinking, working and decision making patterns);
e) Leadership at a physical distance;
f) Communication in a second language.

Data Collection

We have studied two empirical cases about GDT composed of engineers and developers: one belonging to an Italian software company, and the other to a Swedish company working in the automotive sector. We studied software engineers and developers because their work is knowledge intensive, demanding, and individually styled (Barley & Orr, 1997). We refer to the two companies as follows: (1) Delta, with GDT composed of members located in Brescia (Italy), and Tunis (Tunisia) where we have analysed three projects, and; (2) Alpha, with GDT composed of members located in Curitiba (Brazil), Lyon (France) and Gothenburg (Sweden), where we have analysed three projects. In both cases our main interest was in the process of new product development.

We have applied document analysis, participant observations and semi-structured interviews.

a) Semi-Structured Interviews: in order to comprehend structure, project composition, and the context in which products are developed, we began our research by studying two pilot projects inside both of the companies. After having identified the projects in which people work in GDT, we selected 11 developers and engineers in Delta, and we interviewed 8 people in Italy (face-to-face) and 3 people in Tunisia (through video conference). In Alpha, we selected 18 developers and engineers, and we interviewed 6 people in Brazil, 6 in Sweden and 6 in France. All interviews at Alpha were conducted face-to-face. In total we conducted 39 interviews. Each interview was audio recorded and lasted between 1.5 and 2 hours.
b) Observations: in Delta we observed 1 video conference (1.5 hours). In Alpha, we participated in two eight-hour workshops about innovation, 8 hours each one. In both companies we were invited to have lunch and coffee break with the participants, which allowed a more informal and open contact with the attendees. We took notes describing everything that happened, the actions taken by individuals, and the interactions between professionals, while respecting the interviewees' original language. We transcribed our field notes into word files.
c) Archival Data: we consulted Delta procedures, internal reports, newsletters, job descriptions, organizational charts, quality manuals, internal papers, and consultants' reports to provide us with a deeper and richer understanding of the context. In Alpha we consulted: job descriptions, human resources' policies, self-managed teams' institutional presentations, internal reports, press coverage and interviews, company's manuals, Alpha's institutional presentations, internal communications such as magazines, Alpha's intranet and general information available to shareholders on the internet.

The Development of ERP at Delta

Delta is a medium sized Italian company located in Italy and Tunisia which belongs to an international group. Delta develops and commercializes a web-based second generation Enterprise Resource Planning (ERP) for small and medium sized Italian firms. The basic aim of the company is "to offer to our clients solutions, services and innovative technologies to improve business processes and to ensure business competitiveness."

Software packages like ERP are configurable IT, because they typically provide hundreds or even thousands of discrete features and data items that can be combined in multiple ways. Developing ERP software therefore requires advanced technological software skills and a deep knowledge of the potential industrial applications at different clients' sites.

The process of the development of a new ERP is composed of two main tasks: the analysis and the coding work. Let us concentrate our attention on the first task. It is concerned with the development of the main modules (called "domains") of the ERP software. Specifically, it draws up the 'use cases' which are the documents detailing the features and properties of the ERP modules and the ways in which the user is supposed to interact with it. It is precisely this task of the development process which requires different domains of knowledge to be executed. These domains are made up of knowledge concerning the possible industrial applications of the new ERP at different client's sites, and knowledge concerning coding work.

Italian and Tunisian developers and engineers collaborate on: 1) new domains development and 2) maintenance of existing domains. Every domain is associated to a team composed of Italians, or Tunisians, or both. Teams work on multiple projects.

In GDT, the phases of a typical project are sequentially divided between Italy and Tunisia. The Italian management's decisions about domains and customer care calls define the workload. Then Italian analysts define use cases that become the input for Tunisian engineers. A video conference is scheduled to discuss and modify the content of the document and then the coding work is performed by Tunisian engineers, with minimal interactions with Italian analysts. In the final phase, the work is tested in both countries. The overall type of interdependence reflects what Kumar, Von Glinow & Van Fenema (2009) call a "partially sticky form of sequential interdependence".

The Development of Vehicles at Alpha

Alpha is a cross-functional organisation responsible for product planning and purchasing, global vehicle development, global engineering and product range management in the automotive sector. The basic aim of the company is to "propose and develop profitable products to ensure a strong and competitive offer for each company, based upon common vehicle architecture and shared technology". Alpha emphasizes business and customer focus, industrial efficiency and synergies; speed, agility and flexibility; diversity; teamwork; leaders as coaches; and openness to change. It considers employees as key for the success of the organization. In addition to their excellent technical skills, the collaborators are multi-brand and multi-cultural minded.

The projects basically have two axes: an axis of complexity and an axis of length. The organization of the project consists of leaders and of a project head. The project head is the Chef Project Manager (CPM); who is responsible for delivering and implementing the project. The leaders are Project Managers (PMs) that represent every area involved in the project. The team creates its own organization with a basic aim "to manage the project." Very rarely is a project confined to only one area.

The Global Development Process of new products is composed of the following main phases: pre-study, concept study, detailed development, final development, industrialization and commercialisation, and follow-up. Each of these phases of the development process requires different domains of knowledge in order to be accomplished. These domains concern: R&D of the various components of the new truck (cabin, vehicle dynamics, chassis, and electrical systems); the development and manufacturing of truck engines, manufacturing; purchasing; marketing and strategic planning (as shown in Figure 1). These different domains of knowledge are possessed by agents based in Brazil, France and Sweden. The accomplishment of each phase is therefore assigned to different agents so that they act simultaneously, and, in this regard, a particular ICT is used to support the accomplishment of the task: "NetMeeting".

"This project was totally developed with the support of those tools. NetMeeting all the way, almost none of the calls were made without logging onto NetMeeting. [...] It appears on the screen, and the other member is looking at what you are doing, then you pass the mouse to your colleague and he starts scribbling, then you think, "No, no... give me the mouse back, that's not what and where I meant it to be. It is h-e-r-e, look... the radius is here; "aha, on this radius? I haven't thought about that, cool, I'll think about it". Then, half an hour later he calls; "I thought about everything, and let's forget it and do it differently". Everything via NetMeeting. Basically, you feel as if your colleague was sitting next to you".

EMPIRICAL ANALYSIS AND PROPOSITIONS

Following the path outlined by Weber (1904), we have singled out two main conditions that seem to be adequate for the objective possibility of the occurrence of knowledge integration in the creative process of GDT as they develop new products. These conditions are: the adoption of ICTs that mediate relationships and communications among team members geographically dispersed in different nations, and the values orienting the development of new products.

With regard to the first condition, the comparison of the two empirical cases described above provides the following support of its adequacy. In Alpha, *each* phase of the development of a new global product needs the integration and combination of different and complementary domains of knowledge, namely knowledge concerning the R&D of the various components of the new product, the development and manufacturing of engines, general manufacturing, purchasing, marketing and strategic planning.

Figure 1.

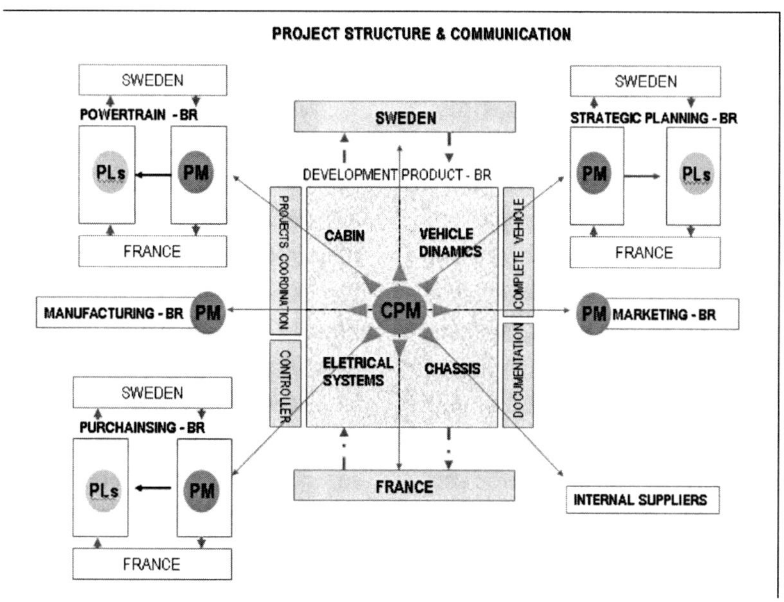

These different kinds of knowledge are possessed by different agents. As a consequence, the accomplishment of each phase of the development process should be assigned to two or more agents so that they can act simultaneously. In the case at hand, the various agents who hold the various domains of knowledge needed for the development of a new product are based in different nations.

Yet the knowledge integration by means of the simultaneous accomplishment of one or more phases of the development process by several agents geographically dispersed in different nations occurs by virtue of the decision to adopt modern ICTs like "Net meeting", and the way this is used by the agents.

In Delta, a phase of the development process involves the drawing up of use cases, which relies on the integration and combination of different and complementary domains of knowledge, namely knowledge concerning the possible industrial applications of the new product at their client's sites and knowledge concerning coding work.

These two domains of knowledge are possessed by different agents based in different nations: the application knowledge is possessed by Italian analysts, and the knowledge concerning coding work is possessed by Tunisian engineers. Therefore, the drawing up of use cases should be assigned to both Italian analysts and Tunisian engineers so that they could act simultaneously. And finally, the relationships and communications between Italian analysts and Tunisian engineers are mediated by ICTs, namely video conference.

However, in Delta, the knowledge integration by means of the simultaneous accomplishment of the task by agents based in different nations does not occur. The drawing up of use cases, in fact, is assigned to both Italian analysts and Tunisian engineers, but these agents act sequentially, not simultaneously. *Firstly* Italian analysts define the document detailing features and properties of the ERP modules, and the way the user is supposed to interact with the program, and *then* use case is discussed with Tunisian engineers and modified accordingly.

The agents act sequentially, not simultaneously, because the analysts and engineers are based in different nations (as in the Alpha case study), and because (differently form the Alpha case study) during the drawing up of use cases their relations and communications are mediated by a video conference, which does not allow fluid communication in real time and does not allow agents to act simultaneously upon the task.

In so far as Italian analysts and Tunisian engineers act sequentially and not simultaneously on the drawing up of use cases, the knowledge integration does not occur within the globally distributed team and the creative process is carried out not by the GDT, but by "individuals".

In light of the above argument, we may summarise as follows the first possible answer to our research question:

Proposition 1.

The decision concerning the adoption and use of Information and Communication Technologies which will mediate relationships and communications among team members geographically dispersed in different nations is a causally adequate condition that enables the occurrence of knowledge integration in the creative process of globally distributed teams while they develop new products.

Let us concentrate on a condition that could account for the different kind of ICT used in the two empirical cases to mediate relationships and communications among members geographically dispersed in different nations. This condition is the consistency between speech (values of the company) and action (technological resources provided to team members).

As has been described above, Alpha emphasizes values such as business and customer focus, industrial efficiency and synergies; speed, agility and flexibility; diversity; teamwork; leaders as coaches; and openness to change. Moreover, it considers employees as key for the success of the organization.

Consistent with these values is Alpha's commitment in constantly updating and making available new technologies to facilitate cross-country communication. A good example of this effort is the use of Live Meeting, a web conference service operated by Microsoft.

"Up to now, NetMeeting has presented some problems, for example: you have the screen up and running, you are using CAD, and you want to show something to your colleague. You should be able to rotate images, but you cannot do that. Now, Live Meeting is a software that allows you to do that, it is like a second generation of NetMeeting. For example, I will be able to work in CAD and in real time; someone in France can see what I am doing. "Look, now I am applying the radius here". "Done" ... the other person in France can see what I have just done and say if it is good enough or not. He can, for example, take control and access the same database. It is cross-ocean interaction...very soon we will not be able to justify business trips any longer. Technological interaction is now immense."

On this basis we may affirm that the organization, in consistence with its values, provides GDT with tools and environments that favour the occurrence of knowledge integration in the creative process of GDT as they develop new products.

Delta also aims to provide its clients with innovative technologies, but unlike Alpha company doesn't supply its team members who are geographically dispersed in different nations with the most innovative technological tools to support their work. Team members, in fact, mainly communicate by video conferences and e-mail. As a consequence, a relevant inconsistency between speech (values of the company) and action (technological resources provided to team members) inhibits the creative process of their GDT.

In accordance with the argument above, we may summarise as follows a possible condition that can account for the decision to adopt ICT that favour the simultaneous accomplishment of one or more tasks of the new products development process by agents based in different nations.

Proposition 1.1.

Consistency between speech (values of the company) and action (technological resources provided to team members) is a causally adequate condition that permits the adoption and use of ICT favouring the occurrence of knowledge integration in the creative process of globally distributed teams as they develop new products.

With regard to the second condition, the comparison of the two empirical cases provides the following support of its adequacy. In the case of Alpha, *each phase* of the development process is oriented by values (or aims) such us profitability, common vehicle architecture, business and customer focus, speed, and industrial efficiency.

This implies that each phase of the development process relies upon heterogeneous domains of knowledge, such as marketing, strategic planning, purchasing, manufacturing, development and manufacturing of truck engines, and R&D of the various components of the new truck.

In so far as these different domains of knowledge are possessed by several agents geographically dispersed in different nations, each phase should be assigned to various agents based in different nations so that they can act simultaneously during the accomplishment of the phase.

However, the development process at Delta is contrarily oriented by the following main value (aim): "to offer to our clients solutions, services and innovative technologies to improve business processes and to ensure business competitiveness."

A closer look at the above statement shows that this value implies that in order to develop new products, and particularly to draw up use cases, knowledge concerning the possible industrial applications of the program at different clients' sites - which is possessed by Italian analysts - is more important than knowledge concerning coding work- which is possessed by Tunisian Engineers.

Accordingly, this value seems to inhibit the integration of different and complementary domains of knowledge by means of the simultaneous accomplishment of the drawing up of use cases by Italian analysts and Tunisian engineers.

With a view to the above argument, we may summarise as follows the second possible answer to our research question:

Proposition 2.

The decision concerning the values (aims) that will drive the development of new products and/or each of its phases is a causally adequate condition that allows the occurrence of knowledge integration in the creative process of GDT as they develop new products.

SOLUTIONS AND RECOMMENDATIONS

The managerial literature stresses the importance of organizational strategic planning, which approaches the sharing of a future vision and the involvement of people with enterprise objectives. More specifically, Malhotra, Majchrzak, Carman & Lott, (2007) highlight that in order to favour a cohesive development of the tasks performed by team members, organizations should articulate a vision for the team, communicate the vision with passion, define an execution plan to support the implementation of the vision, form coalitions of believers, align others behind the vision, and shape a team culture by articulating operating values.

The empirical analysis presented in this contribution provides further support for this argument by highlighting the need for fluid and constant communication of the goals and values of the organization at all levels. In addition, it suggests that the management should clearly define the values within the organizational strategic planning and pay particular attention to the definition of clear and consistent values.

The empirical analysis also suggests that the activities of the new product development process are carried out by agents based in different nations acting simultaneously when the knowledge required to accomplish the activities are geographically dispersed. It also highlights that, with respect to this objective, it is of vital importance that the agents can work in an environment where they have equal access to the most modern information technologies.

FUTURE RESEARCH DIRECTIONS

Future research is encouraged to continue addressing the relevance of knowledge integration in the creative process, taking into account the differences concerning the configuration of the team, and also the team's distribution: whether it is globally distributed, locally distributed or collocated.

Another argument that could be explored is whether (or not) a difference exists when the creative process of GDT concerns new processes rather than new products. As demonstrated by this contribution, information and communication technologies can provide an environment favourable for the creative process of products. But we can say the same when we talk about the abstract process which is to create "processes"?

Kristensson & Norlander (2003) have made an experiment with 96 students in the Cyber/IT Research Laboratory of Karlstad University. Their central aim was to investigate the effects of electronic interaction of groups over individuals' creative performances. The comparison was focused on distributed and co-located teams, and in both situations, the creative process studied concerns product and process. The study provides significant results.

However, we believe that it could be interesting to check the constructs raised not only in laboratory, but also on the field. In fact, regarding this matter it is worth remembering what Kurtzberg &

Amabile (2001) underlined about laboratory experiment concerning teams involved in electronic brainstorming: "These teams had no history with each other, no established patterns of interaction, no knowledge of each other's strengths and weaknesses, and no strong incentive to cooperate or create a mutual understanding that would have helped them as a team at some future time." (p. 288).

CONCLUSION

According to the existing literature, the most important factors that negatively affect creative process in teams are the same factors that distinctively characterize GDT: that is cultural diversity and interpersonal relationships mediated by computer. From this one would presume that GDT experience many difficulties in creating, but our study strongly challenges this conclusion. Indeed, it highlights that the development of new products and the generation of new creative ideas are strictly connected with a successful integration of different and complementary domains of knowledge possessed by team members who are geographically dispersed in different nations.

In this contribution, we have used the term 'knowledge integration' to refer to the situation in which a task to be accomplished requires different domains of knowledge possessed by different agents and its accomplishment is assigned to these agents so that they act simultaneously. Then, following the path outlined by Weber (1904, 1906, 1922), we have compared two empirical cases, thereby ascertaining two adequate conditions for the objective possibility of the occurrence of knowledge integration in the creative process of GDTs as they develop new products.

The first condition is the decision concerning the adoption of the ICT that will mediate the relationships and communications among teams members based in different nations. In this regard, the comparison of the two empirical cases highlights that in situations in which different domains of knowledge are required to accomplish an action and these domains of knowledge are possessed by different agents based in different nations, the choice to adopt more modern ICT, such as NetMeeting rather than video conference, is crucial with respect to the aim of favouring the joint (simultaneous) accomplishment of the task by geographically dispersed agents.

The second condition is the decision concerning the values (aims) that will drive the development of new products and/or each of its phases. Our comparison of the two empirical cases also demonstrates that some company values appear to imply that one or more phases of the development process rely upon different domains of knowledge in order to be accomplished. In this situation, if these different domains of knowledge are possessed by agents based in different nations, those values will favour the assignation of the task to be accomplished to agents geographically dispersed so that they act simultaneously. In this sense, some values seem to favour the occurrence of knowledge integration in the creative process of globally distributed teams.

In conclusion, with respect to the research question *"How and why does knowledge integration occur in the creative process of GDTs as they develop new products"?* We have identified the following causally adequate conditions that enable the occurrence of knowledge integration:

- The decision concerning the adoption and use of Information and Communication Technologies which will mediate relationships and communications among team members geographically dispersed in different nations;
- The decision concerning the values (aims) that will drive the development of new products and/or each of its phases.

With respect to the first condition just outlined above, we have also pointed out that the consistency between speech (values of the company) and action (technological resources provided to team members) can account for the adoption of such ICT that favour the occurrence of knowledge integration in the creative process of GDTs as they develop new products.

REFERENCES

Alavi, M., & Tiwana, A. (2002). Knowledge integration in virtual teams: the potential role of KMS. *Journal of the American Society for Information Science and Technology*, *53*(12), 1029–1037. doi:10.1002/asi.10107

Alves, J., Marques, M. J., Saur, I., & Marques, P. (2007). Creativity and innovation through multidisciplinary and multisectoral cooperation. *Creativity and Innovation Management*, *16*(1), 27–34. doi:10.1111/j.1467-8691.2007.00417.x

Amabile, T. M. (1996). *Creativity in context*. Boulder, CO: Westview Press.

Amabile, T. M., & Gryskiewicz, S. S. (1987). *Creativity in the R&D laboratory*. Greensboro, NC: Center for Creative Leadership.

Argote, L., & Ingram, P. (2000). Knowledge transfer: A basis for competitive advantage in firms. *Organizational Behavior and Human Decision Processes*, *82*(1), 150–169. doi:10.1006/obhd.2000.2893

Baba, M. L., Gluesing, J., Ratner, H., & Wagner, K. (2004). The contexts of knowing: Natural history of a globally distributed team. *Journal of Organizational Behavior*, *25*(5), 547–587. doi:10.1002/job.259

Baer, M., & Oldham, G. R. (2006). The curvilinear relation between experienced creative time preassure and creativity: Moderating effects of openness to experience and support for creativity. *The Journal of Applied Psychology*, *91*(4), 963–970. doi:10.1037/0021-9010.91.4.963

Barley, S. R., & Orr, J. E. (1997). *Between craft and science: Technical settings in U.S.Settings*. London: ILR Press.

Bergh, B. V., & Stuhlfaut, M. (2006). Is advertising creativity primarily an individual or a social process? *Mass Communication & Society*, *9*(4), 373–397. doi:10.1207/s15327825mcs0904_1

Cohen, S. G., & Bailey, D. E. (1997). What makes teams work: group effectiveness research from the shop floor to the executive suite. *Journal of Management*, *23*(3), 239–290. doi:10.1177/014920639702300303

Drucker, P. (2007). *Innovation and Entrepreneurship - Practice and Principles. The classic Drucker Collection*. Oxford, UK: Butterworth-Heinemann.

Ensor, J., Cottam, A., & Band, C. (2001). Fostering knowledge management through the creative work environment: A portable model from the advertising industry. *Journal of Information Science*, *27*(3), 147–155. doi:10.1177/016555150102700304

Gassmann, O. (2001). Multicultural teams: Increasing creativity and innovation by diversity. *Creativity and Innovation Management*, *10*(2), 88–95. doi:10.1111/1467-8691.00206

Gilson, L. L., & Shalley, C. E. (2004). A little creativity goes a long way: An examination of team's engagement in creative processes. *Journal of Management*, *30*(4), 453–470. doi:10.1016/j.jm.2003.07.001

Goh, S. (2002). Managing effective knowledge transfer: An integrative framework and some practice implications. *Journal of Knowledge Management*, *1*(6), 23–30. doi:10.1108/13673270210417664

Grant, R. (1996). Prospering in Dynamically-competitive Environments: Organizational Capability as Knowledge Integration. *Organization Science*, *7*(4), 375–387. doi:10.1287/orsc.7.4.375

Gurteen, D. (1998). Knowledge, creativity and innovation. *Journal of Knowledge Management*, *2*(1), 5–13. doi:10.1108/13673279810800744

Handy, C. (1995). Trust and the virtual organization. *Harvard Business Review*, *73*(3), 40–50.

Hargadon, A. B., & Bechky, B. A. (2006). When collections of creatives become creative collectives: A field study of problem solving at work. *Organization Science*, *17*(4), 484–500. doi:10.1287/orsc.1060.0200

Kao, J. (1996). *Jamming*. London: Haper Collins Business.

Kristensson, P., & Norlander, T. (2003). The creative product and the creative processes in virtual environments. *Creativity and Innovation Management*, *12*(1), 32–40. doi:10.1111/1467-8691.00266

Kumar, K., Von Glinow, M., & Van Fenema, P. (2009). Offshoring and global distribution of work: Implications for task interdependence theory and practice. *Journal of International Business Studies*, *40*(4), 642–667. doi:10.1057/jibs.2008.77

Kurtzberg, T. R., & Amabile, T. A. (2001). From Guilford to creative synergy: Opening the black box of team-level creativity. *Creativity Research Journal*, *13*(3), 285–294. doi:10.1207/S15326934CRJ1334_06

Kylen, S., & Shani, A. B. R. (2002). Triggering creativity in teams: an exploratory investigation. *Creativity and Innovation Management*, *11*(11), 17–30. doi:10.1111/1467-8691.00233

Leenders, R. T. A. J., Van Engelen, J. M. L., & Kratzer, J. (2003). Virtuality, communication, and new product team creativity: A social network perspective. *Journal of Engineering and Technology Management*, *20*, 69–92. doi:10.1016/S0923-4748(03)00005-5

Magnusson, M. G. (2004). Managing the Knowledge Landscape of an MNC: Knowledge Networking at Ericsson. *Knowledge and Process Management*, *11*(4), 262–272. doi:10.1002/kpm.210

Malhotra, A., Majchrzak, A., Carman, R., & Lott, V. (2001). Radical innovation without collocation: A case study at Boeing-Rocketdyne. *MIS Quartely*, *25*(2), 229–249. doi:10.2307/3250930

Mostert, N. M. (2007). Diversity of the mind as the key to successful creativity at Unilever. *Creativity and Innovation Management*, *16*(1), 93–100. doi:10.1111/j.1467-8691.2007.00422.x

Nemiro, J. E. (2004). *Creativity in virtual teams: key components for success. The Collaborative Work Systems Series. Center for the study of work teams*. San Francisco: Pfeiffer.

Nonaka, I. (1994). A dynamic theory of organizational knowledge creation. *Organization Science*, *5*(1), 14–37. doi:10.1287/orsc.5.1.14

O'Dell, C., & Grayson, C. J. J. (1999). Knowledge transfer: Discover your value proposition. *Strategy and Leadership*, 10–15. doi:10.1108/eb054630

Paulus, P. B. (2000). Groups, teams and creativity: the creative potential of idea-generating groups. *Applied psychology. International Review (Steubenville, Ohio)*, *49*(2), 237–262.

Paulus, P. B., Larey, T. S., & Dzindolet, M. T. (2001). Creativity in groups and teams. In Turner, M. E. (Ed.), *Groups at work: Theory and research* (pp. 319–338). Mahwah, NJ: Lawrence Erlbaum Associates Publishers.

Piccoli, G., Powell, A., & Ives, B. (2004). Virtual teams: team control structure, work processes, and team effectiveness. *Information Technology & People*, *17*(4), 359–379. doi:10.1108/09593840410570258

Pozzebon, M., & Pinsonneault, A. (2005). Challenges in conducting empirical work using structuration theory: Learning from IT research. *Organization Studies*, *26*(9), 1353–1376. doi:10.1177/0170840605054621

Prasad, K., & Akhilesh, K. B. (2002). Global virtual teams: What impacts their design and performance? *Team Performance Management: An International Journal*, *8*(5/6), 102–112. doi:10.1108/13527590210442212

Reis, R. S., & Chevalier, F. (2009, August). *Unfolding creative process in development of new products within globally distributed teams*. Paper presented at the meeting of Academy of Management, Chicago, IL.

Revilla, E., & Cury, T. (2008) Antecedents and Consequences of Knowledge Integration in Product Development: An Empirical Evidence. Available at SSRN: http://ssrn.com/abstract=1134491

Schumpeter, J. A. (1942). *Capitalism, Socialism, and Democracy*. New York: Harper.

Weber, M. (1904). Die Objektivität sozial-wissenschaftlicher und sozialpolitischer Erkenntnis. In *Gesammelte Aufsatze zur Wissenschaftslehre*. Tubingen: Mohr.

Weber, M. (1906). Kritische Studien auf dem Gebiet der kultur-wissenschaftlichen Logik. In *Gesammelte Aufsatze zur Wissenschaftslehre*. Tubingen: Mohr.

Weber, M. (1922). *Wirtschaft und Gesellschaft*. Tubingen: Mohr.

ADDITIONAL READING

Ancona, D. G., & Caldwell, D. F. (1992). Demography and design: Predictors of new product team performance. *Organization Science*, *3*(3), 321–341. doi:10.1287/orsc.3.3.321

Andriopoulos, C. (2001). Determinats of organisational creativity: a literature review. *Management Decision*, *39*(10), 834–840. doi:10.1108/00251740110402328

Arnison, L., & Miller, P. (2002). Virtual teams: A virtue for the conventional team. *Journal of Workplace Learning*, *14*(4), 166–173. doi:10.1108/13665620210427294

Brake, T. (2006). Leading global virtual teams. *Industrial and Commercial Training*, *38*(3), 116–121. doi:10.1108/00197850610659364

Brown, S. L., & Eisenhardt, K. M. (1995). Product development: Past research, present findings, and future directions. *Academy of Management Review*, *20*(2), 343–378. doi:10.2307/258850

Chen, M. H., Chang, Y. C., & Hung, S. C. (2008). Social capital and creativity in R&D project teams. *R & D Management*, *38*(1), 21–34.

Chudoba, K. M., Wynn, E., Lu, M., & Watson-Manheim, M. B. (2005). How virtual are we? Measuring virtuality and understanding its impact in a global organization. *Information Systems Journal*, *15*, 279–306. doi:10.1111/j.1365-2575.2005.00200.x

Cramton, C. D. (2001). The mutual knowledge problem and its consequences for dispersed collaboration. *Organization Science*, *12*(3), 346–371. doi:10.1287/orsc.12.3.346.10098

Crossman, A., & Lee-Kelley, L. (2004). Trust, commitment and team working: The paradox of virtual organizations. *Global Networks*, *4*(4), 375–390. doi:10.1111/j.1471-0374.2004.00099.x

DeSanctis, G., & Monge, P. (1999). Introduction of the special issue: Communication process for the virtual organizations. *Organization Science*, *10*(6), 693–703. doi:10.1287/orsc.10.6.693

Dewett, T. (2003). Understanding the relantionship between information technology and creativity in organizations. *Creativity Research Journal, 15*(2-3), 167–182. doi:10.1207/S15326934CRJ152&3_08

Drazin, R., Glynn, M. A., & Kazanjian, R. K. (1999). Multilevel theorinzing about creativity in organizations: A sensemaking perspective. *Academy of Management Review, 24*(2), 286–307. doi:10.2307/259083

Ekvall, G., & Ryhammar, L. (1999). The creative climate: Its determinants and effects at a Swedish University. *Creativity Research Journal, 12*(4), 303–310. doi:10.1207/s15326934crj1204_8

Espinosa, J. A., Slaughter, S. A., Kraut, R. E., & Herbsleb, J. D. (2007). Familiarity, complexity, and team performance in geographically distributed software development. *Organization Science, 18*(4), 613–630. doi:10.1287/orsc.1070.0297

Faraj, S., & Sproull, L. (2000). Coordinating expertise in software development teams. *Management Science, 46*(12), 1554–1568. doi:10.1287/mnsc.46.12.1554.12072

Feldman, D. H., Csikszentmihalyi, M., & Gardner, H. (1994). *Changing the world: A framework for the study of creativity*. Westport, CT: Greenwood Publishing Group Inc.

Fisher, T. (1997). The designer's self-identity – Myths of creativity and the management of teams. *Creativity and Innovation Management, 6*(1), 10–18. doi:10.1111/1467-8691.00044

Ford, C. M. (2000). Creative developments in creativity theory. *Academy of Management Review, 25*(2), 284–285.

Furst, S., Blackburn, R., & Rosen, B. (1999). Virtual teams effectiveness: A proposed research agend. *Information Systems Journal, 9*, 249–269. doi:10.1046/j.1365-2575.1999.00064.x

Gersick, C. J. G. (1988). Time and transition in work teams: Toward a new model of group development. *Academy of Management Journal, 31*(1), 9–41. doi:10.2307/256496

Grevins, J. (2002). *Project virtuality effects on project team processes and project success. Unpuplished doctoral dissertion.* Buffalo: University of New York.

Haner, U. E. (2005). Spaces for creativity and innovation in two established organizations. *Creativity and Innovation Management, 14*(3), 288–298. doi:10.1111/j.1476-8691.2005.00347.x

Holland, S., Gaston, K., & Gomes, J. (2000). Critical success factors for cross-functional teamwork in new product development. *International Journal of Management Reviews, 2*(3), 231–259. doi:10.1111/1468-2370.00040

Horwitz, F. M., Bravington, D., & Silvis, U. (2006). The promise of virtual teams: Identifying key factors in effectiveness and failure. *Journal of European Industrial Training, 30*(6), 472–494. doi:10.1108/03090590610688843

Jackson, P. J. (1999). Organizational change and virtual teams: Strategic and operational integration. *Information Systems Journal, 9*, 313–332. doi:10.1046/j.1365-2575.1999.00066.x

Kang, H. R., Yang, H. D., & Rowley, C. (2006). Factors in team effectiveness: Cognitive and demographic similarities of software development team members. *Human Relations, 59*(12), 1681–1710. doi:10.1177/0018726706072891

Kotlarsky, J., & Oshri, I. (2005). Social ties, knowledge sharing and successful collaboration in globally distributed system development projects. *European Journal of Information Systems, 14*, 37–48. doi:10.1057/palgrave.ejis.3000520

Kratzer, J., Leenders, R. T. A. J., & Van Engelen, J. M. L. (2004). Stimulating the potential: Creative performance and communication in innovation teams. *Creativity and Innovation Management*, *13*(1), 63–71. doi:10.1111/j.1467-8691.2004.00294.x

Malhotra, A., Majchrzak, A., Carman, R., & Lott, V. (2001). Radical innovation without collocation: A case study at Boeing-Rocketdyne. *MIS Quartely*, *25*(2), 229–249. doi:10.2307/3250930

near and far: A comparison between collocated and virtual teams. *Information Technology & People*, *19*(4), 299–322. doi:10.1108/09593840610718018

Nemiro, J. E. (2000). The glue that binds creative virtual teams. In Yogesh Malhotra (Ed.), *Knowledge management and virtual organizations* (101-123). Hershey, PA: Idea Group Publishing.

Nemiro, J. E. (2002). The creative process in virtual teams. *Creativity Research Journal*, *14*(1), 69–83. doi:10.1207/S15326934CRJ1401_6

Nemiro, J. E. (2004). *Creativity in virtual teams*. San Francisco, CA: Pfeiffer.

Oldham, G. R., & Cummings, A. (1996). Employee creativity: Personal and contextual factors at work. *Academy of Management Journal*, *39*(3), 607–634. doi:10.2307/256657

Powell, A., Galvin, J. & Piccoli, G. (2006). Atecedents to team member commitment from

Reis, R. S. (2008). The role of trust in the creative process of globally distributed teams: an exploratory case study. Paper presented at the EGOS 24th Coloquium, Amsterdam, NL.

Rico, R., & Cohen, S. (2005). Effects of task interdependence and type of communication on performance in virtual teams. *Journal of Managerial Psychology*, *20*(3-4), 261–274. doi:10.1108/02683940510589046

Robinson, A. G., & Stern, S. (1997). *Corportae creativity: How innovation and improvement actually happen*. San Francisco: Berrett-Koehler Publishers Inc.

Sole, D., & Applegate, L. (2000). *Knowledge sharing practices and technology use norms in dispersed development teams*. Paper presented at the meeting of International Conference on Information Systems, Brisbane, Queensland, Australia.

Ulrich, K. T., Eppinger, S. D., & Filippini, R. (2004). *Product design and development*. New York: McGraw-Hill Companies Inc.

Van der Smagt, T. (2000). Enhancing virtual teams: social relations v. Communication technology. *Industrial Management & Data Systems*, *100*(4), 148–156. doi:10.1108/02635570010291766

Zaltman, G., Duncan, R., & Holbek, J. (1973). *Innovations and organizations*. New York: Wiley.

KEY TERMS AND DEFINITIONS

Creativity: "The production of novel and useful ideas in any domain" (Amabile, 1996). Creativity refers to the creation of a valuable, useful new product, service, idea, procedure, or process.

Creative Process: Set of activities leading to the generation of ideas that are both novel and appropriate.

Individual Creative Process: Creative process carried out by isolated individuals.

Knowledge Integration: In this chapter this term is used to refer to the situation in which a task to be accomplished requires different domains of knowledge possessed by different agents, and its accomplishment is assigned to various agents so that they can act simultaneously.

Team Creative Process: Members working together in such a manner that they link ideas from sources, delve into unknown areas to find better or unique approaches to a problem, or seek out novel ways of performing a task (Gilson & Shalley, 2004).

Innovation: Concerned with the initial market introduction of a new product or process that either disrupts or entrenches existing competencies. Innovation can be seen as the successful implementation of creative ideas within an organization.

New Product Development Team: Two or more individuals engaged, most of the time, in creating sustaining innovations: new things that make the existing business work better.

Distributed Team: Groups of individuals geographically dispersed in the same country, and collaborating in the execution of a specific project.

GDT - Globally Distributed Team: Work team made up of culturally different members based in two or more nations who share collective responsibilities, brought together by information and communication technologies to jointly accomplish one or more organizational tasks and to reach a common goal in the organization.

Chapter 4
Production Competence and Knowledge Generation for Technology Transfer: A Comparison between UK and South African Case Studies

Ian Hipkin
École Supérieure de Commerce de Pau, France

EXECUTIVE SUMMARY

The ability to execute the physical part of manufacturing will assume greater importance as new technology and knowledge become significant drivers of strategic direction. The case studies described in this chapter address the interaction between technology transfers (TT), production competence and knowledge in enhancing performance in manufacturing organizations. Reference to British and South African case studies provides a useful comparison of production competence in the developed and developing world. In both countries, operators and maintainers lacked detailed knowledge of equipment functionality and performance parameters. United Kingdom (UK) companies demonstrated a deeper understanding of fundamental principles of the underlying production process, enabling them to remedy production deficiencies more thoroughly. South African companies showed greater management commitment to training and group solving approaches.

ORGANIZATION BACKGROUND

New equipment and systems play a vital role as production-oriented firms strive to meet changing customer demands and increased competition. Technology is key to achieving competitive advantage to this and requires an exchange of technical know-how and the transfer of competencies to facilitate organizational learning, consolidation of new technologies, and establishing routines and production skills (Lynskey, 1999). Frequently recurring factors that contribute to poor performance of new technology in both developed and developing countries are inadequate knowledge of plant and processes, managerial commitment,

DOI: 10.4018/978-1-61692-886-5.ch004

information systems, and lack of production competence. This chapter describes an investigation into the impact of such factors on TT in a number of UK and South African manufacturing organizations that have recently acquired new technology, and assesses how similarities and differences in these organizations affect production performance.

SETTING THE STAGE

It has long been suggested that equipment performance is dependent on production competence (Ferdows and De Meyer, 1990; Leonard-Barton, 1995). Employees who understand how something is made are better able to manage manufacturing practices than those who deal only with the underlying symptoms of a process (Gourley, 2006). An environment where knowledge assimilation and sharing generate continuous learning capability lends itself to absorptive capacity (Tu et al, 2006). Early proponents such as Cohen and Levinthal (1990) see absorptive capacity (AC) as the ability of a firm to recognise and utilise the value of new information that comes from continuous learning. Mowery and Oxley (1995) construe AC as a comprehensive set of skills that engages with the tacit component of transferred and imported knowledge. Jones and Craven (2001) emphasise the need for communication networks that diffuse knowledge and technology.

Production competence and knowledge generation may be seen as a function of AC, which Zahra and George (2002) define as: "AC is a set of organizational routines and processes by which firms acquire, assimilate, transform and exploit knowledge to produce a dynamic organizational capability" (p.185). Like many commentators these authors view AC as an essentially organizational phenomenon.

In order to understand a knowledge-based process Rouse and Daellenbach (2002) suggest that we should follow the value generation trail back to its source. This requires "unpacking the process, beginning with performance, then looking for sources of advantage in the form of capabilities and competencies. Unless the sources of firm-specific superiority can be understood, the nature of the competitive advantage is doomed to remain largely a mystery. We will merely have substituted a black box of competencies for the black box of organizations." (p.966) Rouse and Daellenbach (2002) also suggest a theoretical and empirical understanding of what occurs in TT as well as how this takes place:

- Resources, technology and knowledge (tangible and intangible) are bundled, linked and incorporated as management processes.
- Management processes and structural relationships are converted and organised into routines and systems.
- Routines and systems formalise capabilities that lead to production competencies.
- Competencies become the means whereby products and services are generated to provide value and competitive advantage.

Resources and Knowledge

Resource-based paradigms help to explain the success of firms through an analysis of technological skills, complementary assets and routines. Capabilities are clusters of tangible and intangible abilities that span individuals and groups (Coates and McDermott, 2002). Manufacturing firms need a competitive focus to optimise quality, cost and flexibility and develop competencies in the long term, so it is necessary to identify the tangible and intangible knowledge and skills required, and formulate policies to prioritise these. Knowledge further assists designers in understanding the consequences of any changes that may be made in improving product features (Cristofolini et al, 2009).

Knowledge is a useful lens through which to view the development of competencies. Interlinking technical areas allows firms to combine multiple knowledge and functional domains to overcome challenges in implementing new technology. Pools of knowledge lead to competencies so an essential factor in their emergence is an orientation towards learning and developing knowledge. Training and familiarisation are prerequisites for the creation and use of new competencies. Knowledge is a resource that supports a firm's ability to excel in achieving these competitive priorities. Among the reasons for inadequate knowledge is a firm's over-optimism about the knowledge of its own processes, and its inability to build, debug and operate new processes (Bohn, 1994).

Management Support

Firms need management processes that will transcend organizational boundaries and co-ordinate activities across diverse environments and social dynamics (Dekkers, 2009). They benefit from leadership roles played by proactive individuals, boundary spanners and technology champions (Irwin et al, 1998). These facilitate crossing technical boundaries to find multidisciplinary solutions through innovation and bringing together diverse experts. Managers should lead in offering support for new equipment, while being sufficiently flexible to accommodate the emerging behaviour and actions that unfold from operators of the technology. This requires rethinking established processes that affect, or are influenced by, the technology.

Technological competence depends on the ability and contribution of development teams, but managers set the process in motion, provide facilities for training and encourage organizational learning as a means to develop requisite capabilities. Managers should also encourage the process of knowledge acquisition, through "legitimizing familiarisation activities" (Dimnik and Johnston, 1993, p.161). Crossan and Berdrow (2003) emphasise the development of shared understanding amongst individuals and taking coordinated action through mutual adjustment. This will initially be unstructured and informal, but may be institutionalised by embedding the learning that has occurred by into systems, structures and procedures. Within a constantly changing environment, managers must manage the tension between institutionalised learning and the processes of intuitive interpretation and integration.

The more individuals work with colleagues and benefit from physical and psychological proximity, the more they develop mutual trust and are willing to share their specialist knowledge (Uzzi and Lancaster, 2003). The effect is faster knowledge assimilation through observation, the facility to draw on a pool of contextual knowledge, and a shared sense of where expertise lies. Reagans et al (2005) refer to this as a 'relationship-specific' activity where knowledge is assimilated and embedded in the link connecting people performing distinct roles.

Information and Systems

Management scientists and operations researchers have developed decision-making tools for operations and maintenance that are intended to replace subjective decisions with objective ones and assist managerial control through faster, more comprehensive and accurate knowledge of operations, and reducing idiosyncratic judgement and mystique. However, Pintelon and Gelders (1992) claim that little use is made of such techniques and challenge the benefits of a large quantity of explicit data, even if it can be gathered. Researchers can identify some key information systems when assessing TT, but cannot access others that are invisible and buried in the minds of the actors. For example, computerised maintenance management systems have been developed as a form of data repository. Although maintainers perceive the lack of failure data as a barrier to maintenance intervention, it is questionable how much useful data is stored in computerised maintenance systems. Aharoni

(1993) suggests that the diversity of information systems emphasises how the richness in quality and nature of information systems can be imbued with value, context and meaning. However, researchers must assess whether information systems enhance production or support organizational governance through a form of indirect exertion of authority. Even if the systems governing each individual's task within an organizational routine are explicit, the complex totality of the system may well be unknown to most participants, and is therefore tacit. Few systems can accommodate knowledge, know-how, social capital, and other socially complex, difficult-to-transfer data.

Production Competence

Coates and McDermott (2002) define competence as: "a bundle of aptitudes, skills, and technologies that the firm performs better than its competitors." (p.443) The importance of understanding equipment performance is widely supported in the literature on process competencies and organizational learning (Leonard-Barton, 1995; Marmier et al, 2009). Processes dependent on technological subsystems require subtle skills and expertise that are often difficult to codify. The secrets and capabilities of a complex process are understood most effectively through direct, first-hand observation and interaction (Lynskey, 1999).

Galbreath (2005) posits that knowledge is assimilated as capability when familiarity with tangible assets through employee learning, and interaction between management and operational personnel become evident in operational routines. Training can at best only go part-way in imparting this expertise. Assimilation is delayed in conditions of uncertainty when knowledge elements lie beyond an organization's sphere of expertise (Rosenkopf and Nerkar, 2001). This may be due to novelty and layers of complexity which even knowledgeable operators are unable to comprehend (Dehane et al, 2008; Sorenson et al, 2006).

Macher (2006) notes that problem solving is often a trial-and-error exercise where particular parameters are modified, learned about, and improved. A range of solutions is likely to emerge, and their impact may be interpreted in differing ways by multiple stakeholders with possibly contradictory interests. Above all, the success of a technology depends to a large degree on the competencies of personnel, since documented and accessible explicit knowledge may form only a small part of the knowledge that should be shared. The management of collective knowledge in firms and the ability of individuals to work together to convert knowledge into customer value are essential competencies. The challenge for researchers is to develop theory that will "identify unique, proprietary, and special abilities in manufacturing firms and understand how competencies that lead to competitive advantage both occur and are capitalized upon in the firm" (Coates and McDermott, 2002, p.438).

CASE DESCRIPTION

Four organizations that had recently implemented new technology were selected for this study. Broadly comparable firms in two sectors were studied (paper and food): two from the United Kingdom (referred to as PaperUK and FoodUK) and two from South Africa (PaperSA and FoodSA). A brief description of each is provided.

- *PaperUK*: This 30 year old UK paper operated ageing equipment, beset with breakdowns. It produced speciality paper for the printing industry and operated several paper machines, coaters and a finishing department. Production targets were only met with excessive maintenance overtime. Low profitability meant that capital expenditure had been severely limited, so the life of existing equipment was to be extended through maintenance and some small technological advancements. Management had installed a computerised maintenance management system (CMMS), but this was recognised as deficient in its lack of user friendliness. This could not be construed as a knowledge-based operation as the vast quantities of data accumulated over the years were difficult to access, and had little credibility. No direct interface between the CMMS and the production system existed. As part of a long-term plant upgrade a supplier was appointed to install a new coater on one paper machine in close association with a team of paper mill staff from engineering, operations and maintenance. Commissioning of the new coater was accompanied by the installation of a new CMMS, in order to raise the level of knowledge of plant operation and failure data. This represented a substantial upgrade potential for the maintenance function. Reliable information was seen as essential for effective maintenance. The CMMS was to be used to attain the original functionality of equipment, and to ensure higher plant availability. 12 individuals were interviewed in PaperUK.

- *FoodUK*: This UK food company was confronted by tremendous cost competition from global players, and was attempting to find niche markets to build on its quality competitive advantage. Lower operating costs were essential provided quality standards could be maintained. Labour intensive quality checking was no longer sustainable: this was expensive and not sufficiently reliable. Customers had been criticising the company for not meeting its agreed acceptable quality level standards, and several major customers had been lost. In some areas production stoppages could be tolerated depending on downline processes and orders. Opportunistic repairs during production stoppages were becoming the dominant way of doing maintenance, although it was recognised that this was not effective maintenance. A large backlog of data had not been entered into the existing maintenance management system. This food factory installed laser screening equipment to monitor raw materials used in the manufacture of tinned foods and take remedial action to prevent substandard products from entering the manufacturing process. The challenge was to ensure that operators were sufficiently knowledgeable and proficient in utilising the new technology, and to guarantee that strict quality standards were met. 13 individuals were interviewed in Food UK.
- *PaperSA*: The strategy of this South African paper mill was to produce high quality paper at a cost lower than its competitors. Despite an industry characterised by depressed prices, there were profitable markets for the company's products. The company had no knowledge-based system for maintenance since the few historical failure records were based on perceptions of certain members of the engineering and maintenance departments, as well as opportunistic maintenance which was undertaken during scheduled and non-scheduled production stoppages. Maintenance and production staff were required to enter failure and performance data at terminals in the operating control room. This paper mill comprised paper machines, winders, rewinders, core and log handling, and packing machines. Various new projects had been undertaken including an upgrade of the logsaw and associated equipment, and a new packaging machine. The choice of equipment and quality standards was made on the basis of extensive consumer market research which indicated where customer values lay. Significant management was expended in ensuring that operating staff

accepted the new quality standards, which were at times at variance with previous operating practices. 9 individuals were interviewed in PaperSA.

- *FoodSA*: This food-processing factory had pockets of low unit cost competitive advantage brought about by earlier investment in new technologies. The low cost advantage was eroded by fierce competition, and extensive down time on several sorting machines. Maintenance had been predominantly preventive, but management felt that a more structured approach was required, which required information to be recorded in a computerised system. Production departments were to be convinced that maintainers possessed valuable knowledge that would reduce the frequency and duration of production stoppages. Although a number of variables in raw materials influenced the quality of the end product, the level of output variance was limited as quality was strictly in accordance with American fast-food quality. New sorting equipment had been acquired, but no maintenance history was available. The challenge for the company was to acquire sufficient knowledge of the equipment to ensure high levels of available and reliability. 10 individuals were interviewed in FoodSA.

CURRENT CHALLENGES FACING THE ORGANIZATIONS

This section recounts occurrences and experiences in the case organizations in relation to the main points identified in the 'setting the scene' section of this chapter. One of the common features in all case studies was that previous operating practices and equipment were not delivering the required performance. New technology was introduced in each case, but knowledge of new processes was lacking, and managers were not sure how this deficiency should be addressed. Their actions were essentially trial and error, and largely unstructured.

Resources and Knowledge

Resources are the assets owned or controlled by a firm, and include new equipment and knowledge vested in people. In investigating the challenges facing the case organizations, it is instructive to identify subtle knowledge acquisition, working practices and adjustments that maximise the benefits that can be derived from new technology.

New laser equipment at FoodUK replaced manual raw vegetable quality monitoring. Settings by suppliers and plant engineers were based on trial and error. In assessing the quality of the incoming product and monitoring the output, operators were able to use their previous experience of the manual process (tacit knowledge) to set the machine to ensure minimum rejection of acceptable products and maximum rejection of unacceptable products. In order to achieve this, a great deal of management time and effort was dedicated to persuading operators of the value of this knowledge. This was necessary since operators had initially distanced themselves from the technology, expressing the sentiment that the technology had effectively displaced them.

FoodSA also had varying qualities of incoming raw material. Machines were set to allow sizes to vary between limits, depending on customer requirements. Operators followed procedures laid down by supervisors since they were not aware of operating parameters within which outsized materials could be tolerated. FoodSA's explicit instructions were not relaxed to accommodate the type of tacit knowledge demonstrated in FoodUK.

These examples illustrate the importance of delicate adjustments that lie beyond standard operating procedures. According to one respondent at FoodUK the basis for successful use of this equipment resided in 'an uncanny knack' by certain operators to apply incremental fixes and fine adjustments to meet specified quality requirements. Despite their documentation and engineering drawings (explicit knowledge), process engineers were unable to do this. FoodSA's procedures did not encourage the degree of flexibility that would permit some experimentation on the basis of tacit knowledge.

Personnel in the UK companies possessed greater knowledge than their South African counterparts. Although only some events are recounted, it was clear that the experience and initiatives of the UK operators in particular (on the basis of tacit knowledge) added significantly to production output. In neither food company was there any formal approach to knowledge acquisition. PaperSA did make a concerted effort to acquire knowledge of customer quality standards, and to ensure that products adhered to these requirements.

Management Support

Managers in the case studies acknowledged the importance of training staff members involved with operating and maintaining new equipment. PaperUK engaged the supplier of the coater equipment to train operators and maintenance craftsmen. However, the training did not encourage 'ownership' of the plant by the newly constituted team because several features of the coater replicated difficulties with older equipment and operators felt their opinions had not been taken into account when the new coater was specified. Management support was perceived to be wanting. The result was a lack of knowledge assimilation: while knowledge could be acquired through the training, operators did not accept or readily assimilate it.

Much discussion preceded the acquisition of the new equipment at PaperSA, especially regarding internal quality standards. It was possible to meet external requirements but internal standards remained a source of contention. The amount of glue sprayed onto a toilet roll to secure the final sheet (the 'tail') determined how many sheets were glued together while the length of 'tail' affected final packaging. Management established a team of operators, maintenance and process staff to address this type of issue, but only when a champion emerged in the group did team members demonstrate a concerted effort to investigate appropriate specifications for the equipment. In absorptive capacity terms, this reflected assimilation and transformation: the competitive advantage gained from first mover advantage in setting and adhering to appropriate standards was truly transforming.

In this instance the South African managers involved their staff more in discussions of the new technology. This provided a legitimate basis for the group to propose quantitative performance standards which were subsequently accepted as company quality policy, and led to an industry leadership position.

Information and Systems

As part of FoodSA's ISO9000 certification exercise, detailed production procedures were produced and systems established to provide an audit trail for quality inspections. Managers' attention was more directed towards compliance with the system than with production requirements.

PaperSA attempted to make use of its computerised system in determining maintenance intervention. However, existing schedules were done on a conservative time basis. Many failures were prevented, but no one could quantify the costs of such excessive maintenance. The maintenance system was thus of little use: some data existed, but seldom was this analysed to produce information, and no evidence was apparent of a knowledge based maintenance system.

PaperUK had a long-established computerised maintenance system, and managers were adamant that procedures should be based on failure history (data). Failures were recorded, but the data captured was not suitable for establishing maintenance tasks. The emphasis was on data input, rather than ensuring meaningful outputs.

Experiences in the case studies show that information systems were used to little effect. The nature of data collection was suspect, and procedures for analysing information were not satisfactory. While some data was thus collected, it was not adequately assimilated, and failed to transform the maintenance systems of any of the case studies. The findings of this study suggest that all case studies suffered from a lack of accurate data and inability or reluctance to use information systems: operators and maintainers were unable to base remedial action on any documented history and experience.

If time-based maintenance is adopted, then a period of time must elapse before meaningful data can be collected for new equipment. The case studies did not demonstrate that they were going to collect failure data for the new equipment in a meaningful manner, nor that this would be competently analysed.

Production Competence

In order to assess the levels of production competence in the four cases, respondents were asked to provide certain performance related information pertaining to their plant. The items contained in this paper related to the following (the results are summarised in Table 1):

- primary function of the new equipment, with performance standards in terms of output quantity and quality
- details of how non-achievement of the primary function would be detected, and the nature of the consequences
- details of protective systems (safety to humans and equipment protection)
- details of the most severe failures (in terms of cost or safety), with most appropriate remedial action, and the basis for this.

Table 1 shows, for example, that all respondents at FoodUK were able to describe the primary function of the laser technology, and the production supervisor and operator were able to quantify output. However, no one could provide precise quality standards.

The information in Table 1 and interview results confirm that production competence in the UK was greater than in South Africa, but they still demonstrate various degrees of ignorance about production and maintenance issues. While operators were generally familiar with the function and performance of their equipment, maintenance craftsmen were poorly informed. The implication is that those who maintain equipment were not fully aware of its functionality. While most respondents would identify the failure to fulfil the primary function, they were generally unaware of the consequences in the longer term (such as safety, environmental or health implications, costs, effects on customer service). There was a low level of awareness of safety and equipment protection. Respondents were familiar with the most common causes of failure, but none could define precisely the basis on which remedial action was taken (in terms of meeting functionality requirements, assessing root causes of failure, determining consequences of loss of functionality, maintenance tasks and intervals, and so on).

The topics discussed above provide some useful pointers pertaining to TT in the four cases. This section discusses implications of the findings for managers. Table 2 summarises the findings encountered in broad terms.

SOLUTIONS AND RECOMMENDATIONS

The research suggests that the way in which firms manage their resources and competencies will determine the extent to which these will result in enduring sources of competitive advantage through superior performance and functionality. The challenge for managers is to determine what distinctive ways and routines should be harnessed

Table 1. Summary of selected measures of production competence by company and position

FoodUK				
Measure of production competence	Prod supervisor	Operator	Craftsman	Process engineer
Details of primary function Output rate (quantity/time unit) Quantification of quality standards	Yes Yes No	Yes Yes No	Yes No No	Yes Yes Yes
Detection of failure of function Consequences of failure of function	Yes Yes	Yes No	Yes No	No Yes
Safety protective systems Equipment protection systems	Yes No	Yes No	Yes Yes	No Yes
Details of most severe failure Details of appropriate action Basis for remedial action	Yes Yes No	Yes Yes No	Yes Yes No	Yes No No
FoodSA				
Measure of production competence	Prod supervisor	Operator	Craftsman	Process engineer
Details of primary function: Output rate (quantity/time unit) Quantification of quality standards	Yes No No	Yes Yes No	No No No	Yes Yes Yes
Detection of failure of function Consequences of failure of function	Yes No	Yes No	Yes No	No No
Safety protective systems Equipment protection systems	Yes No	Yes No	No Yes	No Yes
Details of most severe failure Details of appropriate action Basis for remedial action	Yes No No	Yes No No	Yes Yes No	Yes No No
PaperUK				
Measure of production competence	Prod supervisor	Operator	Craftsman	Process engineer
Details of primary function: Output rate (quantity/time unit) Quantification of quality standards	Yes Yes Yes	Yes Yes No	Yes No No	Yes Yes Yes
Detection of failure of function Consequences of failure of function	Yes Yes	Yes Yes	Yes No	Yes Yes
Safety protective systems Equipment protection systems	No Yes	Yes Yes	Yes Yes	Yes No
Details of most severe failure Details of appropriate action Basis for remedial action	Yes Yes No	Yes Yes No	Yes Yes No	Yes Yes No
PaperSA				
Measure of production competence	Prod supervisor	Operator	Craftsman	Process engineer
Details of primary function: Output rate (quantity/time unit) Quantification of quality standards	Yes Yes No	Yes No No	Yes No No	Yes No Yes
Detection of failure of function Consequences of failure of function	Yes Yes	Yes No	Yes Yes	Yes Yes
Safety protective systems Equipment protection systems	No Yes	No No	Yes No	No No
Details of most severe failure Details of appropriate action Basis for remedial action	Yes No No	Yes No No	Yes Yes No	Yes No No

to achieve these. Findings show that the UK cases demonstrated greater knowledge and insight of processes, and operators and maintainers had higher levels of production competence than the South Africans. Management intervention in the South African cases was more supportive for TT. Information and systems were of little direct benefit in any of the cases. The transfer of knowledge to competencies is not a one-to-one linear process because of the indeterminate nature of knowledge and the diversity of competencies required for effective utilisation of new technology. Typical competencies include technological skills and complementary assets, as well as routines and capabilities that appear as clusters of tangible and intangible abilities spanning individuals and groups.

Competencies identify the key resources that need to be regenerated, expanded and built on in a firm's future activities, as tacit knowledge can generate more durable advantages (as was found in the UK cases). The feedback loops suggest that competencies help identify the tangible and intangible assets that need to be acquired, and the direct attention that needs to be nurtured and prioritised inside the firm. This is illustrated in Figure 1.

Governance decisions and management support may have both beneficial and detrimental effects. A control spiral can result in yet further control (such as FoodSA's emphasis on procedures and documentation, rather than analysing the broader picture of what the true purpose is of such control). Incremental adjustments in a firm's governance structure may be desirable to permit evolutionary experimentation, encourage individual initiative and enhance organizational learning frameworks, through training, appointment of champions and legitimising activities.

The UK firms encouraged the creation and deployment of knowledge, and the findings suggest that knowledge as a resource accelerated technical progress. The primary motivation for encouraging the emergence and development of pools of new knowledge is therefore to achieve learning and competencies in technology, engineering and processes that will then enable the firm to pursue and develop new opportunities. Respondents in the cases found that technology transfer was frequently hindered by the tacitness of knowledge and complexity in new equipment (thus, the process engineer at FoodUK was unable to arrive at optimal settings).

Table 2. Summary of findings

	FoodUK	PaperUK	FoodSA	PaperSA
1. Resources and knowledge	Tacit knowledge improved performance through an ability to apply subtle changes to process	Operator experience enabled minor adaptations to operating procedures to improve performance	Little scope for enhancing tacit knowledge because procedures were rigidly followed	Specified production processes prevented learning through experimentation. Little tacit knowledge evident
2. Management support	No special support evident to enhance TT	No special support evident to enhance TT	Training and consultation demonstrated management support	Extensive consultation enhanced TT through management support
3. Information and systems	Systems contributed little to TT	Systems contributed little to TT	Systems assisted in monitoring production output	Systems used incorrectly; excessive maintenance increased costs
4. Production competence	Reasonable levels of competence led to meeting desired performance standards in relatively short time	Reasonable operator and craftsman knowledge resulted in good flexibility to meet varying customer demands	Operators and craftsmen not familiar with plant and process. Large wastage of raw materials	Low levels of competence resulted in low initial output and variable quality

Figure 1. Relationship between resources, competence and performance

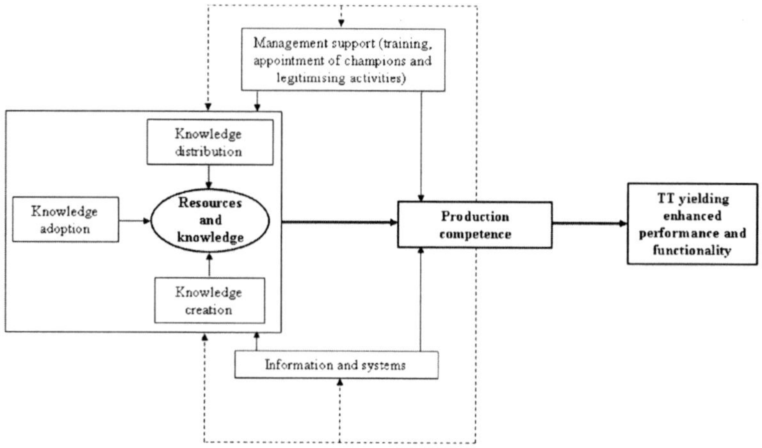

This paper has investigated a number of issues relating to TT in two UK and two South African organizations. Of course, TT encompasses many more concepts than the four identified, and only limited generalisation from four case studies is possible. However, the study has demonstrated that a relationship certainly exists between knowledge, production competence and TT, as illustrated in Figure 1. This becomes apparent when the benefits of greater levels of knowledge are manifested in enhanced production competence, which in turn leads to improved performance and functionality. The UK companies benefited from the knowledge and competences vested within them, when compared to the South African firms. Yet, there is more to TT than knowledge and competences: management support and information systems can play an invaluable important role. Management support was stronger in the South African firms, but information systems did not play a significant part in any of the cases.

Several recommendations are appropriate for managers embarking on successful knowledge-based TT. Managers must develop resources and knowledge if full benefit is to derive from the technology. Management support is a necessary but not a sufficient condition to achieve this. Without specific management attention, information systems are unlikely to assist in knowledge-based decision making, so managers should establish firm processes to collect and analyse data, and utilise this in an analytical manner. Managers should also make determined efforts to enhance production competence. Again, this does not happen on its own, and proactive initiatives are required to enhance an understanding among operators and maintenance staff of how technology operates.

REFERENCES

Aharoni, Y. (1993). In search for the unique: can firm-specific advantages be evaluated? *Journal of Management Studies*, *31*(1), 31–49. doi:10.1111/j.1467-6486.1993.tb00294.x

Bohn, R. E. (1994). Measuring and managing technological knowledge. *Sloan Management Review*, *36*(1), 61–73.

Coates, T. T., & McDermott, C. M. (2002). An exploratory analysis of new competencies: a resource based view perspective. *Journal of Operations Management*, *20*, 435–450. doi:10.1016/S0272-6963(02)00023-2

Cohen, W. M., & Levinthal, D. A. (1990). Absorptive capacity: a new perspective on learning and innovation. *Management Science*, *5*(1), 227–251.

Cristofolini, I., Filippi, S., & Bandera, C. (2009). The role of product feature relations in a knowledge based methodology to manage design modifications for product measurability. *International Journal of Production Research*, *47*(9), 2373–2389. doi:10.1080/00207540701666246

Crossan, M. M., & Berdrow, I. (2003). Organizational learning and strategic renewal. *Strategic Management Journal*, *24*, 1087–1105. doi:10.1002/smj.342

Dahane, M., Clementz, C., & Rezg, N. (2008). Analysis of joint maintenance and production policies under a subcontracting constraint. *International Journal of Production Research*, *46*(19), 5393–5416. doi:10.1080/00207540802273686

Dekkers, R. (2009). Distributed Manufacturing as co-evolutionary system. *International Journal of Production Research*, *47*(8), 2031–2054. doi:10.1080/00207540802350740

Dimnik, T. P., & Johnston, D. A. (1993). Manufacturing managers and the adoption of advanced manufacturing technologies. *Omega*, *21*(2), 155–162. doi:10.1016/0305-0483(93)90048-P

Ferdows, K., & De Meyer, A. (1990). Lasting improvements in manufacturing performance: in search of a new theory. *Journal of Operations Management*, *9*, 168–184. doi:10.1016/0272-6963(90)90094-T

Galbreath, J. (2005). Which resources matter the most to firm success? An exploratory study of resource-based theory. *Technovation*, *25*(9), 979–987. doi:10.1016/j.technovation.2004.02.008

Gourlay, S. (2006). Conceptualizing Knowledge Creation: A Critique of Nonaka's Theory. *Journal of Management Studies*, *43*(7), 1415–1436. doi:10.1111/j.1467-6486.2006.00637.x

Hult, G. T., Ketchen, D. J., Cavusgil, S. T., & Calantone, R. J. (2006). Knowledge as a strategic resource in supply chains. *Journal of Operations Management*, *24*, 458–475. doi:10.1016/j.jom.2005.11.009

Irwin, H., More, E., & McGrath, M. (1998). Relationship management for innovation: the central role of communication in Australia's participation in two high-tech industries. *Technology Analysis and Strategic Management*, *10*(4), 467–481. doi:10.1080/09537329808524328

Jones, O., & Craven, M. (2001). Expanding capabilities in a mature manufacturing firm: absorptive capacity and the TCS. *International Small Business Journal*, *19*(3), 39–55. doi:10.1177/0266242601193003

Katz, R., Rebentisch, E. S., & Allen, T. J. (1996). A study of technology transfer in a multinational cooperative joint venture. *IEEE Transactions on Engineering Management*, *43*(1), 97–105. doi:10.1109/17.491273

Leonard-Barton, D. (1995). *Wellsprings of knowledge*. Boston: Harvard University Press.

Lynskey, M. J. (1999). The transfer of resources and competencies for developing technological capabilities - the case of Fujitsu-ICL. *Technology Analysis and Strategic Management*, *11*(3), 317–336. doi:10.1080/095373299107384

Macher, J. T. (2006). Technological Development and the Boundaries of the Firm: A Knowledge-Based Examination in Semiconductor Manufacturing. *Management Science*, *52*(6), 826–843. doi:10.1287/mnsc.1060.0511

Marmier, F., F., Varnier, C. & Zerhouni, N. (2009). Proactive, dynamic and multi-criteria scheduling of maintenance activities. *International Journal of Production Research*, *47*(8), 2185–2201. doi:10.1080/00207540802311106

McKone, K. E., Schroeder, R. G., & Cua, K. O. (1999). Total productive maintenance: a contextual view. *Journal of Operations Management, 17*, 123–144. doi:10.1016/S0272-6963(98)00039-4

Meredith, J. R. (1987). Automating the factory. *International Journal of Production Research, 25*(10), 1493–1510.

Mowery, D. C., & Oxley, J. E. (1995). Inward technology transfer and competitiveness: the role of national innovation systems. *Cambridge Journal of Economics, 19*, 67–93.

Pintelon, L. M., & van Wassenhove, L. N. (1990). A maintenance management tool. *Omega, 18*, 59–70. doi:10.1016/0305-0483(90)90018-5

Reagans, R., Argote, L., & Brooks, D. (2005). Individual experience and experience working together: predicting learning rates from knowing who knows what and knowing how to work together. *Management Science, 51*(6), 869–881. doi:10.1287/mnsc.1050.0366

Rosenkopf, L., & Nerkar, A. (2001). Beyond local search: boundary spanning, exploration, and impact in the optical disk industry. *Strategic Management Journal, 22*, 287–306. doi:10.1002/smj.160

Rouse, M. J., & Daellenbach, U. S. (2002). More thinking on research methods for the resource-based perspective. *Strategic Management Journal, 23*, 963–967. doi:10.1002/smj.256

Sorenson, O., Rivkin, J. W., & Fleming, L. (2006). Complexity, networks and knowledge flow. *Research Policy, 35*(7), 994–1017. doi:10.1016/j.respol.2006.05.002

Tu, Q., Vonderembse, M. A., Ragu-Nathan, T. S., & Sharkey, T. W. (2006). Absorptive capacity: enhancing the assimilation of time-based manufacturing practices. *Journal of Operations Management, 24*, 692–710. doi:10.1016/j.jom.2005.05.004

Uzzi, B., & Lancaster, R. (2003). Relational embeddedness and learning: The case of bank loan managers and their clients. *Management Science, 49*, 383–399. doi:10.1287/mnsc.49.4.383.14427

Zahra, S. A., & George, G. (2002). Absorptive capacity: a review, reconceptualization, and extension. *Academy of Management Review, 2*(2), 185–203. doi:10.2307/4134351

KEY TERMS AND DEFINITIONS

Technology: May be construed as differentiated knowledge about specific applications: typically the assets that form the basis of a firm's competitiveness.

Production and Process Technology: Embodied in the equipment or the means to produce a specific good, or provide a service. It is manifested in the finished product/service.

Absorptive Capacity: A set of organizational routines and processes by which firms acquire, assimilate, transform and exploit knowledge to produce a dynamic organizational capability (Zahra and George, 2002).

Technology Transfer: The process by which science and technology are diffused throughout human activity. Wherever systematic rational knowledge developed by one group or institution is embodied in a way of doing things by other institutions or groups, we have TT. This can either be transfer from more basic scientific knowledge into technology or adaptation of an existing technology to a new use. TT differs from ordinary scientific information transfer in that to be really transferred, it must be embodied in an actual operation of some kind.

Chapter 5

KM in Higher Learning Institutions:

The Construction and Application of Tacit Knowldege among Malaysain Academicans

Haryani Haron
University Teknologi MARA (UiTM), Malaysia

Rose Alinda Alias
University Technology Malaysia, Malaysia

EXECUTIVE SUMMARY

This study reports the findings from an interpretive exploration of tacit knowledge construction and application among the Malaysian academicians. In the knowledge era, higher learning institutions are increasingly exposed to marketplace pressures similar to other business. Therefore, it is pertinent for higher learning institutions to leverage on their knowledge for sustainability. However, efforts to leverage on tacit knowledge faced difficulties due to lack of understanding of tacit knowledge within its contextual boundary. The ambiguities surrounding tacit knowledge require deeper understanding of its construction and application to manage it as an important resource for academic institutions. The study discovered that tacit knowledge is constructed through implicit learning from the accumulated past experiences. Tacit knowledge is applied by the academicians as their adaptation strategies in appropriate job situation to fulfil the academic excellence norm.

INTRODUCTION

Institutions of higher learning have long been connected with the production of knowledge. In the knowledge era, higher learning institutions expanded their scope of activities in the creation and dissemination of knowledge to include teaching in preparing students to become life-long learners, research to expand human knowledge and creativity and service to universities, national and international community (Metaxiotis and Psarras, 2003). Moreover, as mentioned by Goddard (1998) and Khurana (2007), universities and other higher institutions are recognized to be in the knowledge business and increasingly they are exposed to marketplace pressures in a similar way to other business. In the context of

DOI: 10.4018/978-1-61692-886-5.ch005

Malaysia, the government is seriously pursuing knowledge-based economy. A K-based Economy Strategic Plan is launched and one of the thrust is to develop the public sector into a knowledge-based civil service, which directly affected public institutions of higher learning. It is expected that institutions of higher learning produce quality manpower with a first class mentality. Therefore, to become a knowledge-driven university, higher academic institutions must fulfill the needs of Knowledge Management (KM) which focused on managing their knowledge assets (Mohd Ridzuan Nordin, 2001). However, a study conducted on the implementation of KM in Malaysian higher institutions revealed that KM initiatives are still at the moderate level (Mohd Ghazali Mohayidin et. al, 2007).

The Malaysian government is seriously pursuing knowledge-based economy by laying out seven strategic thrust in its K-based Economy Strategic Plan. One of the thrust is to develop the public sector into a knowledge-based civil service, which directly affected Public Institutions of Higher Learning (PIHL). Therefore, being the producers of human capital, public universities must be prepared to become a knowledge-driven university to face the challenges in the knowledge era.

Tacit knowledge has been credited with the successes of organizations. However, it is an ambiguous construct and has been appropriated diversely in research. The difficulties in managing tacit knowledge for an organization's competitive advantage are due to the lack of understanding of tacit knowledge within its contextual boundary and this is reflected through a survey research by Hijazi and Kelly (2003) that showed that higher learning institutions have some difficulty in understanding the implication of managing knowledge to support business processes.

The study is concerned with the understanding of the tacit knowledge construct among Malaysian academicians. Research on tacit knowledge has been predominantly done in through the knowledge-as-object approach. However, tacit knowledge is a dynamic knowledge that resulted from the ongoing social process of construction and collective action in an organization of an ongoing social accomplishment (Orlikowski, 2002; Aadne, 1996). Therefore, there is a need to explore tacit knowledge through the socially-constructed perspective.

Interpretive inquiry is chosen as it is in-line with the objective of the study which is to understand tacit knowledge, a socially-constructed and contextual knowledge which necessitate subjective method of inquiry to unveil the substance and realities of the research problem. The empirical work was conducted using qualitative strategy of inquiry with Grounded Theory (GT) as the primary overarching approach for data gathering and analysis from a case site.

The objectives of this study are:

1. To describe tacit knowledge construct among the Malaysian academicians.
2. To describe the construction and application of tacit knowledge in the context of Malaysian higher academic institutions.

BACKGROUND

The information age, facilitated by the development of Information Communication Technology, leads to an emergence of a new type of economy called knowledge economy (K-economy). In this era, organizations have to convert themselves into organization of knowledge-specialist in order to survive and remain competitive (Drucker, 1993). Together with globalization, K-economy pushes knowledge to be the main agenda in business survival (Kermally, 1988). Knowledge resides in people and organizations must make more effective use of the knowledge and expertise available within their existing employee base (Hellstrom and Husted, 2004).

Knowledge in people is tacit knowledge, which is the background knowledge of all understanding and expertise, and managing tacit knowledge involved locating and sharing it within the organizations (Stenmark, 2001; Hansen et. al, 1999). However, tacit knowledge is contextual-bounded, it is embedded in specific context that it can only be described and understood in relation to a whole body of knowledge in its context inclusive of both the specific organizational context and specific individual context (Endres et. al., 2007; Li and Gao, 2003). Tacit knowledge has various advantages and potentials to organizations. These include an important role in organizational performance, a key factor in enhancing the quality of strategic decisions making and innovations.

The Ambiguity of Tacit Knowledge

The term tacit knowledge has been appropriated in various situations (Gourlay, 2004). This contributed to the ambiguities surrounding the tacit knowledge construct and is reflected by Gourlay (2006) who summarizes four ambiguities in tacit knowledge, namely the ownership of the tacit knowledge, the way of tacit knowledge acquisitions, the functions of tacit knowledge and the possibility of tacit knowledge articulation. Tacit knowledge has been used in the empirical studies as both articulable and inarticulable knowledge (Gourlay, 2006). The ambiguities of tacit knowledge are also due to the differences in the interpretation of the concept, namely the articulation and ownership of tacit knowledge and the needs for externalization of tacit knowledge (Haldin-Herrgard, 2004).

Disagreements are also prevalent among the authors on whether tacit knowledge resided in an individual, organization (collective) or both individual and organization (Gourlay, 2004). Many scholars regard tacit knowledge as personal, private knowledge at the individual level (Polanyi, 1966; Van Krogh and Roos, 1996; Ambrosini & Bowman, 2001 and Johannessen et. al, 2001; Stenmark, 2000). Haldin-Hergard (2004) concluded from his study on KM literatures that tacit knowledge is highly personal, abstract, and difficult to express and diffuse, acquired through experience and related mainly to practicality.

On the other hand, other scholars including Nonaka and Takeuchi (1995), Spender (1996), Leonard and Sensiper (1998), Choo (2006), and Baumard (1999) treated tacit knowledge as an individual and also collective knowledge. Baumard (1999) defined collective knowledge as knowledge shared among the members of a social entity to differentiate it with individual knowledge. Nonaka and Takeuchi (1995) associated organizational tacit knowledge with knowledge associated within a group of people in an organizational and expand the tacit knowledge construct existence not only in individual levels but also in groups, organizations and inter organizations level. Individual tacit knowledge is expressed through action-based skills while collective tacit knowledge is shared practices and tacit understandings among members of groups working together to perform a task (Choo, 2002). For Lam (2000), individual tacit knowledge which resides in the brain and psychomotor skill of an individual is specialized and domain-specific, and part of the organization's knowledge base, where as collective knowledge is the way knowledge is distributed and shared among members of organizations.

The third view is the collective tacit knowledge which reside in organizational routines, culture and mindset and accessible through interactions among members of the organization (Spender, 1996). For Johannessen et. al. (2001), collective tacit knowledge is organizational tacit knowledge. Swart and Pye (2002) proposed that tacit knowledge is best described as collective tacit knowledge because it is acquired through experience, and the nature of experience is collective. Culture is suggested as a collective form of tacit knowledge by Brockmann and Anthony (2002).

Research Approaches on Tacit Knowledge

The disparity on the conceptualization of tacit knowledge is related to the philosophical assumption and research approach. Two major philosophical assumptions influence the study of KM in the Information Systems (IS) research, the normative or positivist approach, and the interpretive approach. The differences center on the aspects of knowledge metaphor and theoretical assumptions (Schultze and Leidner, 2002). However, more recently, Critical Theory has become another approach in IS research (Stahl, 2008). Specifically, studies focusing on the analyses of organizational knowledge view tacit knowledge either as an objective commodity (positivist approach) or as a socially constructed process (interpretive approach) (Casselman and Samson, 2005).

The positivist perspective focused on the objective and measurable nature of knowledge and operationalized knowledge as an object independent of knower (Orlikowski and Baroudi, 1991). Positivist research focused on knowledge as an object which can be stored, manipulated and transferred to others, centered on the context of problem-solving and decision-making tasks (Schultze and Leidner, 2002). The knowledge-as-object approach is an influential approach in the practitioner and academic literature is popularized by Nonaka as quoted in (Walsham, 2005). In this approach, tacit knowledge is viewed as a discrete type of knowledge possessed by individuals, groups or organizations, which can be converted into explicit knowledge, and shared throughout the organization (Nonaka and Takeuchi, 1995). The articulation of tacit knowledge is difficult but not impossible, according to this approach (Nonaka and Takeuchi, 1995; Sternberg and Horvath, 1999; Hedesstrom and Whitley; 2000). Although the nature of tacit knowledge is sticky to its owner, Szulanski (2003) argues that tacit knowledge can be articulated through proper methodology. Therefore the focus for this approach is the mechanism of transferability for tacit and explicit knowledge (Casselman and Samson, 2005, Hansen et. al., 1999) and to collect, reuse and distribute codified knowledge for an organization's competitive edge (Zack, 1999).

Interpretive research assumed that knowledge is socially constructed through language, consciousness and shared meanings (Orlikowski and Baroudi, 1991). In contrast to the positivist philosophy, interpretive philosophy highlights the dynamic and situated nature of knowledge (Sahay and Robey, 1996). Walsham (1993, 2001) elaborated that interpretive research in IS focused on understanding the context and the process influencing the information system. In contrast with the first approach, the second school of thought disagree with the treatment of tacit knowledge as an object that can be captured, processed and stored similar to explicit knowledge. Tacit knowledge is defined as a knowledge which is impossible to articulate (Polanyi, 1966; Stenmark, 2001; Tsoukas, 2003; Leonard and Sensiper, 1998) and for those belonging to this school of thought, tacit knowledge is no longer tacit if it is articulated, thus focus should be on managing the knowledge resources rather than making tacit knowledge explicit (Stenmark, 2001; Tsoukas, 2003; Leonard and Sensiper, 1998). As such, scholars belonging to this research approach viewed that tacit knowledge cannot be independent of human action, tacit knowledge is associated with something people do, therefore it is more appropriately viewed as the process of knowing and capability to act (Walsham, 2005; Haldin-Herrgard, 2004; Brown and Duguid, 1998; Blackler 1995). Within this perspective, tacit knowledge is an ongoing social accomplishment (Orlikowski, 2002; Aadne, 1996).

The Acquisition of Tacit Knowledge

Tacit knowledge is acquired informally and either unconsciously or semi-consciously (Leonard-Barton & Sensiper, 1998) through personal experiences (Polanyi, 1966, Nonaka and Takeuchi,

1995; Sternberg and Horvath, 1999, Hansen et. al., 1999), collaborative experiences (Nonaka and Takeuchi, 1995) and subconscious integration of explicit and tacit knowledge (Gore and Gore, 1999). Most authors agree that the construction of tacit knowledge is through experiencing and doing in the pursuit of knowledge performing different tasks and duties in different contexts and situations of his life (Wagner and Sternberg, 1985; Koskinen, 2003; Reuber et. al., 1990; Choo, 2006). This view corresponds to Polanyi's (1966) elaboration on the experience which is obtained through repeatedly performing a task in a similar way or experimenting with new approaches to complete a task and new tacit knowledge is acquired when a familiar task is performed in a new way or when a new task is performed through experimentation. Brown and Duguid (1998) echoed the same process of tacit knowledge construction by stressing that tacit knowledge is not only revealed in practice but also created out of practice. In short, tacit knowledge cannot be taught, trained or educated but can be learnt through active contribution of the learner and the process takes time (Brockmann and Anthony, 2002). Other authors like Leonard and Sensiper (1998) and Cook and Brown (1999) added that personal contact and observations of others are also important factors in the construction of tacit knowledge.

The environment in which tacit knowledge can be acquired is either through on-job training or informal learning at workplace (Howells, 1996; Horvath et. al., 1999) such as apprenticeship, discourses and communal story sharing (Choo, 2006). In deliberating on the relationship of experience and tacit knowledge construction, Eraut (2000) associated tacit knowledge acquisition with implicit learning, a kind of unconscious learning from accumulated past personal experiences which form a tacit knowledge base that enables action. Although tacit knowledge is constructed through experience, it is the ability to learn from experience is important rather than the length of experience (Sternberg et. al., 2000). Learning is a social action and requires interaction, and individual learning involves reflections on actions for improvement and gaining new knowledge (Haldin-Herrgard, 2000). Johnson et. al. (2002) summarized that tacit knowledge builds up on personal skill, through social activity, informal discussion of work problems and internalization. Sharing the same view, McInerney (2002) stressed that tacit knowledge is also created not only by individual experiences but also by the networks of people who meet and work with each other and cause knowledge to be shared and created.

The Manifestation of Tacit Knowledge

Tacit knowledge is a hidden construct which cannot be seen but can be observed through its effect or manifestation (Polanyi, 1966). Tsoukas (2003) believed that tacit knowledge can only be displayed and manifested in an individual's action and cannot be captured, translated or converted. The possession of tacit knowledge of an individual remains hidden until it is used to perform a task, understand something, make decisions or solve problems, thus tacit knowledge is a characteristics of an individual that influences his/her behavioral potential (Hunt, 2003). Polanyi's (1966) description of tacit knowledge as "knowing more than we can tell" pointed out that tacit knowledge is evidence through an individual's action but not through verbalization of what a person's know. Based on Polanyi's (1966) concept of knowledge as a justified personal belief that increases an individual's capacity to take effective action, Connell et al. (2003) argued that tacit knowledge is essential in performing actions. Hedesstrom and Whitley (2000) did an extensive study on tacit knowledge and action, concluded that tacit knowledge plays an important role in polymorphic actions, actions which take its shape from society, and action which is based on the behavior and appropriateness in the society.

Application of tacit knowledge is not only through actions in social circumstances, but also through action-based skills (Choo, 2006; Stenmark, 2000). Howells (1996) connects tacit knowledge directly with performance skill. Stenmark (2000) in recognizing that expertise is a quality highly dependent on tacit knowledge stressed that expertise can only be observed and recognized through its resulting actions, and Gourlay (2006) referred to this kind of expertise as the internalized expertise.

As for Koskinen (2003), application of tacit knowledge is in the form of evaluations, attitudes, views, commitments and motivation. Gore and Gore (1999) added that tacit knowledge expresses itself not only through perception of an individual but also in the behavior of individuals, which is in line with Schmidt and Hunter (1993) that tacit knowledge reflects an individual understanding of the requirements in job context and suitable behavior in various situations within the working setting. Tacit knowledge application can also be an ability in working as a group (Collins, 2001), or ability in solving complex problems (Gore and Gore, 1999). Van Krogh and Roos (1996) believe that tacit knowledge manifests in relationship, attitudes, information flows, and ways of making decisions that shape people's dealings with each other.

Interdependence of Individual and Organization Tacit Knowledge

Differences in opinion by authors in interpreting tacit knowledge are mostly found in the ontological assumptions of tacit knowledge. Although most authors agree that tacit knowledge is personal (Polanyi, 1966; Davenport and Prusak, 1998; Gore and Gore, 1999; Stenmark, 2001), it has no relevance unless it is captured for the benefit of organization (McInnerney, 2002). Organizational knowledge is the collective knowledge made of individuals' experience in an organization (Nonaka and Takeuchi, 1995; Baumard, 1999) and if it is used and shared, it will give the competitive advantage for the organization (Dixon, 2000).

Furthermore, individual and organizational knowledge are distinct but interdependent because individuals must refresh their knowledge base by interacting with those who have work-specific skill and expertise to create valuable "organizational knowledge" (Bhatt, 2002). Organizational knowledge is collective tacit knowledge which resides in organizational culture, routines and organizational experiences which become the context of interaction among the individuals in the organization (Cavusgil, 2003; Bhatt, 2002), but if individuals do not posses necessary skill and knowledge, their interaction will not create valuable organizational knowledge of organizational learning.

The Context-Bounded Nature of Tacit Knowledge

Organizational knowledge is a wide-scope concept as it involves people and context (Guzman and Wilson, 2005; Nonaka and Takeuchi, 1995). Tacit knowledge is a part of organizational knowledge which resides in people and is embedded in specific context. Thus it is a contextual-bound knowledge that it can only be described and understood in relation to a whole body of knowledge in its context, such as specific organizational context or specific individual context (Endres et. al., 2007; Li and Gao, 2003). According to Nonaka and Takeuchi (1995), tacit knowledge is context-specific, and they define tacit knowledge as involving both cognitive and technical elements. The cognitive elements center on mental models which includes schemata, paradigms, beliefs, and viewpoints that provide perspectives to perceive and define their world. By contrast, the technical element of tacit knowledge covers concrete know-how, crafts, and skills that apply to specific contexts.

Tacit knowledge is contextual-bounded as it enables the knower to take actions that are situated in particular social and physical circumstances

(Stenmark, 2000). Fernie et. al (2003) conducted a study on the utilization of knowledge in the industrial sectors concluded that context is an important issue because knowledge has to be extracted from one context and adapted to another context, a process referred to recontextualization. Maqsood et. al. (2004) added that the same hold true for tacit knowledge, capturing the context of tacit knowledge's acquisition is an important factor in investigating tacit knowledge. Furthermore, recontextualization is clearly reflected in the manifestation of tacit knowledge construct, which is through the actions performed in a particular context. Evans et al. (2004) expressed that tacit knowledge is contextually bounded, as actions taken are situated in a particular circumstance, thus the manifestation of tacit knowledge is situation-specific which may contribute success in one context but not necessarily in another.

According to Kakabadse et. al. (2003), historical background and culture contextualize tacit knowledge. As concluded by Koskinen (2001) in his study, the usefulness and utilization of tacit knowledge always depends on the context and situation where it is used. In addition, context also plays an important role in tacit knowledge proliferation (D'Eredita and Barreto, 2006), and it plays an important role when investigating learning processes (Brézillon, 1999), the acquisition process of tacit knowledge.

Tacit Knowledge Conceptual Definition

In order to explore tacit knowledge through a social construction perspective, a conceptual definition is developed. Since knowledge cannot be treated independently from the process which it is acquired (Ancori et. al., 2000), the construction of tacit knowledge is an important aspect in exploring tacit knowledge. Tacit knowledge is a hidden construct which cannot be verbalized, but can only be observed through its manifestations (Polanyi, 1966). This rationalizes the need to understand the application of tacit knowledge in the whole process of exploring tacit knowledge. Both construction and manifestation processes are main features considered in conceptually defining the tacit knowledge besides its ontological dimensions.

Firstly, tacit knowledge is acquired from personal experience in a particular context. Experience within a particular context like job domain plays an important role in the implicit learning which contributes to tacit knowledge construction. Implicit learning is learning from experience or practice, and it involves transformation of knowledge into situation appropriate form (Eraut, 2000). The accumulated knowledge forms the knowledge base of an individual, and the application of this knowledge becomes informal and automatic.

Secondly, tacit knowledge manifests through actions in a particular context. Tacit knowledge is closely associated to action, as it takes the form of "know-how". It is the knowledge that guides an individual's action in a given situation. Fenstermacher (2005) elaborates those actions enabled by tacit knowledge are physical actions, thinking and a combination of physical and thinking. Tacit knowledge increase is related to an individual's capacity to take actions which requires either physical skills and competencies or cognitive/intellectual activity or both (Connell et. al., 2003). Thus, the manifestations of tacit knowledge are actions that are reflective and judgmental in nature. In short, tacit knowledge assists in performing actions, the act of knowing.

Thirdly, tacit knowledge is conceptualized as the knowledge which is owned by both individuals and organizations. They are distinct but interdependent from each other. Individuals with knowledge and skill interact with others within the organization setting and their interactions create valuable organizational knowledge, and at the same time it enriches the individual knowledge.

A conceptual framework for the exploration of tacit knowledge is illustrated based on the conceptual definition. Figure 1 portrays the conceptual framework for the research. It is based on the socially-constructed approach of tacit knowledge, taking into consideration of Eraut's (2000) perspective of implicit learning in the acquisition process, Polanyi's (1966) perspective of tacit knowledge manifestation and Nonaka's (1995) idea of contextual knowledge.

The framework suggests that tacit knowledge is constructed through the implicit learning from the experience accumulated from the job domain. Although the value of experience is important to tacit knowledge construction, it is the ability to learn from the experience from the on-job-exposure and appropriately apply to job problems and situations or implicit learning is the factor which enhances the tacit knowledge. Individuals are different with respect to what and how much they learn from experience, and this becomes the basis in the differences between experts and novices. Experts are introspective about their knowledge and have the ability to learn better from previous experience.

The framework also depicts the ontological dimension of tacit knowledge as owned both by individual and organization. Both individual and organization tacit knowledge are distinct but mutually enabling each other. As tacit knowledge guides the action of an individual, it is externalized through individual's action in a given situation. Actions can either be physical actions, cognitive/ intellectual activity or a combination of both.

PROBLEM STATEMENT

The creation and dissemination of knowledge had long been the social role of institutions of higher learning. In the knowledge era, higher learning institutions role are no longer just to provide knowledge to the students, but also to manage and collaborate the existing knowledge towards becoming a learning organization. PIHL must respond to their changing role in a knowledge-based society and need to be consciously managing their knowledge assets. The knowledge assets in institutions of higher learning include the academicians which own a reservoir of tacit knowledge.

While tacit knowledge is credited for all its potentials and values, it is also an ambiguous construct in an organization. The ambiguities surrounding the tacit knowledge necessitate a deeper understanding of the construct so that it can be

Figure 1. Conceptual framework for the study

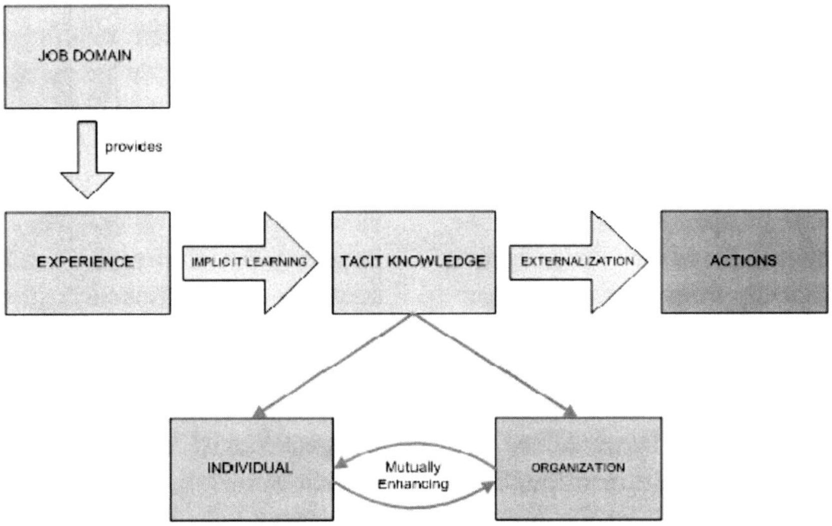

managed as a source of competitive advantage. Researches on tacit knowledge are predominantly done through the knowledge-as-object approach. However, tacit knowledge is a dynamic knowledge resulted from the ongoing social process of construction and collective action in an organization therefore there is a need to explore tacit knowledge through the social construction perspective.

This research aims to understand tacit knowledge among Malaysian academicians through a socially-constructed perspective. The social construction of tacit knowledge involves two important processes, which are its construction and application. This rationalizes the need to understand the construction and application of tacit knowledge in the whole process of exploring tacit knowledge.

RESEARCH METHODOLOGY

IS research is founded on three main paradigms, positivist, interpretivist and critical research (Klein and Myers, 1999). However, other researchers expanded and proposed four types of paradigms, positivist, post-positivist (also referred to as realism), constructivist (also known as interpretivist) and critical theory (Creswell, 2003; Denzin and Lincoln, 1998; Tashakkori and Teddlie, 1998; Guba and Lincoln, 1994). All four paradigms have advantages and limitations however this study follows the interpretive inquiry because it is in-line with the objective of the study and is able to provide a more meaningful research outcome through a richer and contextualized view of the phenomenon and also knowledge enrichment for the researcher.

Qualitative approach is chosen and grounded theory (GT) is the selected qualitative inquiry the research because there is no *apriori theory* on tacit knowledge in academic domain and furthermore it provides the rigor in the exploring tacit knowledge. GT provides the rigor in the data collection and analysis process (Hughes and Howcroft, 2000). The other strength in applying GT is that the theory is generated from data collected in the area of interest (Glaser and Strauss, 1967), thus it is a theory constructed from reality.GT is used as the overarching methodology which drives data collection and data analysis activities within a case site.

Data Collection

The data collection in this study involves the process of eliciting tacit knowledge from the expert domain. The case site is one of the premiers PIHL in Malaysia, with 3 satellite campuses, 14 branch campuses and 8 city campuses. It has 25 faculties and offered over 200 academic programs spreading over science and technology, social science, humanities and business management. Therefore, it met the criteria in providing a multitude of different elements in the context of an academician's working environment that leads to a richer pool of tacit knowledge. Like other public universities in Malaysia, the activities of the case site were coordinated by the Ministry of Higher Education. The Ministry provides the leadership and guidance in policy matters to universities in-line with the country's aspirations, towards the National Vision of 2020.

The basic premise of the study is experts have more tacit knowledge than novice therefore the population frame of the study are the Professors. Professors are the experts in academic institutions; they are in the highest rank in an academic environment and are recognized as scholars in academic domain. The second factor considered is experience. Experience is vital in acquisition and use of tacit knowledge. In the context of PIHL, experiences in teaching, research and services play critical role in the selection of samples to ensure that the success of capturing tacit knowledge. The Professors were selected according to their experience in teaching, research and community and university services.

Theoretical sampling guides the process of selecting participants and directs the data collection process. Overall, the participants are mature academicians with age groups above 45years old. All of them have the experience in teaching, research and community and university services. A total of 24 participants were interviewed, comprising of 5 females and 21males. Other characteristics of the participants were also noted, with two of them exceeding their retirement age of 55 years old, 3 of the participants have the experience of working in other government agencies prior to being an academician and 22 of them were awarded with outstanding service at least once during their services. Two of them were awarded with state honorary titles during their service. In terms of their academic discipline, there are five participants from the medical school, four from the engineering disciplines, and the 15 are from business management, arts, pharmacy, sciences and others.

Data are collected during a period of 9 months. Semi-structured in-depth interviews are conducted which lasted between 50 to 90 minutes, in private and mutually agreed location and time. Critical incidence technique is incorporated in the interviews, where participants were asked to recall specific situations or incidents in their professional life that shaped their understanding, perception, views and behavior as a scholar. Among the issues discussed during the interviews were the actions taken to overcome or handle the situations and the underlying reasons for their actions. They were also asked to elaborate on the critical elements that contributed to their success in their career. Interview protocol is developed to ensure consistencies throughout data collection.

Interviews are tape-recorded to ensure all the details are captured. In minimizing the tape-recording drawbacks, assurance of the confidentiality on the issues discussed is made before the interview. The validation process is done immediately after the interviews ended through paraphrasing the interviewee's comments in a scenario which consists of situations, actions taken and reasons for the actions taken.

Data Analysis

All interviews are transcribed verbatim after each interview by the researcher. Field notes which consisted of ideas which the researcher noted during data collection process for the purpose of reorienting the researcher in the analysis stage were also made after each interview. Data is analyzed based on the GT method of analysis where three levels of coding are performed, namely the open coding, axial coding and selective coding. After conducting 24 interviews, 149 codes are obtained in the open coding stage. In determining the coding reliability, an independent expert was consulted and Kappa value is calculated. Kappa is the measure classically used for summarizing agreement beyond chance, had a value of 0.91, which signified high agreement in the coding with the expert, as Kappa has a value ranging from 0 to 1, where 0 represents no agreement beyond chance and 1 as perfect agreement.

In the axial and selective coding stage, paradigm modeling, an analytical tool is used to identify the key themes and cluster them into broad themes which eventually summarize and categorize the emerging strategy in a structured form of causal condition →phenomenon →context →intervening condition→ action strategies → consequences. During the whole analysis, the qualitative analysis computer software, QSR NVivo 2.0, is used to organize the vast amount of data and to support the coding process. In the open coding phase, NVivo is used to label the concepts with code words using the free nodes. Later as more concepts are generated, certain concepts are grouped together under a higher order concept, leading to the discovery of categories. The context of the incidences becomes the property in the grouping of the concepts. Constant comparative method, a major feature of grounded theory is applied in comparing the context of the new concept generated with the other existing ones for new categories. In this process, NVivo assisted in organizing the concepts into trees. Each tree node represents the categories. Categories are named according to the academic institution context to establish relevant meanings to guide the researcher.

FINDINGS

Twenty-six categories emerged from open coding process. In the axial coding, paradigm modeling is used to establish relationships between categories. In the process of determining the emerging phenomenon from the data, careful examination for repeated patterns of events and happenings are performed. The repeated patterns describe how the participants adapt to the academic excellence norms arising from the recently introduced key performance indicators for academicians, known as the 7Ps requirements, which are teaching, research, publication, writing, public service, consultancy, and management. Academicians used and applied their knowledge from previous experience in adapting to the new academic excellence norm. This resulted in the internalization of knowledge (tacit) by the academicians. Therefore knowledge internalization is identified as the phenomenon, and all other categories are organized related to the phenomenon.

The paradigm model discovered from axial coding is shown in Table 1.

In reflecting a more dynamic illustration on the findings of GT, a theoretical model is developed as in Figure 2.

A story line is developed to describe the relationships of all categories centered on the discovered phenomenon, as shown in Table 2.

TACIT KNOWLEDGE IN THE CONTEXT OF MALAYSIAN ACADEMICIANS

Tacit knowledge exists in the PIHL domain as knowledge internalized by the academicians in their work practices. In the context of Malaysian PIHL, basically the workload of an academician centres on three main academic activities, namely teaching, research and services. However, the seven performance indicators (7Ps) detailed out seven academic activities, namely teaching, research, publication, writing, public services, consultancy and management, to be fulfilled by an academician. Therefore, the work practice of academicians is centred around the seven academic activities listed on the performance indicators.

Table 1. Paradigm model of knowledge internalization

Phenomenon	Knowledge Internalization
Causal Condition	Academic Excellence
Context	Proliferation of Academic Demands Increase Need of Social Solidarity Transition in Work Practice
Intervening Conditions	Institutional Culture Institutional Setting
Action Strategies	Affirming Intellectual Capabilities Capitalizing on Self-Knowledge Building Interpersonal Relationship
Consequences	1. Self-fulfillment 2. Self-worthiness 3. Career Advancement 4. Legitimate Practices 5. Personal Development 6. Authority

Figure 2. Theoretical model for knowledge internalization

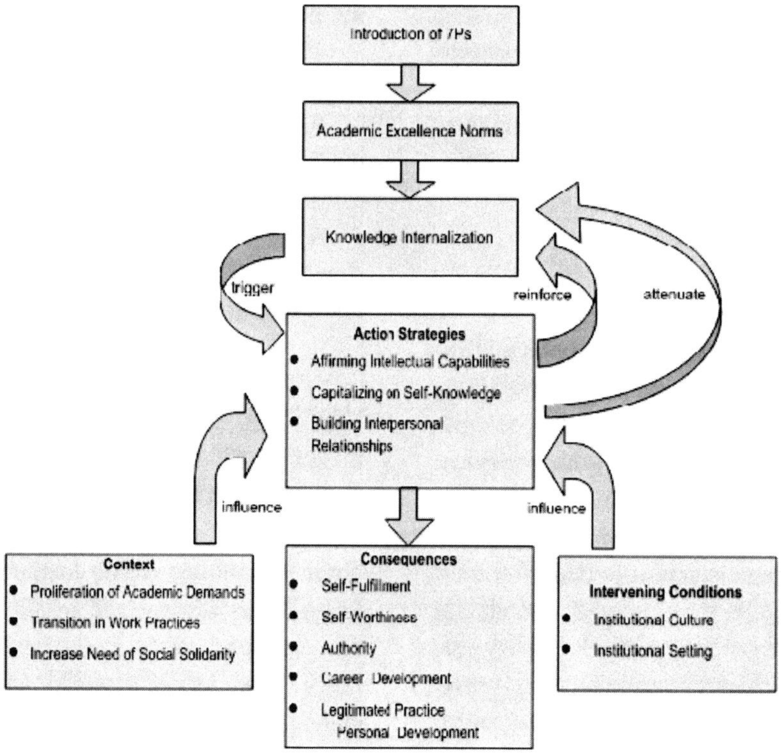

Findings revealed that tacit knowledge in academic domain is important for an academician to achieve his/her goals. Tacit knowledge has goal-attainment values and is the know-how in achieving such goals and this is reflected when academicians applied their tacit knowledge to strategize their actions in order for them to reach their intended goal of fulfilling the 7Ps which resulted in the internalization of knowledge (tacit) by the academicians. Internalization of knowledge is the implicit process of applying the accumulated knowledge learned from previous experience when faced with challenges in situations related to their work practices. Challenges in academicians' work practice include fulfilling the proliferations of academic demands, fulfilling the increased needs of social solidarity and adapting to the work practice transition. In attenuating the effects from the proliferations of academic demands, fulfilling the increased needs of social solidarity and adapting to work practice transitions, academicians internalized their knowledge to build their intellectual worthiness, in interacting with people, which resulted in influencing others to work for them, and in distinguishing own needs thus creating meanings for their academic activities.

In short, internalization of knowledge is reflected when academicians applied their knowledge (tacit) to devise their action strategies in overcoming a particular challenge they encounter in their work practices. Three categories of action strategies are discovered from the data, namely affirming intellectual capabilities, capitalizing on self knowledge and building interpersonal relationships. Affirming intellectual capabilities are actions within the intellectual capabilities context. Capitalizing self knowledge consists of actions within the personal context while building interpersonal relationships are actions within the social context.

Table 2. The storyline for knowledge internalization

Knowledge Internalization among the Academicians

Basically the workload of an academician can be divided into three, teaching, research activities and university-community services. However, as a guideline, the Ministry and public higher institutions agreed on the seven performance indicators (7Ps) for the academicians. The introduction of the 7 Ps requirements gives rise to the academic excellence norms which is characterized by the proliferation of academic demands, transition in work practice and the increase in needs for social solidarity. Academicians adapted to the norms by striving for scholarly achievements and acting befitting the concept of a university, which ultimately led them to view and do things using the knowledge they possess.

The proliferation of academic demands resulted in the feeling of intellectual inadequacies among the academicians. They also experience the complexities of their intellectual capabilities worthiness such as "unappreciated" and "unnoticed" intellectual capabilities. Their work practices changes due to the changing focus of their academic activities. The new academicians' work environment is characterized by competing priorities due to institutional expectations and scholarly works, besides the existing work practice norms. They experience the transition of work practice from mainly teaching to fulfilling other scholarly works such as research, publications and services. Transition of work practice affected the academicians whereby dilemma arises when there are conflicts between institutional expectation and personal goals or when self-principles or self-values are in conflict with institutional policies or norms. In another context, to fulfill the 7Ps, academicians realize the need to strategize by performing team-work and collective action. Their interactions occur within diverse groups of people, varying from different backgrounds and academic position. The maintenance of social solidarity becomes a necessity in order to conform to the new norm.

Adding to these complexities are the institutional culture and institutional settings which at times counters the scholarly developments. An issue in the public institutions of higher learning is the constraint of monetary and resources. In the context and within the conditions mentioned above, academicians perform adaptive responses to attenuate them. Within their capacity, based on their knowledge and accumulated experience that they devise their action strategies. They capitalize on their self knowledge, build interpersonal relationship and affirm their intellectual capabilities to adapt to the new demands.

Affirming their intellectual capabilities involves utilizing their knowledge to strengthen, highlight, assert and position their intellectual in accordance to the new norm. Capitalizing self knowledge is utilizing their reservoir of knowledge in knowing their own strengths and weakness to evaluate, motivate, stay focused, distinguish own self besides expressing their self-principles and self-values. Building interpersonal relationships is the strategy used in adapting to the increase needs of social solidarity. Academicians use their knowledge to adapt to the social context in performing their professional demands. Utilizing their knowledge, they also reconcile their differences with others, and motivate others, created a sense of connectedness besides negotiating the institutional politics. The strategies reinforce the knowledge internalization process among the academicians which enables them to create meanings of their academic activities, thus giving them self-fulfillment and personal development. Through building interpersonal relationship, the adoption of a shared reference frame leads to mutual agreements and legitimated practice, thus sanctioning the others towards working for the academicians. Affirming intellectual capabilities allows the academicians to highlight their self-worthiness within and outside the institutions. Proven by their academic capabilities, they become an authority in their discipline, which give them unintended power base and career development.

In performing their actions, academicians draw upon their reservoir of internalized knowledge, suggesting that actions taken are aligned to the type of knowledge internalized. As conceptualized in this research, tacit knowledge is knowledge internalized through past experience. Since actions are manifestations of tacit knowledge, this research categorizes tacit knowledge according to the types and context of actions involved. Therefore, in affirming their intellectual capabilities, academicians draw upon their intellectual-affirmation tacit knowledge. Whereas in capitalizing self tacit knowledge, they draw upon their self tacit knowledge and in building interpersonal relationships they draw upon their social tacit knowledge. Nevertheless, the researcher acknowledged that the three types of tacit knowledge are mutually enhancing.

In conclusion, three types of tacit knowledge are discovered, intellectual affirmation tacit knowledge, self tacit knowledge and social tacit knowledge. Knowledge is considered intellectual-affirmation tacit knowledge if it is related to actions in building intellectual capabilities worthiness. Self tacit knowledge is knowledge related to actions which are individually-centered and are performed in distinguishing one self-needs. Social tacit knowledge is knowledge related to actions which involve interaction with others.

Intellectual-affirmation tacit knowledge is the knowledge of understanding behaviors relating to intellectual capabilities. The know-how in this dimension of tacit knowledge is the know-how to strengthen academic capabilities, know-how to assert academic capabilities, know-how to highlight academic capabilities and know-how to position academic work in relation to industries. Self tacit knowledge is the knowledge of understanding behaviors relating to one-self. The know-how in this dimension of tacit knowledge includes the know-how to motivate own self, know-how to stay focused to achieve personal target, know-how to evaluate own self, know-how to distinguish self needs and know-how to express self principle, self beliefs and self value. Social tacit knowledge reflects the understanding of human behaviors and how these behaviors affect an academician maneuvers in fulfilling the 7Ps requirements. Thus, social tacit knowledge is the knowledge in interacting with people. The know-how in social tacit knowledge are know-how to adapt to social context, know-how to reconcile differences with others, know-how to negotiate institutional politics, know-how to create a sense of connectedness and know-how to motivate people.

The Application Process of Tacit Knowledge in Malaysian Academia

Through the socially-constructed perspective, tacit knowledge manifestation is displayed through actions of an individual in some organizational situations. In the context of Malaysian PIHL, the introduction of 7Ps create situations which include proliferation of academic demands, increased need of social solidarity and transition of work practices. Therefore the application of tacit knowledge among the Malaysian academicians is centered towards their actions in to the new situation, the academic excellence norm.

The application of tacit knowledge in Malaysian academia corresponds to the academicians' goals and values. The actions that they perform are guided by their values and norms and also the intended results and effects. As much as the academicians intend to fulfill the 7Ps, the underlying "Eastern values" which includes conflict avoidance, harmony-seeking, relationship-oriented and respect for authority and seniority are visible in their actions. Two other values influence the actions of the academicians, namely religious values and nationalistic values.

Three categories of actions representing the academicians' maneuvers are discovered in this research, namely affirming intellectual capabilities, capitalizing self- knowledge and building interpersonal relationships. Affirming intellectual capabilities represents actions performed by the academicians in proving their intellectual worthiness. In building their intellectual worthiness, their actions are centered around conflict-avoidance and relationship-oriented, which is reflected through actions taken such as collaboration and networking with people. Religious value is also prevalent when actions taken are regarded as fulfilling the religious obligation such as sharing intellectual ideas and pursuing knowledge.

Meanwhile, capitalizing self knowledge represents actions performed by the academicians for their personal fulfillment and development. Eastern values, religious and nationalistic values are all prevalent in the way academicians capitalize their self knowledge. Religious values influence their actions in which they believed that they are accountable for their decisions and actions in the world and also in the hereafter. The non-confrontation value is reflected when academicians relied on their own strengths and weaknesses in encountering lack of support for scholarly works from the working environment. Nevertheless, there are traces of more aggressive actions especially in expressing their self-principles, where they will stand by their principles although they are against the majority. However, these actions are expressed in a subtle manner such as voicing the disagreement in a "closed-situation". Nationalism is also reflected when academicians capitalize their self knowledge as they associated their actions for the reasons of correcting social injustice and carrying out the government agenda. Academicians believe that they are the agents in the implementation of government agenda.

Lastly, building interpersonal relationships represents actions in interacting with other members of the institutions. In interacting with people, a majority of the academicians resort to conflict avoidance actions such as sharing responsibilities rather than enforcement. Legitimation of an action is done either through consensus or directives from the top management. They also incorporate their religious beliefs that people are created with certain advantages and disadvantages thus leading them to value others. The actions they took are also associated with the loyalty to the institution.

The Construction Process of Tacit Knowledge in Malaysian Academia

The construction process of tacit knowledge is important in understanding tacit knowledge. Job domain provides the environment for on-the-job training and informal learning which is vital for tacit knowledge construction. On-the-job training and experience provides the context for acquiring new knowledge for the academicians. Informal or implicit learning is the non-formal learning which occurs when academicians perform their job or learn from their previous experience.

The relationship of experience and tacit knowledge is associated with knowledge internalization, a process of implicit learning from the accumulated past experiences. In exploring the role of experience in tacit knowledge acquisition, examination of qualitative data reflected that experience from the training process, either before or after entering the workforce is used in taking actions in the work situations. The training process in specific area of expertise promotes the implicit learning of the experience. In this research, length of service, which reflects the length of experience, did not contribute to tacit knowledge construction. It is an indication that years of experience are less important to tacit knowledge construction than the ability to learn from experience. In the context of the academicians, the ability to acquire on-the-job exposure and apply it to job situations is more relevant to tacit knowledge construction.

Summarizing the tacit knowledge construction process, academicians are exposed to different work-related situations and their capabilities to learn from the situations vary, and these are the two factors which distinguish an academician from others. Although tacit knowledge is acquired through experience, the operational contribution of experience to tacit knowledge lies in an academician's ability to learn and apply it in an appropriate situation.

FUTURE RESEARCH DIRECTIONS

This study discovers tacit knowledge in academic domain through an academician's own experience. Although the research discovers the relationship between experience and tacit knowledge construction and application, a further study focusing on examining and identifying strategies for creating tacit knowledge in the academic domain is strongly recommended. Among the strategies that can be explored are strategies in sensitizing academicians to learning opportunities and lessons learned. A future research on strategies in applying knowledge acquired to their future work problems is recommended.

Secondly, research on exploring how organizational culture contributed to tacit knowledge construction and application are recommended to provide understandings on tacit knowledge from the organizational perspectives. Besides organizational culture, individual personalities and cognitive styles are also associated with individuals' reactions to a problem. They influenced individuals' perception on taking actions in a situation. Hence, a further study on the relationships between personality or cognitive style and tacit knowledge is worth pursuing. The relationship between personalities or cognitive styles with tacit knowledge would allow insights on the propensity to acquire and use knowledge.

Thirdly, two of the three dimensions of tacit knowledge discovered from this research resemble the social and emotional quotients of an individual. A further research on tacit knowledge using the meta-cognitive approach is recommended for further insights and understanding on tacit knowledge. Another recommendation for further research is to develop a concrete approach for measuring tacit knowledge in academic domain. The ability to measure the level of tacit knowledge is useful for organizations to locate its knowledge resources. This will also provides an academician with the knowledge of their own strength and weaknesses. A future research on measuring tacit knowledge can benefit organizations in evaluating knowledge transferred among their workers, before and after knowledge management initiatives.

CONCLUSION

This study is concerned with the exploration of the socially-constructed notion of tacit knowledge among Malaysian academicians. A research review of relevant research approaches and theoretical lens on tacit knowledge resulted on the decision on an open method for exploration of tacit knowledge. The empirical work was conducted using qualitative approach. GT is selected as the overarching approach for data gathering and analysis from a case site. The qualitative approach allowed two levels of understandings on tacit knowledge in academic domain. The first understanding was an interpretive understanding of tacit knowledge in academic domain discovered from the field data collected, through the GT approach. The second level of understanding was the conceptual understanding of the construction and application of tacit knowledge in academic domain.

This study concluded that tacit knowledge in Malaysian academia is knowledge internalized in the academicians' work practice. In the context of Malaysian PIHL, basically the workload of an academician centres on three main academic activities, namely teaching, research and services. However, the seven performance indicators (7Ps) detailed out seven academic activities, namely teaching, research, publication, writing, public services, consultancy and management, to be fulfilled by an academician. Therefore, the work practice of academicians is centred around the seven academic activities listed on the performance indicators.

Tacit knowledge has goal-attainment values and is the know-how in achieving such goals. Tacit knowledge is used by the academicians to strategize their actions in order for them to reach their intended goal of fulfilling the 7Ps.

Tacit knowledge is constructed through the implicit learning of the experience accumulated. The ability of an academician to learn from experience is the most important factor in tacit knowledge construction. The application of tacit knowledge among Malaysian academicians is through the internalization of knowledge to devise their action strategies in overcoming a particular challenge they encounter in their work practices. The application of tacit knowledge is reflected through the academicians' maneuvers in new academic excellence norms. This study does not only inform the social construction behind tacit knowledge but also it presents a particular interpretation of interpretive research.

REFERENCES

Aadne, J. H., Von Krogh, G., & Roos, J. (1996). Representationism: The Traditional Approach to Cooperative Strategies. In Von Krogh, G., & Roos, J. (Eds.), *Managing Knowledge Perspectives on Cooperation and Competition* (pp. 9–31). London: Sage.

Ambrosini, V., & Bowman, C. (2001). Tacit Knowledge: Some Suggestions for Operationalization. *Journal of Management Studies*, *38*, 811–829. doi:10.1111/1467-6486.00260

Baumard, P. (1999). *Tacit knowledge in organization*. London: Sage.

Best, J. W., & Kahn, J. V. (1998). *Research in Education*. Boston: Library of Congress Cataloguing-in-Publication data.

Bhardwaj, M., & Monin, J. (2006). Tacit to Explicit: An Interplay Shaping Organization Knowledge. *Journal of Knowledge Management*, *10*(3), 72–85. doi:10.1108/13673270610670867

Bhatt, G. D. (2002). Management Strategies for Individual Knowledge and Organizational Knowledge. *Journal of Knowledge Management*, *6*(1), 31–39. doi:10.1108/13673270210417673

Blackler, F. (1995). Knowledge, Knowledge Work and Organization: An Overview and Interpretation. *Organization Studies*, *16*, 1021–1046. doi:10.1177/017084069501600605

Brézillon, P. (1999). Context in problem solving: A survey. *The Knowledge Engineering Review*, *14*(1), 1–34. doi:10.1017/S0269888999141018

Brockman, E. N., & Anthony, W. P. (2002). Tacit Knowledge and Strategic Decision Making. *Group & Organization Management*, *27*(4), 436–455. doi:10.1177/1059601102238356

Brown, J. S., & Duguid, P. (1998). Organizing Knowledge. *California Management Review*, *40*(3), 90–111.

Casselman, R. M., & Samson, D. (2005). Moving Beyond Tacit and Explicit: Four Dimensions of Knowledge. *Proceedings of the 38th Hawaii International Conference on System Sciences*. 3 – 6 January. Hawaii: IEEE Computer Society Press.

Cavusgil, S. T., Calantone, R. J., & Zhao, Y. (2003). Tacit Knowledge Transfer and Firm Innovation Capability. *Journal of Business and Industrial Marketing*, *18*(1), 6–21. doi:10.1108/08858620310458615

Choo, C. W. (2006). *The Knowing Organization How Organizations Use Information to Construct Meaning, Create Knowledge and Make Decisions* (2nd ed.). New York: Oxford University Press.

Choo, W. C. (2002). Sensemaking, Knowledge Creation, and Decision Making. In Choo, C. W., & Bontis, N. (Eds.), *The Strategic Management of Intellectual Capital and Organizational Knowledge* (pp. 79–88). New York: Oxford University Press.

Collins, H. M. (2001). Tacit Knowledge, Trust, and the Q of Sapphire. *Social Studies of Science*, *31*(1), 71–85. doi:10.1177/030631201031001004

Connell, N. A., Klein, J. H., & Powell, P. L. (2003). It's Tacit Knowledge But Not As We Know It: Redirecting the Search for Knowledge. *The Journal of the Operational Research Society*, *54*, 140–152. doi:10.1057/palgrave.jors.2601444

Cook, S. D., & Brown, J. S. (1999). Bridging Epistemologies: The Generative Dance between Organization Knowledge and Organizational Knowing. *Organization Studies*, *10*(4), 381–400. doi:10.1287/orsc.10.4.381

Creswell, J. W. (2003). *Research Design: Qualitative, Quantitative, and Mixed Methods Approaches*. London: Sage.

D'Eredita, M. A., & Barreto, C. (2006). How Does Tacit Knowledge Proliferate? An Episode-Based Perspective. *Organization Studies*, *27*(12), 1821–1841. doi:10.1177/0170840606067666

Davenport, T. H., & Prusak, L. (1998). *Working Knowledge – How Organizations Manage What They Know*. New York: Harvard University Press.

Denzin, N. K., & Lincoln, Y. S. (1998). *The Landscape of Qualitative Research*. London: Sage.

Dixon, N. M. (2000). *Common Knowledge: How Companies Thrive by Sharing What They Know*. Boston: Harvard Business School Press.

Drucker, P. F. (1993). *The post capitalist society*. New York: HarperBusiness.

Endres, M. L., Endres, S. P., Chowdhury, S. K., & Alam, I. (2007). Tacit Knowledge Sharing, Self-Efficacy Theory, and Application to the Open Source Community. *Journal of Knowledge Management*, *11*(3), 92–103. doi:10.1108/13673270710752135

Eraut, M. (2000). Non-formal Learning and Tacit Knowledge in Professional Work. *The British Journal of Educational Psychology*, *70*, 113–136. doi:10.1348/000709900158001

Evans, K., Kersh, N., & Kontiainen, S. (2004). Recognition of Tacit Skills: Sustaining Learning Outcomes in Adult Learning and Work Re-entry. *International Journal of Training and Development*, *8*(1), 54–72. doi:10.1111/j.1360-3736.2004.00196.x

Fenstermacher, K. D. (2005). The Tyranny of Tacit Knowledge: What Artificial Intelligence Tells Us About Knowledge Representation. *Proceedings of the 38th Hawaii International Conference on System Sciences.* 3 – 6 January. Hawaii: IEEE Computer Society Press.

Fernie, S., Green, S. D., Weller, S. J., & Newcombe, R. (2003). Knowledge Sharing: Context, Confusion and Controversy. *International Journal of Project Management*, *21*(3), 177–186. doi:10.1016/S0263-7863(02)00092-3

Glaser, B. G., & Strauss, A. L. (1967). *The Discovery of Grounded Theory: Strategies for Qualitative Research*. Mill Valley, CA: Sociology Press.

Goddard, A. (1998). Facing Up to Market Forces. *Times Higher Education Supplement*, *13*(November), 6–7.

Gore, C., & Gore, E. (1999). Knowledge Management: The Way Forward. *Total Quality Management*, *10*(4-5), 554–560.

Gourlay, S. (2004). Tacit Knowledge: The Variety of Meaning in Empirical Research. *Proceedings of the Fifth European Conference on Organizational Knowledge, Learning and Capabilities.* 2-3 April. Innsbruck, Austria.

Gourlay, S. (2006). Towards Conceptual Clarity for 'Tacit Knowledge': A Review of Empirical Studies. *Knowledge Management Research and Practice.*, *4*(1), 60–69. doi:10.1057/palgrave.kmrp.8500082

Guba, E., & Lincoln, Y. (1994). Competing paradigms in qualitative research. In Denzin, N., & Lincoln, Y. (Eds.), *Handbook of qualitative research* (pp. 105–177). Thousand Oaks, CA: Sage.

Guzman, G. A., & Wilson, J. (2005). The "Soft" Dimension of Organizational Knowledge Transfer. *Journal of Knowledge Management*, *9*(2), 59–74. doi:10.1108/13673270510590227

Haldin-Herrgard, T. (2000). Difficulties in Diffusing of Tacit Knowledge in Organization. *Journal of Intellectual Capital*, *1*(4), 357–363. doi:10.1108/14691930010359252

Haldin-Herrgard, T. (2004). Diving Under the Surface of Tacit Knowledge. *Proceedings of Fifth European Conference on Organizational Knowledge, Learning and Capabilities.* 2-3 April. Innsbruck, Austria.

Hansen, B. H., & Kautz, K. (2005). Grounded Theory Applied- Studying Information Systems Development Methodologies in Practice. *Proceedings of the 38th Hawaii International Conference on System Sciences.* 3 – 6 January. Hawaii: IEEE Computer Society Press.

Hansen, M. T., Nohria, N., & Tierney, T. (1999). What's Your Strategy for Managing Knowledge? *Harvard Business Review*, (March-April): 106–116.

Hedesstrom, T., & Whitley, E. A. (2000). What is meant by tacit knowledge? Towards a better understanding of the shape of actions. *Proceedings of the 8th European Conference on Information Systems*. 3-5 July. Vienna, Austria: ECIS 2000, 46-51.

Hellstrom, T., & Husted, K. (2004). Mapping Knowledge and Intellectual Capital in Academic Environments. *Journal of Intellectual Capital*, *5*(1), 165–180. doi:10.1108/4691930410512987

Hijazi, S., & Kelly, L. (2003). Knowledge Creation in Higher Education Institutions: A Conceptual Model. *Proceedings of Association of Small Computer Users in Education (ASCUE) Conference.* 8-12 June. South Carolina, United States of America.

Horvath, J. A., Forsythe, G. B., Bullis, R. C., Sweeny, P. J., Williams, W. M., & McNally, J. A. (1999). Experience, Knowledge, and Military Leadership. In Sternberg, R. J., & Horvath, J. A. (Eds.), *Tacit Knowledge in Professional Practice* (pp. 39–58). Mahwah, NJ and London: Lawrence Erlbaum Associates.

Howells, J. (1996). Tacit Knowledge, Innovation and Technology Transfer. *Technology Analysis and Strategic Management*, *8*, 91–106. doi:10.1080/09537329608524237

Hughes, J., & Howcroft, D. A. (2000). Grounded Theory: Never Knowingly Understood. *Information Systems Research*, *4*(1), 181–1974.

Hunt, D. P. (2003). The Concept of Knowledge and How to Measure It. *Journal of Intellectual Capital*, *4*(1), 100–112. doi:10.1108/14691930310455414

Johannessen, J. A., Olaisen, J., & Olsen, B. (2001). Mismanagement of Tacit Knowledge: The Importance of Tacit Knowledge, The Danger of Information Technology, and What To Do About It. *International Journal of Information Management*, *21*, 3–20. doi:10.1016/S0268-4012(00)00047-5

Johnson, J. M. (2002). In depth interviewing. In Gubrium, J. F., & Holsteins, J. A. (Eds.), *Handbook of Interview Research: Context and Methods* (pp. 103–120). Thousand Oaks, CA: Sage.

Kakabadse, N. K., Kakabadse, A., & Kouzmin, A. (2003). Reviewing the Knowledge Management Literature: Towards a Taxonomy. *Journal of Knowledge Management*, *7*(4), 75–91. doi:10.1108/13673270310492967

Kermally, S. (1998). *New Economy Energy: Unleasing Knowledge for Competitive Advantage*. Sussex, England: John Wiley and Sons.

Khurana, R. (2007). *From Higher Aims to Hired Hands: The Social Transformation of American Business Schools and the Unfulfilled Promise of Management as a Profession*. Princeton: Princeton University Press.

Klein, H. K., & Myers, M. D. (1999). A Set of Principles for Conducting and Evaluating Interpretive Field Studies in Information Systems. *Management Information Systems Quarterly*, *23*(1), 67–93. doi:10.2307/249410

Koenig, M. (2002). The Third Stage of KM Emerges. *KMWorld.*, *11*(3), 20–21.

Koskinen, K. U. (2003). Evaluation of Tacit Knowledge Utilization in Work Units. *Journal of Knowledge Management*, *7*(5), 67–80. doi:10.1108/13673270310505395

Lam, A. (2000). Tacit Knowledge, Organizational Learning and Societal Institutions: An Integrated Framework. *Organization Studies*, *21*(3), 487–513. doi:10.1177/0170840600213001

Leonard, D., & Sensiper, S. (1998). The Role of Tacit Knowledge in Group Innovation. *California Management Review*, *40*(3), 112–132.

Li, M., & Gao, F. (2003). Why Nonaka Highlights Tacit Knowledge: A Critical Review. *Journal of Knowledge Management*, *7*(4), 6–14. doi:10.1108/13673270310492903

Maqsood, T., Finegan, A. D., & Walker, D. H. (2004). Biases and Heuristics in Judgment and Decision Making: The Dark Side of Tacit Knowledge. *Issues in Informing Science and Information Technology.*, *1*, 295–301.

McInerney, C. (2002). Knowledge Management and the Dynamic Nature of Knowledge. *Journal of the American Society for Information Science and Technology*, *53*(12), 1009–1018. doi:10.1002/asi.10109

Metaxiotis, K., & Psarras, J. (2003). Applying Knowledge Management in Higher Education: The Creation of a Learning Organisation. *Journal of Information and Knowledge Management.*, *2*(4), 1–7.

Mohayidin, M. G., Man, N., Kamaruddin, N., & Margono, M. I. (2007). The Application of Knowledge Management in Enhancing the Performance of Malaysian Universities. *Electronic Journal of Knowledge Management*, *5*(3), 301–312.

Mohd Ridzuan Nordin. (2001). The Challenge for Higher Education in Malaysia in an Era of a Knowledge-Based Economy. In Abdulai, D. N. (Ed.) *Malaysia and the K-Economy: Challenges, Solutions and the Road Ahead.* (pp. 1-30). Subang Jaya: Pelanduk Publications (M) Sdn. Bhd.

Nonaka, I., & Takeuchi, H. (1995). *The Knowledge-Creating Company*. Oxford, UK: Oxford University Press.

Orlikowski, W. J. (2002). Knowing in Practice: Enacting a Collective Capability in Distributed Organizing. *Organization Science*, *13*(3), 249–273. doi:10.1287/orsc.13.3.249.2776

Orlikowski, W. J., & Baroudi, J. J. (1991). Studying Information Technology in Organizations: Research Approaches and Assumptions. *Information Systems Research*, *2*, 1–28. doi:10.1287/isre.2.1.1

Polanyi, M. (1966). *Personal Knowledge: Towards a Post-Critical Philosophy*. Chicago: The University of Chicago Press.

Reuber, A. R., Dyke, L. S., & Fischer, E. M. (1990). Using Tacit Knowledge Methodology to Define Expertise. *Proceedings of ACM SIGBDP Conference on Trends and Directions in Expert Systems*. November 1990. Orlando, Florida. ACM, (pp. 262-274).

Sahay, S., & Robey, D. (1996). Organizational Context, Social Interpretation, and The Implementation and Consequences of Geographic Information Systems.Accounting. *Management and Information Technology.*, *6*(4), 255–282. doi:10.1016/S0959-8022(96)90016-8

Schmidt, F. L.& Hunter, J. E. (1993). Tacit Knowledge, Practical Intelligence, General Mental Ability, and Job Knowledge. *Current Directions in Psychological Science: A Journal of American Psychological Society.* *2*(1), 8-9.

Schultze, U., & Leidner, D. E. (2002). Knowledge management in IS research. *Management Information Systems Quarterly*, *26*(3), 213–242. doi:10.2307/4132331

Spender, J. C. (1996). Competitive Advantage from Tacit Knowledge? Unpacking The Concept and Its Strategic Implication. In Moingeon, B., & Edondson, A. (Eds.), *Organizational Learning and Competitive Advantage* (pp. 56–73). London: Sage.

Stahl, B. C. (2008). *Information Systems: Critical Perspectives*. Routledge.

Stenmark, D. (2000). Turning Tacit Knowledge Tangible. *Proceedings of the 33rd Hawaii International Conference on System Sciences*. 4-7 January. Maui, Hawaii: IEEE Computer Society Press.

Stenmark, D. (2001). Leveraging Tacit Organisational Knowledge. *Journal of Management Information Systems*. 17(3), (Winter 2000-2001), 9-24.

Sternberg, R. J., Forsythe, G. B., Hedlund, J., Horvath, J. A., Wagner, R. K., & Williams, W. M. (2000). *Practical Intelligence in Everyday Life*. New York: Cambridge University Press.

Sternberg, R. J., & Horvath, J. A. (Eds.). (1999). *Tacit Knowledge in Professional Practise: Researcher and Practitioner Perspectives*. Mahwah, NJ: Lawrence Erlbaum Associated Inc.

Swart, J., & Pye, A. (2002). Conceptualizing Organization Knowledge as Collective Tacit Knowledge: A Model of Redescription. *Proceedings of the 3rd European Conference on Organization Knowledge, Learning and Capabilities.* 5 – 6 April. ALBA, Athens, Greece.

Szulanski, G. (2003). *Sticky Knowledge: Barriers to Knowing in the Firm*. London: Sage.

Tashakkori, A., & Teddlie, C. (1998). *Mixed Methodology: Combining Qualitative and Quantitative Approaches*. London: Sage.

Tsoukas, H. (2003). Do We Really Understand Tacit Knowledge? In Easterby-Smith, M., & Lyles, M. A. (Eds.), *The Blackwell Handbook of Organizational Learning and Knowledge Management* (pp. 410–427). Oxford, UK: Blackwell Publishing Ltd.

Von Krogh, G., & Roos, J. (Eds.). (1996). *Managing Knowledge Perspectives on Cooperation and Competition*. London: Sage.

Wagner, R. K., & Sternberg, R. J. (1985). Practical Intelligence in Real-World Pursuits: The Role of Tacit Knowledge. *Journal of Personality and Social Psychology*, *49*, 436–458. doi:10.1037/0022-3514.49.2.436

Walsham, G. (1993). *Interpreting Information Systems in Organisations*. Chichester, UK: Wiley.

Walsham, G. (2001). *IT in a Global Context*. Chichester, UK: Wiley.

Walsham, G. (2005). Knowledge Management Systems: Representation and Communication in Context. *An International Journal on Communication. Information Technology and Work.*, *1*(1), 6–18.

Zack, M. (1999). Managing Codified Knowledge. *Sloan Management Review*, *40*(4), 45–58.

ADDITIONAL READING

Abdullah, A. (Ed.). (1993). *Understanding the Malaysian Workforce – Guidelines for Managers*. Kuala Lumpur: Malaysian Institute of Management.

Boiral, O. (2002). Tacit Knowledge and Environmental Management. *Long Range Planning*, *35*, 291–317. doi:10.1016/S0024-6301(02)00047-X

Boisot, M. (2002). The Creation and Sharing of Knowledge. In Choo, C. W., & Bontis, N. (Eds.), *The Strategic Management of Intellectual Capital and Organizational Knowledge* (pp. 65–78). New York: Oxford University Press.

Chen, W. S., & Hirschheim, R. (2004). A Paradigmatic and Methodological Examination of Information Systems Research From 1991 to 2001. *Information Systems Journal*, *14*, 197–235. doi:10.1111/j.1365-2575.2004.00173.x

Choo, C. W., & Bontis, N. (Eds.). (2002). *The Strategic Management of Intellectual Capital and Organizational Knowledge*. New York: Oxford University Press.

Creswell, J. W. (2007). *Qualitative Inquiry & Research Design* (2nd ed.). Thousand Oaks, CA: Sage.

Denzin, N. K. (1978). *The Research Act: A Theoretical Introduction to Sociological Methods* (2nd ed.). New York: McGraw-Hill.

Drucker, P. (1999). Knowledge-worker Productivity: The Biggest Challenge. *California Management Review*, *41*(2), 79–85.

Easterby-Smith, M., Thorpe, R., & Lowe, A. (2002). *Management Research: An Introduction* (2nd ed.). London: Sage.

Galliers, R. D. (Ed.). (1992). *Information Systems Research: Issues, Methods, and Practical Guidelines*. Oxford, UK: Blackwell.

Garcia, B. (2006). Learning Conversation: Knowledge, Meanings and Learning Networks in Greater Manchester. *Journal of Knowledge Management*, *10*(5), 99–109. doi:10.1108/13673270610691215

Germain, L. S., & Quinn, D. M. (2005). Investigation of Tacit Knowledge in Principal Leadership. *The Educational Forum*, *70*(1), 75–90. doi:10.1080/00131720508984873

Glaser, B. G. (1978). *Theoretical Sensitivity: Advances in the Methodology of Grounded Theory*. Mill Valley, CA: Sociology Press.

Glaser, B. G. (1992). *Basics of Grounded Theory Analysis*. Mill Valley, CA: Sociology Press.

Glaser, B. G. (1998). *Doing Grounded Theory: Issues and Discussions*. Mill Valley, CA: Sociology Press.

Goddard, A. (1998). Facing Up to Market Forces. *Times Higher Education Supplement*, *13*(November), 6–7.

Guba, E., & Lincoln, Y. (1998). Competing paradigms in qualitative research. In Denzin, N., & Lincoln, Y. (Eds.), *The Landscape of Qualitative Research: Theories and Issues* (pp. 195–220). Thousand Oaks, CA: Sage.

Johnson, B., Lorenz, E., & Lundvall, B. E. (2002). Why All This Fuss About Codified and Tacit Knowledge? *Industrial and Corporate Change*, *11*(2), 245–262. doi:10.1093/icc/11.2.245

Jones, S., & Hughes, J. (2001). Understanding IS evaluation as a complex social process: a case study of a UK local Authority. *European Journal of Information Systems*, *10*, 189–203. doi:10.1057/palgrave.ejis.3000405

Kolodner, J. L. (1983). Towards an Understanding of the Role of Experience in the Evolution from Novice to Expert. *International Journal of Man-Machine Studies*, *19*, 497–518. doi:10.1016/S0020-7373(83)80068-6

Leonard, N., & Insch, G. S. (2005). Tacit Knowledge in Academia: A Proposed Model and Measurement Scale. *The Journal of Psychology*, *139*(6), 495–512. doi:10.3200/JRLP.139.6.495-512

McElroy, M. W. (2003). *The New Knowledge Management*. Burlington, MA: Butterwoth-Heinemann.

Metaxiotis, K., & Psarras, J. (2003). Applying Knowledge Management in Higher Education: The Creation of a Learning Organisation. *Journal of Information and Knowledge Management.*, *2*(4), 1–7.

Metcalfe, A. M. (Ed.). (2005). *Knowledge Management and Higher Education: A Critical Analysis*. Hershey, PA: Idea Group Publishing.

Mingers, J. (2000). The Contribution of Critical Realism as an Underpinning Philosophy for OS/MS and Systems. *The Journal of the Operational Research Society*, *51*(11), 1256–1270.

Mingers, J., & Willcocks, L. (2004). (Eds.) *Social Theory and Philosophy for Information Systems*. Chichester, UK: Wiley.

Ministry of Higher Education. (2006). *Report by the Committee to Study, Review and Make Recommendations Concerning the Development and Directions of Higher Education in Malaysia*. [Report]. University Publication Center (UPENA), University Teknologi MARA: Shah Alam.

Myers, M. D. (1997). Qualitative Research in Information Systems. *Management Information Systems Quarterly*, *21*(2), 241–242. doi:10.2307/249422

Nestor-Baker, N. S. (2002). Knowing When to Hold 'Em and Fold 'Em: Tacit Knowledge of Place-Bound and Career-Bound Superintendents. *Journal of Educational Administration*, *40*(3), 230–256. doi:10.1108/09578230210427163

Nestor-Baker, N. S., & Hoy, W. K. (2001). Tacit Knowledge or School Superintendents: Its Nature, Content, and Meaning. *Educational Administration Quarterly*, *37*(1), 59–86. doi:10.1177/00131610121969253

Ning, Y., Fan, Z., & Feng, B. (2005). Motivation Factors That Make Knowledge Workers Share Their Tacit Knowledge in Universities: An Empirical Research. *Services Systems and Services Management.*, *2*, 923–927.

Nonaka, I., & Konno, N. (1998). The Concept of "BA" – Building a Foundation for Knowledge Creation. *California Management Review*, *40*(3), 40–54.

Pozzali, A. (2007). *Tacit Knowledge, Implicit Learning and Scientific Reasoning. Mind & Society. (Online First Status)*. SpringlerLink.

Renken, J., & Moswetsi, W. (2006). Experiences in Interpretive Information Systems Research: Investigating E-commerce Adoption in the Botswana Defence Force. In Ruth, A. (Ed.), *Quality and Impact of Qualitative Research* (pp. 104–115). Brisbane: Institute for Integrated and Intelligent Systems, Griffith University.

Rose Alinda Alias. (1997). *A Multiple Perspective Exploration of Information Systems Quality in a Malaysian Context*. Salford, UK: Doctor Philosophy, University of Salford.

Rowley, J. (2000). Is Higher Education Ready for Knowledge Management? *International Journal of Educational Management*, *14*(7), 325–333. doi:10.1108/09513540010378978

Sarker, S., Lau, F., & Sahay, S. (2001). Using an Adapted Grounded Theory Approach for Inductive Theory Building About Virtual Team Development. *The Data Base for Advances in Information Systems*, *32*(1), 38–56.

Schultze, U., & Leidner, D. E. (2002). Knowledge management in IS research. *Management Information Systems Quarterly*, *26*(3), 213–242. doi:10.2307/4132331

Silverman, D. (1985). *Qualitative Methodology and Sociology*. London: Gower.

Smit, J., & Bryant, A. (2000). *Grounded Theory Method in IS Research: Glaser Vs Strauss* [Working Paper]. Leeds, UK: University of Leeds.

Smithson, S. (1991). Continuing Different Approaches. In Nissen, H. E., Klein, H. K., & Hirschiem, R. (Eds.), *Information Systems Research: Contemporary Approaches and Emergent Traditions* (pp. 365–369). Amsterdam: Elsevier.

Snowden, D. (2002). Complex Acts of Knowing: Paradox and Descriptive Self-Awareness. *Journal of Knowledge Management*, *6*(2), 1–14. doi:10.1108/13673270210424639

Somech, A., & Bogler, R. (1999). Tacit Knowledge in Academia: Its Effects on Student Learning and Achievement. *The Journal of Psychology*, *133*(6), 605–616. doi:10.1080/00223989909599766

Spender, J. C. (1996). Competitive Advantage from Tacit Knowledge? Unpacking The Concept and Its Strategic Implication. In Moingeon, B., & Edondson, A. (Eds.), *Organizational Learning and Competitive Advantage* (pp. 56–73). London: Sage.

St. Germain, L., & Quinn, D. M. (2005). Investigation of Tacit Knowledge in Principal Leadership. *The Educational Forum*, *70*(1), 75–89. doi:10.1080/00131720508984873

Sternberg, R. J. & Grigorenko, E. L. (2001). *Practical Intelligence and the Principal*. [Technical Report]. Office of Educational Research and Improvement (OERI): United States Department of Education.

Stewart, T. A. (2001). *The Wealth of Knowledge: Intellectual Capital and the Twenty-first Century Organization*. New York: Random House Inc.

Steyn, G. M. (2004, Summer). Harnessing the Power of Knowledge in Higher Education. *Education*, *124*(4), 615–632.

Stones, R. (2005). *Structuration Theory*. New York: Palgrave Macmillan.

Strauss, A. L., & Corbin, J. (1990). *Basics of qualitative research: grounded theory procedures and techniques*. Thousand Oaks, CA: Sage.

Strauss, A. L., & Corbin, J. (1998). *Basics of qualitative research: grounded theory procedures and techniques* (2nd ed.). Thousand Oaks, CA: Sage.

Sveiby, K. E. (1997). *The New Organizational Wealth*. San Francisco: Berrett-Koehler Publishers, Inc.

Tschannen-Moran, M., & Nestor-Baker, N. (2004). The Tacit Knowledge of Productive Scholars in Education. *Teachers College Record*, *106*(7), 1484–1511. doi:10.1111/j.1467-9620.2004.00387.x

Tsoukas, H., & Vladimirou, E. (2001). What is Organizational Knowledge? *Journal of Management Studies*, *38*(7), 973–993. doi:10.1111/1467-6486.00268

Urquhart, C. (2001). An Encounter with Grounded Theory: Tackling the Practical and Philosophical Issues. In Trauth, E. (Ed.), *Qualitative Research in IS: Issues and Trends* (pp. 104–140). Hershey, PA: Idea Group Publishing.

Van Krogh, G., Ichiro, K., & Nonaka, I. (2000). *Enabling Knowledge Creation: How to Unlock the Mystery of Tacit Knowledge and Release the Power of Innovation*. New York: Oxford University Press.

Walsham, G. (1995). Interpretive Case Studies in IS Research: Nature and Method. *European Journal of Information Systems*, *4*(2), 74–81. doi:10.1057/ejis.1995.9

Walsham, G., & Han, C.-K. (1993). Information Systems Strategy Formation and Implementation: The Case of a Central Government Agency. *Accounting. Management and Information Technologies*, *3*(3), 191–209. doi:10.1016/0959-8022(93)90016-Y

Wiig, K. M. (2002). Reinventing Business with New Generation Knowledge Management. [July, Singapore.]. *Keynote Address in KM Asia, 2002*, 16–18.

Wong, W. L. P., & Radcliffe, D. F. (2000). The Tacit Nature of Design Knowledge. *Technology Analysis and Strategic Management*, *12*(4), 493–512. doi:10.1080/713698497

KEY TERMS AND DEFINITIONS

Tacit Knowledge: Knowledge internalized by the academicians in their work practices in strategizing towards goal accomplishment.

Knowledge Management: The process of leveraging knowledge and expertise in an organization for competitive advantage.

Intellectual Affirmation Tacit Knowledge: Knowledge of understanding behaviors relating to intellectual capabilities;.related to actions in building intellectual capabilities worthiness.

Self Tacit Knowledge: Knowledge of understanding behaviors relating to one-self; related to actions which are individually-centered and are performed in distinguishing one self-needs.

Social Tacit Knowledge: Knowledge in interacting with people; related to actions which involves interaction with others.

Knowledge: Understanding acquired through formal education or experience.

Implicit Learning: Non-formal learning, learning by experience.

Expert: A person who posses knowledge and skill from years of experience.

Chapter 6
Supply Chain Knowledge Integration in Emerging Economies

Ryan Atkins
Queen's University Belfast, Northern Ireland

EXECUTIVE SUMMARY

Firms in developed economies have been purchasing from firms in emerging economies for years, but they typically purchase low value, low margin items. Opportunities have arisen for emerging economy manufacturers to supply higher value products and services to firms in developed economies. In today's global, knowledge based economy, suppliers must have cutting edge knowledge, and they must constantly upgrade their knowledge to maintain a competitive advantage. When supplying high-value products, complex knowledge based interactions between the buyer and supplier are necessary. A conceptual model is developed in this chapter, proposing that if suppliers in emerging economies are committed to long term relationships with developed economy customers, they can increase the level of knowledge integration in the relationship, and in turn, improve performance. The primary contribution of this chapter is to show that firms in emerging economies can achieve sales growth by becoming critical links in today's global, knowledge based supply chains.

INTRODUCTION

The globalization of supply chains has provided tremendous opportunities for manufacturers in emerging economies to improve their sales and profitability. Responding to these opportunities, however, can be a significant challenge. Firms in developed economies have been purchasing from manufacturers in less advanced economies for many years because of the lower cost of labor and other inputs to production. These purchases have typically been limited to low value products, but more recently they have begun to include more complex, high value products due to the improving capabilities of the emerging economy firms (Koudal & Engel, 2007). Major companies in North America and Europe purchase almost 50% of their raw materials, semi-finished goods, and

DOI: 10.4018/978-1-61692-886-5.ch006

finished goods from abroad, so there are many opportunities for emerging economy manufacturers to capitalize on their cost advantage and increase sales (Cudahy et al., 2008). One of the top priorities of emerging economy manufacturers for the upcoming years is the ability to sense and respond to growth opportunities (Pandey, 2010). To achieve these opportunities, suppliers must often meet strict customer demands, which can require significant amounts of knowledge exchange between the firms. Research has begun to investigate the effects of knowledge exchange and learning in cross-border buyer-supplier relationships, and it has shown that significant value can be generated from these interactions (Cheung et al., In Press). The objective of this chapter is to provide insight into how firms in emerging economies can take advantage of growth opportunities through supply chain relationships. A conceptual model is proposed, explaining how long term buyer-supplier relationships affect knowledge integration and sales growth. The next sections provide a literature review, followed by the theoretical framework and model, and finally a discussion of the recommendations that emerge from this research.

LITERATURE REVIEW

Competition among manufacturers has evolved in recent years to be supply chain vs. supply chain instead of firm vs. firm (Handfield & Bechtel, 2002). The Supply-Chain Council (2008) defines a supply chain as a network of firms that encompasses every effort involved in producing and delivering a final product from the supplier's supplier to the customer's customer. In the past, efficiency was often gained through vertical integration, with firms designing and manufacturing their complete products. The quintessential example of vertical integration is Ford's River Rouge complex, in which raw materials were brought into one end of the plant and finished vehicles came out of the other end, and nearly every other step was performed in the plant, from steelmaking to electricity generation (Fine, 1998). This method worked well for standardized, mass produced products. Today, customers demand flexibility and responsiveness, which limits the ability of one firm to meet all possible customer needs. Manufacturers have responded to the call for flexibility and responsiveness in part by relying on supply chain management (Richardson, 1993). By performing only the functions for which they can add significant value to the end product and outsourcing the functions for which they are at a comparative disadvantage, manufacturers have become more flexible and innovative, while reducing costs. A supply chain focus enables the individual firms in the chain to become more specialized, thus making the chain as a whole more efficient (Prahalad & Hamel, 1990). Although the management of the manufacturing process is more complex due to the additional parties involved and their potentially conflicting goals, a firm's supply chain can be a significant source of value and competitive advantage if managed successfully (Porter, 1985).

Suppliers have taken on greater responsibility with the shift towards supply chain based competition. Instead of purchasing only raw materials and low value products, many firms today purchase high value components or complete subsystems, which often represent over 50% of the value of the final product (Holcomb & Hitt, 2007). The quality and cost of the buyer's end product can be directly and significantly impacted by the quality and cost of the supplied component, so it is imperative that suppliers provide superior products (Liker & Choi, 2004). Suppliers can also play a significant role in the success of the end product by participating in the product design process (Petersen et al., 2005). If a supplier provides products with quality or safety problems, there can be significant repercussions for the buying firm. High profile recalls have significantly damaged the reputations and stock prices of firms such as

Toyota and Mattel, whose suppliers provided them with faulty accelerator pedals and toys with lead paint, respectively (Deloitte Touche Tohmatsu, 2008; Motavalli, 2010).

To better manage the supply chain, many firms have built long term, collaborative relationships with a limited number of suppliers. Japanese automakers led the way in improving quality and performance through supplier relationships in the 1980's (Richardson, 1993). While US firms were attempting to achieve low costs by increasing competitive pressure on a large number of suppliers (Ellram & Carr, 1994), Japanese firms were using long-term relationships with a smaller number of suppliers (Lyons et al., 1990). The longer duration of these relationships enabled the firms to collaborate to develop more value (Richardson, 1993). The improved coordination and economies of scale that come from consolidating purchases with a smaller number of suppliers are beneficial for cost reduction, while the increased level of collaboration provides opportunities to generate value (Handfield et al., 1999).

Manufacturers in emerging economies are becoming important members of global supply chains. Despite the advances that have been made in supply chain management practices, cost remains a critical factor in purchasing decisions, so there are tremendous opportunities for suppliers in emerging economies. Emerging economy suppliers have long been able to provide inputs at a lower cost than developed economy suppliers due to the relatively low cost of labor, capital, and other inputs compared to developed countries. These purchases have typically been limited to low margin, commodity products that are manufactured using standardized equipment and skills (Steinfeld, 2004). Today, however, opportunities have arisen as the cost of global transportation and communication has decreased, while in countries such as China and India, the workforces are becoming more educated and infrastructure is improving (Kouvelis & Niederhoff, 2007). Firms in developed economies are turning to firms in emerging economies to provide a greater portion of the value of their final products. Some firms are even outsourcing their core manufacturing processes to focus more on the service and innovation side of the business (Li et al., 2008). Contract manufacturers in China, for example, perform much of the manufacturing and assembly of the Apple iPhone, while Apple focuses on their core strengths: design and marketing (Burrows, 2005).

Emerging economy suppliers must overcome many challenges to take advantage of their potential opportunities. Despite the low cost of resources in emerging economies, the additional costs to ship the products, manage the supply chain, and handle potential risks can outweigh the initial cost benefits. Developed economy buyers must spend money on supplier selection, testing products, monitoring performance, and managing contracts to ensure that they receive high quality products on time and at a low cost (Deloitte Touche Tohmatsu, 2008). When a supplier is located on the other side of the world, problems can arise due to transportation delays and customs holds, while communication problems due to language barriers, time zone issues, and cultural issues can limit the buyer's visibility over the supplier's processes. Firms in developed economies who have outsourced operations have had their expectations met less than 25% of the time, prompting many firms to move operations back to their home country or to another country that is closer to home (Cudahy et al., 2008). It can be especially difficult to overcome these obstacles given the fact that many of the products being purchased from emerging economies are relatively low end or mature products that can be made by a large number of competitors, and for which cost is the primary differentiating factor (Steinfeld, 2004).

Buyers and suppliers can achieve superior performance by building and exchanging supply chain knowledge. Knowledge is created when decisions are made cross-functionally (Dyer & Nobeoka, 2000), when there are networks, teams, and an open organizational design (Nonaka, 1994), and through the use of quality management practices (Linderman et al., 2004; Molina et al., 2007). An organic flow and integration of knowledge creates stores of knowledge among the individuals involved that can be tapped for making decisions and formulating operational strategies (Germaine et al., 2001; Hult et al., 2006). Buyer-supplier relationships can be an external source of knowledge (Paiva et al., 2008). Selnis and Sallis (2003) introduced the concept of relationship learning, which is joint activity between a buyer and supplier to share and interpret information. Cheung et al. (In Press) extended this concept to international relationships, and linked cross-border relationship learning to enhanced competitiveness. Knowledge of the supply chain can be a competitive weapon at the supply chain and firm levels (Craighead et al., 2009). Acquiring and using knowledge has been linked to reductions in cycle time (Hult et al., 2007). Information visibility in the supply chain can promote the efficient use of resources and can improve operational performance (Barratt & Oke, 2007), while poor communication was identified by as the biggest contributor to the failure of buyer-supplier partnerships (Ellram, 1995).

THEORETICAL FOUNDATION AND CONCEPTUAL MODEL

To address how firms in emerging economies can use supply chain relationships to improve knowledge management and performance, a conceptual model, grounded in the knowledge based view of the firm (KBV), is presented below. First a review of knowledge and the KBV is provided, followed by the research propositions and two illustrative examples of firms in emerging economies that have benefitted from supply chain relationships.

Theoretical Foundation

Knowledge is a complex topic with varying definitions. A general definition of knowledge is 'that which is known', or more specifically in the context of business, knowledge is credible information, held by individuals, that can be applied to productive activities (Grant, 1996b). Knowledge is often divided into tacit and explicit knowledge. Tacit knowledge is highly subjective information that is understood by an individual, while explicit knowledge refers to facts and data that can be documented (Kogut & Zander, 1992). The role of the firm in knowledge management has also been debated. Some researchers argue that firms own knowledge as a socially constructed asset (Spender, 1996), that the firm's role is to acquire, distribute, interpret, and store information (Huber, 1991), and to embody the firm's recombined knowledge into new innovative products (Nonaka, 1994). Others view knowledge as an asset owned by the individuals within an organization (Simon, 1991; Nahapiet & Ghoshal, 1998), with the firm playing a more passive role by providing socialization and interaction mechanisms for the creation, storage, and transferral of knowledge among individuals (Simon, 1991) or providing integration mechanisms for the creation and use of knowledge (Grant, 1996a).

This chapter addresses knowledge as one of the limited resources of a firm, so the KBV is an appropriate theoretical framework. The KBV is based in the logic of the resource based view of the firm (RBV), which is built upon the assumption that resources are heterogeneously divided among firms and are imperfectly mobile among firms (Penrose, 1959; Wernerfelt, 1984), and those resources that are valuable and rare can provide a competitive advantage (Barney, 1991). The KBV extends the RBV by adding the assumptions that most differences in firms can be broken down into knowledge differences, and that the productive capability of firms emerges more from a firm's intellectual assets than its physical assets (Conner & Prahalad, 1996; Grant, 1996b).

The role of the firm and of supply chains, according to the KBV, is to integrate the knowledge of individuals into productive activities.

Knowledge integration is defined as accessing and utilizing the knowledge of individuals for productive activities (Grant, 1996a). Integration does not imply a transfer of knowledge from one individual to another, but rather the diverse knowledge of individuals is brought together through routines in which each individual performs a specific portion of a greater task. Grant (1996a) argues that firms can gain a competitive advantage by integrating knowledge more efficiently, with greater scope, and more flexibly than competitors. This view of knowledge is appropriate for supply chain management because in a diversified supply chain, in which the productive activities are distributed among firms, the individuals who hold the knowledge of the products are also divided among the firms (Becker & Zirpoli, 2003). Production requires a large scope of specialized knowledge, and it is not realistic for a firm to have specialized knowledge for every task, from how to manufacture nuts and bolts to how to develop software for the firm's computers (Grant & Baden-Fuller, 2004). Individuals and firms are limited in their attention and capacity to hold knowledge, so specialization ensures that they can be experts in a particular field (Prahalad & Hamel, 1990). When individuals and firms specialize, they can be more efficient in the creation and storage of knowledge (Simon, 1991). The challenge, then, is that all of the distributed knowledge of individuals cannot be gathered together for a single decision maker (von Hayek, 1937). Manufacturing knowledge has become decentralized as firms have focused on their core competencies and outsourced non-core activities to suppliers (Becker & Zirpoli, 2003). Buyer-supplier relationships are used to match the firm's knowledge base with the knowledge required to fulfil opportunities (Grant & Baden-Fuller, 2004). More than simple transactions, these relationships often require complex interactions on issues such as designing, manufacturing, transporting, distributing, selling, and marketing of products (Cudahy et al., 2008).

Conceptual Model

A conceptual model is proposed applying the KBV to supply chain management, and in particular to emerging economy firms as suppliers to developed economy customers. The basic premise of the model is that through stronger buyer-supplier relationships based on long term commitment the level of knowledge integration in a relationship between the firms can be increased, which in turn leads to opportunities to increase sales. The model is shown in Figure 1.

Research Propositions

Long Term Commitment and Knowledge Integration

Long term relationships are a foundation of supply chain management. Long term commitment is defined as the desire to develop a stable relationship, a willingness to make short-term sacrifices to maintain the relationship, and a confidence in the stability of the relationship (Anderson & Weitz, 1992). Firms that are committed to a long term relationship can achieve benefits such as increased relational behaviour (Lusch & Brown, 1996) and a reduced risk of opportunism (Williamson, 2008). When a firm is committed to a long term relationship, the benefits from the relationship warrant

Figure 1. Conceptual model

maximum efforts at maintaining the relationship (Morgan & Hunt, 1994).

In the context of this chapter, knowledge integration is defined as the extent to which an emerging economy supplier makes an effort to ensure the compatibility of decisions between the firm and its developed economy customer. Grant (1996a) defined knowledge integration as the access and use of the knowledge of individuals. In an operations context, knowledge integration has been associated with cross-functional activities, team meetings, dedicated personnel, and the development of idiosyncratic routines and networks of individuals and firms to store and exchange knowledge and to ensure the consistency of decisions among individuals and groups (Germain et al., 2001; Selnis & Sallis, 2003). The definition provided in this chapter considers the access to and use of the knowledge of both firms through interfirm interactions to make decisions that consider the needs and abilities of both firms. Grant (1996a) proposed knowledge integration measures of efficiency, scope, and flexibility. While the current definition of knowledge integration does not formally include these three characteristics, they are useful in considering the impact of knowledge integration.

Long term commitment can contribute to increased levels of knowledge integration by improving the efficiency of communication and reducing inefficient practices. Emerging economy suppliers are often located thousands of miles from the buyer and may pose an increased risk of quality or safety problems, so developed economy firms must spend significant time and effort searching for suppliers, monitoring supplier performance, and establishing relationship governance (Grover & Malhotra, 2003). If a relationship is expected to continue, uncertainty is reduced, and search and negotiation efforts are minimized, thus allowing the firms to focus on working together to achieve productive goals (DeToni & Nassimbeni, 1999). Exchange partners are also more likely to expend the effort to share information when they expect the relationship to continue beyond a single, low cost purchase (Moberg et al., 2002).

Long term commitment promotes stable, long term relationships. The repetition and improvement of tasks over time causes the individuals involved to develop common knowledge and shared interpretations, thus ensuring that they know where to find information when needed, and enabling them to understand each other more fully when they communicate (Anderson & Weitz, 1992; Paiva et al., 2007). Accurate communication is particularly important when there are cultural barriers in addition to geographic distance (Anderson & Weitz, 1992; Ring & van de Ven, 1994). Grant and Baden-Fuller (2004) argue that interorganizational relationships are less efficient at integrating knowledge than firms because the stability within a firm enables stable integrating routines to be developed; however, if firms are committed to a long term relationship, they can approach the stability within a firm and can develop stable knowledge integration routines.

When a relationship is expected to continue, the firms involved are more likely to prepare for the future of the relationship by sharing a broad scope of information and using that information to make decisions. To realize the potential value of an ongoing relationship, firms must often exchange information regarding current and future needs, including quality and delivery information, expected changes in technology, markets, and competitors, and any other information deemed necessary (Madhok & Tallman, 1998; Moberg et al., 2002). These exchanges may extend to complex technological assistance if the investment is expected to pay off in the long run (Modi & Mabert, 2007; Noordewier et al., 1990). Firms that work with a small number of suppliers in long term relationships tend to include a broader coverage of information in their interactions (Larson, 1992). Suppliers in emerging economies in particular may benefit from promoting long term commitment because developed economy firms are looking to find and retain responsible exchange partners in emerging economies that can adapt with market changes over time (Deloitte Touche Tohmatsu, 2008).

If a relationship is expected to last a long time, the firms are more likely to be flexible to make decisions that are beneficial to both parties in the relationship. Recognizing the fact that long term relationships are more effective than short term relationships, firms may make efforts to use cooperative behaviors to maintain the stability of a relationship when they expect that the relationship will continue into the future (Lusch & Brown, 1996). They will not just focus on short term opportunities, but will plan for the future of the relationship so they do not lose the value associated with the ongoing relationship (Madhok & Tallman, 1998). Fulfilling new business opportunities may require large investments from buyers and suppliers. A commitment to the long term success of a relationship provides an incentive to work together to make decisions in the best interest of both parties (Hoetker, 2005; Teece et al., 1997). It can be a challenge for a developed economy buyer to differentiate between emerging economy suppliers, but if a supplier is committed to the success of a relationship with a buyer, the buyer will be more likely to consult the supplier for the new opportunity rather than moving on to a new supplier.

Proposition 1. Emerging economy suppliers that display greater levels of long term commitment to their developing economy customers are more likely to have greater levels of knowledge integration in the relationships

Knowledge Integration and Sales Growth

Sales growth is an important measure for emerging economy firms. If a firm is able to differentiate its products or provide additional value to customers in some other way, it will be reflected in increased sales (Henderson & Cockburn, 1994). Sales growth, therefore, reflects the firm's ability to take advantage of new opportunities. Operational performance measures such as cost and quality are important in supply chain management, (Ward et al., 1998), but cost is more appropriately designated as an order qualifier than an order winner in emerging economies, and quality should be reflected in the sales growth measure because poor quality should prevent sales from increasing over time.

Knowledge integration can contribute to sales growth by improving the firm's ability to provide value to its customers. This value can come in many forms, such as assuring the customer that their product is of an adequate quality (Deloitte Touche Tohmatsu, 2008), producing a more high tech product, or moving up the value chain to take on more assembly steps from the final manufacturer. The efficient use of knowledge can help to prevent or at lease provide early warning regarding quality problems and delays in shipments, which can increase costs and reduce customer satisfaction (Paulraj et al., 2008). Efficient knowledge integration can also ensure that information regarding changes is transmitted quickly and accurately so that the firms have enough time to make the appropriate responses in terms of production schedules, inventory levels, logistical planning, and changes to products and processes (Fugate et al., 2008).

Integrating a large scope of knowledge can make a firm's capabilities more difficult for competitors to imitate (Dierickx & Cool, 1989; Lippman & Rumelt, 1982), thus giving the firm a more sustainable competitive advantage (Barney, 1991). A broad scope of knowledge integration gives a firm a bigger and less easily imitable knowledge base from which to continuously improve and derive a competitive advantage (McEvily et al., 2000). A firm that draws upon information from external sources should develop solutions to problems that are at least as good if not better than those solutions that would be made without the external knowledge (Siemsen et al., 2008). Innovation tends to occur at boundaries, so the more interpersonal connections there are and the more points are addressed between firms, the more opportunities can be addressed (Carlile, 2004; Nahapiet & Ghoshal, 1998). When products or markets are complex, the knowledge that is relevant to present or future value is not always clear, so a larger scope of knowledge integration can ensure that the necessary knowledge is available to make mutual adjustments to achieve success (Carlile & Rebentisch, 2003). The efficient sharing of information can also lead to operational performance improvements, which can improve the reputation of the supplier and potentially lead to increased sales over time. Sharing information related to production schedules and inventory information can improve operational measures such as inventory (Lee, 2004) and quality and cost performance (Dyer & Hatch, 2006; Sanders, 2008). An emerging economy supplier that can be depended upon to provide high quality products on time can charge a premium price to its developed economy suppliers by reducing the buyer's level of risk (Accenture, 2007).

In the current business environment, technology changes quickly and product life cycles have shortened, so existing relationships will not last long unless firms are flexible enough to adjust to the necessary changes. Suppliers must be flexible to keep up with external changes in products, processes, and technology when deemed necessary according to knowledge received from customers (Henderson & Cockburn, 1994). As firms become increasingly interdependent, their changes must become more coordinated (Thompson, 1967). Knowledge integration between firms can keep both firms abreast of market changes so they continue to reflect the current competitive environment (Heide & John, 1992). Incremental knowledge improvements can be made when firms work together to find solutions to new problems and introduce new products, thus providing more opportunities to increase sales (Helfat & Raubitschek, 2000).

Proposition 2. Emerging economy suppliers who have greater levels of knowledge integration with developed economy customers are more likely to have positive sales growth

Empirical Examples

Two examples are provided of firms in emerging economies that have achieved significant sales growth in part by becoming integral links in global supply chains. The first example is Hon Hai Precision Industry (Hon Hai), a Taiwanese electronics manufacturer. Hon Hai, which is also known by its international trade name, Foxconn, began supplying plastic connectors to TV and computer manufacturers in the 1970's, and since then has become 'the biggest electronics company you never heard of' (Dean, 2007). Taiwan's economy today is one of the world's most technologically advanced, but in the 1970's, was an emerging economy. Hon Hai excelled in the emerging economy of Taiwan in the 1970's and 1980's due in part to the relationships with its customers. One of the keys to Hon Hai's strategy was to nurture its customer relationships by committing resources to the relationships and adapting to meet the needs of its customers (Dean, 2007). By maintaining close working relationships with its customers, Hon Hai was able to adapt to new technologies and demands as they arose, and is now the supplier of many of the most popular electronics products, such as Apple's iPhone (Hughes, 2009). Hon Hai's commitment to understanding customer needs continues today, with their business model that they call eCMMS, which describes the company's core strengths: Components, Module, Move, Services. This principle represents the firm's ability to provide any element of the value chain that a customer requires, from basic component manufacturing to complete design and manufacturing (fih-foxconn.com, 2007). Working together with customers, Hon Hai has evolved from supplying inexpensive plastic components to being on the cutting edge of high tech knowledge for over two decades.

The second example, IDS Group (IDS), is a China-based distributor that that has revolutionized the role of distributors in supply chain management in Southeast Asia. While much of this chapter has focused on manufacturing firms, IDS illustrates how services are critical to adding value to a manufacturer's products. Distributors in China in the late 1990's were known to be very inefficient, with costs almost double those of the western economies as a percentage of sales (Idsgroup.com, 2010). Typical distributors in China attempted to push products out onto the market, but often ended up with excess inventory, for which they had to cut prices and margins (Chang & Phi, 2007). IDS was launched in 1999 to provide integrated marketing, logistics, and manufacturing services, which the company refers to as 'Value-Chain Logistics', instead of the typical distribution services. The IDS business model includes building relationships with manufacturers and customers in order to gain knowledge of products and markets, so that rather than pushing products onto the market, it can help manufacturers make the right products and market them in the right way to increase sales and profitability. By building stronger relationships with manufacturers and customers, IDS has been able to gather and exchange more knowledge, which has allowed the company to provide superior value to its clients' supply chains (Chang & Phi, 2007; Idsgroup.com, 2010).

DISCUSSION

In today's globalized economy, emerging economies have made enormous steps forward in terms of improving infrastructure, wages, and living conditions (Lee & Lee, 2007). The rising tide of these economies can potentially lift all boats, but there is risk that some firms will run aground if they do not keep up with the changes in the economy. This research contributes to the discussion of how emerging economy manufacturers can achieve success in today's global, knowledge based economy. As the workforces in emerging economies have become more educated, the firms have developed the basic knowledge needed to compete, such as engineering and management skill (Steinfeld, 2004). Gaining tacit knowledge of products, processes, and technology, however, can be more complicated. Collaborative relationships with customers can ensure the supplier understands its customers' needs in the short and long term. Over time, products, processes, and technology evolve, so firm's must also evolve to remain competitive (Teece et al., 1997). Suppliers must evolve to meet their customers' needs or risk losing them. Committing to long term relationships and openly exchanging knowledge with customers can help emerging economy manufacturers satisfy customer demands.

There are potential risks that must be considered when building supply chain relationships. Developed economy firms must consider ways to deal with the potential risks of purchasing from emerging economy firms, such as delays and quality problems (Lee & Lee, 2007). Some firms have dealt with these issues by gaining more control through vertical integration, but that option is often not feasible (Peleg-Gillai, 2007). Long term buyer-supplier relationships can provide many of the benefits of vertical integration, but with less investment and financial risk for the buyer. One of the primary benefits of not vertically integrating is the flexibility to change suppliers if necessary. This benefit for the buyer is a risk for the supplier. It takes time to develop effective relationships, and because many firms in emerging economies do not have experience in a long term buyer-supplier relationship with a multinational company, the process may be unsuccessful (Li & Atuahene-Gima, 2001, Liu et al., 2008). Investments in human and financial resources are required to take advantage of opportunities. Any investments into the relationship made by the supplier may become less valuable if the buyer would choose to end the relationship, so additional transaction costs may be incurred by the supplier to protect its investments (Williamson, 1981). Any proprietary knowledge shared by the supplier is at risk of being leaked to competitors, so the benefits and costs of exchanging knowledge with customers must be considered carefully.

There are also long term risks that must be considered. As conditions improve in emerging economies, wages and other costs can increase over time and reduce the cost advantage held by these firms (Lee & Lee, 2007). If a manufacturer attempts to compete solely on cost, they will face the risk of losing business to lower cost countries or competitors with superior products. If, on the other hand, the firm makes improvements to its products, in terms of technology, quality, safety, and other factors, competitors may also make similar changes, and these conditions may no longer serve as order winners, but only as order qualifiers. The firms may end up working harder and spending more money to stay in the same place relative to competitors (Porter, 1996). Knowledge, however, may be even more important in this situation, as knowledge promotes innovation, which may become necessary to maintain a competitive advantage and continue to increase sales (Nonaka, 1994).

Aside from working on external relationships, firms must also do internal work to improve performance. Relationships may provide a firm with knowledge of opportunities, but may not help the firm achieve those opportunities. Superior operational abilities are necessary to manufacture products that are acceptable in developed economies. Some customers may be willing to offer development support to suppliers, but the investment must be worthwhile to the customer (Krause et al., 2007). In addition to the external knowledge integration, internal knowledge integration can help a firm provide its customers with products and resources that are valuable and can provide a competitive advantage (Paiva et al., 2008).

The model proposed in this chapter is aimed at emerging economy firms, but is also applicable to firms in developed economies. Distance and culture can be problems that lead developed economy firms to bring production back to their home country whether or not the supplier is in an emerging economy (Cudahy et al., 2008). The development of the Boeing 787 highlights these problems with suppliers in Europe and Japan (Bailey & Clark, 2008). Toyota's 2009 and 2010 recalls are linked to products developed and manufactured in Asia and the US (Bunkley & Maynard, 2010). Due to the high profile nature of these issues, developed economy firms may be reluctant to outsource product development and may purchase more components domestically. In the long run, these problems may be forgotten, and the low cost of products from emerging economies will continue to tempt firms to purchase from emerging economies. The manufacturers that are able to satisfy their developed economy customers can continue to increase sales and profitability.

FUTURE RESEARCH DIRECTIONS

This chapter provides insight into knowledge management in emerging economy supply chains, but there are some limitations to the research that provide opportunities for future research. The context of this research is manufacturing firms, but there are also opportunities for firms in emerging economies to take on more service related roles in the supply chain, as referenced in the IDS Group example. The model may also be relevant for firms that supply pure services such as customer service and support, as these types of services are increasingly being outsourced to India and other emerging economies (Balakrishnan et al., 2008). These services require little more than an office with a phone and internet connection to get started. Because of the low startup costs, the key differentiator among these firms could be the knowledge of the individuals involved and the relationship with their customers.

The conceptual model can also be applied to the distribution end of the supply chain. As the wealth of emerging economies has grown, these countries have become large consumers (Lee & Lee, 2007). Some notable examples are China, which is the world's largest cellular telephone market, and Mexico, which is the world's second largest soft drink market (Accenture, 2007). There are opportunities for local firms to distribute products for manufacturers who have limited knowledge of the local laws and markets. In this type of relationship, a firm can apply its knowledge of local issues and represent the customer-facing end of the supply chain (Balakrishnan et al., 2008). Firms in developed economies are increasingly outsourcing their manufacturing to be located in the same region as these new markets, so there are opportunities for research on supply chains that are wholly in emerging economies as well as those that serve emerging economies, but involve developed economies (Lee & Lee, 2007).

The performance measure used in this chapter is sales growth, which is appropriate due to the rapid growth of emerging economies, but over time, growth will slow down. As costs rise and competition increases, sales growth may slow down, and cost reduction may become more important to maintain profitability. Research into internal operations management practices will become more important as cost becomes a greater priority.

CONCLUSION

Firms in emerging economies are faced with many opportunities and challenges. This research points to the fact that emerging economy manufacturers can achieve the opportunities and overcome the challenges by building relationships with developed economy customers. A supplier that commits to a long term relationship with its customers signals that they are interested in making the relationship successful, and as a result, knowledge integration between the firms can be more successful. Knowledge integration with developed economy customers can help an emerging economy manufacturer satisfy its customers, fulfil new opportunities, and achieve sales growth. Long term buyer-supplier relationships have proven to be beneficial to firms in developed economies, and by adopting these practices, emerging economy manufacturers can achieve similar success.

REFERENCES

Accenture (2007). *The Rise of the Multi-Polar World.* Retrieved February 10, 2010, from www.accenture.com.

Anderson, E., & Weitz, B. (1992). The use of pledges to build and sustain commitment in distribution channels. *JMR, Journal of Marketing Research*, *29*(1), 18–34. doi:10.2307/3172490

Anderson, J. C., & Narus, J. A. (1990). A model of distributor firm and manufacturer firm working partnerships. *Journal of Marketing*, *54*(1), 42–58. doi:10.2307/1252172

Artz, K. W., & Brush, T. H. (2000). Asset specificity, uncertainty and relational norms: An examination of coordination costs in collaborative strategic alliances. *Journal of Economic Behavior & Organization*, *41*, 337–362. doi:10.1016/S0167-2681(99)00080-3

Bailey, J., & Clark, N. (2008). Supplier problems led to new delay of Boeing 787. Retrieved February 10, 2010, from www.nytimes.com.

Balakrishnan, K., Mohan, U., & Seshadri, S. (2008). Outsourcing of front-end business processes: Quality information and customer contact. *Journal of Operations Management*, *26*(2), 288–302. doi:10.1016/j.jom.2007.08.001

Barney, J. (1991). Firm resources and sustained competitive advantage. *Journal of Management*, *17*(1), 99–120. doi:10.1177/014920639101700108

Barratt, M., & Oke, A. (2007). Antecedents of supply chain visibility in retail supply chains: A resource-based theory perspective. *Journal of Operations Management*, *25*(6), 1217–1233. doi:10.1016/j.jom.2007.01.003

Becker, M. C., & Zirpoli, F. (2003). Organizing new product development: Knowledge hollowing-out and knowledge integration – the FIAT Auto case. *International Journal of Operations & Production Management*, *23*(9), 1033–1061. doi:10.1108/01443570310491765

Braunscheidel, M. J., & Suresh, N. C. (2009). The organizational antecedents of a firm's supply chain agility for risk mitigation and response. *Journal of Operations Management*, *27*(2), 119–140. doi:10.1016/j.jom.2008.09.006

Bunkley, N., & Maynard, M. (2010). With recall expanding, Toyota gives an apology. Retrieved February 10, 2010, from www.nytimes.com.

Burrows, P. (2005). *Commentary: Apple's Blueprint for Genius*. Retrieved on February 10, 2010, from www.businessweek.com.

Cai, S., & Yang, Z. (2008). Development of cooperative norms in the buyer-supplier relationship: The Chinese Experience. *Journal of Supply Chain Management*, *44*(1), 55–70. doi:10.1111/j.1745-493X.2008.00045.x

Cannon, J. P., Achrol, R. S., & Gundlach, G. T. (2000). Contracts, norms, and plural form governance. *Academy of Marketing Science*, *28*(2), 180–194. doi:10.1177/0092070300282001

Cannon, J. P., & Perreault, W. D. (1999). Buyer-seller relationships in business markets. *JMR, Journal of Marketing Research*, *36*(4), 439–460. doi:10.2307/3151999

Carlile, P. R. (2004). Transferring, translating, and transforming: An integrative framework for managing knowledge across boundaries. *Organization Science*, *15*(5), 555–568. doi:10.1287/orsc.1040.0094

Carlile, P. R., & Rebentisch, E. S. (2003). Into the black box: The knowledge transformation cycle. *Management Science*, *49*(9), 1180–1195. doi:10.1287/mnsc.49.9.1180.16564

Chang, B., & Phi, J. (2007). The IDS story: Reinventing distribution through supply chain logistics. In Lee, H. L., & Lee, C.-Y. (Eds.), *Building Supply Chain Excellence in Emerging Economies*. New York: Springer.

Cheung, M.-S., Myers, M. B., & Mentzer, J. T. (in press). Does relationship learning lead to relationship value? A cross-national supply chain investigation. *Journal of Operations Management*.

Conner, K. R., & Prahalad, C. K. (1996). A resource-based theory of the firm: knowledge versus opportunism. *Organization Science*, *7*(5), 477–501. doi:10.1287/orsc.7.5.477

Craighead, C. W., Hult, G. T. M., & Ketchen, D. J. (2009). The effects of innovation–cost strategy, knowledge, and action in the supply chain on firm performance. *Journal of Operations Management*, *27*(5), 405–421. doi:10.1016/j.jom.2009.01.002

Cudahy, G., Mulani, N., & Cases, C. (2008). Mastering global operations in a multipolar world. *Supply Chain Management Review*, *12*(2), 22–29.

Dean, J. (2007). The Forbidden City of Terry Gou. *Wall Street Journal, 11 Aug 2007, A1*.

Deloitte Touche Tohmatsu. (2008). *Innovation in emerging markets: Managing product sourcing risks in emerging markets.* Retrieved February 10, 2010, from www.deloitte.com.

Dierickx, I., & Cool, K. (1989). Asset stock accumulation and sustainability of competitive advantage. *Management Science*, *35*(12), 1504–1511. doi:10.1287/mnsc.35.12.1504

Dyer, J. H., & Hatch, N. W. (2006). Relation-specific capabilities and barriers to knowledge transfers: Creating advantage through network relationships. *Strategic Management Journal*, *27*(8), 701–719. doi:10.1002/smj.543

Dyer, J. H., & Nobeoka, K. (2000). Creating and managing a high-performance knowledge-sharing network: The Toyota case. *Strategic Management Journal*, *21*(3), 345–367. doi:10.1002/(SICI)1097-0266(200003)21:3<345::AID-SMJ96>3.0.CO;2-N

Ellram, L. (1995). Partnering pitfalls and success factors. *International Journal of Purchasing and Materials Management*, *31*(2), 36–44.

Ellram, L., & Carr, A. (1994). Strategic purchasing: A history and review of the literature. *International Journal of Purchasing and Materials Management*, *30*(2), 10–18.

Fih.foxconn.com (2007). *FIH: Vertical Integration.* Retreived February 10, 2010, from fih.foxconn.com.

Fine, C. (1998). *Clockspeed: Winning Industry Control in the Age of Temporary Advantage.* Cambridge: Perseus Publishing.

Fugate, B. S., Stank, T. P., & Mentzer, J. T. (2008). Linking improved knowledge management to operational and organizational performance. *Journal of Operations Management*, *27*(3), 247–264. doi:10.1016/j.jom.2008.09.003

Ganesan, S. (1994). Determinants of long-term orientation in buyer–supplier relationships. *Journal of Marketing*, *58*(2), 1–19. doi:10.2307/1252265

Germain, R., Droge, C., & Christensen, W. (2001). The mediating role of operations knowledge in the relationship of context with performance. *Journal of Operations Management*, *19*, 453–469. doi:10.1016/S0272-6963(00)00067-X

Grant, R. M. (1996a). Prospering in dynamically-competitive environments: Organizational capability as knowledge integration. *Organization Science*, *7*(4), 375–387. doi:10.1287/orsc.7.4.375

Grant, R. M. (1996b). Towards a knowledge-based theory of the firm. *Strategic Management Journal,* 17(Winter Special Issue), 109-122.

Grant, R. M., & Baden-Fuller, C. (2004). A knowledge accessing theory of strategic alliances. *Journal of Management Studies*, *41*(1), 61–84. doi:10.1111/j.1467-6486.2004.00421.x

Grover, V., & Malhotra, M. K. (2003). Transaction cost framework in operations and supply chain management research: theory and measurement. *Journal of Operations Management*, *21*, 457–473. doi:10.1016/S0272-6963(03)00040-8

Gulati, R., & Sytch, M. (2007). Dependence asymmetry and joint dependence in interorganizational relationships: Effects of embeddedness on a manufacturer's performance in procurement relationships. *Administrative Science Quarterly*, *52*, 32–69.

Handfield, R. B., & Bechtel, C. (2002). The role of trust and relationship structure in improving supply chain responsiveness. *Industrial Marketing Management*, *31*, 367–382. doi:10.1016/S0019-8501(01)00169-9

Heide, J. B., & John, G. (1992). Do norms matter in marketing relationships? *Journal of Marketing*, *56*(2), 32–44. doi:10.2307/1252040

Heide, J. B., & Miner, A. S. (1992). The shadow of the future: Effects of anticipated interaction and frequency of contact on buyer-seller cooperation. *Academy of Management Journal*, *35*, 265–291. doi:10.2307/256374

Helfat, C. E., & Raubitschek, R. S. (2000). Product sequencing: Co-evolution of knowledge, capabilities and products. *Strategic Management Journal*, *21*(10-11), 961–979. doi:10.1002/1097-0266(200010/11)21:10/11<961::AID-SMJ132>3.0.CO;2-E

Henderson, R. M., & Clark, K. B. (1990). Architecturai innovation: The reconfiguration of existing product technologies and the failure of established firms. *Administrative Science Quarterly*, *35*, 9–30. doi:10.2307/2393549

Henderson, R. M., & Cockburn, I. (1994). Measuring competence? Exploring firm effects in pharmaceutical research. *Strategic Management Journal*, *15*(Winter), 63–84. doi:10.1002/smj.4250150906

Hill, C. W. L. (1990). Cooperation, opportunism, and the invisible hand: Implications for transaction cost theory. *Academy of Management Review*, *15*(3), 500–513. doi:10.2307/258020

Hoetker, G. (2005). How much you know versus how well I know you: Selecting a supplier for a technically innovative component. *Strategic Management Journal*, *26*(1), 75–96. doi:10.1002/smj.453

Holcomb, T. R., & Hitt, M. A. (2007). Toward a model of strategic outsourcing. *Journal of Operations Management*, *25*(2), 464–481. doi:10.1016/j.jom.2006.05.003

Huber, G. P. (1991). Organizational learning: The contributing processes and the literatures. *Organization Science*, *2*(1), 88–115. doi:10.1287/orsc.2.1.88

Hughes, N. (2009). *Foxconn reportedly tapped to manufacture next-gen iPhone*. Retrieved February 10, 2010, from www.appleinsider.com.

Hult, G. T. M., & Ketchen, J. D. J., S. T. & Arfelt, M. (2007). Strategic supply chain management: Improving performance through a culture of competitiveness and knowledge development. *Strategic Management Journal*, *28*, 1035–1052. doi:10.1002/smj.627

Hult, G. T. M., Ketchen, J. D. J., Cavusgil, S. T., & Calantone, R. J. (2006). Knowledge as a strategic resource in supply chains. *Journal of Operations Management*, *24*(5), 458–475. doi:10.1016/j.jom.2005.11.009

Idsgroup.com (2009). *Profile*. Retreived February 10, 2010, from idsgroup.com.

Jap, S. D. (1999). Pie-expansion efforts: Collaboration processes in buyer-supplier relationships. *JMR, Journal of Marketing Research*, *36*(November), 461–475. doi:10.2307/3152000

Jap, S. D., & Anderson, E. (2003). Safeguarding interorganizational performance and continuity under ex post opportunism. *Management Science*, *49*(12), 1684–1701. doi:10.1287/mnsc.49.12.1684.25112

Kogut, B., & Zander, U. (1992). Knowledge of the firm, combinative capabilities, and the replication of technology. *Organization Science*, *3*(3), 383–397. doi:10.1287/orsc.3.3.383

Kogut, B., & Zander, U. (1996). What firms do? Coordination, identity, and learning. *Organization Science*, *7*(5), 502–518. doi:10.1287/orsc.7.5.502

Koudal, P., & Engel, D. A. (2007). Globalization and emerging markets: The challenge of continuous global network optimization. In Lee, H. L., & Lee, C.-Y. (Eds.), *Building Supply Chain Excellence in Emerging Economies*. New York: Springer.

Kouvelis, P., & Niederhoff, J. (2007). On the globalization of operations and supply chain strategies: A conceptual framework and its application. In Lee, H. L., & Lee, C.-Y. (Eds.), *Building Supply Chain Excellence in Emerging Economies*. New York: Springer.

Krause, D. R., Handfield, R. B., & Tyler, B. B. (2007). The relationships between supplier development, commitment, social capital accumulation and performance improvement. *Journal of Operations Management, 25*(2), 528–545. doi:10.1016/j.jom.2006.05.007

Larson, A. (1992). Network dyads in entrepreneurial settings: A study of the governance of exchange relationships. *Administrative Science Quarterly, 37*, 76–104. doi:10.2307/2393534

Lee, H. L. (2004). The triple-A supply chain. *Harvard Business Review, 82*(10), 102–112.

Lee, H. L., & Lee, C.-Y. (Eds.). (2007). *Building Supply Chain Excellence in Emerging Economies*. New York: Springer.

Li, H., & Atuahene-Gima, K. (2001). Product innovation strategy and the performance of new technology ventures in China. *Academy of Management Journal, 44*(6), 1123–1134. doi:10.2307/3069392

Li, Y., Liu, Y., Li, M., & Wu, H. (2008). Transformational offshore outsourcing: Empirical evidence from alliances in China. *Journal of Operations Management, 26*, 257–274. doi:10.1016/j.jom.2007.02.011

Liebeskind, J. P. (1996). Knowledge, strategy, and the theory of the firm. *Strategic Management Journal,* 17(Winter Special Issue), 93-107.

Liker, J. K., & Choi, T. Y. (2004). Building deep supplier relationships. *Harvard Business Review, 82*(12), 104–113.

Linderman, K. W., Schroeder, R. G., Zaheer, S., Liedtke, C., & Choo, A. S. (2004). Integrating quality management practices with knowledge creation processes. *Journal of Operations Management, 22*, 589–607. doi:10.1016/j.jom.2004.07.001

Lippman, S. A., & Rumelt, R. P. (1982). Uncertain imitability: An analysis of interfirm differences in efficiency under competition. *The Bell Journal of Economics, 13*(2), 418–438. doi:10.2307/3003464

Liu, Y., Luo, Y., & Liu, T. (2008). Governing buyer–supplier relationships through transactional and relational mechanisms: Evidence from China. *Journal of Operations Management, 27*(4), 294–309. doi:10.1016/j.jom.2008.09.004

Lusch, R. F., & Brown, J. R. (1996). Interdependency, contracting, and relational behaviour in marketing channels. *Journal of Marketing, 60*, 19–38. doi:10.2307/1251899

Madhok, A., & Tallman, S. B. (1998). Resources, transactions and rents: Managing value through interfirm collaborative relationships. *Organization Science, 9*(3). doi:10.1287/orsc.9.3.326

McEvily, S. K., Das, S., & McCabe, K. (2000). Avoiding competence substitution through knowledge sharing. *Academy of Management Review, 25*(2), 294–311. doi:10.2307/259015

Moberg, C. R., Cutler, B. D., Gross, A., & Speh, T. W. (2002). Identifying antecedents of information exchange within supply chains. *International Journal of Physical Distribution and Logistics Management, 32*(9/10), 755–770.

Modi, S. B., & Mabert, V. A. (2007). Supplier development: Improving supplier performance through knowledge transfer. *Journal of Operations Management, 25*(1), 42–64. doi:10.1016/j.jom.2006.02.001

Molina, L. M., Llorens-Montes, J., & Ruiz-Moreno, A. (2007). Relationship between quality management practices and knowledge transfer. *Journal of Operations Management*, *25*, 682–701. doi:10.1016/j.jom.2006.04.007

Morgan, R. M., & Hunt, S. D. (1994). The commitment-trust theory of relationship marketing. *Journal of Marketing*, *58*, 20–38. doi:10.2307/1252308

Motavalli, J. (2010). *Accelerator Pedal Supplier in Toyota's Recall Has Many Customers*. Retreived on February 10, 2010, from www.nytimes.com.

Nahapiet, J., & Ghoshal, S. (1998). Social capital, intellectual capital, and the organizational advantage. *Academy of Management Review*, *23*(2), 242–266. doi:10.2307/259373

Nelson, R. R., & Winter, S. G. (1982). *An Evolutionary Theory of Economic Change*. Cambridge, MA: Harvard University Press.

Nonaka, I. (1994). A dynamic theory of organizational knowledge creation. *Organization Science*, *5*(1), 14–37. doi:10.1287/orsc.5.1.14

Noordewier, T. G., John, G., & Nevin, J. R. (1990). Performance outcomes of purchasing arrangements in industrial buyer-vendor relationships. *Journal of Marketing*, *54*, 80–93. doi:10.2307/1251761

Paiva, E. L., Roth, A. V., & Fensterseifer, J. E. (2008). Organizational knowledge and the manufacturing strategy process: A resource-based view analysis. *Journal of Operations Management*, *26*(1), 115–132. doi:10.1016/j.jom.2007.05.003

Pandey, A. C. (2010). *Y2010 & Ahead – value chain trends in emerging economy.* Retrieved February 15, 2010, from www.infosysblogs.com.

Paulraj, A., Lado, A. A., & Chen, I. J. (2008). Inter-organizational communication as a relational competency: Antecedents and performance outcomes in collaborative buyer-supplier relationships. *Journal of Operations Management*, *26*(1), 45–64. doi:10.1016/j.jom.2007.04.001

Peleg-Gillai, B. (2007). Esquel Group: Going beyond the traditional approach in the apparel industry. In Lee, H. L., & Lee, C.-Y. (Eds.), *Building Supply Chain Excellence in Emerging Economies*. New York: Springer.

Penrose, E. (1959). *The theory of the growth of the firm*. New York: Wiley.

Petersen, K. J., Handfield, R. B., & Ragatz, G. L. (2005). Supplier integration into new product development: coordinating product, process and supply chain design. *Journal of Operations Management*, *23*(3-4), 371–388. doi:10.1016/j.jom.2004.07.009

Porter, M. E. (1985). *Competitive advantage: Creating and sustaining superior performance*. New York: Free Press.

Porter, M. E. (1996). What is strategy? *Harvard Business Review*, *74*(6), 61–78.

Prahalad, C., & Hamel, G. (1990). The core competencies of the corporation. *Harvard Business Review*, *68*(3), 79–91.

Richardson, J. (1993). Parallel sourcing and supplier performance in the Japanese automobile industry. *Strategic Management Journal*, *14*(5), 339–350. doi:10.1002/smj.4250140503

Ring, P. S., & van de Ven, A. H. (1994). Developmental processes of cooperative interorganizational relationships. *Academy of Management Review*, *19*(1), 90–118. doi:10.2307/258836

Sanders, N. R. (2008). Pattern of information technology use: The impact on buyer–suppler coordination and performance. *Journal of Operations Management*, *26*, 349–367. doi:10.1016/j.jom.2007.07.003

Selnes, F., & Sallis, J. (2003). Promoting relationship learning. *Journal of Marketing*, *67*(July), 80–95. doi:10.1509/jmkg.67.3.80.18656

Siemsen, E., Roth, A. V., & Balasubramanian, S. (2008). How motivation, opportunity, and ability drive knowledge sharing: The constraining-factor model. *Journal of Operations Management*, *26*, 426–445. doi:10.1016/j.jom.2007.09.001

Simon, H. A. (1991). Bounded rationality and organizational learning. *Organization Science*, *2*(1), 125–134. doi:10.1287/orsc.2.1.125

Spender, J.-C. (1996). Making knowledge the basis of a dynamic theory of the firm. *Strategic Management Journal,* 17(Winter Special Issue), 45-62.

Steinfeld, E. S. (2004). China's shallow integration: Networked production and the new challenges for late industrialization. *World Development*, *32*(11), 1971–1987. doi:10.1016/j.worlddev.2004.04.003

Teece, D. J., Pisano, G., & Shuen, A. (1997). Dynamic capabilities and strategic management. *Strategic Management Journal*, *18*(7), 509–533. doi:10.1002/(SICI)1097-0266(199708)18:7<509::AID-SMJ882>3.0.CO;2-Z

The_Supply_Chain_Council. (2008). Retrieved 11/1/2008, from http://www.supply-chain.org

Thompson, J. D. (1967). *Organizations in Action*. New York: McGraw-Hill.

Ward, P. T., McCreery, J. K., Ritzman, L. P., & Sharma, D. (1998). Competitive priorities in operations management. *Decision Sciences*, *29*(4), 1035–1046. doi:10.1111/j.1540-5915.1998.tb00886.x

Wernerfelt, B. (1984). A resource based view of the firm. *Strategic Management Journal*, *5*(5), 171–180. doi:10.1002/smj.4250050207

Williamson, O. E. (1981). The economics of organization: The transaction cost approach. *American Journal of Sociology*, *87*(3), 548–577. doi:10.1086/227496

Williamson, O. E. (2008). Outsourcing: Transaction cost economics and supply chain management. *Journal of Supply Chain Management*, *44*(2), 5–16. doi:10.1111/j.1745-493X.2008.00051.x

Young, G., Sapienza, H., & Baumer, D. (2003). The influence of flexibility in buyer–seller relationships on the productivity of knowledge. *Journal of Business Research*, *56*, 443–451. doi:10.1016/S0148-2963(01)00243-0

Zand, D. E. (1972). Trust and managerial problem solving. *Administrative Science Quarterly*, *17*(2), 229–239. doi:10.2307/2393957

KEY TERMS AND DEFINITIONS

Supply Chain: A network of firms that encompasses every effort involved in producing and delivering a final product from the supplier's supplier to the customer's customer

Value Chain: A network of firms and activities concerned with the marketing, design, production, delivery, and support of a product or service

Knowledge: Broadly defined as any credible information, held by individuals, that can be applied to productive activities

Knowledge Based View: A theory that argues that most important resource in firms is the knowledge of the individuals in the firm and that most differences in firms can be represented as differences in knowledge

Knowledge Integration: The access and utilization of the knowledge of individuals in a productive activity

Long Term Commitment: The desire to develop a stable relationship, a willingness to make short-term sacrifices to maintain the relationship, and a confidence in the stability of the relationship

Vertical Integration: The degree to which a firm owns its upstream suppliers and its downstream buyers

Chapter 7
A Social and Technical Investigation of Knowledge Utilization from a Repository Knowledge Management System

Kamla Ali Al-Busaidi
Sultan Qaboos University, Oman

EXECUTIVE SUMMARY

The need for developing countries to empower themselves through knowledge management cannot be underestimated. Knowledge utilization is the process where the benefits of knowledge management systems (KMS) are actually realized. This study investigated the effects of knowledge quality, system level, system form, service quality, management support, knowledge trust and rewards policy on knowledge utilization behavior, and the benefits of knowledge utilization from a repository KMS in an Omani organization. Findings suggested that determinants of knowledge utilization were not related to system technical characteristics or information technology service quality; they were related to knowledge quality, management support, knowledge trust, and rewards policy. Results also suggested that knowledge utilization results in individual benefits. Findings implied to practitioners that the development of high-level KMS does not guarantee their success. Organizations need to establish a knowledge-oriented culture and develop standards that ensure high knowledge quality to promote KMS and knowledge utilization behavior.

INTRODUCTION

The need for developing countries to empower themselves through knowledge management (KM) cannot be underestimated. Knowledge plays a major role in individuals' learning and decision making capabilities. It improves an individual's ability to take an effective action (Huber, 1991; Nonaka, 1994). Thus, KM has become one of the main imperatives of knowledge-based economy. KM is driven by increasing knowledge domain complexity, accelerating marketing volatility, intensifying speed of responsiveness and diminishing individual experience (Becerra-Fernandez et al., 2004).

DOI: 10.4018/978-1-61692-886-5.ch007

Knowledge management systems (KMS) are IT-based systems that support and boost the organizational processes of knowledge creation, storage/retrieval, transfer, and application (Alavi & Leidner 2001). KMS also are defined as "the applications resulting from the synergy between the latest technologies and social/structural mechanisms" (Becerra-Fernandez et al., 2004, p.7). They offer organizations an efficient and effective way to create knowledge repositories that can be utilized to improve employees' performance and productivity. Two common KMS models are repository KMS and network KMS (Alavi 2000; Davenport & Prusak, 1998). Empirical studies have indicated that the use of KMS results in several benefits for organizations such as improved operating performance (Jennex & Olfman, 2006; Liu & Tsai, 2007), learning and innovative performance (Chang & Lee, 2007; Jiang & Lia, 2008). Moreover, KMS improve developing nations' and their organizations' efforts to manage their knowledge, and consequently build their human resources (World Bank, 1998; World Bank, 2003).

However, KM and KMS benefits can be actually achieved through knowledge utilization (Alavi, 2000; Scholl et al., 2004).Hence, knowledge utilization is critical to the success of repository KMS. Knowledge utilization is the application of the stored knowledge to solve daily work problems and make decisions. Knowledge utilization, however, is inhibited by several corporate cultural issues such as "Not-Invented here (NIH) syndrome" (Katz & Allen, 1982), lack of time, and risk aversion (Davenport & Prusak, 1998). Thus, knowledge utilization from repository KMS is a social and technical process. An effective knowledge utilization process will be achieved only by an effective integration of technical and social factors. The need for this investigation has been stressed in both management and knowledge management systems literatures (O'Dell & Grayson, 1998; Scholl et al., 2004).

This study aimed to provide a better understanding of the factors that improve the individuals' knowledge utilization in a petroleum company in Oman. In Oman, KMS is emerging; several organizations (e.g. Sultan Qaboos University, Ministry of Education, Petroleum Development of Oman etc) have recognized the importance of KMS and created KMS tools such as portals to integrate knowledge and information from internal and external sources. The development of knowledge repositories is becoming imperative for petroleum companies. In the oil industry, market volatility is accelerating, speed of responsiveness is intensifying and individual experience is diminishing. For instance, in the oil refining industry, there is a shortage of qualified and knowledgeable personnel who are ranging from 30 to 50 years old, and 60 percent of plant engineers will retire by 2010 according to the journal of petroleum technology (Clark, 2005). "Oil exploration can cost millions of dollars a day, and anything that makes it more efficient has magnified value" (Steinberg, 2002, p.45). The use of KMS may result in some efficiency and financial gains for petroleum organizations. For instance, Shell Company saved $200 million in 2000 by implementing discussion groups and knowledge repositories where more than 10,000 engineers can share ideas and solve problems (King, 2001; Steinberg, 2002). Also, British Petroleum (BP) Company is implementing computer-supported visualization environments to share know-how knowledge to promote teamwork, innovation and creativity; BP is expected to improve its productivity by four percent per year from this KMS implementation (Barrow, 2001). Thus, individual knowledge utilization through KMS leads to financial benefits. Investigating the factors that influence the knowledge utilization behavior is, therefore, essential to promote these emerging KMS and ensure their successful deployment in Omani organizations.

However, there are very limited studies that investigate KMS deployment in developing countries specifically in Oman. Information technology(IT) managers in Oman perceived knowledge culture, organizational infrastructure, technical infrastructure, management support, vision clarity and economic return as critical factors for the deployment of KMS in Omani organizations according to a recent study (Al-Busaidi & Olfman, 2005). However, IT managers, in this study, did not perceive rewards policy as a valuable factor for KMS success. Also, a couple of studies in developing countries such as Kuwait (Al-Athari & Zairi, 2001) and Malaysia (Chong, 2006; Syed-Ikhsan, & Rowl, 2004) illustrated that several social and technical factors lead to a successful KMS deployment. Nevertheless, the determinants of an information system usage are best investigated at individual users' level (DeLone & McLean, 2003). However, there are limited studies that are focused on KMS users (Kankanhalli & Tan, 2004) and clear measurements of KMS users' satisfaction are still not well established (Ong & Lai, 2007). In a qualitative exploratory study, Al-Busaidi et al. (2007) unveiled some determinants of knowledge utilization and knowledge sharing behaviors.

Consequently, this study empirically examined the social and technical factors that affect the knowledge utilization from repository KMS. It specifically investigated the effects of knowledge quality, system quality (form and level), service quality, management support, knowledge trust, and rewards policy on knowledge utilization behavior. It also examined the benefits that individuals gain from utilizing knowledge from a repository KMS.

The next section discusses the background literature of KMS, knowledge utilization, repository KMS, KMS technical success factors and KMS social success factors. The literature section is followed by the study framework and hypotheses, methodology, analysis and conclusion sections respectively.

BACKGROUND LITERATURE

Knowledge Utilization

Knowledge management systems (KMS) are IT-based systems that support and boost the organizational processes of knowledge creation, storage/retrieval, transfer, and application (Alavi & Leidner 2001). Simply KMS are systems that manage knowledge throughout the organization. Organizational KMS store structured or unstructured explicit knowledge from internal or external sources (Davenport & Prusak, 1998; Turban et al., 2001).

Knowledge utilization is one of the main knowledge management processes (Alavi, 2000; Davenport & Prusak, 1998). Knowledge management includes the generation of knowledge from databases, individuals or group of individuals. Once the knowledge is generated, it is codified and stored in a storage area to establish an organizational memory or a knowledge repository that can be accessed for later use and distribution. Knowledge utilization is the application of the stored knowledge to solve daily work problems and make decisions. Some studies, such as Alavi and Leidner (2001), Gold et al. (2001) and Becerra-Fernandez et al. (2004), named knowledge utilization as knowledge application.

The utilization and application of knowledge creates value for individuals and organizations. Empirical studies have confirmed this assertion. Liu (2003) empirically found that knowledge application, as a dimension of KMS use, improves individual learning. Jennex and Olfman (2006) found that the utilization of KMS results in improved individual productivity in terms of decision making, root cause analysis, problem resolution, timeliness, and operability assessment documentation; this improved individual productivity further positively impacts organizational productivity. Al-Busaidi et al.'s (2007) exploratory study revealed that some of the individual benefits of knowledge utilization are enhanced knowledge, improved performance, improved decision making.

Repository KMS

Repository KMS is one of two common KMS models (Alavi 2000; Davenport & Prusak, 1998). The objective of the repository model is to codify the organization's explicit knowledge. The repository model is the widespread form of KMS initiatives in organizations. The mostly used technologies for the development of repository KMS are relational databases and document management systems. One main advantage of repository KMS is that it enhances organizational memory (OM), which is general, explicit and articulated knowledge of the organization. Accordingly, repository KMS assists in efficiently storing and reapplying workable solutions.

Repository KMS codifies knowledge from internal sources and external sources. Internal knowledge may include reports, document templates, memos, methodologies, manuals, product and marketing knowledge, and internal best practices; whereas external knowledge includes competitive intelligence, industry trends, customer knowledge, and professional and trade journal publications (Alavi 2000). The two major processes for the development of an effective knowledge repository are content structure and content quality structure (Debowski, 2006). Content structure includes knowledge map, source description, source identification and user interface analysis; while content quality structure includes content evaluation, content maintenance and content archives.

The network model of KMS, compared to the repository model, does not codify knowledge, but instead transfer knowledge among individuals in organizations mainly through person-to-person contacts. It also refers to a system that stores sources of knowledge (experts' profiles) such as "yellow pages", or/and a system that provides communication channels for knowledge exchange such as "video conferencing" (Alavi 2000; Davenport & Prusak, 1998).

Technical Success Factors

Technical factors, here, refer to the characteristics of the information system and the IT staff service to the end users. Generally, the deployment of a successful KMS requires several technical factors. Practically, the knowledge utilization process involves several technical tasks; individuals perform knowledge search and retrieval tasks to utilize knowledge from the repository KMS. Several theoretical frameworks have been proposed in the literature for the success factors of an information system (IS). Based on DeLone and McLean 2003 IS success model, the system use is determined by system quality, information quality and service quality.

Applying DeLone and McLean's Model, Jennex and Olfman (2002) proposed that system quality for KMS can be measured by "level" and "form", while information (knowledge) quality is measured by "richness" and "linkage". System level includes ease of search, speed of retrieval, and completeness of the search function. System form refers to the degree to which information/knowledge is online and accessible through a single interface. System form defines the level of computerization of the knowledge base and integration of the organization's knowledge sources. Linkage is related to the completeness, accuracy, and currency of linkage to experts, while richness is related to the completeness, accuracy, and timeliness of knowledge.

Social Success Factors

Social factors refer to the dimensions of corporate culture. Corporate culture plays a key role in the success of KMS. Schein defined culture as "the way we do things around here" (1992, p. 12). Culture values shape an organization's norms and practices, which consequently influence employees' behaviors such as knowledge utilization (De Long & Fahey, 2000).

Several dimensions of knowledge culture have been highlighted by various theoretical and qualitative studies (Davenport & Prusak, 1998; De Long & Fahey, 2000; Krogh, 1998; O'Dell & Grayson, 1998). Some of these dimensions are management support, rewards, and trust. Few KMS studies have included a cultural construct in their models. This study aimed to provide a better understanding of the dimensions of knowledge culture that improve the effectiveness of individuals' knowledge utilization in a repository KMS. It specifically investigated the effects of management support, rewards policy, and trust. More discussion on these social factors is provided in the framework section.

FRAMEWORK AND HYPOTHESES

Framework Development

This study adopted the DeLone and McLean 2003 IS success model to investigate the determinants of the knowledge utilization from a repository KMS. For information quality and system quality factors, this study adopted Jennex and Olfman's proposed factors for the KMS context. Thus, this study investigated the information richness, system level and system form factors. For clarity, in this study, "richness" was called "knowledge quality". Jennex and Olfman's "linkage" construct was not included here because the investigated system does include the experts' profiles on its knowledge base.

However, the knowledge utilization is also a social process, thus incorporating social factors on this investigation is very essential to get a better understanding of this phenomenon. The most cited social factors on the KM literature are management support, trust and rewards policy. As a behavior, knowledge utilization is mostly inhibited by the "Not-Invented here (NIH) syndrome" (Husted, & Michailova, 2002). In most situations, individuals are not willing to utilize knowledge from repositories because it is not their own knowledge. Some users do not trust others' knowledge and others may think it is more impressive to create their own knowledge than using others'. Other reasons that inhibit knowledge utilization are lack of time and risk aversion (punishment for making a mistake) (Davenport & Prusak, 1998). Consequently, several social factors are needed to eliminate these inhibitors and hence motivate knowledge utilization. Management support is very important to clarify and acknowledge the importance of KMS and knowledge utilization to the organization's success. Management support is also important to provide individual users time to utilize knowledge. Trust is a critical factor for knowledge utilization. Knowledge-trust improves the reliability of the knowledge or/and its source and reduces the individual's fear of utilizing knowledge. Rewards policy is another important factor that motivates KMS users to utilize knowledge. Figure 1 shows the study framework.

Knowledge Quality

Knowledge quality includes completeness, accuracy, sufficiency, currency, timeliness, relevance, understandability and contextuality of the knowledge. Completeness, accuracy and timeliness are important information measurements as indicated by many IS researchers. KMS researchers also proposed the importance of these measurements in the KMS context (Jennex & Olfman, 2002; Maier, 2002). Understandability is also important; having a thesaurus in any knowledge repository is important to clarify the meaning of stored knowledge, and improves its understandability by knowledge utilizers (Davenport & Prusak, 1998). Also, the application of specific knowledge to a new situation depends on the relevance and contextuality of the applied knowledge to the new situation. The context is what differentiates knowledge from information. These different

Figure 1. The study's framework for knowledge utilization in repository KMS

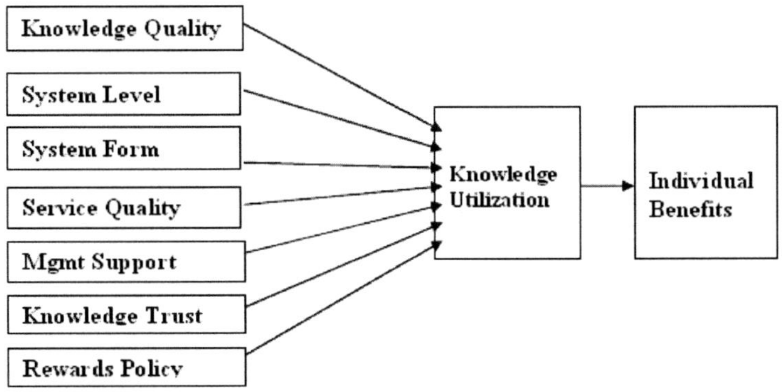

attributes of knowledge contribute to its richness. Thus, rich knowledge quality is essential to knowledge utilization. From these arguments, the following hypothesis was derived:

Hypothesis 1. Higher knowledge quality improves knowledge utilization from a repository KMS.

System Quality

System quality includes two constructs: system level and system form. The KMS search/retrieval system level is very important for KMS utilization. The KMS should have effective search/retrieval capabilities that enable users to find needed documents (Davenport & Prusak, 1998; Lehaney et al., 2004). Thus, an effective and efficient KMS search-level enables faster and easier knowledge extraction. Having all the needed documents accessible in a single integrated interface is also important for knowledge utilization (Jennex & Olfman, 2002; Lehaney et al., 2004). An integrated KMS enables more effective and efficient knowledge extraction. Thus, we hypothesized the following:

Hypothesis 2. Higher system level improves knowledge utilization from a repository KMS.

Hypothesis 3. Higher system form improves knowledge utilization from a repository KMS.

Service Quality

Service quality refers to the quality of IS staff support to the system's end-users. It is measured by the following indicators: reliability, responsiveness, assurance, and empathy (based on Kettinger and Lee (1994)), and training. Users have similar criteria for evaluating service quality regardless of system type (Parasuraman et al., 1985). DeLone and McLean (2003) indicated that ignoring service quality endangers IS effectiveness measurements. Service quality is also important for evaluating KMS success (Maier, 2002); service quality in terms of reliability, responsiveness, assurance, and empathy is essential to motivate KMS users' participation. Also, training is needed to improve the success of an information system (Turban et al., 2001). Thus, we hypothesized the following:

Hypothesis 4. Higher service quality improves knowledge utilization from a repository KMS.

Management Support

Management support involves clarifying the objective and the importance of a KMS, and encouraging end-users (Davenport & Prusak, 1998). Management's open approval and acknowledgement of knowledge exchange reduces individuals' fear about using others' knowledge. It assures employees that using others' knowledge is part of the organization's culture, and encourages employees to feel okay about using others' knowledge. Management support is extremely critical to endorse the KMS and consequently change employees' attitudes. In the Arab culture, managers are recognized as high authority (Ali, 1990), and their support for KMS projects, which are emerging systems, certainly enhances employees' confidence to utilize the stored knowledge for problem solving and decisions making. The World Bank report (1998) indicated the importance to have a plan for KMS. Igbaria (1990) found that management support of end-users improves computer usage. Therefore, the following hypothesis was proposed:

Hypothesis 5. Higher management support improves knowledge utilization from a repository KMS.

Knowledge Trust

Trust is highlighted as a crucial factor for knowledge exchange and collaboration (Davenport & Prusak, 1998; O'Dell & Grayson, 1998). Typically, trust is defined as a set of mutual expectations shared by people involved in collaboration and exchange (Zucker, 1986). Mutual trust in knowledge exchange brings people closer and reduces their individual fears. In knowledge exchange, trust can be interpreted in different ways. From the knowledge utilizer's perspective, trust may be also interpreted in several ways: as source trust (Sussman & Siegal, 2003), or knowledge trust or credibility (Dennis, 1996)). In this paper, trust is referred to as "knowledge trust" (credibility), adopted based on Dennis's (1996) information credibility: the degree of truthfulness participants attributed to information contributed by others. Trust is crucial for knowledge utilization (Davenport & Prusak, 1998; Husted & Michailova, 2002). Knowledge-trust reduces knowledge utilizers' fears of using others' knowledge. Consequently, the following hypothesis is proposed:

Hypothesis 6. Higher knowledge trust improves knowledge utilization from a repository KMS.

Rewards Policy

Rewards refer to "non trivial" monetary and non-monetary incentives. Rewards policy encourages employees to spend time and make the effort to exchange knowledge, that is, to use and share knowledge (Davenport & Prusak, 1998). In this study, a rewards policy is specifically investigated form the perspective of knowledge utilization. The rewards-policy is a critical success factor of KMS deployment because unlike other IS projects, the breadth and depth of a KMS project is based on the participation of the employees to create and use the knowledge that is stored in these systems. Without good incentives employees will be reluctant to exchange knowledge (O'Dell & Grayson, 1998). Therefore:

Hypothesis 7. More effective rewards policies improve knowledge utilization from a repository KMS.

Individual Benefits

Individuals may gain several benefits from utilizing knowledge from repository KMS to solve business problems and make decisions. The main benefit of knowledge utilization for individuals is individual productivity which is indicated by improvement on individuals' decision making and innovation capabilities (Davenport & Prusak, 1998; Jennex & Olfman, 2006; Liu, 2003). More specifically, productivity improvement means that individuals will improve their judgments and skills, which will help them, make better decisions and accomplish their work more efficiently. Consequently:

Hypothesis 8. Higher knowledge utilization of a repository KMS results in higher individual benefits.

METHODOLOGY

Sample

The study sample includes 104 employees in a major private petroleum company in Oman. The company accounts for about 90 percent of the country's crude-oil production and nearly all of its natural-gas supply. Oil is the major industry in Oman. Based on recent statistics, most of the employees (3784 staff) of the company are local, which represent 82 percent of the total employees in the company. The participants of this survey are KMS users of a specific organizational knowledge management system in this company.

The organization developed this KMS because of business, technological and cultural factors. The objective of the organization was to enhance the transparency and the accessibility of the organization's information and knowledge throughout the organization, so employees are able to access it from anywhere. The system is a mean to transfer information/knowledge within one department or across departments. For example, petroleum engineers across several oil fields can use the system to share or retrieve common problems' solutions. Also information/knowledge can be shared across several departments such as between personnel and finance departments or drilling department and geophysicists or petroleum engineers.

According to the IT department, this investigated system was a web-centric application, with strong integration with the MS-Office suite and mail. It enabled employees to store search and retrieve organizational documents, information and knowledge. The system was a purchased software package from an international organization. Any employees in the organization can voluntarily access the system from the organization's web home page.

Research Design

Data was collected through a survey questionnaire from employees in a petroleum company in Oman. The employees are users of an organizational KMS; the applicable sample was randomly selected from the organization's email list. The study sample was initially invited through an email by an official employee in the human resources department at the participating organization. The study was conducted in English (the typical medium of business activities in Oman).

Questionnaire

The questionnaire contained the constructs to be measured for quantitative analysis, along with 10 demographic questions (e.g., gender, age, degree, KMS experience, work experience, and job function). Construct measurements items were phrased according to a 7–point Likert scale. For the study's independent constructs, the scale was defined as follows: 1= strongly disagree, 2= disagree, 3= somewhat disagree, 4= neither agree nor disagree, 5= somewhat agree, 6= agree, 7= strongly agree. For the dependent constructs, the scale is defined as follows: 1= Never, 2= Very infrequently, 3= infrequently, 4= Sometimes, 5= frequently, 6= Very frequently, 7= Always. A "Not applicable" option was also given for all constructs to ensure that individuals' ratings are valid responses.

To test this study's theoretical model, the questionnaire had 42 indicators that formed the independent and dependent constructs. Some of the measurements were modified from previous studies. Knowledge quality, system level, system form, knowledge utilization and individual benefits were modified from Liu (2003), and Jennex and Olfman (2002). Service quality was modified from Kettinger and Lee (1994). Knowledge trust was modified from Dennis (1996). The new self-constructed measurements (management support and rewards-policy) were developed based on the relevant literature and following the construction recommendations of Moore and Benbasat (1991). Appendix A illustrates the study measurements.

DATA ANALYSIS & RESULTS

Analysis Methodology

Data was analyzed by the PLS-Graph 3.0 software. PLS is a variance-based structural equation model that allows path analysis of models with latent variables.

Participants' Profile

More than 80 percent of the participants were male, and all but three were at least 26 years old. About 86 percent of participants had at least two years of KMS-use experience. Omanis represented 73 percent of the participants. About 56 percent of the participants were group leaders, project managers or department heads. About half of the participants were engineers. Nineteen percent were analysts, and 13 percent were consultants. The majority of the participants had at least a Bachelors degree: four percent of participants had PhD, 25 percent had Master degree, 14 percent had postgraduate diploma, 51 percent had Bachelors degree, and 10 percent had diploma.

Reliability and Validity

According to PLS, the reliability of the measurements is assessed through internal consistency reliability, and the validity is assessed by the average variance extracted (AVE). AVE refers to the amount of variance a latent variable, captures from its indicators. The recommended level for internal consistency reliability is at least 0.70, whereas for AVE, it is at least 0.50 (Chin, 1998). Table 1 shows that the study constructs' reliability and AVE are above the recommended levels.

Model Evaluation and Hypotheses Testing

PLS evaluates the predictive relevance of a structural model by the R-square values for the dependent latent variables and paths coefficients. T-tests were used to test the model hypotheses. To test the significance of the PLS estimates of path coefficients, the bootstrapping technique was utilized with a re-sampling of 200, which was suggested by PLS-Graph user's guide.

Table 2 shows that knowledge quality ($\beta=0.392$; p-value = 0.005), management support ($\beta=0.187$; $p = 0.05$), knowledge trust ($\beta=0.219$; $p = 0.05$) and rewards policy ($\beta=0.172$; $p = 0.1$) are the only significant factors of knowledge utilization. System level, system form and service quality are

Table 1. Constructs reliability and AVE

Construct	Total items	Reliability	AVE
Management (MG)	4	0.926	0.760
Knowledge Quality (KQ)	7	0.957	0.759
System Level (SL)	4	0.953	0.835
System Form (SF)	2	0.804	0.677
Service Quality (SVQ)	5	0.940	0.757
Rewards Policy (RW)	2	0.943	0.892
Knowledge Trust (KT)	3	0.951	0.866
Knowledge Utilization (KU)	6	0.923	0.665
Individual Benefits (IB)	9	0.969	0.747

not significant predictors of knowledge utilization. Knowledge utilization was also a significant predictor of individual benefits (β=0.869; p = 0.005).

Thus, hypotheses H1, H5, H6, H7, and H8 were supported, but hypotheses H2, H3, H4 were not supported. Even though detecting the effect of service quality on system use has been problematic (DeLone and McLean, 2003), system quality factors have been long believed to be major determinants of any system use. Table 2 also shows the R^2 values of the independents constructs. The model explains 50 percent of variance in knowledge utilization, and 75.5 percent in individual benefits.

CONCLUSION

Discussion of Findings

Knowledge is a powerful intangible resource that enables individuals and organizations to improve their learning and decision-making processes and consequently achieve competitive advantages in the knowledge-based economy. The need for developing countries and their organizations, including petroleum companies, to empower themselves and achieve competitive advantage through knowledge management (KM) cannot be underestimated. Several global petroleum companies such as BP and Shell have gained some financial benefits from the deployment of KMS. Knowledge utilization is the process where the benefits of KMS are realized. This study aimed to develop a social and technical framework for knowledge utilization from repository KMS in an Omani organization.

This study found that the factors that significantly affect knowledge utilization were, in order of their contributions, knowledge quality (β = 0.392; p = 0.005), knowledge trust (β = 0.219; p =0.05), management support (β = 0.187; p = 0.05), and rewards policy (β = 0.172; p = 0.1). Surprisingly, the system level (β = 0.007), system form (β = 0.013), and service quality (β = -0.077) were found to be insignificant. This indicates that the most important issue for utilizing knowledge from KMS is the quality of knowledge and confidence in the knowledge. It seems that once individuals highly value the knowledge, and the management supports and rewards its utilization, then system quality and IT service quality are not critical factors for their knowledge utilization.

Also, this study empirically detected significant individual benefits resulting from knowledge utilization (β= 0.869; p = 0.005). Respondents

Table 2. Model evaluation statistics

Construct	Mean	R-Square	Path coefficient (β)	Significance level (α)	Hypothesis
Knowledge quality (KQ)	2.89	NA	0.392	0.005	H1: Supported
System level (SL)	2.67	NA	0.007	NS	H2: Rejected
System Form (SF)	2.93	NA	0.013	NS	H3: Rejected
Service Quality (SVQ)	4.25	NA	-0.077	NS	H4: Rejected
Management (MG)	4.41	NA	0.187	0.05	H5: Supported
Knowledge Trust (KT)	2.89	NA	0.219	0.05	H6: Supported
Rewards Policy(RW)	2.33	NA	0.172	0.1	H7: Supported
Knowledge Utilization (KU)	3.84	0.50	0.869	0.005	H8: Supported
Individual Benefit (IB)	3.69	0.755	NA	NA	NA

NS = Not Significant NA = Not Applicable

perceived that higher knowledge utilization leads to higher individual benefits (performance and innovation), and utilizing knowledge from KMS improves their performance in completing work-related tasks and enlightening their ways of thinking.

This study showed that the development of a knowledge-oriented culture is very significant on the success of KMS use consistent with a number of studies in developing countries (e.g. Al-Athari and Zairi (2001) and Syed-Ikhsan and Rowland (2004)). The significance of management support on the success of IT deployment was highly supported by several studies from Arab countries such as Ahmed and Hegazy (2006) and Khalfan and Alshawaf (2004). The significant of management support is also consistent to an earlier study conducted by the researcher on KMS success factors in Omani organizations from the IT managers' perspective. However, this study showed that individual knowledge utilizers consider rewards policy as a valuable strategy unlike the IT managers in the earlier study. The significance of rewards policy is also consistent with a study conducted in Malaysian context (Yahya and Goh, 2001).

Limitations and Future Research

This study had some limitations. First, the study was limited to a specific KMS's users in one company in one country. Generally the Arab countries share the same cultural values; nevertheless, more research in this area is needed from different countries to establish a solid framework on KMS evaluation. Second, the sample of this study was also from one organization. Investigating one system may add some control to the study, but limits its generalization. The model can now be tested with more organizations and systems, and across several countries. Future research may also conduct this investigation through longitudinal study to understand whether knowledge utilization is improved by the independent variables suggested in this study and/or by the benefits achieved through knowledge utilization. In addition, future studies should carry out detailed investigations of the individual and organizational benefits of KMS and knowledge utilization.

Implications for Practice

Despite limitations mentioned above, this study offered some implications for KMS practitioners and researchers. This study provided practitioners and researchers an initial framework and operational measurements to evaluate KMS use in this under investigated geographical area. It also provided some insights on the determinants of KMS use in an Omani context, which might be applied to the rest of the Middle East countries as they generally share similar cultural values.

As indicated above, developing countries and their organizations, including petroleum organizations, are in great need for KM and KMS to empower themselves and create competitive advantage. This study confirmed that the deployment of KMS results in individual benefits. Also, the study confirmed that the adoption of organizational KMS in terms of knowledge utilization requires the development of a suitable organizational culture consistent with Al-Busaidi et al.'s (2007) exploratory study. It illustrated that developing a high-level knowledge management system does not guarantee its success. Management support is extremely crucial to endorse a KMS initiative, clarify its objective to end users, encourage end users, and most importantly provide them the sufficient time to use it. Top management support is extremely critical in the Arab culture as mangers are considered a high authority, and individuals value the use of personal knowledge more than others' knowledge. Moreover, the establishment of rewards policy is essential to promote the utilization

Like Al-Busaidi et al.'s (2007) study, this study showed that the quality and the credibility of the knowledge stored in KMS are vital for knowledge utilization behavior. High knowledge quality may improve knowledge credibility (or trust), and, consequently knowledge utilization. Thus, developing standards that ensure high knowledge quality is crucial to improve users' confidence and consequently knowledge utilization behavior.

Finally, the study illustrated to organizations that it is worth it to establish organizational KMS. The study showed that the utilization of knowledge from KMS repository improves individuals' performance and productivity, which consequently may lead to organizational benefits.

REFERENCES

Ahmed, A., & Hegazy, K. (2006). Knowledge management perception in the Middle Eastern region: an empirical investigation within Egypt context. *International Journal of Management Practice*, *2*(2), 109–126.

Al-Athari, A., & Zairi, M. (2001). Building benchmarking competence through knowledge management capability: An empirical study of the Kuwaiti context. *Benchmarking: An International Journal*, *8*(1), 70–80. doi:10.1108/14635770110383489

Al-Busaidi, K. A., & Olfman, L. (2005). An investigation of the determinants of knowledge management systems success in Omani organizations. *Journal of Global Information Technology Management*, *8*(3), 6–27.

Al-Busaidi, K. A., Olfman, L., Ryan, T., & Leroy, G. (2007). Revealing the antecedents and benefits of KMS use: An exploratory study in petroleum company in Oman, *The Electronic Proceeding of the 9th International Conference on Decision Support Systems*, Kolkata, India.

Alavi, M. (2000). Managing organizational knowledge. In Zmud, R. (Ed.), *Framing the domains of IT management* (pp. 29–50). Cincinnati, OH: Pinnaflex.

Alavi, M., & Leidner, D. (2001). Review: Knowledge management and knowledge management systems: Conceptual foundations and research issues. *Management Information Systems Quarterly*, *25*(1), 107–136. doi:10.2307/3250961

Ali, A. J. (1990). Management theory in a transitional society: The Arab's experience. *International Studies of Management and Organization*, *20*(3), 7–35.

Becerra-Fernandez, I., Gonzalez, A., & Sabherwal, R. (2004). *Knowledge management: Challenges, solutions and technologies*. New Jersey: Pearson Education Inc.

Chang, S., & Lee, M. (2007). The effects of organizational culture and knowledge management mechanisms on organizational innovation: An empirical study in Taiwan. *The Business Review-Cambridge*, *7*(1), 295–301.

Chin, W. W. (1998). The partial least square approach to structural equation modeling. In G.A Marcoulides (Ed), *Modern methods for business research*(pp. 295-336). Mahawah, London, UK: Lawrence Erlbaum Associates.

Chong, S. C. (2006). KM critical success factors: A comparison of perceived importance versus implementation in Malaysian ICT companies. *The Learning Organization*, *13*(3), 230–256. doi:10.1108/09696470610661108

Clark, S. (2005). Refinery expertise preservation. *Control Engineering*, *52*(8), IE5.

Davenport, T. H., & Prusak, L. (1998). *Working knowledge*. Boston, MA: Harvard Business.

De Long, D. W., & Fahey, L. (2000). Diagnosing cultural barriers to knowledge management. *The Academy of Management Executive*, *14*(4), 113–127.

Debowski, S. (2006). *Knowledge management*. Australia: Wiley.

DeLone, W., & McLean, E. R. (2003). The DeLone and McLean model of information systems success: A ten-year update. *Journal of Management Information Systems*, *19*(4), 9–30.

Dennis, A. R. (1996). Information exchange and use in group decision making: You can lead a group to information, but you can't make it think. *Management Information Systems Quarterly*, *20*(4), 433–457. doi:10.2307/249563

Gold, A. H., Malhotra, A., & Segars, A. H. (2001). Knowledge management: An organizational capabilities perspective. *Journal of Management Information Systems*, *18*(1), 185–214.

Huber, G. (1991). Organizational learning: The contributing processes and the literatures. *Organization Science*, *2*(1), 88–115. doi:10.1287/orsc.2.1.88

Husted, K., & Michailova, S. (2002). Knowledge sharing in Russian companies with Western participation. *Management International*, *6*(2), 17.

Igbaria, M. (1990). End-user computing effectiveness: A structural equation model. *Omega*, *18*(6), 637. doi:10.1016/0305-0483(90)90055-E

Jennex, M., & Olfman, L. (2002). Organizational memory/knowledge effects on productivity, A longitudinal study. *Proceedings of the 35th Hawaii International Conference on System Sciences(HICSS'02), Vol.6*(pp. 1316 – 1316). Hawaii: HICSS.

Jennex, M., & Olfman, L. (2006). A model of knowledge management success. *International Journal of Knowledge Management*, *2*(3), 51–68.

Jiang, X., & Lia, Y. (2008). An empirical investigation of knowledge management and innovative performance: The case of alliances. *Research Policy*, *38*(2), 358–368. doi:10.1016/j.respol.2008.11.002

Kankanhalli, A., & Tan, B. (2004). A review of metrics for knowledge management systems and knowledge management initiatives. *Proceedings of the 37th Annual Hawaii International Conference on System Sciences (HICSS'04),* vol. 8.(pg.8). Hawaii: HICSS.

Katz, R., & Allen, T. (1982). Investigating the not invented here (NIH) syndrome: A look at the performance, tenure and communication patterns of 50 R&D projects groups. *R & D Management*, *12*(1), 7–19. doi:10.1111/j.1467-9310.1982.tb00478.x

Kettinger, W. J., & Lee, C. C. (1994). Perceived service quality and user satisfaction with the Information services function. *Decision Sciences*, *25*(5/6), 737–765. doi:10.1111/j.1540-5915.1994.tb01868.x

Khalfan, A., & Alshawaf, A. (2004). Adoption and implementation problems of e-banking: A study of the managerial perspective of the banking industry in Oman. *Journal of Global Information Technology Management*, *7*(1), 47–64.

King, J. (2001). Shell strikes knowledge gold. *Computerworld*, *1*(3), 34.

Krogh, G. (1998). Care in knowledge creation. *California Management Review*, *40*(3), 133–153.

Lehaney, B., Clarke S., Coakes, E, and Jack Gillian. (2004). *Beyond knowledge management*. Hershey, PA,: Idea Group Publishing.

Liu, P., & Tsai, C. (2007). Effect of knowledge management systems on operating performance: An empirical study of hi-tech companies sing the balanced scorecard approach. *International Journal of Management*, *24*(4), 734–744.

Liu, S. C. (2003). *A study of factors that facilitate use of knowledge management systems and the impact of use on individual learning. Unpublished* doctoral dissertation, Claremont Graduate University, Claremont, CA.

Maier, R. (2002). *Knowledge Management Systems: Information and communication technologies for knowledge management*. Berlin: Springer.

Moore, G., & Benbasat, I. (1991). Development of an instrument to measure the perceptions of adopting an information technology innovation. *Information Systems Research*, *2*, 192–222. doi:10.1287/isre.2.3.192

Nonaka, I. (1994). A dynamic theory of organizational knowledge creation. *Organization Science*, *5*(1), 14–37. doi:10.1287/orsc.5.1.14

O'Dell, C., & Grayson, C. J. (1998). If only we knew what we know: Identification and transfer of internal best practices. *California Management Review*, *40*(3), 14–37.

Ong, C., & Lai, J. (2007). Measuring user satisfaction with knowledge management systems: scale development, purification, and initial test. *Computers in Human Behavior*, *23*(3), 1329–1346. doi:10.1016/j.chb.2004.12.012

Parasuraman, A., Zeithaml, A. V., & Berry, L. L. (1985). A conceptual model of service quality and its implications for future research. *Journal of Marketing*, *49*(4), 41–50. doi:10.2307/1251430

Schein, E. (1985). *Organizational culture and leadership*. San Francisco, CA: Jossey-Bass.

Scholl, W., Konig, C., & Meyer, B., & Heisig, p. (2004). The future of knowledge management: An international Delphi study. *Journal of Knowledge Management*, *8*(2), 19–35. doi:10.1108/13673270410529082

Steinberg, D. (2002). Share knowledge (ignorance costs millions) Shell international exploration and production. *Ziff Davis Smart Business*, *15*(4), 45.

Sussman, S. W., & Siegal, W. S. (2003). Informational influence in organizations: An integrated approach to knowledge adoption. *Information Systems Research*, *14*(1), 47–65. doi:10.1287/isre.14.1.47.14767

Syed-Ikhsan, S. O., & Rowland, F. (2004). Benchmarking knowledge management in a public organization in Malaysia. *Benchmarking: An International Journal*, *11*(3), 236–266. doi:10.1108/14635770410538745

Turban, E., McLean, E., & Whetherbe, J. (2001). *Information Technology for Management: Making connections for strategic advantage*. New York: John Wiley and Sons Inc.

World Bank. (1998). *World development report 1998/1999: Knowledge for development*. Retrieved May 20, 2005, from http://www.worldbank.org/wdr/wdr98/contents.htm

World Bank. (2003). *Technical cooperation program brief on GCC*. Retrieved May 20, 2005, from http://web.worldbank.org/WBSITE/EXTERNAL/COUNTRIES/MENAEXT/BAHRAINEXTN/0,menuPK:312668~pagePK:141132~piPK:141107~theSitePK:312658,00.html

Yahya, S., & Goh, W. (2002). Managing human resources toward achieving knowledge management. *Journal of Knowledge Management*, *6*(5), 457–468. doi:10.1108/13673270210450414

Zucker, L. G. (1986). Production of Trust: Institutional sources of economic structures, 1840-1920. In Shaw, B., & Cummings, L. (Eds.), *Research in Organizational Behavior* (*Vol. 8*, pp. 53–111). Greenwich, CT: JAI Press.

ADDITIONAL READING

Abu-Zed, E. (2002). A knowledge management reference model. *Journal of Knowledge Management*, *6*(5), 486–499. doi:10.1108/13673270210450432

Additively, A., Maurer, M., & Li, W., Went ling, T., & Steadman, R. (2006). Cultural influences on knowledge sharing through online communities of practice. *Journal of Knowledge Management*, *10*(1), 94–107. doi:10.1108/13673270610650139

Al-Busaidi, K. A., & Olfman, L. (2005). Knowledge Management Systems Success Factors: An investigation in Middle-Eastern Organizations. *Proceedings of the 16th Information Resources Management Association International Conference,* San Diego, California, USA.

Argot, L., & Civilly, B. & Reagan's, R. (2003). Managing knowledge in organizations: an integrative framework and review of emerging themes. *Management Science, 49*(4), 571–582. doi:10.1287/mnsc.49.4.571.14424

Barrow, D. (2001). Sharing know-how at BP Amoco. *Research Technology Management, 44*(3), 18–25.

Bell, M., & Album, M. (1999). Knowledge systems and technological dynamism in industrial clusters in developing countries. *World Development, 27*(9), 1715. doi:10.1016/S0305-750X(99)00073-X

Bhatt, G. (2001). Knowledge management in organizations: examining the interactions between technologies, techniques and people. *Journal of Knowledge Management, 3*(2), 94–109.

Chon, C. W., Chon, S. C., & Hang, P. Y. P. (2006). KM implementation in Malaysian telecommunication industry: an empirical analysis. *Industrial Management & Data Systems, 106*(8), 1112–1132. doi:10.1108/02635570610710782

Chow, C., Deng, F., & Ho, J. (2000). The openness of knowledge sharing within organizations: a comparative study in the United States and the People's Republic of China. *Journal of Management Accounting Research, 12*, 65–95. doi:10.2308/jmar.2000.12.1.65

Delouse, K., & Venially, G. (2005). Securing knowledge in organizations: lessons from the defense and intelligence sectors. *International Journal of Information Management, 25*(1), 85–98. doi:10.1016/j.ijinfomgt.2004.10.007

Doctor, G. (2008). Determining the number of simultaneous users of an institutional knowledge repository at a management institute in India. *Vine, 38*(3), 334–347. doi:10.1108/03055720810904844

Eftekharzadeh, R. (2008). Knowledge management implementation in developing countries: An experimental study. *Review of Business, 28*(3), 44–58.

Goel, R., & Rustagi, N. (2006). Increasing competitiveness through knowledge transfer in developing countries. *Advances in Competitiveness Research, 14*(1), 77–89.

Hegde, D., & Shapira, P. (2007). Knowledge, technology trajectories, and innovation in a developing country context: evidence from a survey of Malaysian firms. *International Journal of Technology Management, 40*(4), 349. doi:10.1504/IJTM.2007.015757

Hicks, R., Dattero, R., & Galup, S. (2007). A metaphor for knowledge management explicit islands in a tacit sea. *Journal of Knowledge Management, 11*(1), 5–16. doi:10.1108/13673270710728204

Jennex, M. (2008). Exploring system use as a measure of knowledge management success. *Journal of Organizational and End User Computing, 20*(1), 50–63.

Lee, H., & Choi, B. (2003). Knowledge management enablers, processes, and organizational performance: An integrative view and empirical examination. *Journal of Management Information Systems, 20*(1), 179–228.

Lytras, M. D. (2008). *The open knowledge society.* Berlin: Springer-Verlag.

Marwick, A. D. (2001). Knowledge management technology. *IBM Systems Journal, 40*(4), 814–830. doi:10.1147/sj.404.0814

Mayasandra, R. N., & Pan, S. L. (2005). Governing and creating a KM platform for strategic success: a case study of Wipro Technologies, India. In Menkhoff, T., Evers, H., & Chay, Y. W. (Eds.), *Governing and managing knowledge in Asia* (pp. 259–275). Singapore: World Scientific Publishing.

Mullin, R. (1998). Knowledge management: A cultural evolution. *The Journal of Business Strategy*, *17*(5).

Narteh, B. (2008). Knowledge transfer in developed-developing country interfirm collaborations: a conceptual framework. *Journal of Knowledge Management*, *12*(1), 78–91. doi:10.1108/13673270810852403

Nayir, D., & Uzunçarsili, U. (2008). A cultural perspective on knowledge management: the success story of Sarkuysan company. *Journal of Knowledge Management*, *12*(2), 141–155. doi:10.1108/13673270810859578

Nevo, D., & Chan, Y. E. (2007). A Delphi study of knowledge management systems: Scope and requirements. *Information & Management*, *44*, 583–597. doi:10.1016/j.im.2007.06.001

Persad, D., Quach, U., & Throsteinsdottir, H. (2006). Enabling knowledge societies in developing countries: the example of genomics. *International Journal of Biotechnology*, *8*(1), 4.

Srinivasan, R. (2004). Knowledge Architectures for Cultural Narratives. *Journal of Knowledge Management*, *8*(4), 65–74. doi:10.1108/13673270410548487

Sveiby, K. (2001). A knowledge based theory of the firm guide in strategy formulation. *Journal of Intellectual Capital*, *2*(4), 344–350. doi:10.1108/14691930110409651

Tiwana, A. (2002). *The knowledge management toolkit: Orchestrating IT, strategy, and knowledge platforms*. Upper Saddle River, NJ: Prentic Hall PTR.

Tiwana, A. (2004). An empirical study of the effect of knowledge integration on software development performance. *Information and Software Technology*, *46*(13), 899–906. doi:10.1016/j.infsof.2004.03.006

Wang, J., Peters, H., & Guan, J. (2006). Factors influencing knowledge productivity in German research groups: lessons for developing countries. *Journal of Knowledge Management*, *10*(4), 113–126. doi:10.1108/13673270610679408

Wu, J.-H., & Wang, Y.-M. (2006). Measuring KMS success: A respecification of the DeLone and McLean's model. *Information & Management*, *43*, 728–739. doi:10.1016/j.im.2006.05.002

Zhu, Z. (2004). Knowledge management: Towards a universal concept or cross-cultural contexts? *Knowledge Management Research and Practice*, *2*(2), 67–79. doi:10.1057/palgrave.kmrp.8500032

KEY TERMS AND DEFINITIONS

Knowledge: Understanding gained through experience; the capability to use information to make actions.

Explicit Knowledge: Knowledge that can be easily verbalized and communicated to others.

Knowledge Management: The process of creating, capturing, integrating and disseminating knowledge throughout the organization to add value to the organization.

Knowledge Utilization: The application of the stored knowledge to solve daily work problems and make decisions.

Knowledge Management Systems: Systems that manage knowledge throughout the organization; they are systems that create, integrate, store, retrieve, transfer and distribute knowledge thought the organization.

Organizational Memory: Explicit and articulated knowledge of the organization.

Repository Knowledge Management System: A type of knowledge management systems that codifies the organization's explicit knowledge for later reuse and distribution.

Technical KMS Success Factors: The characteristics of the KMS that determine its success.

Social KMS Success Factors: The dimensions of the corporate culture that determine the success of KMS.

APPENDIX

Table 3. Operational measurements of the study constructs

Construct	Operational measurements (indicators)	Source
Knowledge Quality (KQ)	Your KMS provides information/ knowledge that is: KQ1. Accurate. KQ2. Relevant. KQ3. Adequate for you to complete work-related tasks. KQ4. Contextual, easy-to-apply. KQ5. Up-to-date. KQ6. Complete. KQ7. Presented in a good format.	Modified from Liu (2003); originally based on Jennex & Olfman (2002)
System Level (SL)	Your KMS search/retrieval function is: SL1. Complete. SL2. Reliable. SL3. Quick. SL4. Easy-to-understand.	Modified from Liu (2003); originally based on Jennex & Olfman (2002)
System Form (SF)	Whenever you do an online search: SF1. You can find most of the organizational information/ knowledge. SF2. You need to access more than one system to locate the needed information/ knowledge.	Modified from Liu (2003) & originally based on Jennex & Olfman (2002)
Service Quality (SVQ)	The IT support staff: SVQ1. Does what it promises to do on time. SVQ2. Have the knowledge to answer your questions. SVQ3. Provides adequate training for your system use. SVQ4. Understands your specific needs. SVQ5. Does its best to respond as soon as possible when you have a problem.	Modified from Kettinger & Lee (1994)
Management Support (MG)	Your senior managers: MG1.Emphasize to you the importance of knowledge management to the company's success MG2.Clearly state the objective of the KMS to you. MG3.Encourage you to exchange (use and share) knowledge. MG4.Provide you time to use, create and share knowledge	Self-constructed based on Davenport & Prusak (1998) and Igbaria (1990)
Rewards Policy (RW)	Your organization: RW1.Rewards knowledge utilization. RW.2.Provides non-monetary incentives (e.g. certificate) to knowledge utilizers.	Self-constructed based on Davenport & Prusak(1998)
Knowledge Trust (KT)	KT1: You believe in everything you use from the KMS. KT2.The knowledge/information you use is truthful. KT3.The knowledge/information you use is reliable.	Dennis (1996);
Knowledge Utilization (KU)	How frequently do you use knowledge/information from KMS to: KU1. Solve work-related problems. KU2. Perform work-related decision-making. KU3. Question existing rules and routines. KU4. Search for and explore alternatives. KU5. Assist you in a work-related task. KU6. Acquire knowledge/information of a new job position.	Modified from Liu (2003); based on Jennex & Olfman (2002).
Individual Benefits (IB)	Because you use knowledge/information from KMS that was contributed by others, how frequently does it: IB1. Improve the quality of your decisions at work. IB2. Faster your decision-making at work. IB3. Reduce time you spend on problem solving at work. IB4. Improve your performance. IB5. Improve your sense of autonomy. IB6. Improve your experience of knowledge/information transfer and reuse. IB7. Enlighten you to new ways of thinking. IB8. Help you to detect work-related problems. IB9. Make you more innovative and creative.	Modified from Liu (2003); originally based on Jennex & Olfman (2002)

Section 2
Knowledge Sharing and Transfer

Chapter 8
Knowledge Creation and Sharing in the Malaysian Housebuilding Industry:
Improving the Housing Delivery System

Nor'Aini Yusof
Universiti Sains Malaysia, Malaysia

Mohd Wira Mohd Shafiei
Universiti Sains Malaysia, Malaysia

EXECUTIVE SUMMARY

This chapter examines how knowledge management (KM) develops within the Malaysian housebuilding industry. The objectives of this chapter are twofold. First, it explains how innovation is able to encourage or influence housing developers in generating the new Build Then Sell (BTS) system. Second, it identifies the characteristics of successful adopters so that other developers can imitate them and funnel resources to those areas that successful adopters emphasise. Focus group and semi-structured interviews were conducted to obtain in-depth responses from one house buyers' organisation and successful BTS adopters. The effort to implement BTS is seen as being the result of a strategic investment in new knowledge by developers responding to the pressure they have experienced on the part of the house buyers and the government. Discussion and the sharing of both explicit and tacit knowledge between the main stakeholders have generated the finer details of the BTS. A strong learning culture and the development of KM tendencies also lead to successful BTS implementation.

INTRODUCTION

Contemporary housing development industry is characterised by dynamic markets, continuous technological advancement, and increasingly discerning house buyers. To cope with these trends, housing developers must become more flexible, and one certain way for them to do so is to strengthen their potential to learn as organisations. Thus, for enlightened companies, knowledge and the flow of information become the essential organisational drivers of innovative ideas and value.

DOI: 10.4018/978-1-61692-886-5.ch008

The Knowledge Management (KM) approach can be used to help housing developers to focus more on their particular organisational processes and the creation of new knowledge to remain one step ahead of their competitors. By implementing KM, these companies can shift from management based on compliance to a management system based on self-control and self-organisation. This change towards a more aggressive business mindset can then lead these companies to shift from the utilisation of already known knowledge to the creation of new knowledge. This shift in strategy had been described by Binney (2001) as the movement from pure technology KM applications to the inclusion of process applications.

The new housing delivery system known as the Build-Then-Sell (BTS) system is a manifestation of KM in progress. The system can be viewed as the newly created new knowledge that can improve the product offerings of developers for the existing housing market. Initially proposed by consumer associations to protect the rights of house buyers, the BTS system has been slowly embraced by some property developers as an innovative product that can help them to increase their market share. The system has undergone a series of metamorphoses since it was originally developed that have made it more attractive to housing developers. It has been argued that the BTS system was mainly implemented by those housing developers with a strong learning culture and well developed KM tendencies. This evidence encourages the implementation of KM in organisations.

This chapter examines how KM develops within housing developers' organisations in Malaysia by exploring their readiness to adopt the new BTS system. The chapter is based on a study done by the authors who investigate housing developers' organisational capabilities in trying to implement the BTS system. The BTS system generates benefits not only for customers but also for the developers themselves and the housing industry as a whole. The objectives of this chapter are twofold. First, this chapter explains how innovation is able to encourage or influence housing developers in generating knowledge-based products, processes, or services. The second objective is to identify the characteristics of successful adopters so that other developers can imitate and funnel resources to those areas that adopters successfully practice. We argue that a strong learning culture and well developed KM tendencies will lead to successful BTS implementation. Successful BTS adopters are likely to become points of reference for other organisations that help to facilitate the implementation of BTS in the housing industry.

In order to obtain in-depth information about the BTS system as perceived from house buyers' point of view, a focus group interview was carried out with one house buyers' organisation. House buyers' organisation is one of the main proponents of BTS implementation in Malaysia. The focus group interview was attended by the chairman, secretary and the treasurer of the association and three members of the research team. The discussion was held at the association's main office and lasted approximately three hours.

A series of interviews was also conducted with housing developers that have successfully implemented the BTS system. A snowballing method was employed to locate these developers. This type of sampling method was employed because it is extremely difficult to locate the developers who have experience adopting the BTS system. Referrals from the initial developer were used to locate additional developers. The initial developer was identified based on its self-identification in media reports as a BTS adopter. In total, three BTS-adopting developers' organisations participated in semi-structured interviews held between July and August of 2008. Preliminary telephone conservations were conducted with the developers to secure their participation and identify those individuals who are most knowledgeable, have the most experience and are in the best position to provide information regarding the implementation of the BTS system in their housing projects.

Phone calls were then made to the identified individuals, followed by official confirmation letters. Interviews were conducted to obtain in-depth responses from the respondents regarding the issues surrounding BTS implementation. The meetings were held face-to-face at the participants' office or at a location determined by the participants. An interview guide was developed to minimise variation and add flexibility to the discussion. The key themes of the interviews were the main drivers for BTS implementation and their experience in implementing the BTS system. Each interview was tape-recorded and subsequently transcribed in full by student assistants into word documents as quickly as possible. The transcriptions were then cross-checked by the interviewers with written notes taken during the meetings and to ensure accuracy. The data were analysed manually using content analysis following Miles and Huberman (1994). The transcribed data were categorised using several codes. After eliminating or combining code categories, the final codes were organised according to common themes to identify patterns and highlight important ideas.

The chapter begins by highlighting the problems in Malaysia's housing industry and the need for a new housing delivery system. It then describes the new BTS system in depth and discusses how innovation promotes the generation of knowledge-based products, processes, and services. It will also emphasise the importance of experience in readying housing developers to adopt BTS. Conclusion and directions for future research are presented at the end of the chapter.

PROBLEMS IN THE MALAYSIAN HOUSING INDUSTRY

Serious concerns have been voiced about the performance of the Malaysian housing industry, which is based on the decades-old practice of "Sell Then Build" (STB), which allows developers to sell houses and collect payments before construction is completed. Complaints abound regarding abandoned housing projects, housing oversupply and poor workmanship. Abandoned projects are projects in which all work and construction activity has stopped for a period of no less than six continuous calendar months even when sale and purchase agreements have been signed (MHLG, 1999). Surprisingly, data on abandoned housing projects is scanty. Most statistics only cover projects abandoned by private developers in Peninsular Malaysia. The first case of an abandoned housing project was reported in 1983 during an economic downturn that resulted in a huge property overhang. In 1986, 126 private housing projects involving 14,568 units and 6,834 house buyers failed to be completed as scheduled (Raman, 1997). After 1990, 'project accounts' were introduced to help resolve the problem of abandoned housing projects. However, the introduction of project accounts and the end of the recession did not end the problem of abandoned projects. Between 1990 and 2002, 80,070 house buyers became victims of abandoned housing projects involving 544 private housing projects and 125,649 housing units valued at RM 9.5 billion (MHLG, 2003). The Ministry of Housing and Local Government (MHLG) revealed that while some of these projects could be salvaged, others could not be revived at all. As of April 2009, 146 housing projects were still abandoned (Hassanuddin, 2009). The causes of abandoned housing projects are various, but one of the main factors that encourage developers to stop their projects is STB, which allows anybody to become a developer and provides prospective developers with an easy way out; they can simply abandon their projects when they face problems (Mohd. Hanipah, *et al.* 2005; HBA, 2005). Abandoned projects place a particular burden on buyers, who not only make up-front payments for the new houses but also continue to pay rent for their existing accommodations during the construction process (Raman, 1997).

In addition to the issue of abandoned projects, there is also a problem of housing oversupply. Official figures published by the National Property Information Centre (NAPIC) on housing oversupply are disturbing. In 2000, out of 5,338,000 housing units supplied in Malaysia, including existing and committed housing units, the country required only 3,941,000 units to fill its housing needs. Thus, around 1,396,000 housing units that were approved by planning authorities in that year were in fact oversupply. In an efficient housing delivery system, market forces push the housing market toward a point of equilibrium, balancing supply and demand (Eisendiel, 1997). However, the preceding discussion suggests that the STB system has contributed to an imbalance between demand and supply in the Malaysian housing market. Speculation by housing developers, weaknesses in the housing forecast process and ineffective procedures for assessing new development applications may have contributed to this imbalance (Yusof, et al., 2009). Under the STB, only 40% of a project's houses need to be sold during the first launch, leading developers to begin development even if there is an oversupply of houses in the area. The problem is compounded by a lack of rigorous market studies by developers and the current approval criteria set by local authorities, which are based on to what extent the developers comply with regulations and by-laws rather than on the local demand for housing.

The current STB system has also encouraged shoddy workmanship. Shoddy workmanship has occurred as a result of poor supervision, the use of substandard materials and negligence by the parties concerned (Ong, 1997). A typical sale and purchase agreement signed between developers and house buyers requires that the developer adhere to good, fair construction standards. However, such requirements seem to have had little impact on the actual quality of completed houses. Given that houses built under the STB system are sold before completion, developers are believed to have less incentive to provide quality workmanship (Chau *et al.*, 2007). Developers do not monitor their housing projects diligently because they have received payment for houses still under construction. In addition, model houses, which have been widely used by developers to market their houses, have often been a poor indicator of the quality of actual units purchased and built (HBA, 2005). The quality of the workmanship, the finish, and the fittings provided in housing units often differ substantially even between different units of the same project in the same phase of construction (ibid). There are vast quality differences between units at different phases of construction and between different projects. There is a dearth of concrete data on the incidence of shoddy workmanship in the Malaysian housing industry. However, reports made to consumer associations and those highlighted by the mass media have provided ample examples. Every year, HBA statistics show that not less than 7% of house buyers complain about shoddy workmanship and defects. The percentage may be small, but studies in other countries reveal that the cost for defect rectification is around 5% of the construction contract value (Mills, *et al.*, 2009; Josephson and Hammarlund, 1999). This finding shows that defects are costly and cannot be ignored.

Considering the above-mentioned problems, it would appear that something must be done to ensure the quality completion of housing projects and to protect the rights of house buyers. Many have argued that an overhaul of the present concept of housing delivery in Malaysia is long overdue and essential to protect the reputation of the industry. This idea suggests feasibility of introducing a new housing delivery system commonly known as the BTS system.

THE NEW BTS SYSTEM

The proposed solution is a move towards a system that allows houses to be sold only after completion and aims to promote better quality housing and to provide greater protection for house buyers. The new BTS system was introduced in April 2007 by the government alongside the existing STB system. The BTS system will require housing developers to find alternative sources of project financing rather than relying on advance payments from house buyers through pre-sale financing. Thus, under the new system, developers need to be financially strong before carrying out housing development. In addition, the BTS system will cause developers to be more cautious with regard to completion time and the quality of the houses that they develop, which means that these developers will need to be better organised, with improved working practices and optimised project delivery systems (ibid). Nonetheless, the BTS system has the potential to benefit not only house buyers but also developers and the housing industry in general. Under the new system, buyers are able to examine finished houses and their surroundings before deciding to buy, and they are also able to move in soon after purchase and thereby avoid buying an abandoned project (HBA, 2003). Developers can benefit as well because the actual construction costs are known at the time of sale, and thus, the selling prices of housing units under BTS reflect actual costs and market conditions (Kasi, 1992). It is also argued that the BTS system would likely give rise to stronger housing development companies that would invest in advanced technologies and efficient construction methods for better quality control to persuade purchasers to buy their products (Yusof et al., in press). This practice would raise the standard of construction and the quality of houses produced by the industry as a whole.

In order to encourage developers to implement the new BTS system, the government has offered incentives to developers of BTS projects: fast-track approval, exemption from building low-cost housing and exemption from paying deposits of RM 200,000 (40,728.5 Euro) for a developer's license. Unfortunately, not many developers have taken advantage of the government's offer (Star, June 2008). This finding is not surprising; the conventional STB system has proven profitable to developers for more than 40 years, and few developers would risk a new, unproven system when given the option of continuing to use a profitable existing one. Even in other countries worldwide, evidence suggests that the housebuilding industry is not keen to adopt new systems or products. Among the reasons for this phenomenon include fear of changes in attitude and the perception that change can influence employees and clients in an undesirable way (Barlow, 1999). In addition, the temporary nature of housing projects and the involvement of sub-contractors and other external consultants (Asibong & Barlow, 1997) also make the adoption of new systems or products challenging within the industry. In several developed countries, expenditure on research and development in the housing industry is also minimal (Gibb, 1999). Does the same scenario apply in Malaysia? In other words, are Malaysian housing developers are capable or ready to adopt the new BTS system?

Interestingly, some Malaysian developers have already implemented BTS in their housing development projects. Bandar Tasik Semenyih Group (BTSG), Guthrie Properties and Worldwide Holdings Ltd are a few housing developers that claim in media reports to be BTS adopters. For example, in 2004, BTSG undertook a 16-acre project, intending to build 136 residential properties: bungalows, semi-detached units and link houses. They were to use the BTS system in Kajang Township. Encouraged by the success of the first phase of the project, the BTSG recently began Phase Two using the same new BTS system. Phase Two will entail building 19 units of double-story bungalows and another 58 units of double-story semi-detached houses. In addition, Guthrie Properties launched its first BTS development in 2005. The development consists of 160 link housing units with prices ranging from RM 420,000 (82,380 Euro) per unit to RM 620,000 (121,610 Euro) per unit. These BTS projects are relatively small in scale, and all have already been completed and are sold out. Some were even completed ahead of schedule.

Nevertheless, housing developers differ in terms of their size, the location of their operations and their number of years in business. It is suspected that not all developers will be able to adopt BTS. The effort to implement BTS is seen as a result of strategic investments in new knowledge (in this case, BTS) by developers responding to the pressure from the government and house buyers. Investments in new knowledge and increasing knowledge levels among managers and employees pertaining to the BTS system have resulted in these developers' launching new BTS houses that show better-quality and conform to house buyers' needs, thus allowing them to gain competitive advantage. Therefore, there is a need to ascertain the characteristics of those developers that are most likely to succeed under a BTS system so that other developers also adapt themselves to this mode of house construction.

THE ROLE OF INNOVATION IN GENERATING THE BTS SYSTEM

The literature on innovation is well established. Innovation can generally be described as the adoption of an idea, technology, process or product that is new to an organisation (Herkema, 2003). Since it involves the adoption of an entity that is new to the organisation, innovation adoption will definitely occur in a process whereby knowledge is acquired, shared and distributed among organisation members. Gloet and Terziovski (2004) describes this process as KM, by which innovation is created, developed and implemented in an environment that is conducive to sharing knowledge. According to Cavusgil *et al.* (2003), knowledge-sharing is a pivotal factor for organisations seeking to foster or support innovation. Herkema (2003) goes further in her definition of innovation by stating that innovation is the adoption of an idea or behaviour that is new to the organisation. The innovation can be a new product, a new service or a new technology. Innovation is related to change, which can be radical or incremental. These definitions of innovation have put into context the origin of BTS, its development and application in the developed countries, its proposed implementation by house buyers and the Malaysian government, and its adoption by a few housing developers.

It has been generally accepted by the stakeholders in the Malaysian housing industry that BTS was originally conceived by the house buyers associations. Due to the housing developers' incompetence and their inability to fulfil their promises, the associations suggested that major improvements be made to solve the problem. As one respondent from a house buyers' organisation described its efforts in lobbying for BTS:

'Realising the challenges we are up against are too great, we made use of every opportunity that we have had to expound on our views, to lobby... We even gave a briefing to the chairman of the national public account committee on why we feel that implementing BTS is the right thing to do. And he was very sympathetic to us. Immediately the following day, he made a very strong press comment calling for the implementation of BTS.'

They argued that the roots of the problems prevalent in the housing development industry can be traced to the existing housing delivery system. The widely used STB system makes house buyers assume all of the major risks. This lopsided structure means that for over six decades, the housing developers were operating in a market that was always favourable to them. It can be postulated that the low-risk exposure and high profit margin that the market could offer them had made the housing developers complacent. Without pressure to perform exceptionally well to stay in the business, many housing developers simply stuck to tried and tested strategies to continue profiting from the housing market. It can be argued that this low-pressure environment led to the stagnation of new ideas and products in the industry. In the end, it was the customers who were disadvantaged in this traditionally oriented and backward culture.

Housing developers must accept the fact that over the years, the nature of the industry has been changed by the speed of innovation, which has been precipitated by rapidly evolving technology, shorter new housing product lifecycles, and a higher rate of development of new housing concepts. The industry as a whole must ensure that its product offerings are innovative to face more discerning house buyers and changing customer needs. In the case of BTS innovation, the push towards implementation in the Malaysian housing industry actually came from the house buyers. The housing developers must accept that the housing market is slowly changing from a seller's market to a buyer's market. The change in the industry landscape began with the 1997 Asian economic crisis, which led to property glut in the housing market. This excess in housing supply turned the industry structure on its head by increasing the bargaining power of house buyers. This major change only strengthened the calls by house buyers for the implementation of a new, innovative housing delivery system: the BTS.

It is not surprising that when the government announced the implementation of the BTS system, this house buyers' organisation was very pleased. Nevertheless, it was mindful of the time lag factor: the policy had to be streamlined with the existing law before it could become effective. One respondent said:

'We were very glad in June 2006, when Najib Abdul Razak (then the Deputy Prime Minister) made the announcement that BTS would run concurrently with the present STB system for a trial period of two years. But because the law has not changed... this is only a policy in its beginning stages.... So even if the developers want to implement BTS, they cannot do so. To implement [BTS] you have to have the law intact.'

Afraid to change the status quo, the housing industry did not respond enthusiastically to the proposed implementation of BTS. In order to maintain the push towards BTS, the government took advantage of the house buyers' strong market position by formally incorporating the system into the Malaysian housing delivery system. Despite the slowness of the industry to implement the new BTS system, a few housing developers demonstrated their high level of market elasticity by taking up the new system as a challenge and then integrating innovative features into it. Talking about some developers' bold decision to implement BTS, one respondent revealed,

'Even S (a housing development company) got its own version of BTS—i.e., the 20:80 BTS system... S has already implemented the 20:80 program, meaning that you pay 20 percent (as the) down payment... After that, there will be no more payments until the project is completed.'

Another respondent continued in the same vein:

'There are also some developers who give buyers [an option of] a 10 percent down payment, and during the duration of the construction, the developer would service the bank interest for the house buyers. There are some developers making that [kind of offer], who service the interest, and it is equivalent to 10:90... meaning that you don't have to pay any interest during the duration of construction. Instead, the developer is making the arrangements with the banker... that is, during construction; the developers pay the loan interest (of house buyers) directly to the banker.'

Commenting on these efforts by housing developers, the respondent went on and said:

'...so it is all in the making... in one form or another, this Build Then Sell will be seen.'

These companies recognised that the evolved KM process that had given birth to BTS is extremely dependent on innovation to make it successful. The original BTS system was proposed by the house buyers through their representatives at the house buyers' organisation, which had observed the success of such systems in developed nations. However, there is no guarantee that this system would work in the Malaysian context. Thus, the complexity of the house buyers' BTS needed to be unravelled and then reconstructed using innovative inputs from the housing developers.

As previously explained, BTS innovation was initiated by the house buyers, the government, and then, later in the BTS development process, the housing developers. BTS innovation allows collaboration among these important stakeholders. They worked together to achieve shared business and economic objectives, producing benefits for all stakeholders. This collaboration, in turn, plays an especially significant role in the transfer of both tacit and explicit knowledge related to BTS. Initially, the house buyers associations only managed to engage the government as part of this collaboration. Without the other main player in the equation, i.e. the housing developers, the transfer of tacit and explicit knowledge was fairly limited.

It is widely accepted that the stronger the relationship is between collaboration partners, the greater the extent of the tacit knowledge transfer will be (Scarbrough, 2003). Some innovative housing developers were able to see the benefits of BTS and began to become involved in the collaboration between the house buyers and the government. The transfer of tacit knowledge that had happened between the house buyers and housing developers was evidenced by the cooperation between a house buyers' organisation and a number of small property developers who were interested in implementing BTS in their projects. Having understood each other's needs and requirements, both the house buyers' organisation and the housing developers embarked on the next stage of their plan of action: to work out the finer details of the system. Depicting how the system was first begun, one respondent said,

'We helped out this first-time developer.... They heard about us proposing the 10:90. They came to us and asked us to help with how to go about [instituting] this 10:90. And we helped them out. At that time, the regulations were not there.... So this developer came out, he had this proposal wanting to implement the 10:90 but had no idea how to work it out. And we worked with this developer.'

As the BTS is a new policy, the housing developers in particular were not sure of the needs of the house buyers with regard to asking price, methods of payment, and amortisation period. There had to be a convergent point where both the house buyers and housing developers could understand one another's needs and requirements. Even though the HBA had outlined the needs of the house buyers, the developers still insisted on having meetings with the house buyers just to ensure that they understood the buyers' requirements clearly. By the same token, it was clear that the developers used the meetings to communicate their tacit knowledge to house buyers' organisation, especially regarding the scenario of the housing market in Malaysia. It had long been argued by researchers that many developers based their marketing decisions on hunches and guesses rather than on intelligence drawn from marketing research. Cadman and Austin-Crowe (1983) argue that, despite the need for a proper market analysis prior to development, many projects have been developed without sufficient study of the market. Cognisant of these facts, the housing developers used the series of meetings that they had with the house buyers' organisation to communicate their intuitions, guesses, and hunches to the house buyers so that the latter understood what they could expect from the former under the BTS system.

When the government announced the use of BTS in June 2006, the Ministry of Housing and Local Government had not yet finalised major parts of the BTS implementation plan. This lack of finalisation was the main obstacle that hindered the implementation of BTS system. After much lobbying, one of the respondents said,

'...the law has been changed along the way; they brought the matter before the parliament. We were there when it was debated. And on 12 April 2007, the bill was passed and the Housing Developers Act was amended.'

Despite the official announcement made by the government regarding the implementation of BTS, many housing developers and house buyers complained that the system was ineffectual. After the announcement, it was discovered that the schedule of payment upon which the delivery system is based had not been amended. To resolve these problems, all of the stakeholders involved-namely, the house buyers' organisation, the government and the housing developers-had a succession of meetings. One of the main outcomes of these meetings was the agreement regarding the schedule of payment to be used in this system. Said one respondent, describing the involvement of the three parties (housing developers, the house buyers' organisation and the government),

'...you cannot simply change the schedule of payment... So we [the housing developer and the house buyers' organisation] went to the Ministry of Housing and local government, to this special unit.... and we got the schedule of payment changed.... and it then became the new schedule for 10:90. This is the pilot program (for BTS).... and the developer launched the project in February 2007. So this is one of those innovative developers which does not bother about Housing Act being changed or what, but they took the initiative and go for it.'

In short, the schedule of payment, known as Schedule G in the BTS system, is a testament to explicit knowledge creation and sharing in action involving the three main stakeholders.

THE IMPORTANCE OF EXPERIENCE IN CREATING THE HOUSING DEVELOPERS' READINESS

Now that the housing developers are showing more interest and are beginning to adopt BTS in their housing delivery process, the next challenge in the development of BTS is the implementation of the system. No system can be claimed as a success if it is not rigorously tried and tested in various scenarios and situations. In the case of BTS, a number of housing developers have already implemented the system with mixed results. The majority of the BTS adopter-developers have managed to improve their revenue by implementing the system. However, there are some cases of housing developers who have claimed that they end up with unsold houses as a result of BTS system implementation. Nevertheless, after further inquiries, it was discovered that these developments were simply because the development projects were not selling well.

A case study commissioned by the government was conducted on three housing developers that had gainfully and successfully implemented BTS. These companies were happy to share their experiential learning process with the other stakeholders in the BTS system (the government) because they strongly believed in the sharing of tacit and explicit knowledge necessary to make the system a success. The companies are based in three different geographical settings: the northern, central, and southern parts of Peninsular Malaysia.

The findings of the study indicate that these three companies share a common characteristic: all of them have a structured land banks policy. This result means that the companies have a long-term strategy in the housing development industry because they know that once they complete one project, there will be another one started as a result of a steady stream of land bank purchases. Table 1 summarises the findings regarding these housing developers.

As shown in Table 1, the three developers that have successfully used BTS models all possess their own land banks as a result of a land banks policy. Interestingly, one developer (Developer B) intimated that it had only ventured into the housing development industry because it had been holding a vast tract of matured land banks that had originally been intended for more than 10 years to become agricultural developments. It could be said that this company demonstrated its willingness to change its strategies and policies according to the needs of the purchasing market. This inclination was made possible by a knowledge support structure available within the company whose main role is to help decision-makers to identify the activities that the company will be involved in and the resources required. Nonetheless, the main driver of the company's willingness to change its strategies was perhaps the company's leadership aspirations and ability. Debowski (2006) terms this type of leadership as that of the 'knowledge leader' whose role is to provide strategic visions and guidance and to identify priorities among changes to take place.

In addition, none of these three developers relied entirely upon bank financing. Instead, they made use of their own capital during the early development stages. A representative of Developer B commented that the bank was not supportive in giving a loan to finance the company's project. Moreover, he said,

'We encountered a lot of trouble dealing with banks, they don't understand the concept. I feel that the reason why they are not supportive is that they don't gain [from it].'

A representative of Developer C also stated that he tried to avoid obtaining a loan from the bank because he found that the bank was 'unfriendly' in its reception of his company's BTS ideas. This finding supports the argument that banks are not yet doing much to encourage developers to adopt BTS.

Nevertheless, some small developers are still able and motivated to adopt 100 percent BTS or the 10:90 BTS model. Developer B consciously implemented a BTS system to penetrate the housing market. Cognisant of its limitations as a

Table 1. Housing projects of BTS adopters

	Developer A	**Developer B**	**Developer C**
Location of Project (District, State)	Tanjung Bungah, Penang	Hulu Langat, Selangor	Muar, Johor
Type of Houses Built	Condominium	Double-story terrace houses and shop lots	Single-story semi-detached and single story bungalows
Number of Units	104 units	77 units of terrace housing and 12 units of shop lot	12 semi-detached units and 10 bungalows
Types of Financing	Mix of own capital and bank loans	Used own funds in the earlier period; later used bank loans	Retained profit and financing from co-operative
Possess Own Land Bank	Yes	Yes	Yes
Received Incentives	No	No	No
Approval of Local Authority	More than 12 months	Nearly one year	Usually 12 months
Target Market	Middle-end to high-end	Middle-income	Government servants (middle-income)

small-sized and relatively new housing developer on the supplier side of the market, it decided to jump on the new bandwagon of BTS to build up its reputation with the buying public. Developer B's representative noted that

'We went for this 10:90 BTS because we have no track record and no references—we are indeed a new player in this industry. For example, would you buy from me if you didn't know me at all? So by adopting this concept, we hope to instil confidence in the buyer.'

The remarks made by Developer B show that increasing customer demand and changing market needs are another important driver of the implementation of the BTS system. Although the driving force is coming from outside the company, it is ultimately creating the need for change and increasing companies' tendency to implement the BTS system.

As mentioned earlier, to promote the concept and implementation of BTS among the housing developers, the government offered them a number of BTS-related incentives. Fast-track approval, exemptions from building low-cost housing and exemptions from paying RM 200,000 (40,728.5 Euro) deposits for a developer's license are incentives offered to encourage developers to implement the BTS system and thus stimulate the supply of BTS houses in the market. None of the three developers received any incentives from the government, simply because they had started their projects before the incentives were announced. This finding shows that these innovative companies are already self-motivated to use a new system that they thought would give them a competitive advantage against their rivals. Their capacity to change in response to external demand without relying too much on the help of the government or financial institutions signals these companies' adaptability with regard to the new system. According to Dennison and Mishra (1995), this adaptability comes from the norms and beliefs developed within a company, which in turn translates into catering customer demand or transmitting signals from the environment into product or process change.

Developer A's representative brought up an interesting point about the quality of its BTS projects. The company discovered that the quality of construction under the BTS system is appreciably better than that of STB houses. Developer A was able to fairly compare the quality of both BTS and STB previous projects because the same contractor was used for each project. Developer A maintained that the monitoring system for BTS projects is much more stringent than that for STB projects because many precautions need to be taken before delivering completed houses to BTS buyers. House buyers also submitted fewer complaints regarding defects in the BTS houses.

Table 2 shows the number of defects reported in a 6-month period for the STB and BTS projects of Developer A. It is obvious that the number of defects detected under the BTS project was much lower than that under the STB project during the same time period. This result suggests that BTS is effective in motivating developers to be more cautious about the quality of houses built.

Developer A revealed that better housing quality is first achieved under the BTS system through the selection of appropriate individuals. To ensure that the project team has the ability to integrate

Table 2. Defects reported in 6 months for two different projects of developer A

Month	1	2	3	4	5	6	Total Defects
Project A (STB)	47	64	70	59	59	37	336
Project B (BTS)	11	14	10	17	19	13	84

knowledge from various sources, the conventional approach to employee selection was revised. The required knowledge and skills were identified in advance, and an improved approach was adopted that allows individuals with appropriate competencies and attitudes to be employed or selected from different units to that particular BTS project team. This approach helped to create the environment that the developer desired, one that is conducive to implementing new ways of doing things (in this case, stricter monitoring of project work).

Additionally, a system is developed in which experience from past projects can be learned from and shared among the parties involved. Experience from past projects and lessons learned were systematically collected from the respective parties and kept in databases. These databases come from the weekly project meetings held by the project team and feedback received from the house buyers or the residents of past projects. The databases provide guidelines for future projects and are easily accessible to all individuals involved.

Thirdly, a reward system aimed at promoting lessons learned and knowledge sharing is also in place. Employees who have been able to acquire the new knowledge to be learned have been rewarded, as have employees who are able to share the new knowledge they have acquired, either through bonuses or through promotions to higher position.

According to Developer A, the above-mentioned efforts have helped the firm to successfully implement the BTS system with better quality houses and thus to gain the element of competitive advantage crucial for its survival.

CONCLUSION AND FUTURE RESEARCH DIRECTIONS

BTS is a system that has been developed through knowledge sharing among the three main stakeholders in the housing development industry: house buyers, the government, and housing developers. The push towards BTS was initiated by representatives of house buyers at the house buyers' organisation who were seeking to protect the legal rights of these house buyers. Over time, the government and then housing developers became part of the team. Ideas and input from these stakeholders through discussions and the sharing of both explicit and tacit knowledge made the system more innovatively generated and acceptable to all stakeholders. The meetings and discussions that the house buyers, the government and the housing developers had in the early stages of the BTS implementation plan generated the finer details of the BTS, especially regarding the schedule of payment (Schedule G) to the housing developers. Finally, knowledge or experience sharing amongst the members of the developer's company is also important for the successful implementation of the new BTS system.

The study adds to our limited knowledge on how a housing delivery system is improved through knowledge creation and sharing in the Malaysian housebuilding industry. Nevertheless, the results of the study should be interpreted in light of the limitations posed by the data collection techniques employed in the study. Because the data were gathered through focus group and semi-structured interviews with one house buyers' organisation (which represents the house buyers' views) and only three developers, it is difficult to generalise the results to other countries, especially emerging economies where KM is still new. Therefore, we call for more empirical research to explore and test whether knowledge sharing among the key stakeholders and a strong learning culture and developed KM tendencies will lead to successful innovation implementation. In addition, many past studies have proved that effective leadership and employee skills are crucial factors in the successful implementation of innovation in an organisation. A strong leadership commitment at top levels encourages other members of an organisation to join in the change process, thus eliciting their knowledge of business processes and leading to the successful adoption of new ideas (Kearns and Lederer, 2004). On the other hand, employee skills are also crucial. Lim (2003) stresses that knowledge and skills of an organisation's workforce are among the most important factors contributing to the organisation's ability to implement new products, processes, or services. Siemieniuch and Sinclair (2004) suggest that individuals responsible for developing and implementing new ideas must possess appropriate knowledge, skills and expertise, either via formal training or through experiential learning opportunities made available within the organisation. Because these factors are not fully investigated, we suggest additional studies concerning the influence of leadership commitment and employee interest on knowledge that will help developers to implement the new BTS system or other types of innovation. This chapter also indicates the need for holistic studies on the role of KM in other stakeholder organisations within the housing industry that influence the implementation of the new system, product or process. Research into ways to enhance knowledge development, including house buyer empowerment and higher levels of professionalism on the part of the approving and monitoring authorities, would be valuable.

ACKNOWLEDGMENT

This research forms part of a larger project that investigates the readiness of housing developers and financial institutions to implement the BTS system in Malaysia. It is funded by the National Property Research Coordinator (NAPREC) of Malaysia. We also wish to acknowledge the two anonymous reviewers and the editor of the book for their constructive comments, which have greatly improved the chapter.

REFERENCES

Asibong, C., & Barlow, J. (1997, April). *Barriers to innovation and change in the housebuilding industry*. Paper presented at the Housing Studies Association Conference, York, England.

Barlow, J. (1999). From craft production to mass customization. Innovation requirements for the UK Housebuilding Industry. *Housing Studies*, *14*(1), 23–42. doi:10.1080/02673039982984

Binney, D. (2001). The knowledge management spectrum - understanding the KM landscape. *Journal of Knowledge Management*, *5*(1), 33–42. doi:10.1108/13673270110384383

Cadman, D., & Austin-Crowe, L. (1983). *Property Development* (2nd ed.). London: Spon.

Cavusgil, S. T., Calantone, R. J., & Zhao, Y. (2003). Tacit knowledge transfer and firm innovation capability. *Journal of Business and Industrial Marketing*, *18*(1), 6–21. doi:10.1108/08858620310458615

Chau, K. W., Wong, S. K., & Yiu, C. Y. (2007). Housing quality in the forward contracts market. *The Journal of Real Estate Finance and Economics*, *34*, 313–325. doi:10.1007/s11146-007-9018-x

Debowski, S. (2006). *Knowledge Management*. Milton, QLD: John Wiley & Sons.

Denison, D. R., & Mishra, A. K. (1995). Toward a theory of organisational culture and effectiveness. *Organization Science*, *6*(2), 204–223. doi:10.1287/orsc.6.2.204

Eisendiel, V. (1997, May). *Towards a Sustainable Housing Strategy*. Paper presented at the National Housing Convention. Kuala Lumpur, Malaysia.

Gibb, K. (1999). Regional Differentiation and the Scottish private housebuilding sector. *Housing Studies*, *14*(1), 43–56. doi:10.1080/02673039982993

Gloet, M., & Terziovski, M. (2004). Exploring the relationship between knowledge management practices and innovation performance. *Journal of Manufacturing Technology Management, 15*(5), 402–409. doi:10.1108/17410380410540390

Hassanuddin, T. (2009). *House Buyer's Awareness: the case of Penang, Malaysia*. Unpublished M. Sc. Project Management dissertation, Universiti Sains Malaysia, Malaysia.

Herkema, S. (2003). A complex adaptive perspective on learning within innovation projects. *The Learning Organization, 10*(6), 340–346. doi:10.1108/09696470310497177

House Buyers Association. (2003), Adopt build-then sell variant. *NST-PROPERTY*. Retrieved January 11, 2007 from http://www.hba.org.my/articles/buyer_ watch/2003/adopt_build.htm

House Buyers Association. (2005, June). *Sell then Build or Build then Sell or 10:90 Variant Models for the Housing Delivery System*. Paper presented at a Housing Workshop, Kajian Dasar Perumahan Negara, Penang, Malaysia.

Josephson, P. E., & Hammarlund, Y. (1999). The causes and costs of defects in construction: A study of seven building projects. *Automation in Construction, 8*(6), 681–687. doi:10.1016/S0926-5805(98)00114-9

Kasi, P. (1992, April). *Pros and Cons towards Build Then Sell concept*. Paper presented at the Seminar of the Build then Sell Housing Development Approach Towards 2020, Kuala Lumpur, Malaysia.

Kearns, G. S., & Lederer, A. L. (2004). The impact of industry contextual factors on IT focus and the use of IT for competitive advantage. *Information & Management, 41*(7), 899–919. doi:10.1016/j.im.2003.08.018

Lim, C. Y. (2003). *Factors contributing to E-readiness of SMEs*. Unpublished MBA thesis, Universiti Sains Malaysia, Malaysia.

Miles, M. B., & Huberman, A. M. (1994). *Qualitative data analysis* (2nd ed.). Newbury Park, CA: Sage Publication.

Miller, J. G. (1978). *Living Systems*. New York: McGraw-Hill.

Mills, A., Love, P. E. D., & Williams, P. (2009). Defect costs in residential construction. *Journal of Construction Engineering and Management*, (January): 12–16. doi:10.1061/(ASCE)0733-9364(2009)135:1(12)

Ministry of Housing & Local Government. (1999). *Housing in the New Millenium. A Malaysian Perspective*. Article 6. Retrieved March 5, 2007 from http://www.kpkt.gov.my/jpn/artikel/arti04.htm

Ministry of Housing & Local Government. (2003). *Number of Projects, Houses, House Buyers and the Estimated Value of Abandoned Housing Projects in Malaysia*. Retrieved March 5, 2007, from http://www.kpkt.gov.my/statistik/ perangkaan2002/kandungan.htm.

Mohd. Hanipah, B., Abd Rahman, M. Y., & Ishak, Z. (2005, September). *The Role of Financial Support and Strength in Ensuring the Completion of Housing Development Projects in Malaysia*. Paper presented at Inaugural Meeting & Seminar of the Real Estate Educators & Researchers Association Malaysia, Universiti Teknologi Mara City Campus, Kuala Lumpur, Malaysia.

Ong, S. E. (1997). Building defects, warranties and project financing from pre-completion marketing. *Journal of Property Finance, 8*(1), 35–51. doi:10.1108/09588689710160507

Raman, M. (1997). Housing and the Concerns of Consumers. In Berhad, C. (Ed.), *Housing The Nation: A Definitive Study* (pp. 709–741). Kuala Lumpur: ISIS.

Scarbrough, H. (2003). Knowledge Management, HRM and the Innovation Process. *International Journal of Manpower*, *24*(5), 501–516. doi:10.1108/01437720310491053

Siemieniuch, C. E., & Sinclair, M. A. (2004). A framework for organisational readiness for knowledge management. *International Journal of Operations & Production Management*, *24*(1), 79–98. doi:10.1108/01443570410511004

Star (2008). Slow response from Developers. 4 June 2008, pg 5.

Yusof, N., Mohd Shafiei, M. W., & Yahya, S. (2009). *Assessment of the Readiness of Private Housing Developers and Financial Institutions to the BTS concept*. Unpublished Final Report. NAPREC Research Grant.

Yusof, N., Mohd Shafiei, M. W., Yahya, S., & Ridzuan, M. (2010). Strategies to implement the "build then sell" housing delivery system in Malaysia. *Habitat International*, *34*, 53–58. doi:10.1016/j.habitatint.2009.06.001

ADDITIONAL READING

Ball, M. (2003). Markets and the Structure of the Housebuilding Industry: An International Perspective. *Urban Studies (Edinburgh, Scotland)*, *40*(5-6), 897–916. doi:10.1080/0042098032000074236

Beckman, T. J. (1999). The current state of knowledge management. In Liebowitz, J. (Ed.), *The Knowledge Management Handbook*. Boca Raton, FL: CRC Press.

Cagamas Berhad. (1997). *Housing The Nation: A Definitive Study*. Kuala Lumpur: ISIS.

Kamaruddeen, A. M., Yusof, N., & Said, I. (2010). Innovation and Innovativeness: Difference and Antecedent Relationship. *The Icfai University Journal of Architecture*, *2*(1), 1–13.

Mohd Shafiei, M. W., Yusof, N., & Yahya, S. (2008, April). *Assessing Property Developers' Readiness in the Implementation of Build Then Sell System: the proposed model*. Paper presented at International Real Estate Research Symposium (IRERS), PWTC Kuala Lumpur.

Naim, M., Naylor, J., & Barlow, J. (1999). Developing Lean and Agile Supply Chains in the UK Housebuilding Industry. [Berkeley, CA: University of California.]. *Proceedings, IGLC-7*, 159–170.

Nonaka, I., & Takeuchi, H. (1995). *The Knowledge-Creating Company*. New York: Oxford University Press.

Pan, W. (2010). Strategies for managing innovation in UK housebuilding. *Engineering, Construction, and Architectural Management*, *17*(1), 78–88. doi:10.1108/09699981011011339

Teece, D. J. (1998). Research directions for knowledge management. *California Management Review*, *40*(3), 289–292.

Tsoukas, H. (1996). The firm as a distributed knowledge system: a constructionist perspective. *Strategic Management Journal*, *17*, 11–25.

Yusof, N. (2007). *Pemaju Swasta dan Perumahan Kos Rendah*. Penang, Malaysia: Penerbit Universiti Sains Malaysia.

Yusof, N., Mohd Shafiei, M. W., Said, I., & Zainul Abidin, N. (2010). Factors Influencing Firms' Readiness Towards Innovation In House Building Industry: A Multi-Dimensional Construct. *International Journal of Organizational Innovation*, *2*(3), 74–88.

Yusof, N., Mohd Shafiei, M. W., & Yahya, S. (2007). Build Then Sell Models for Housing Industry: a review. *Journal of Valuation and Property Services*, *7*(1), 1–20.

Yusof, N., Mohd Shafiei, M. W., Yahya, S., & Fen, N. L. (2009). Adopting a new housing delivery system: the case of private housing developers in Malaysia. *Fifth International Conference on Construction in the 21st Century (CITC-V), May 20-22, Istanbul, Turkey,* (pp. 710-717).

KEYWORDS AND DEFINITIONS

Knowledge Management: The targeted coordination of knowledge as a factor of production and the management of the organisational environment to support individual knowledge transfer and the subsequent creation of collective knowledge.

Knowledge Sharing: An activity through which knowledge is exchanged among all the relevant stakeholders in any organisation of project situations.

Innovation: Innovation can generally be described as the adoption of an idea, technology, process or product that is new to an organisation.

Housing Delivery System: The procurement of houses, from the signing of the Sales and Purchase Agreements to the handover stage.

Housing Industry: An economic sector that concerns the supply and demand in the housing market.

Stakeholders: The parties involved that form the basis of the housing industry.

Knowledge Creation: Formation of new ideas through interactions between the tacit and explicit knowledge in individual human minds.

BTS Adopter: A party which implement build then sell housing delivery system

Build Then Sell (BTS): A housing delivery system whereby the housing developers must complete the houses before being able to sell them to the buyers.

Sell Then Build: A housing delivery system that allows housing developers to sell houses and collect payments before housing completion.

Chapter 9
Factors Affecting Knowledge Sharing Practice in Malaysia: A Preliminary Overview

Zawiyah Mohd Yusof
Universiti Kebangsaan Malaysia, Malaysia

Mohd Bakhari Ismail
Universiti Kebangsaan Malaysia, Malaysia

EXECUTIVE SUMMARY

The need for the adoption of knowledge management practice, particularly knowledge sharing is pressing in any modern organisation today in order to ensure their survival in this highly competitive world. It is evidence that knowledge sharing has given untold advantages to organisations such as the ability to make good decisions and formulate new strategies. This in turn enables organisations to remain competitive in their respective industry. Knowledge bears no value unless it is shared and well disseminated, particularly among workers within the organisation. Despite its advantages, knowledge sharing seems to be taken lightly by many organisations in Malaysia and only a few studies have been carried out. Therefore, the authors take an initiative to identify factors affecting knowledge sharing in Malaysian public sector. An extensive literature review has been conducted in this area in order to determine such factors. Synthesizing from the literatures, this chapter proposes a theoretical framework that takes into consideration the individual, organizational and technological dimensions that might affect knowledge sharing practise in Malaysian public sector. It is not the intention of this chapter to include the findings of this study and therefore the current study only develops a framework based on literature review. The findings of the study will be reported upon completion of the entire study.

DOI: 10.4018/978-1-61692-886-5.ch009

INTRODUCTION

Malaysia, as many other countries, is always concerned with the efficiency and effectiveness of its public service. As the government is striving towards becoming a developed nation in the year 2020, the role of public agencies is seen crucial to materialise the vision. Thus, the efficiency, speedy, transparency, openness, resourcefulness, integrity and accountability of public agencies have always been important issues (Ali, 2006). It is indicated statistically by the Public Complaints Bureau (PCB) of Malaysia that on average, 50 per cent of complaints received by the PCB were on the failures or delays in attending or responding to the needs of customers (Ninth Malaysia Plan, 2006). Though there are many factors contributing to this problem, but it has been identified that lack of information and knowledge sharing become the prime reason for the situation.

It has been argued that knowledge management could improve both the speed and the quality of public service delivery (Wigg, 1999). On the contrary, organisations hardly embark on the endeavour as they are not convinced of the advantages of sharing their knowledge and expertise. This is due to the fact that people are not willing to share knowledge unless it is useful and beneficial to them (Ryu, Hee & Han, 2003). Furthermore, it is simply difficult for any one person to share knowledge (Davenport & Prusak, 1998).

It is the aim of this paper to extensively and critically review the existing research in the area in order to determine the factors that contribute to the seemingly low acceptance of *knowledge sharing*. Finally, a theoretical framework of factors that influence knowledge sharing in public sectors is proposed. The results from the proposed model will not be presented here as the empirical study yet to be completed.

KNOWLEDGE SHARING AS A DIMENSION TO KNOWLEDGE MANAGEMENT

Knowledge management is a fast growing discipline with various issues yet to be tested, resolved and discovered (Beckman, 1999). Knowledge management is a process of identifying, organizing and managing knowledge resources (Al-Hawamdeh, 2003). It involves creating, generating, capturing, storing and using of knowledge to support and improve individual performance (Kinney, 1998; Davenport & Prusak, 1998). Despite the wide area covered by knowledge management field, Al-Hawamdeh (2003) suggests that there are at least five important dimensions to knowledge management activities namely: knowledge capture, knowledge creation, knowledge use (leverage), knowledge sharing, and knowledge retention. This paper focuses on the knowledge sharing dimension of knowledge management.

What Is Knowledge Sharing?

Knowledge sharing is a process that takes place between individuals (Ryu, Hee & Han, 2003). This process is not visible as it cannot be observed physically (Lee, 1989). In a broader perspective, knowledge sharing refers to the communication of all types of knowledge including explicit knowledge (information, know-how and know-who) and tacit knowledge (skills and competency) (Al-Hawamdeh, 2003). However sharing knowledge does not merely mean giving something to someone, or getting something from someone (Senge, 1990). This process occurs when an individual is interested to help others in developing a new capability for action. Thus, knowledge sharing is perceived to be only taken place if there is willingness of individuals in an organization to share the knowledge they created or possessed (Gibbert & Krause, 2002). Knowledge sharing is also defined as a deliberate act that make knowledge reusable by other people through knowledge transfer (Lee &

Al-Hawamdeh, 2002). Van den Hooff et al.(2003) added a new dimension to the above definition by asserting that knowledge sharing is a process in which individuals exchange knowledge (tacit or explicit) and together create new knowledge. It is also noted that knowledge sharing involves the dissemination of information and knowledge to the entire organization or department (Yang, 2004).

Despite various definitions of knowledge sharing, in this paper, knowledge sharing is defined as a process that involves individuals in public sector to share knowledge (tacit or explicit) with the intention to increase the performance and the quality of public service delivery.

Knowledge Sharing: Why Is It Importance?

Knowledge is considered critical and crucial to organizations (Ives, Torrey & Gordon, 2000) but it has little value unless it is shared (Small & Sage, 2006). Knowledge-based Theory of the Firm (Grant, 1996; Spender 1996) posits that knowledge is the most strategically significant resource of an organization to gain competitive advantage and increase organizational performance. Although knowledge belongs to individuals, it is only expressed when people cooperate in social communities such as groups, organizations and networks. In this environment, knowledge sharing exists.

In organization, knowledge sharing could serve four functions as follows (Zhang, Li & Shi (2005):

i) increases intellectual capital structure in organizations. Al-Hawamdeh (2003) asserts many industries nowadays are facing aging population and as a result it reduces the number of workforce. This situation requires knowledge which has been stored for years to be codified and transferred before they retire.
ii) changes individual competitiveness into organizational competitiveness, minimize organizational dependency on individual and reduce the possibility of lost employee because of changing place of work;
iii) changes organizational competitiveness into individual competitiveness in which individual can gain knowledge from organizational repository. This will increase individual competitiveness;
iv) the cost to gather knowledge in organization will be reduced compare to those available in the market.

STUDIES ON KNOWLEDGE SHARING IN PUBLIC SECTOR

Apparently, nowadays, the public organizations globally begin to acknowledge the importance of knowledge management. Knowledge management has been acknowledged to significantly contribute to making good decision, formulating good policies and better service delivery (Thomas, 2005). In public sector, knowledge sharing seems to be important to increase public sector performance and deliver better service to the public. However, study in this area is rather limited (McAdam & Reid, 2000) particularly in the developing countries like Malaysia (Syed Omar & Rowland, 2004).

Studies on knowledge management in public organizations have been carried out by a number of researchers like Liebowitz & Chen (2003); Shields, Holden & Schmidt (2000), Lee & Al-Hawamdeh (2003) and Wiig (2002). In Malaysia, studies on knowledge management in public sector have been conducted by Quin, Yusoff & Hamdan (2005) and Salleh & Syed Ahmad (2005). However, there were only three studies related to knowledge sharing in public organizations in Malaysia. The studies were carried out by Syed Omar (2005), Supar, Ibrahim, Mohamed, Yahya & Abdul (2005) and Ahmad, Sharom & Abdullah (2006).

Syed Omar (2005) conducted a study on knowledge transfer performance in a ministry in Malaysia. He studied the relationship between organizational elements and knowledge transfer performance. Syed Omar (2005) developed a model (Figure 1) to study how the organizational culture, organizational structure, technology, human resource and political directive affect knowledge transfer performance in a ministry.

Syed Omar (2005) model hypothesised that knowledgetransfer performance is influenced by sharing culture, individualistic, document confidentiality status, communication flow, ICT infrastructure, ICT tools, ICT know-how, posting, training, staff turnover and directive from politician. The results of the study show that organizational culture, individualistic, training, ICT know-how have significant relationship with knowledgeperformance transfer. However communication flow, ICT infrastructure and posting have no significant relationship with knowledge-performance transfer.

Supar et al. (2005) studied on factors affecting knowledge sharing in three selected higher institution and its impact on performance. The objective of the study is to understand the factors affecting knowledge sharing among academicians. The factors identified are culture (socialibility, solidarity, power distance), information technology (availability of IT infrastructure, availability of IT infrastructure for the purpose of knowledge sharing, expert vs, distributed model, problem of codification), communications (trust, face-to-face communication, reciprocity, reputation, altruism, acknowledgement), organizational support, rewards, mentoring, knowledge sharing to be included as part of work process). The results of the study indicates that management support is the most significant predictor to knowledge sharing followed by solidarity, expert vs. distributed model, knowledge sharing to be part of work process and the availability of IT infrastructure. Other factors have no significant impact on knowledge sharing among academicians in those three selected higher educations. The study also indicates that knowledge sharing has a significant impact on performance.

Ahmad, Sharom & Abdullah (2006) conducted a study on knowledge sharing in public sectors from business process perspective. They studied the knowledge sharing behaviour of public sectors in terms of individual attitudes, organizational environment, reward system and information and communication technology (ICT) within the system. The study looks into the business process management which embedded in a knowledge sharing behaviour. The preliminary results of the study indicate that all the dimensions contributed to the knowledge sharing environment in Malaysian public sector. Individual attitude was found to be the highest contributor.

From the studies mentioned above, it shows that study on knowledge sharing in public sector is limited. Why study on knowledge sharing in public organization is under researched? One of the reasons could be due to the status that public sector is a non-profit organization (Syed Omar & Rowland, 2004) which does not stress the importance of knowledge sharing. It shows that the adoption of the knowledge sharing practice in public sector is not pressing as compared to its private counterparts. Nevertheless, knowledge sharing is crucial to improve performance, increase customer and employee satisfactions (Pan & Scarbrough, 1999) whether in private or public sectors. Hence, to fill the gap, this study focuses on the knowledge sharing in public organizations in Malaysia.

PROPOSED MODEL FOR KNOWLEDGE SHARING

There are many factors that influence knowledge sharing in organization and a few models have been developed by previous researchers. For instance, Riege (2005) has listed three dozens factors or barriers that affect knowledge sharing behaviour in organization and divides them into 3 categories namely: individual, organizational and technology. Lee & al-Hawamdeh (2002) suggest that actors, channels, organizational elements and environment are factors that affect knowledge sharing behaviour. Taylor & Wright (2004) suggest six factors that influence knowledge sharing behaviour namely change readiness, innovative culture, clarity of responsibility, quality of information, strategic connection and accountability. Obviously, there are many factors that influence knowledge sharing practice.

Using Syed Omar (2005) model as the basis for the proposed framework as shown in Figure 2, this study focuses on how organizational and technological factors affect knowledge sharing quality among Malaysian public sector officers. In Syed Omar's model, sharing culture, individualistic, document confidentiality status, communication flow are organizational factors. Whereas ICT infrastructure, ICT tools, ICT know-how are technological factors. Meanwhile posting, training, staff turnover are considered human resource factors. Directive from politician is environmental factor.

In this study, modification is made to the Syed Omar's model using Orlikowski (1992) structurational model of technology as a basis. In her model, Orlikowski consider people, organization and technology are interrelated. Human create technology and decision makers to an organization. Technology facilitates human to do jobs. Organizational values influence how people interact with technology. Interaction between technology, people and organization is determined by social process, organizational structure and workers' activities with technological support (Orlikowski, 1992).

Based on Orlikowski's model of technology, Syed Omar's model lacks of individual factors. Thus, this paper considers individual factors as

Figure 1. Syed Omar (2005) knowledge transfer performance model

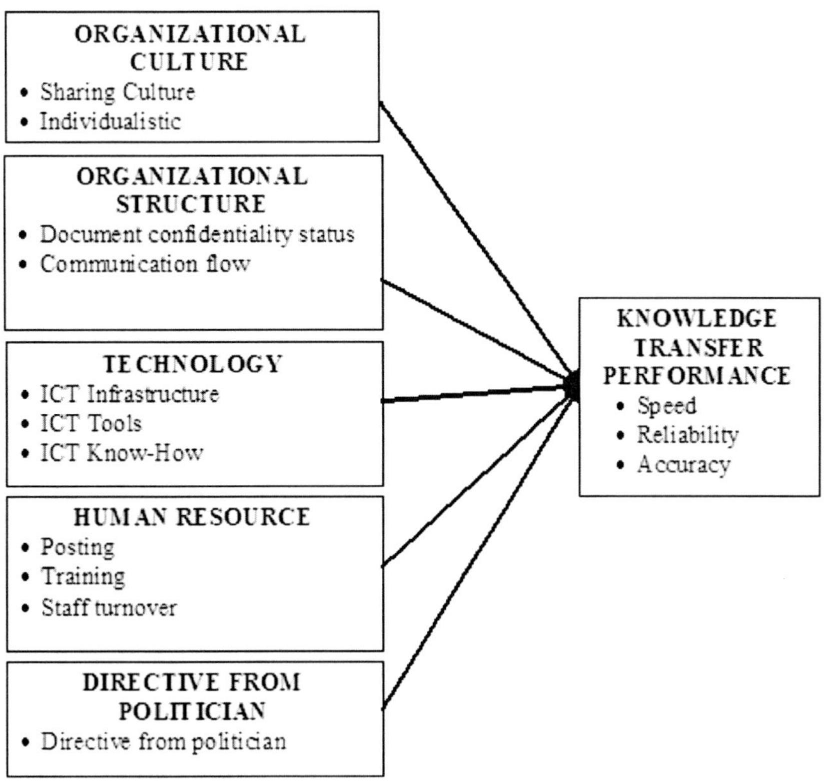

another potential predictor of knowledge sharing behaviour in public sector. The role of human is important in donating and using knowledge(Van den Brink, 2003). An effective knowledge management requires organizations to change their ways of doing business and people is the key of any effective change (Davis, 1998). Therefore, individual or human factors are important to implement knowledge sharing in any organization. Many researchers combined socio-psychological theories in developing theoretical framework to study knowledge sharing in organization (Samieh, 2007). Many researchers studied factors influencing knowledge sharing from organizational and technological perspective but less emphasis on individual perspective (Samieh, 2007). Many researchers agree that knowledge sharing depends on individual criteria such as experience, values, motivation and trust (Lin 2007). Therefore this study combines the individual factors together with organizational and technological factors in studying knowledge sharing in public organization in Malaysia. Other elements are not included in this study because of time and financial constraints.

The common question raised in knowledge management is the quality of knowledgeshared (Larsson & Ohlin 2002). Therefore, this study considers knowledge sharing quality instead of knowledgetransfer as dependent variable because quality knowledgebecomes important attention in a matured community (Chiu et al. 2006). Syed Omar focused on knowledgetransfer which involves one way interaction whereas knowledge sharing involves two ways or parties who are the

Figure 2. Proposed theoretical framework

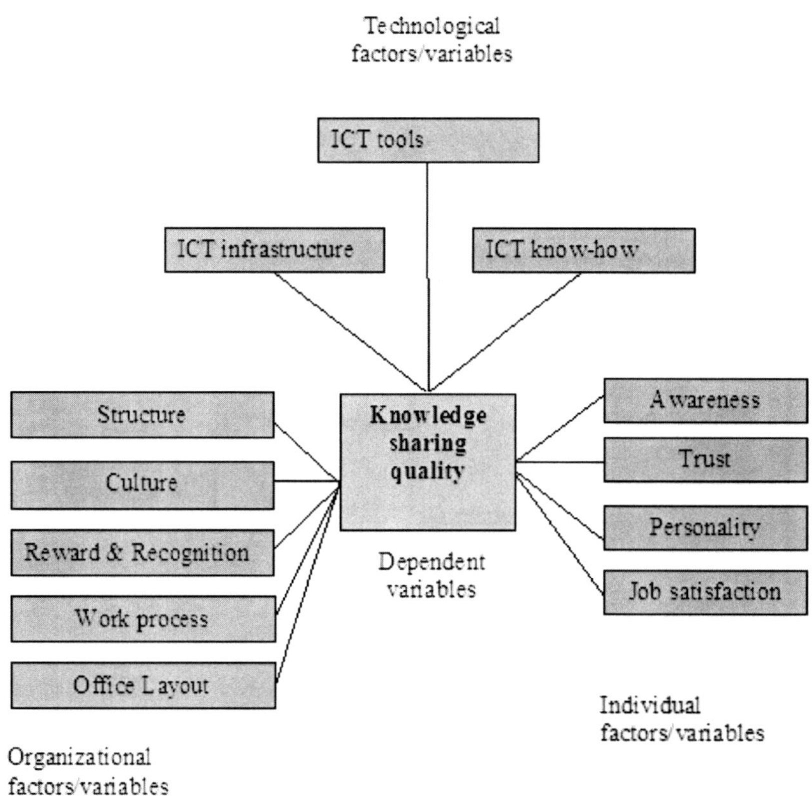

donors and the receivers. Furthermore this study focuses not only on knowledge sharing but also the quality of the knowledgeshared. The knowledgethat is shared should be useful for decision making, performance improvement and better service delivery. Employees should not spend time sharing knowledgewhich is not relevant to their jobs at workplace. It wastes time and does not benefit to the organization. In a matured community, people are more concerned about the quality of information and knowledgethey received. Although knowledge sharing can be divided into formal and informal knowledge sharing, what is important is the quality of the knowledgethat can complement each other (Zahra, Neubaum & Larraneta 2007). Therefore, organization should take both types of quality knowledge sharing to maximise achievement through quality knowledge management (Nidumolu et al. 2001; Orlikowski 2002).

Syed Omar's (2005) model is chosen and adapted because the study is done in Malaysia though only tested on only one public organization. Testing the model on a few public organizations is deemed necessary to draw generalization. Supar et al. (2005) is not chosen because the study was conducted on higher institutions not government organizations. While Ahmad et al. model was not considered because the focus in on the business process perspective rather than public service. Based on Syed Omar's (2005) model and upon critically synthesising literatures in the area such as Orlikowski's (1992) model of technology, Lee & Al-Hawamdeh (2002) actor's framework and a study by Van den Brink (2003), the authors suggest a theoretical framework as shown in Figure 2. The proposed framework is considered comprehensive as it includes combination of individual, organizational and technological factors as predictors of knowledge sharing quality. Factors affecting knowledge sharing at individual level must be understood in order for organizations to fully leverage their knowledge-based assets (Sharratt & Usoro, 2003). In terms of organizational factors, it is important to be considered because knowledge sharing is a process between individuals (Ryu et al., 2003) that takes place in organizations (Disterer, 2001; Van den Brink, 2003). Meanwhile, the role of information and communication technology is important to facilitate knowledge sharing process.

Individual Factors

Four components of individual dimension that affect knowledge sharing are proposed. These are: awareness (Lee & Al-Hawamdeh, 2002); trust (Sharratt & Usoro, 2003); personality (Awad & Ghaziri, 2004); and job satisfaction (Engstrom, 2003). Unawareness represents the first phase of knowledge sharing initiative in organizations without knowledge sharing process in place (Van den Brink, 2003). Awareness about the important of knowledge sharing is considered as an attitude that every employee should have including the top management.

Knowledge sharing is also facilitated by reciprocal and trust amongst members in a community (Scarbrough & Swan, 2001). Trust is described as an expression of confidence between several parties during whatever exchange, which means confidence that does not harm or risk through other parties' action, or confidence that is not exploited by any party (Jones & George, 1998). So, trust is the key to knowledge transfer (Davenport & Prusak, 1998).

According to Awad & Ghaziri (2004), personality is one of the impediments of knowledge sharing and employees who are extroverts, self confidence, feel secured have more tendency to share their experience and knowledge compared to those who are introverts, self-centred of security conscious. An individual personality can be characterized through his values, attitude, mood and emotion (Van den Brink 2003). In this study, personality is defined as employee's attitude whether extrovert, confident and feel secure to share knowledge compare to those who are introvert, self-centred and cautious (Awad & Ghaziri, 2004).

An employee should feel satisfied with his daily jobs in order to be in knowledge transfer environment (Engstrom, 2003) Through community of practise, employees share ideas and best practise to increase job satisfaction and overall team performance (*Socitm Insight,* 2003).

Organizational Factors

There are five variables identified under organizational dimension. These are organizational structure (Lee & Al-Hawamdeh 2002; Syed Omar & Rowland, 2004; Sharrat & Usoro, 2003), organizational culture (Lee & Al-Hawamdeh 2002; Syed Omar & Rowland, 2004; Sharrat & Usoro, 2003), reward and recognition, (Lee & Al-Hawamdeh 2002), work process (Lee & Al-Hawamdeh, 2002), and office layout (Lee & Al-Hawamdeh, 2002).

Organizational structure refers to how people and task in an organization is arranged to ensure the work done (*Encyclopaedia of Management*, 2000). Traditionally, public sector organizational structures are compartmentalized and this complicates the information and knowledge sharing between units and different levels in organizations (Cong & Pandya, 2003). Organizational structure defines how task, responsibilities, coordination and communication flow are managed (Van den Brink, 2003).

Organizational culture is one of the biggest challenges to knowledge sharing (Skyrme, 1997). Organizational culture means beliefs or values that are shared (Van den Brink 2003). Long (1997) explains organizational culture in terms of values, norms and practises.

Rewards could be in terms of both monetary and non monetary incentives (Bartol & Srivastava, 2002). But Davenport & Prusak (1998) suggest that monetary value such as financial rewards, salary increment and the like should be used to encourage and create a consistent knowledge sharing.

Knowledge management process similar to knowledge sharing should be included in work process (Davenport & Prusak 2000). Realising this, many organizations around the world had and are trying to introduce effective knowledge management in their work process (Chaudhry, 2005). According to Andersson (2000), one should be able to contribute knowledge as part of their work process. Larsson & Ohlin (2002) believe that the implementation of knowledge management initiatives (such as knowledge sharing) should be, if possible, integrated into current work process.

Davenport & Prusak (2000) suggest that corporate planner, architects, academics and executives should give consideration and creative thought to the issue of office design which hinder corporate world citizens from working with knowledge. It has becoming more important to design offices that could encourage socialization between employees to transfer knowledge (Arora, 2002). Furthermore, it is argued that office layout could encourage social interaction among employee (Lee & Al-Hawamdeh, 2002).

Technological Factors

Based on Orlikowski's model (1992), Van den Brink (2003) defines technology as software and hardware (ICT) that people in organizations use in order to perform their task. The main role of ICT in knowledge sharing is '*to connect people with other people or with explicit knowledge* (Van den Brink, 2003). Although real knowledge sharing has little to do with hardware or technology (Wenger & Synder 2000), three variables are considered to be included in the study. These are ICT tools, ICT infrastructure and ICT know-how (Syed Omar & Rowland, 2004).

In addition, effective knowledge management depends on the readiness of employees to share knowledge through computer facilities that could be accessed by other employees in the organization. ICT infrastructure is needed to support knowledge creation, knowledge structure, knowledge penetration and knowledge use (Van den Brink, 2003).

Apparently, ICT tools play important role in knowledge management (Smith, 2001) and the functionalities could be divided into five segments (Anderson & Smith, 1998). These are office applications (e-mail, messaging, calendaring and scheduling), groupware (databases, application sharing and electronic meeting systems), document systems (digital documents), work process systems (workflow management systems, process support systems and e-forms) analytical systems (decision support systems and data warehouse) and knowledge systems (portals, e-learning and knowledge sharing). ICT tools that exist in public organizations could help and facilitate employees to share knowledge (Syed Omar & Rowland, 2004). Also, sufficient and suitable ICT training to all employees have a positive relationship with knowledge creation and knowledge transfer.

Knowledge Sharing Quality

Much of the previous studies were focussing to only knowledge sharing behaviour. The quality of the knowledge shared was never become the concern. This study concentrates on the quality of knowledge shared among individuals in organization. Quality knowledge becomes central as community become mature (Chiu et al. 2006). In this study, the quality of knowledge shared is measured based on McKinney et al. (2002), DeLone & McLean (2003), Chiu et al. (2006) as the approach is relevant, easy to understand, accurate, complete, can be trusted and timely.

EVALUATION

The proposed theoretical framework provides the ground work for the next phase of the study. The framework has several advantages compared to previous knowledge sharing models related to public sector. Firstly, this model uses Structurational Model of Technology (Orlikowski 1992) as a basis which considers three important categories namely individual factors, organizational factors and technological factors. The combination of these three factors can be considered comprehensive because they are interrelated to each other. Secondly, factors like awareness, personality, job satisfaction and office layout have not been studied empirically particularly in public sector. Thirdly, important factors such as trust, organizational culture, organizational structure, rewards and recognition, ICT know-how, ICT infrastructure and ICT tools were not left out because they were empirically tested in previous researches and found to be important predictors to knowledge sharing. Finally, this paper focuses on knowledge sharing quality rather than knowledge sharing behaviour because quality knowledgeis important for performance improvement, decision making and better service delivery.

FUTURE RESEARCH DIRECTIONS

Methodologically, this study uses triangulation approach which combine both questionnaire and interview. Triangulation is an approach in which multiple observations theoretical perpsectives, sources of data and methodologies are combined (Bryman, 1996). Once the technique is identified, the next step is to develop questionnaires and interview questions to collect data. To ensure the validity and reliability of the instruments, pre-test and pilot study has to be conducted. Officers from three selected government agencies in Putrajaya, the administrative capital of Malaysia, has been selected as unit of analysis. Self-administered questionnaires are to be distributed to respondents from those agencies. The survey is complemented by face-to-face interview with a few government officers from those agencies. The interview could unearth factors that cannot be expressed through survey (Sekaran, 2005)

This study limits its investigation on factors affecting knowledge sharing quality to only three government agencies. As such the findings could not be generalised to wider public agencies. Future research ought to consider bigger samples involving more agencies to be more meaningful.

In this study, both questionnaire and interview are used to collect data. It is suggested that in future, another technique like participation observation technique should be considered to yield better results as the last technique would be able to collect detail information about the quality of knowledge sharing.

Research should also be extended to knowledge sharing between organisations rather than to only knowledge sharing within organisation. Samples should also comprise of senior officers and supporting staff so as to enable vertical knowedge sharing to be explored. As it is now, this study limits its sample to only officers from middle management group.

CONCLUSION

It is imperative to note that for public organizations to fully leverage the knowledge of their employees, they must first understand the factors that make their employees share knowledge. Sharing knowledge is an 'unnatural' act (Al-Hawamdeh 2003) because knowledgeis considered as 'power syndrome' (Davenport & Prusak 1998; Lee & Al-Hawamdeh 2000). If a worker considers knowledgeas a power, this will make knowledge sharing unnatural act in most organizations (Mitchell, 2006). Therefore knowledge sharing has to be nurtured. To encourage knowledge sharing, factors that influence the behaviour have to be identified.

Effective knowledge management requires a combination of many elements such as technology, human resource, structure and organisational culture (Donoghue et al. 1999). Thus, this study proposed a theoretical framework which comprises twelve factors seemingly to influence the practice of knowledge sharing in Malaysian public sector. Those factors are grouped into three categories namely individual, organizational and technological factors as suggested by Donoghue et al. (1999). The constructs have been operationally defined based on the scope of the research. The operational definition provides the foundations for empirical testing of the research model in the subsequent phase of the study. Groundwork is laid for a follow up study that will test and validate this framework. The findings of the study are yet to be finalised.

REFERENCES

Ahmad, H., Sharom, N., & Abdullah, C. S. (2006). *Knowledge Sharing Behaviour in the public sector: the business process management perspectives*. Retrieved from http://www.kmice.edu.my/files/kmice06.PDF.

Al-Hawamdeh, S. (2003). *Knowledge management: Cultivating knowledge professionals*. Oxford, UK: Chandos Publishing.

Ali, J. (2006). *Mengurus sumber manusia*. Dewan Bahasa dan Pustaka, Kuala Lumpur.

Andersson, K. (2000). *Knowledgetechnology applications for knowledge management*. Sweden: Tesis Dr. Falsafah. Uppsala University.

Arora, R. (2002). Implementing KM - a balances score card approach. *Journal of Knowledge Management*, *6*(3), 240–249. doi:10.1108/13673270210434340

Awad, E. M., & Ghaziri, H. M. (2004). *Knowledge management*. Upper Saddle River, NJ: Pearson Education Inc.

Bartol, K. M., & Srivastava, A. (2002). Encouraging knowledge sharing: The role of organizational reward systems. *Journal of Leadership & Organizational Studies*, *9*(1), 64–76. doi:10.1177/107179190200900105

Beckman, T. J. (1999). The current state of knowledge management . In *Knowledge management handbook*. New York: CRC Press.

Chaudhry, A. S. (2005). *Knowledge sharing practices in Asian institutions: a multi perspective from Singapore*. Retrieved from http://archive.ifla.org/IV/ifla71/papers/066e-Chaudhry.pdf.

Chiu, C. M., Hsu, M. H., & Wang, E. T. G. (2006). Understanding knowledge sharing in virtual communities: An integration of social capital and social cognitive theories . *Decision Support Systems*, *42*(3), 1872–1888. doi:10.1016/j.dss.2006.04.001

Cohen, W. M., & Levinthal, D. A. (1990). Absorptive Capacity: A new perspective on learning and innovation . *Administrative Science Quarterly*, *35*(1), 128–152. doi:10.2307/2393553

Cong, X. dan Pandya, K.V. (2003). Issues of knowledge management in the public sector . *Electronic Journal of Knowledge Management*, *1*(2), 25–33.

DeLone, W. H., & McLean, E. R. (2003). The DeLone and McLean model of information systems success: a ten year update . *Journal of Management Information Systems*, *19*(4), 9–30.

Davenport, T. H., & Prusak, L. (1998). *Working knowledge: How organizations manage what they know*. Boston: Harvard Business School Press.

Davenport, T. H., & Prusak, L. (2000). *Working knowledge: How organizations manage what they know*. Boston: Harvard Business School Press.

Disterer, G. (2001). *Individual and social barriers to knowledge transfer.* Retrieved from http://www.hicss.hawaii.edu/HICSS_34/PDFs/OSTTA01.pdf.

Engstrom, T. E. J. (2003). Sharing knowledge through mentoring. *Performance Improvement*, *42*(8), 36–42. doi:10.1002/pfi.4930420811

Gibbert, M., & Krause, H. (2002). Practice exchange in a best practice market place . In Davenport, T. H., & Gilbert, J. B. (Eds.), *Knowledge management case book: Siemen Best Practices*. Erlangen, Germany: Publicis Corporate Publishing.

Grant, R.M. (1996). Toward a Knowledge-based theory of the firm. *Strategic Management Journal,* 17, Winter Special Issue, 109-122.

Hawamdeh, S. (2003). *Knowledge management cultivating knowledge professionals*. Oxford, UK: Chandos Publishing.

Ives, W., Torrey, B., & Gordon, C. (2000). Knowledge sharing is human behavior . In Morey, D., Maybury, M., & Thuraisingham, B. (Eds.), *Knowledge Management: Classic and Contemporary Works*. Massachusetts: Massachussetts Institute of Technology Press.

Jones, G. R., & George, J. M. (1998). The experience and evolution of trust: Implications for cooperation and teamwork . *Academy of Management Review*, *23*(3), 531–546. doi:10.2307/259293

Larsson, K. J., & Ohlin, L. (2002). *Implementing a knowledge sharing initiative – principles for success*. Sweden: Tesis Sarjana. Linkoping Institute of Technology.

Lee, C. K., & Al-Hawamdeh, S. (2002). Factors impacting knowledge sharing . *Journal of Information and Knowledge Management*, *1*(1), 49–56. doi:10.1142/S0219649202000169

Lee, A. S. (1989). A scientific methodology for MIS studies. *Management Information Systems Quarterly*, *13*(1), 33–50. doi:10.2307/248698

Liebowitz, J., & Chen, Y. (2003). Knowledge sharing proficiencies: the key to knowledge management . In Hosapple, C. W. (Ed.), *Handbook on knowledge management 1: Knowledge Matter*. Berlin: Springer-Verlag.

McAdam, R., & Reid, R. (2000). A comparison of public and private sector perception and use of knowledge management. *Journal of European Industrial Training*, *24*(6), 317–329. doi:10.1108/03090590010346424

McKinney, V., Yoon, K., & Zahedi, F. M. (2002). The measurement of web-customer satisfaction: An expectation and disconfirmation approach. *Information Systems Research*, *11*(3), 296–315. doi:10.1287/isre.13.3.296.76

Mitchell, H. J. (2006). Knowledge Sharing- The value of story telling. *International Journal of Organisational Behaviour*, *9*(5), 632–641.

Orlikowski, W. J. (1992). The duality of technology: Rethinking the concept of technology in organizations. *Organization Science*, *3*(3), 398–427. doi:10.1287/orsc.3.3.398

Pan, S. L., & Scarbrough, H. (1999). Knowledge management in practice: An exploratory case study. *Technology Analysis and Strategic Management*, *11*(3), 359–374. doi:10.1080/095373299107401

Ninth Malaysia Plan 2006-2010. (2006). *Retrieved from*http://www.pmo.gov.my/ RancanganWeb/Rancangan1.nsf/fec3c825c96ab6be-482571a20024010b/ 520459cd0c6d953848257 1a90028cf68?OpenDocument [15 November 2006].

Riege, A. (2005). Three-dozen knowledge sharing barriers managers must consider. *Journal of Knowledge Management*, *9*, 18–35. doi:10.1108/13673270510602746

Ryu, S., Hee, H. S., & Han, I. (2003). Knowledge sharing behavior of physicians in hospitals . *Expert Systems With Application*, *25*, 113–122. doi:10.1016/S0957-4174(03)00011-3

Salleh, K. dan Syed Noh, S.A. (2005). *Knowledge management in the local authorities – a suitable platform for e-government.* Retrieved from http://www.kmice.edu.my.

Scarbrough, H., & Swan, J. (2001). Knowledge-communities and innovation . In Husyman, M., & Van Baalen, P. (Eds.), *Trends in communication cpecial issues on communities of practice.* Amsterdam: Boom.

Senge, P. (1990). *The fifth discipline: the art and practice of the learning organization* (1st ed.). Currency Doubleday. [s.l.]

Sharratt, M., & Usoro, A. (2003). Understanding knowledge-sharing in online communities of practice . *Electronic Journal of Knowledge Management*, *1*(2), 187–196.

Shields, R., Holden, T., & Schmidth, R. A. (2000). *A critical analysis of knowledge management initiatives in the Canadian public service: the impact of a knowledge-based economy on work in the public service, the virtual of expertise and knowledge-based economy on work in the public service, the virtual of expertise and knowledge.* Retrieved from http://www.carleton.ca/innovation/ km-fed.pdf.

Skyrme, D. J. (1997). *From information to knowledge management: Are you prepared?* Retrieved from http://www.skyrme.com/pubs/on97full.htm.

Small, C. T., & Sage, A. P. (2006). Knowledge management and knowledge sharing: A review. *Information Knowledge Systems Management*, *5*, 153–169.

Socitm Insight. (2003). *Rediscovering knowledge: an overview of knowledge sharing in the public sector*. retrieved from http://www.socitm.gov.uk.

Supar, N., Ibrahim, A. A., Mohamed, Z. A., Yahya, M., & Abdul, M. (2005). *Factors affecting knowledge sharing and its effects on performance: a study of three selected higher academic institutions.* Retrieved from http://www.kmice.edu.my.

Syed Omar, S. S. I. (2005). *Knowledge Management in a public organisation: A Study on the performance of knowledge transfer in the Ministry of Entrepreneur Development of Malaysia*. unpublished PhD thesis University of Lougborough United Kingdom.

Syed Omar, S. S. I., & Rowland, F. (2004). Benchmarking knowledge management in a public organisation in Malaysia. *Benchmarking: An International Journal*, *11*(3), 238–266. doi:10.1108/14635770410538745

Quin, T. Y., Yusof, M., & Hamdan, A. R. (2005). *Knowledge management readiness in organization: a case of public sector in Malaysia*. Retrieved from http://www.ickm.edu.my.

Taylor, W. A., & Wright, G. H. (2004). Organizational Readiness for Successful Knowledge Sharing: Challenges for Public Sector Managers. *Information Resources Management Journal*, *17*(2), 22–37.

Thomas, E. J. (2005). *Knowledge management in the public and private sector: a synthetic analysis the contemporary literature*. unpublished Master Thesis, School of Information Sciences and Technology, The Pennsylvania State University.

Van den Brink, P. (2003). *Social, organizational and technological conditions that enable knowledge sharing*. unpublished PhD Thesis, Faculty of Technology, Policy and Management, Delft University of Technology, Holland.

Van den Hooff, B., Elving, W. J. L., Meeuwsen, J. M., & Dumoulin, C. M. (2003). Knowledge Sharing in Knowledge Communities . In Huysman, M. H., Wulf, V., & Wenger, E. (Eds.), *Communities and Technologies*. Deventer: Kluwer Academic Publishers.

Wenger, E., & Synder, W. (2000). Communities of practice: the organizational frontier. *Harvard Business Review*, *78*(1), 139–145.

Wiig, K. M. (1999). Introducing knowledge management into the enterprise . In Liebowitz, J. (Ed.), *Knowledge management handbook*. New York: CRC Press.

Wiig, K. M. (2002). Knowledge management in public administration . *Journal of Knowledge Management*, *6*(3), 224–239. doi:10.1108/13673270210434331

Yang, J. T. (2004). Job-related knowledge sharing: comparative case studies . *Journal of Knowledge Management*, *8*(3), 118–126. doi:10.1108/13673270410541088

Zhang, L., Li, J., & Shi, Y. (2005). Study on improving efficiency of knowledge sharing in knowledge-intensive organizations. Retrieved from http://www.springerlink.com/content/f4nh617100065813/fulltext.pdf (accessed 7 March 2007).

KEY TERMS AND DEFINITIONS

Awareness: to what extent an officer's understands knowledge management, knowledge sharing and benefits he/she can get when sharing knowledge.

Trust: the confidence level of an officer on his/her colleagues' expertise and help, and his/her colleagues do not exploit him/her for their own interest.

Personality: to what extent an officer finds himself/herself as extrovert or introvert, self confidence or self-centered and secure or security conscious.

Job Satisfaction: the satisfaction level of an officer on daily works, job completion and colleagues' help to overcome problem.

Organizational Culture: to what extent an officer feels organizational culture encourage knowledge sharing, helping each other, knowledge dissemination and teamwork.

Organizational Structure: to what extent an officer feels document confidentiality hinder information retrieval, organizational structure hinder information flow and bureaucratic structure hinder knowledge sharing

Reward and Recognition: to what extent an officer feels incentives given by organization in terms of monetary, promotion and performance appraisal influence his/her knowledge sharing behaviour.

Work Process: to what extent an officer feels procedures, levels, routines and policies in doing daily works affect his/her knowledge sharing behaviour.

Office Layout: to what extent and officer feels his/her superiors, colleagues and subordinates' offices, physically open or close, affect his/her knowledge sharing behaviour.

ICT Infrastructure: to what extent an officer feels ICT infrastructure available in his/her office such as computers, internet and computer networks affect his/her knowledge sharing behaviour.

ICT Tools: to what extent an officer feels ICT tools such as groupware and computer-based information systems affect his/her knowledge sharing behaviour.

ICT Know-How: to what extent an officer feels ICT training is sufficient and facilitate him/her to share knowledge.

Chapter 10
The Influence of Emotional Intelligence on Tacit Knowledge Sharing in Service Organizations

Abdul Kadir Othman
Universiti Teknologi MARA (UiTM), Malaysia

Hazman Shah Abdullah
Universiti Teknologi MARA (UiTM), Malaysia

EXECUTIVE SUMMARY

The challenge of knowledge management is to generate, share, exploit and renew knowledge faster than competitors. To achieve sustainable competitive advantage, the knowledge management strategy is to develop and share tacit knowledge so that organizational capabilities rest in many, if not all, employees. In team-work environment, emotional intelligence, trust and organizational citizenship behavior are crucial to tacit knowledge sharing. Emotional intelligence as the ability to control one's and others' emotions, among other abilities, will facilitate the sharing of high value tacit knowledge, of work and clients, among team members. The road to a trusting relationship is marked by inter-personal understanding and empathic communication between members leading potentially to organizational citizenship behavior. Organizational citizenship behavior which manifests the employees' commitment to the job and organization is a significant mediator in tacit knowledge sharing. However, the role of emotional intelligence in facilitating tacit knowledge sharing is constrained by the type of service environment, i.e. professional, service shop and mass service which limits the scope for tacit knowledge sharing. This chapter offers several testable propositions to explain the influence of emotional intelligence on tacit knowledge sharing in service businesses.

INTRODUCTION

Emerging economies are the countries that have restructured their economies to allow greater opportunities in trade, technology transfer and foreign direct investments. China, India, Indonesia, Brazil and Russia are among the biggest emerging economies while countries like Malaysia, Mexico, Argentina, South Africa, Poland, Turkey, and South Korea are also considered as emerging economies. These countries have made critical moves from a developing country to emerging

DOI: 10.4018/978-1-61692-886-5.ch010

economy due to rapid economic growth. One of the fundamental elements to sustain the economic growth is to properly manage knowledge in organizations since this serves as the catalyst for innovation and creativity (Rebernik & Sirec, 2007). At corporate level, competitive advantage can be obtained when a company values its tacit knowledge (Alwis & Hartmann, 2008) as inimitable resource, which can differentiate a company from the rest.

Realizing the huge benefits that can be obtained from tacit knowledge sharing, authors and researchers have attempted to investigate how tacit knowledge can be shared among the organizational members. However, technology driven paradigm of knowledge management has failed to deliver the promised benefits (Gilmour, 2003). Hansen, Nohria and Tierney's (1999) codification strategy is not a viable option for firms with high tacit knowledge. Of course, this strategy is very successful when dealing with explicit knowledge. Tacit knowledge sharing, as the essence of the knowledge management implementation, is much related to the human aspects specifically the attitudes and behaviors of organizational members (Gilmour, 2003). Although research efforts have been directed at describing the human role in knowledge sharing (Haesli & Boxall, 2005), none has come out with a generic explanation of emotion and how it affects tacit knowledge sharing among organizational members particularly in service organizations. Therefore, this chapter is meant to bring together previous works on EI and tacit knowledge sharing into a cohesive framework with service types as moderators and team affiliation and organizational citizenship behaviors (OCB) as the mediators.

DEFINITION OF KNOWLEDGE

Davenport and Prusak (1998) defined knowledge as "a fluid of framed experience, values, contextual information, and expert insights that provides a framework for evaluating and incorporating new experiences and information. It originates in and is applied in the minds of knowers". Nonaka and Takeuchi (1995) defined knowledge in a broader perspective which is "a dynamic human process of justifying personal belief toward the truth". Two types of knowledge reside in an organization – explicit and tacit knowledge (Nonaka, 1991). Explicit knowledge is knowledge that can be codified therefore it is easily shared and communicated via organization databases, hence making it available to all members of the organizations. Explicit knowledge also refers to rationalized knowledge which is general, context independent, standardized, and public (Weiss, 1999).

Tacit knowledge comprises mental models, beliefs and persuasions of each employee that are so deeply rooted within the individual and is difficult to express in words. Tacit knowledge also refers to embedded knowledge which is context dependent, narrowly applicable, personalized, and may be personally or professionally sensitive (Weiss, 1999). It sometimes refers to practical intelligence (Wagner & Sternberg, 1985), common sense (Sternberg, 1985) or street smart (Horvath *et al.*, 1994). Tacit knowledge is seldom communicated and hence, seldom shared because it is subjective and intuitive in nature and it is lost when the employee possessing it leaves the organization. Examples of individual tacit knowledge are self-motivation, adaptability and flexibility and customer service experience, which generally refer to the blend of complex individual's abilities to accomplish work effectively.

With regard to knowledge sharing, Nonaka and Takeuchi (1995) have proposed a model describing how tacit knowledge can be transformed into explicit knowledge and shared in the organizations. Knowledge conversion model (also known as SECI Model), comprises four modes of conversions: socialization, externalization, combination and internalization. Socialization is the process by which an individual's tacit knowledge is shared with another individual through observation, imitation and practice, etc. Externalization is the process by which tacit knowledge is converted into explicit knowledge by using metaphors and analogies that are known to be inaccurate to reproduce knowledge, thus causing dialogue and collective reflection that serve as the foundation of externalization. Combination is the exchange process by which an individual's explicit knowledge is transferred to others by means of documents, lectures, meetings, telephone communication, etc. Internalization is the process of converting explicit knowledge into tacit knowledge through which the incorporated explicit knowledge in one individual is converted into tacit knowledge under the influence of his mental model. Knowledge conversion model is the basis of recent knowledge management theories and practices.

Tacit knowledge sharing has become the heightened interest of researchers over the years to divulge the fact of how it can be shared among the organizational members because of its importance in improving organizational performance (Harlow, 2008). The difficulties in sharing of tacit knowledge are associated with the difficulties in expressing or documenting such knowledge, the language issues, time, value as well as distance (Haldin-Herrgard, 2000). Besides, there are also factors that prevent individuals from sharing tacit knowledge, which include the lack of confidence, anxiety, unwillingness, confusion, and being carried away by strong feelings (Harvey & Butcher, 1998). Summarizing the existing work on knowledge sharing, Ipe (2003) has developed a cohesive framework highlighting the importance of three factors that affect knowledge sharing – nature of knowledge, motivation to share and opportunities to share. Since the focus of the chapter is on tacit knowledge sharing, motivation to share is the most relevant factor that affects knowledge sharing. Others are beyond the control of the employees.

Motivation to Share

Previous works have established that knowledge sharing is mostly influenced by trust and commitment (Morgan & Hunt, 1994), which refer to intrinsic motivation. Research has fairly established that intrinsic motivation can enhance knowledge sharing (Cruz, Perez, & Cantero, 2009). Organizational commitment, defined as an individuals' willingness to commit extra effort to their workplace job (Meyer & Allen, 1997), is expected to be associated with the willingness to contribute and receive knowledge (Nonaka, 1994). Individuals who are emotionally attached to their organization are likely to share their knowledge if the work environment is favorable where such behavior is appreciated and where their knowledge will be actually used and eventually be valuable to their organization (Lin, 2007).

Trust, on the other hand, is an expression of faith and confidence that a person or an institution will be fair, reliable, ethical, competent, and non-threatening (Morgan & Hunt, 1994). A lot of research has been conducted to investigate the role of trust in facilitating inter- and intra-organizational cooperation which includes knowledge sharing (Nonaka, 1991). Individuals are willing to share knowledge if they believe that their coworkers are trustworthy and faithful and would also do the same thing to them as well as others (Lin, 2007). Trust may reduce perceived uncertainty, facilitate risk-taking behaviors, and foster a positive orientation (Morgan & Hunt, 1994) that accordingly enhances their willingness to share tacit knowledge with their coworkers (Lin, 2007).

Individuals have two contradictory instincts – hoarding or sharing. They hoard if they feel threatened, fearful, distrustful or insecure, urging them to hold on tightly to whatever they possess. On the contrary, they share if they feel their contribution matters (Goman, 2007). Group membership plays an important role in deciding to share knowledge or not. The better a group of people knows each other, the more the people in the group will call on each other's knowledge (Dixon, 2002). In Asian culture and most of the emerging economies, group membership is considered as a source of identity, protection and loyalty, and in exchange for such loyalty, information can be expected to be shared within the group (Hutchings & Michailova, 2004). People are unwilling to share knowledge with those they perceived are not the group members (Yao, Kam, & Chan, 2007). Besides, personal networking has been eventually increasing in importance in business success as well as in knowledge sharing (Hutchings & Michailova, 2004).

To summarize, knowledge sharing takes place when there are networks and communities of people who are bound together by traditions of trust, mutual respect, etc. Emotional intelligence (EI), as the abilities to perceive emotions, to access and generate emotions so as to assist thought, to understand emotions and emotional knowledge, and to reflectively regulate emotions so as to promote emotional and intellectual growth, can facilitate the formation and maintenance of networks of people or teams and hence enhancing tacit knowledge sharing amongst them. Organizational members who possess high EI are expected to have high willingness to share tacit knowledge because of their heightened understanding of the organizational needs as well as the needs of those around them.

DEFINITION OF EI

EI is defined as a set of abilities to perceive emotions, to access and generate emotions so as to assist thought, to understand emotions and emotional knowledge, and to reflectively regulate emotions so as to promote emotional and intellectual growth (Mayer & Salovey, 1997). The detail descriptions of each dimension of EI are included in the table below (Table 1).

EI AND KNOWLEDGE SHARING IN EMERGING ECONOMIES

One of the elements to facilitate the economic growth in emerging economies is technology transfer but it can only be materialized if the com-

Table 1. EI dimensions/branches and descriptions of exemplary skills

Branches	Descriptions
Perceiving Emotion	Ability to identify emotion in one's physical and psychological states Ability to identify emotion in other people Ability to express emotion s accurately and to express needs related to them Ability to discriminate between accurate/honest and inaccurate/dishonest feelings
Using Emotions to Facilitate Thought	Ability to redirect and prioritize thinking on the basis of associated feelings Ability to generate emotions to facilitate judgment and memory Ability to capitalize on mood changes to appreciate multiple points of view Ability to use emotional states to facilitate problem solving and creativity
Understanding Emotions	Ability to understand relationships among various emotions Ability to perceive the causes and consequences of emotions Ability to understand complex feelings, emotional blends, and contradictory states Ability to understand transitions among emotions
Managing Emotions	Ability to be open to feelings, both pleasant and unpleasant Ability to monitor and reflect on emotions Ability to engage in, prolong, or detach from an emotional state Ability to manage emotions in oneself Ability to manage emotions in others

Source: Salovey, Kokkonen, Lopes, & Mayer (2004)

panies have properly institute sound knowledge management practices in their organizations. Technology transfer involves the sharing of explicit as well as implicit knowledge from the owner to the recipient. For example, Perodua, a Malaysian car manufacturer, has established relationship with Japanese counterpart, Daihatsu Motor Co., to facilitate technology transfer in car manufacturing industry. The explicit part of the technology transfer (e.g. machineries, procedures, systems) can be realized once both parties have reached a mutual agreement. Nevertheless, the implicit part or tacit knowledge (e.g. the technical know-how) can only be shared when there is an employee exchange program from both companies as part of the training efforts. At this juncture, Malaysian Perodua employees may be sent to Japan for training while the Japanese Daihatsu employees may be sent to Malaysia to assist Perodua employees in understanding the whole process of car making. The process is similar in service organizations. Therefore, knowledge sharing either explicit or implicit is crucial for companies to achieve sustained economic growth.

However, the process is not always easy because many factors might influence the success or failure of the technology transfer especially with regard to the transfer of implicit or tacit knowledge, for examples, cultural differences, and personality traits of participants. Some countries are seen as individualist but some are perceived as collectivist. Within the countries, there are individuals who are idiocentric (similar to people in individualist cultures) as well as those who are allocentric (similar to people in collectivist cultures) (Triandis, 2006). The cultural differences might affect the smoothness of the technology transfer process. In terms of personality traits, individuals are unique with regard to their varying levels of extraversion, openness-to-experience, conscientiousness, neuroticism and agreeableness (Costa & McCrae, 1995; John & Srivastava, 1999). Some traits facilitate knowledge sharing but others might impede the sharing process. These factors, among others, have to be considered in examining tacit knowledge transfer particularly in service organizations.

EI and Collaboration/Team Work

Team has been defined as working with co-workers and team members to achieve success of the firm (Welbourne, Johnson, & Erez, 1998). The importance of team role in organization as well as in research has recently been realized and given a special attention due to its effect in creating a synergy in work outcomes. However, individuals differ from each other in terms of culture and personality. Some individuals are idiocentric but others are allocentric. Similarly, some people are high in openness-to-experience but some are high in neuroticism. These individual differences impede the development of work team. Here, EI comes into play. A number of studies have established how EI can be used to enhance team performance. Rice (1999) and Feyerherm and Rice (2002) suggest that EI abilities are inevitable in obtaining effective team performance. Stough and De Guara (2003) discover that emotional control is positively related to the ability to work as a team member towards the success of the organization. Perlini and Halverson (2006) establish both intrapersonal competency and general mood predict number of National Hockey League points and games played. Jordan, Ashkanasy, Hartel and Hooper (2002) found that high EI teams operated at high level of performance throughout the study period but low EI teams performed lower at the beginning of the study period but managed to match the performance of the high EI teams at the end of the study period. The findings suggest that EI is effective at facilitating team performance. Jordan and Troth (2004), on the other hand, found that EI predicted successful problem solving of cognitive task in teams, and contributed to integrative conflict resolution styles and team performance. The findings are consistent with their earlier studies (Jordan & Troth, 2002). These studies have highlighted the importance of EI in generating effective teams, which can be realized through such practices as empathy, cooperation, collaborative efforts in problems solving and conflict avoidance.

EI has also been proven to enhance employees' commitment (Wong & Law, 2002) and trust (social and emotional contract) (Mayer & Beltz, 1998) – the core elements in knowledge sharing; therefore, having high EI team members can ensure the success of knowledge sharing as mentioned by Druskat and Wolff (2001) that EI of a group produces trust, identification and efficacy between group members and this in turn creates an atmosphere that is conducive towards a group's efficiency and creativity. It is suggested that team task, team structure and processes, along with the physical, virtual and social work spaces as well as the organizational context play important roles in enabling knowledge work productivity (Bosch-Sijtsema, Ruohomaki, & Vartiainen, 2009). It is evidenced in research that knowledge sharing is likely to occur if team members are emotionally welcome as valuable contributors to a common task (Sackmann & Friesl, 2007) by encouraging positive shared emotions, which can lead to group effectiveness through broadening and building interactions among group members (Rhee, 2005). EI is also practical to augment interpersonal relationships (Schutte *et al.*, 2001), which can enhance tacit knowledge sharing specifically in team and in organization as a whole. Therefore, the following proposition is developed:

Proposition 1. EI enhances tacit knowledge sharing by facilitating stronger team affiliation.

EI and OCB

Organizational citizenship behavior (OCB) is defined as discretionary behavior on the part of a person that is believed to promote directly the effective functioning of an organization (Organ, 1988). OCB consists of behaviors that go beyond specific role requirements, with the stipulation that such behaviors are performed voluntarily without expectation of rewards (Brief & Motowidlo, 1986). They include altruism (helping others), conscientiousness (going beyond the minimum role requirements), courtesy (preventing work-related problems from occurring), sportsmanship (tolerating unfavorable circumstances without complaining) and civic virtue (participating in, involving in, and concerning about the life of the company) (Podsakoff, MacKenzie, Moorman, & Fetter, 1990). Knowledge sharing, although is part of pro-social organizational behaviors (Lin, 2007), which captures the general propensity of people anticipating good consequences not only for themselves, but also for their co-workers and organization (Brief & Motowidlo, 1986) it is not necessarily synonymous with OCB. To be considered as OCB, the behaviors should be performed both spontaneously and voluntarily. Knowledge sharing, on the other hand, is often subject to managerial exhortations and organizational reward structures. Managers are now expected to coach, mentor or guide their subordinates to use and share knowledge across functional boundaries. It is increasingly becoming involuntary behavior. Moreover, it is assessed as part of overall performance management scheme. However, employees with high OCB tendency are expected to engage in tacit knowledge sharing based on the general premise that those with high OCB tendency are willing to go beyond their minimum role requirement helping others with work-related problems as well as pursuing organizational interests rather than personal interests.

As mentioned earlier, people are different with reference to their culture and personality traits. Studies have established that EI reduces the effects of having negative traits, such as low openness-to-experience and high neuroticism (Othman, 2010). Employees with low openness-to-experience and high neuroticism are perceived as low performers. However, when they are equipped with EI their perceived performance increases to the extent comparable with the performance of high openness-to-experience counterparts and low neuroticism counterparts. With regard to cultural differences, EI or specifically cultural intelligence can assist the individuals with different cultural upbringing to fit with the environment through actions such as suspending judgment until enough information is gathered, paying attention to the situation, cross cultural training, and matching personal and organizational attributes (Triandis, 2006).

EI, as the abilities to perceive emotions, to access and generate emotions so as to assist thought, to understand emotions and emotional knowledge, and to reflectively regulate emotions so as to promote emotional and intellectual growth, can be applied to enhance OCB of organizational members in a number of ways. First, employees with high EI are able to neutralize negative emotions or to turn negative emotions into positive ones. Positive emotions (e.g. enthusiasm and pride) is related to OCB whereas negative emotions (e.g. anger and anxiety) are related to counterproductive work behavior (Spector & Fox, 2002). Second, employees with high EI can "feel" the emotions of others. They have the ability to empathize the pain and pleasure experienced by their colleagues. Abraham (1999) claims that EI may enhance certain pro-social behaviors, one of which is assisting co-workers with personal matters. Those with high EI tend to exhibit high OCB. And third, employees with high EI are able to observe the consequences of their actions from different perspectives: personal, group as well as organizational. Day and Carroll (2004) established that EI is unrelated to individual-level OCB, but related to group-level OCB. Côté and Miners (2006) also discover a similar association between EI and organization-level OCB. Employees with high EI are inclined to pursue organizational interests rather than their personal interests. Research has also established the association between OCB and knowledge sharing (Kim & Gong, 2009; Yu, Lu, & Liu, 2010). These findings, although inconclusive, suggest that EI enhances pro-social behaviors or OCB, which subsequently contribute to tacit knowledge sharing among organizational members. Thus, the following proposition is suggested:

Proposition 2. EI increases the propensity of tacit knowledge sharing as a result of employees' willingness to engage in voluntary behaviors (OCB).

SERVICE TYPOLOGY

The diversity of service sector makes it difficult to make useful generalizations concerning the management of service organizations. Therefore, classification of services has greatly contributed to this area. The purpose of service typology is to address the complexities of services by developing service criteria that reflect core service aspects that go beyond narrow industry boundaries (Cook, Goh, & Chung, 1999). Realizing the need to develop a practical service typology, Silvestro, Fitzgerald, Johnson and Voss (1992), among others, proposed a classification scheme based on business processes, bringing together several previously suggested classification schemes into a cohesive framework. Based on data from 11 service industries, Silvestro *et al.* (1992) suggested three types of service processes; professional service, service shops, and mass services. The classification is based on six dimensions; people versus equipment, level of customization, extent of employee/customer contact, level of employee discretion, value added in back office versus front office, and product versus process focus. The service process types are as follows:

- Professional service – Organizations with few transactions, highly customized, process oriented, with a long contact time. Value-added process occurs in the front office where considerable judgment is applied in meeting customer needs. Examples are consultants, doctors and architects.
- Service shop – A category which falls between professional and mass services with

levels of classification dimensions falling between the other two extremes. Examples are retail banks, rental services and hotels.

- Mass service – Organizations with many customer transactions, limited contact time, little customization, product-oriented with most of the value being added in the back office and little judgment applied by the front office staff. Examples are telecommunication, bus services and fast foods.

The Role of Service Typology in Affecting EI and Knowledge Sharing

Although the evidence supports that EI enhances tacit knowledge sharing among team members, service type moderates this relationship. In professional service, tacit knowledge sharing is crucial in achieving team effectiveness. Service providers engage in extended interactions with the customers, use personal judgment and discretion in providing customized service to the customers. There is a demand to share their personal experiences so that others can follow their success efforts (Weiss, 1999). Having high EI, professional service providers are able to develop better relationship with colleagues, thus facilitating tacit knowledge sharing as trust is necessary for effective knowledge sharing to take place. In emerging economies, businesses are becoming to a greater extent professionally oriented. Services are changing to become increasingly customized. For example, insurance companies, because of the fierce competition in the industry, offer various products to cater for varying needs of the customers. The insurance agents have to be, not only knowledgeable, but also flexible in dealing with customers. Companies in emerging economies, especially Asian countries are having advantages in facilitating tacit knowledge transfer because research has recently established that, although the results are inconclusive, knowledge transfer is subject to national cultural characteristics which are power distance, performance orientation, in-group collectivism, and uncertainty avoidance (Wilkesmann, Fischer, & Wilkesmann, 2009). Therefore, to facilitate knowledge sharing, they have to be equipped with high EI to effectively perform their duties. Besides, they also need high EI to be readily available to share their experiences with co-workers especially in team environment for the benefits of the organization as a whole.

However, in service shops, since the services are more standardized and there is lack of interaction, EI abilities are moderately needed. Similarly, tacit knowledge sharing, although important, is not as critical as in professional service. Tacit knowledge sharing together with conformance to standard operating procedures (SOP) might contribute to team effectiveness. Usually, there are some guidelines stipulated by the management about teamwork. However, having high EI is encouraged so that tacit knowledge sharing can be enhanced to achieve team effectiveness.

In mass service, the employees are offering standardized and routine services with minimum interaction with others. Although working in team is encouraged, the management has already determined every aspect of team to function effectively. Rules and regulations are established, roles and functions of members are predetermined, and methods are fixed. Therefore, in mass service, because of structured team role, the need for EI is the least significant for the employees to achieve effective team role. Similarly, tacit knowledge sharing is the least important since the nature of work does not require employees to learn from others' personal experience. Therefore, the following proposition is formed:

Proposition 3. The role of EI in facilitating tacit knowledge sharing among team members is the most significant in professional service, moderate in service shop and the least significant in mass service due to the different scope for tacit knowledge sharing across service types (it is wide in professional service, moderate in service shop and narrow in mass service).

It is proposed that EI enhances tacit knowledge sharing as a result of employees' willingness to engage in voluntary behaviors (Tschannen-Moran & Nestor-Baker, 2004). Again, the relationship is moderated by types of services by which the employees are associated. A study conducted by Turnipseed (1996) establishes that work environment that consists of involvement, peer cohesion, supervisor support, autonomy, task orientation, clarity, innovation and physical comfort is positively related to OCB. In professional service (described as having favorable work environment), services are customized and dependent on employees' discretion and judgment. The work environment encourages the employees to engage in OCB. Since tacit knowledge sharing is critical for work accomplishment in professional service due to the service nature, EI comes into play to facilitate the sharing. The call for EI is escalating in emerging economies due to ever-increasing competition intensity. Since human capital is considered as a source of competitive advantage, businesses are looking for the right formula for employees to voluntarily engage in OCB. Research has established that employees with high EI abilities are capable of aligning their interests with those of the organization and are capable of anticipating their future with the organization (Othman, 2010). Besides, they are capable of empathizing others' work-related problems and consequently, are willing to offer their assistance (Abraham, 1999). Therefore, EI is crucial in fostering tacit knowledge sharing among employees in professional service setting through employees' willingness to engage in OCB.

In service shops, since the services are more standardized and they require less interaction with the customers as well as other employees, going above the call of duties e.g., helping others and promoting the company might not be as imperative as in professional service. Therefore, in service shop, because of moderate work environment, employees' OCB level is also moderate. The need for tacit knowledge sharing is also moderate. Hence, the role of EI in facilitating knowledge sharing in service shop lessens with reference to the nature of service.

In mass service (described as having unfavorable work environment), offering standardized and routine service does not significantly require employees' discretion and judgment as well as extended interaction with others. Therefore, going above the call of duty i.e. engage in OCB is the least significant in mass service. Conforming to the standard operating procedures (SOP) is the norm and sharing of tacit knowledge seems unnecessary. Likewise, due to nature of service, the role of EI is the least significant in promoting OCB and subsequently enhancing tacit knowledge sharing as suggested by Zapf (2002) that doing "object-related work" does not require effective emotion management. For this reason, the following proposition is put forth.

Proposition 4. The role of EI in facilitating tacit knowledge sharing through employees' willingness to engage in OCB is the most significant in professional service, moderate in service shop and the least significant in mass service due to the different scope for tacit knowledge sharing across service types (it is wide in professional service, moderate in service shop and narrow in mass service).

Figure 1 below illustrates the proposed relationships between EI and tacit knowledge sharing and the role of mediators – team affiliation, and OCB, as well as moderators – service types (professional service, service shop and mass service) in affecting the relationships.

FUTURE RESEARCH DIRECTIONS

The chapter is meant to bring together the works on EI and knowledge management into a cohesive framework of how EI can facilitate tacit knowledge sharing in service organizations especially in professional service. Future research should focus on testing the propositions developed in

Figure 1. A proposed framework

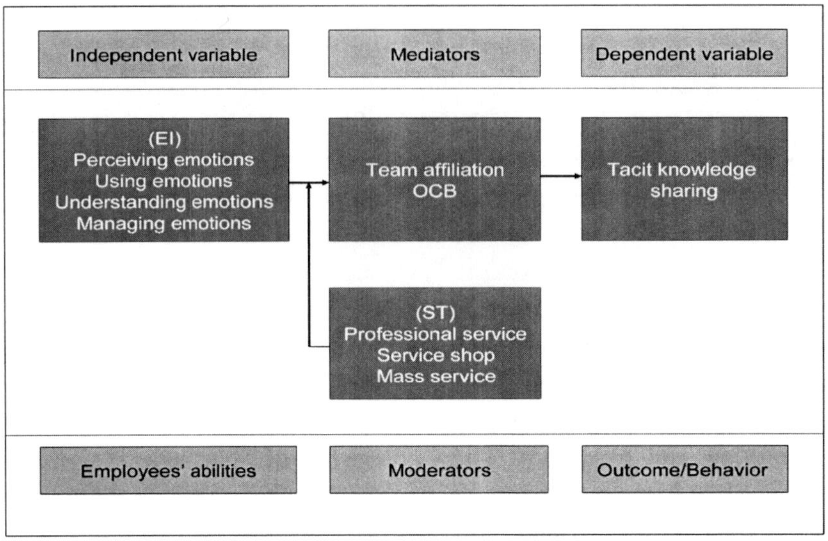

this chapter. Besides, several factors such as leadership, organizational culture and personality traits of service providers should be reviewed and considered since these factors are expected to significantly influence knowledge sharing practices in organizations.

With regard to the scales to measure the constructs, researchers who are interested in this area should be highly cautious to select the most relevant and validated scales. To date, there are abundant available measures, some of which are empirically validated but others are not. Selecting empirically validated measures ensure the validity and reliability of the research findings, which can be used with less skepticism.

Since the proposed model includes testing the mediating and moderating effects of the variables concerned, it is suggested that future research should use Structural Equation Modeling (SEM) to clearly identify the relationships among the variables in each model (based on service types). To permit the analyses, enough data should be collected and be divided into three categories according to service types; professional, service shop and mass service. Then, the comparison among the models could be conducted and respective model for each service type established.

CONCLUSION

In emerging economies, tacit knowledge sharing has become a central issue in most organization because of its role in contributing to competitive advantage. Tacit knowledge sharing is a dynamic process as it depends on individual' attributes and social relationships. Besides, based on the nature of services (e.g. standardization versus customization), it is critical in professional service, moderate in service shop and the least significant in mass service. In professional service, having a successful tacit knowledge sharing requires service providers to be equipped with high EI abilities so that they can interact effectively among team members and engage in pro-social behaviors. The EI abilities have been demonstrated to enhance team performance and to encourage pro-social behaviors among organizational members and these practices will facilitate tacit knowledge sharing in the organizations.

Managerial Implication

Organizations/managers must have a policy that calls for or facilitates the sharing of knowledge. If what we have argued in the chapter is to be true, managers must not leave this critical behavior to chance – as OCB may or may not happen. They should take control of developing and also ensuring the sharing of knowledge especially of the tacit type both for spreading the capabilities across the organization and also develop "sustainable" (as opposed to fragile) sources of strength. To take control means to organize, monitor, review and enhance. A policy to this end will signal to all staff the institutional importance of tacit knowledge sharing. Too many organizations are still focused on ensuring the explicit knowledge is captured, promoted, renewed and protected. Tacit knowledge sharing is highly valued but still an uncharted territory for managers. There is much rhetoric than focused action on tacit knowledge sharing. So, raising issue on the institutional radar is the first call for action.

A policy without a strategy for action is pure politics. An HR strategy should be devised to ensure that tacit knowledge sharing becomes a deliverable for the HR managers. This is in keeping with the most strategic HRM observations of the new roles for HR managers – to be an administrative expert, change champion, employee champion and business/strategic partner. This HR strategy must recognize also that the issue of tacit knowledge sharing is a k-worker issue probably most felt in professional service organizations as argued in the chapter. The strategy should therefore identify the strategic positions that offer most impact. The chapter offers an approach to looking at the scope for this strategy to work. Professional services offer the best scope for tacit knowledge sharing and hence, require greater need for a policy and managerial action.

Tacit knowledge sharing should be part of the attributes/criteria that organization sets for this category of staff i.e. professional service staff. The empirical indicators should be developed and used as a guide to observing and appraising this attribute. Regular feedback on the tacit knowledge sharing should be provided to help the sharing process. When a criterion becomes clear (communicated in words and actions), accepted (by all as necessary), valuable (because it enhances) and durable (remains a policy over time across different leaders/managers), it becomes a culture of the organization. Culturally conditioned behaviors are as powerful if not more forceful than contracts in enforcing rules of conduct.

Managers should also understand and develop the individual and group competencies that drive or facilitate the tacit knowledge sharing. This chapter offers one such driver i.e. EI. EI is a key ability that organizations must develop because tacit knowledge sharing is a social exchange process that takes place under a mutually trusting environment. People with great EI can regulate themselves and also the people environment, thus, creating conditions suitable for the excise of tacit knowledge sharing. Therefore, as much as functional and technical skills must be developed and renewed, EI should be given equal attention and time as it acts as a great facilitator of the strategic knowledge management and HR changes organizations seek.

REFERENCES

Abraham, R. (1999). Emotional intelligence in organizations: A conceptualization. *Genetic, Social, and General Psychology Monographs*, *125*(2), 209–225.

Alwis, S.-R., & Hartmann, E. (2008). The use of tacit knowledge within innovative companies: Knowledge management in innovative enterprises. *Journal of Knowledge Management*, *12*(1), 133–147. doi:10.1108/13673270810852449

Bosch-Sijtsema, P. M., Ruohomaki, V., & Vartiainen, M. (2009). Knowledge work productivity in distributed teams. *Journal of Knowledge Management*, *13*(6), 533–546. doi:10.1108/13673270910997178

Brief, A. P., & Motowidlo, S. J. (1986). Prosocial organizational behaviors. *Academy of Management Review*, *11*, 710–725. doi:10.2307/258391

Cook, D. P., Goh, C.-H., & Chung, C. H. (1999). Service typologies: A state of the art survey. *Production and Operations Management*, *8*(3), 318–339. doi:10.1111/j.1937-5956.1999.tb00311.x

Costa, P. T., & McCrae, R. R. (1995). Domains and facets: Hierarchical personality assessment using the Revised NEO Personality Inventory. *Journal of Personality Assessment*, *64*, 21–50. doi:10.1207/s15327752jpa6401_2

Côté, S., & Miners, C. T. H. (2006). Emotional intelligence, cognitive intelligence, and job performance. *Administrative Science Quarterly*, *51*(1), 1–28.

Cruz, N. M., Perez, V. M., & Cantero, C. T. (2009). The influence of employee motivation on knowledge transfer. *Journal of Knowledge Management*, *13*(6), 478–490. doi:10.1108/13673270910997132

Davenport, T. H., & Prusak, L. (1998). *Working knowledge: How organizations manage what they know*. Boston: Harvard Business School Press.

Day, A. L., & Carroll, S. A. (2004). Using an ability-based measure of emotional intelligence to predict individual performance, group performance, and group citizenship behaviors. *Personality and Individual Differences*, *36*(6), 1443–1458. doi:10.1016/S0191-8869(03)00240-X

Dixon, N. M. (2002). The neglected receiver of knowledge sharing. *Ivey Business Journal* (March), 35-40.

Druskat, V. U., & Wolff, S. B. (2001). Building the emotional intelligence of groups. *Harvard Business Review*, *79*(3), 80–90.

Feyerherm, A. E., & Rice, C. L. (2002). Emotional intelligence and team performance: The good, the bad and the ugly. *The International Journal of Organizational Analysis*, *10*(4), 343–362. doi:10.1108/eb028957

Gilmour, D. (2003). How to fix knowledge management. *Harvard Business Review*, *81*(10), 16–17.

Goman, C. K. (2007). The hoarding/sharing instinct. *Knowledge Management Review*, *9*(6), 6–6.

Haesli, A., & Boxall, P. (2005). When knowledge management meets HR strategy: an exploration of personalization-retention and codification-recruitment configurations. *International Journal of Human Resource Management*, *16*(11), 1955–1975. doi:10.1080/09585190500314680

Haldin-Herrgard, T. (2000). Difficulties in diffusion of tacit knowledge in organizations. *Journal of Intellectual Capital*, *1*(4), 357–365. doi:10.1108/14691930010359252

Hansen, M. T., Nohria, N., & Tierney, T. (1999). What's your strategy for managing knowledge? *Harvard Business Review*, *77*(2), 106–116.

Harlow, H. (2008). The effect of tacit knowledge on firm performance. *Journal of Knowledge Management*, *12*(1), 148–163. doi:10.1108/13673270810852458

Harvey, P., & Butcher, D. (1998). Those who make a difference: Developing business through developing individuals. *Industrial and Commercial Training*, *30*(1), 12–15. doi:10.1108/00197859810197690

Horvath, J. A., Williams, W. M., Forsythe, G. B., Sweeney, P. J., Sternberg, R. J., McNally, J. A., et al. (1994). *Tacit knowledge in military leadership: A review of the literature* (Technical Report 1017). Alexandria, VA: United States Army Research Institute for the Behavioral and Social Sciences.

Hutchings, K., & Michailova, S. (2004). Facilitating knowledge sharing in Russian and Chinese subsidiaries: The role of personal networks and group membership. *Journal of Knowledge Management*, *8*(2), 84–94. doi:10.1108/13673270410529136

Ipe, M. (2003). Knowledge sharing in organizations: A conceptual framework. *Human Resource Development Review*, *2*(4), 337–359. doi:10.1177/1534484303257985

John, O. P., & Srivastava, S. (1999). The Big Five trait taxonomy: History, measurement and theoretical perspectives. In Pervin, L. A., & John, O. P. (Eds.), *Handbook of personality: Theory and research* (pp. 102–138). New York: Guilford Press.

Jordan, P. J., Ashkanasy, N. M., Hartel, C. E. J., & Hooper, G. S. (2002). Workgroup emotional intelligence development and relationship to team process effectiveness and goal focus. *Human Resource Management Review*, *12*(2), 195–214. doi:10.1016/S1053-4822(02)00046-3

Jordan, P. J., & Troth, A. C. (2002). Emotional intelligence and conflict resolution: Implications for human resource development. *Advances in Developing Human Resources*, *4*(1), 62–79. doi:10.1177/1523422302004001005

Jordan, P. J., & Troth, A. C. (2004). Managing emotions during team problem solving. *Human Performance*, *17*(2), 195–218. doi:10.1207/s15327043hup1702_4

Kim, H., & Gong, Y. (2009). The roles of tacit knowledge and OCB in the relationship between group-based pay and firm performance. *Human Resource Management Journal*, *19*(2), 120–139. doi:10.1111/j.1748-8583.2009.00093.x

Lin, C.-P. (2007). To share or not to share: Modeling tacit knowledge sharing, its mediators and antecedents. *Journal of Business Ethics*, *70*, 411–428. doi:10.1007/s10551-006-9119-0

Mayer, J. D., & Beltz, C. M. (1998). Socialization, society's "emotional contract," and emotional intelligence. *Psychological Inquiry*, *9*(4), 300–304. doi:10.1207/s15327965pli0904_12

Mayer, J. D., & Salovey, P. (1997). What is emotional intelligence? In Salovey, P., & Sluyter, D. (Eds.), *Emotional development and emotional intelligence: Education implication*. New York: Basic Books.

Meyer, J. P., & Allen, N. J. (1997). *Commitment in the workplace: Theory, research and application*. Thousand Oaks, CA: Sage.

Morgan, R. M., & Hunt, S. D. (1994). The commitment-trust theory of relationship marketing. *Journal of Marketing*, *58*, 20–38. doi:10.2307/1252308

Nonaka, I. (1991). The knowledge creating company. *Harvard Business Review*, (November/December): 96–104.

Nonaka, I. (1994). A dynamic theory of organizational knowledge creation. *Organization Science*, *5*(1), 14–37. doi:10.1287/orsc.5.1.14

Nonaka, I., & Takeuchi, H. (1995). *The knowledge creating company: How Japanese companies create the dynamics of innovation*. New York: Oxford University Press.

Organ, D. W. (1988). *Organizational citizenship behaviors: The good soldier syndrome*. Lexington, MA: Lexington Books.

Othman, A. K. (2010). *The influence of emotional intelligence on job performance in service businesses*. Unpublished PhD thesis, Universiti Teknologi MARA, Shah Alam.

Perlini, A. H., & Halverson, T. R. (2006). Emotional intelligence in the National Hockey League. *Canadian Journal of Behavioural Science*, *38*(2), 109–120. doi:10.1037/cjbs2006001

Podsakoff, P. M., MacKenzie, S. B., Moorman, R. H., & Fetter, R. (1990). Transformational leader behaviors and their effects on followers' trust in leader, satisfaction, and organizational citizenship behaviors. *The Leadership Quarterly*, *1*, 107–142. doi:10.1016/1048-9843(90)90009-7

Rebernik, M., & Sirec, K. (2007). Fostering innovation by unlearning tacit knowledge. *Kybernetes*, *36*(3/4), 406–419. doi:10.1108/03684920710747039

Rhee, S.-Y. (2005). *How shared emotions among group members influence group effectiveness? The role of broadening-and-building interactions.* Unpublished PhD dissertation, The University of Michigan.

Rice, C. L. (1999). *A quantitative study of emotional intelligence and its impact on team performance.* Unpublished master's thesis, Pepperdine University, Malibu, CA.

Sackmann, S. A., & Friesl, M. (2007). Exploring cultural impacts on knowledge sharing behavior in project teams: results from a simulation study. *Journal of Knowledge Management*, *11*(6), 142. doi:10.1108/13673270710832226

Salovey, P., Kokkonen, M., Lopes, P. N., & Mayer, J. D. (2004). Emotional Intelligence: What do we know? In Manstead, A. S. R., Frijda, N., & Fischer, A. (Eds.), *Feelings and Emotions: The Amsterdam symposium* (pp. 321–340). Cambridge, UK: Cambridge University Press.

Schutte, N. S., Malouff, J. M., Bobik, C., Coston, T. D., Greeson, C., & Jedlicka, C. (2001). Emotional Intelligence and Interpersonal Relations. *The Journal of Social Psychology*, *141*(4). doi:10.1080/00224540109600569

Silvestro, R., Fitzgerald, L., Johnson, R., & Voss, C. (1992). Towards a classification of service processes. *International Journal of Service Industry Management*, *3*(3), 62–76. doi:10.1108/09564239210015175

Spector, P. E., & Fox, S. (2002). An emotional-centered model of voluntary work behavior: Some parallels between counterproductive work behavior and organizational citizenship behavior. *Human Resource Management Review*, *12*, 269–292. doi:10.1016/S1053-4822(02)00049-9

Sternberg, R. J. (1985). *Beyond IQ: A triarchic theory of human intelligence*. Cambridge, UK: Cambridge University Press.

Stough, C., & De Guara, D. (2003). Examining the relationship between emotional intelligence and job performance. *Australian Journal of Psychology*, *55*, 145.

Triandis, H. C. (2006). Cultural intelligence in organizations. *Group & Organization Management*, *31*(1), 20–26. doi:10.1177/1059601105275253

Tschannen-Moran, M., & Nestor-Baker, N. (2004). Tacit knowledge of productive scholars in education. *Teachers College Record*, *106*(7), 1484–1511. doi:10.1111/j.1467-9620.2004.00387.x

Turnipseed, D. (1996). Organization citizenship behaviour: An examination of the influence of the workplace. *Leadership and Organization Development Journal*, *17*(2), 42–47. doi:10.1108/01437739610111222

Wagner, R. K., & Sternberg, R. J. (1985). Practical intelligence in real-world pursuits: The role of tacit knowledge. *Journal of Personality and Social Psychology*, *49*(2), 436–458. doi:10.1037/0022-3514.49.2.436

Weiss, L. (1999). Collection and connection: The anatomy of knowledge sharing in professional service firms. *Organization Development Journal*, *17*(4), 61–77.

Welbourne, T. M., Johnson, D. E., & Erez, A. (1998). The role-based performance scale: Validity analysis of a theory-based measure. *Academy of Management Journal*, *41*(5), 540–555. doi:10.2307/256941

Wilkesmann, U., Fischer, H., & Wilkesmann, M. (2009). Cultural characteristics of knowledge transfer. *Journal of Knowledge Management*, *13*(6), 464–477. doi:10.1108/13673270910997123

Wong, C.-S., & Law, K. S. (2002). The effects of leader and follower emotional intelligence on performance and attitude: An exploratory study. *The Leadership Quarterly*, *13*(3), 243–274. doi:10.1016/S1048-9843(02)00099-1

Yao, L. J., Kam, T. H. Y., & Chan, S. H. (2007). Knowledge sharing in Asian public administration sector: The case of Hong Kong. *Journal of Enterprise Information Management*, *20*(1), 51–69. doi:10.1108/17410390710717138

Yu, T.-K., Lu, L.-C., & Liu, T.-F. (2010). Exploring factors that influence knowledge sharing behavior via weblogs. *Computers in Human Behavior*, *26*(1), 32–41. doi:10.1016/j.chb.2009.08.002

Zapf, D. (2002). Emotion work and psychological well-being: A review of the literature and some conceptual considerations. *Human Resource Management Review*, *12*, 237–268. doi:10.1016/S1053-4822(02)00048-7

KEY TERMS AND DEFINITIONS

Implicit Knowledge: Mental models, beliefs and ideas of each employee that are so deeply rooted within the individual and are difficult to express in words.

Explicit Knowledge: Knowledge that can be codified therefore it is easily shared and communicated via organization databases, thus making it available to all members of the organizations.

Tacit Knowledge Sharing: The process of sharing tacit knowledge that is deeply embedded in individual employees and difficult to be expressed in words.

Emotional Intelligence: The abilities to perceive, understand, and regulate emotion to promote emotional and intellectual growth.

Service Typology: Classification of services according to their specific characteristics; namely people versus equipment, level of customization, extent of employee/customer contact, level of employee discretion, value added in back office versus front office, and product versus process focus.

Team Work: Working together with co-workers and team members, toward success of the firm.

Organizational Citizenship Behavior: Employee's discretionary behavior to promote the effective functioning of an organization.

Chapter 11
Transfer Knowledge Using Stories:
A Malaysian University Case Study

Khairul Shafee B. Kalid
Universiti Teknologi PETRONAS, Malaysia

EXECUTIVE SUMMARY

Storytelling is part of our daily life. Everybody tells stories everyday without even realizing it. Storytelling has always been used in knowledge transfer. This chapter attempts to explore the usage of knowledge-embedded stories in a Malaysian institute of higher learning and will consider the understanding of organization members concerning storytelling in knowledge management and the facilitators and barriers of implementing storytelling. The significant of this study is that it provides insights on the culture of storytelling as a method of knowledge transfer in a Malaysian institute of higher learning. This study examines how other organizations are establishing storytelling practices and the way knowledge-embedded stories are being used. This study explores the practice of storytelling as a knowledge transfer medium and explores the practicality of using stories in the organization and the employees' perception of the usage of stories to transfer knowledge. This case reveals that storytelling is regarded as an acceptable approach in knowledge transfer. It is demonstrated through gathering sessions that stimulate storytelling. The case also presented context-specific factors that influence KM storytelling.

INTRODUCTION

Storytelling is an ancient and traditional way of passing on complex and multidimensional information and ideas through narratives (Sole, 2002). It exists in many domains and in many forms. From an artistic point of view, storytelling is the art of using language, vocalization, and/or physical movement and gesture to reveal the elements and images of a story to a specific, live audience. A central, unique aspect of storytelling is its reliance on the audience to develop specific visual imagery and detail to complete and co-create the story (National Storytelling Association, 1997).

Organizational story is defined as a detailed narrative of past management actions, employee

DOI: 10.4018/978-1-61692-886-5.ch011

interactions, or other intra- or extra-organizational events (Swap, et al 2001). Organization storytelling is used for many purposes such as promoting values, communicating change and stimulating product innovation (Groff & Jones, 2003). Linde (2001) conducted a case study on the use of narratives in an insurance company. The insurance company used the stories to convey its organization values and identity, inducting new staffs and inspiring existing staffs. This is done by telling the stories of the company's founder, staffs telling stories about their early days, career and careers of exemplary staffs. Katzeff & Ware (2006) conducted a case study on the usage of storytelling to capture experience in a temporary but recurring organization. The organization in the study is a festival organization whereby it is established for a temporary period of time and got dissolved after the festival ends. In this type of organization, the employees are also temporary but recurring. The authors presented a storytelling application to capture the employees' personal accounts of their work in the organization in story form. The stories are usually personal stories derived from personal experiences.

From knowledge management (KM) perspective, storytelling is seen as a traditional way of passing on complex multi-dimensional information and ideas through narrative (Sole, 2002). Gabriel (2000) regards storytelling as 'an art of weaving, of constructing, the product of intimate knowledge'. Groce (2004) defines storytelling as orally communicating ideas, beliefs, personal stories and life-lessons which insinuate the element of tacit knowledge in stories. Stories can come in different forms and are labeled as cases, anecdotes, examples, histories or simple experience (Sole, 2002).

Storytelling as a Knowledge Transfer Approach

A great deal of research and literature has been dedicated to the role that storytelling in effective knowledge transfer. This can be seen from the work on the role of stories among disparate project members (MacLeod & Davidson, 2007; Nielsen & Madsen, 2006), the role of stories in representing tacit knowledge (LeBlanc & Hogg, 2006) and also the tools used to create and construct organizational stories (Appan, Sundaram, & Birchfield, 1999; Bradner, Kellogg, & Erickson, 1999; Santoro & Brezillon, 2005). However, there is a lack of study that focus on the perception of knowledge users towards the usage of storytelling in knowledge management and the factors that influence the use of stories as a knowledge transfer medium.

The knowledge transfer factors yielded in existing knowledge transfer studies are broad factors. Joshi, Sarker & Sarker (2007) conducted a study that examines the factors associated with knowledge transfer in information system development project teams. The study focuses on the team member's absorptive capacity, motivation, communication among members, group culture and group cohesion on knowledge transfer. Zhihong & Tao (2009) investigated the factors that influence informal knowledge transfer. They proposed that knowledge transfer opportunities, motives and capacity are the decisive influential factors to informal knowledge transfer among individuals. Khamseh & Jolly (2008) identify and classify knowledge transfer factors in strategic alliances. They identified four categories of knowledge transfer factors which are the characteristics of knowledge, factors related to absorptive capacity, the reciprocal behavior of the partners and finally, the nature and form of alliance activity. Since one of the purposes of storytelling is to transfer knowledge, we believe that the knowledge transfer factors are applicable to storytelling but with other additional factors which are storytelling specific.

Case Study Objectives

This purpose of the case study is to address the lack of literature on storytelling practices in organizations and its factors that influence the use of stories for knowledge transfer. The case study demonstrates storytelling practices in the organization; therefore, it provides an overall view on how storytelling is practiced and the factors that influence storytelling. This includes the storytelling sessions, the use of stories and the factors that affect storytelling. The insights in this case study will prepare other organizations in coordinating and promoting storytelling sessions. This case study attempts to explore the practice of storytelling as a mean of transferring knowledge in UTP, the unit of analysis of this study. UTP, a private Malaysian university is selected as the case study unit of analysis due to researcher's attachment to the university.

This case study covers formal and informal storytelling sessions. The formal storytelling sessions held in UTP are described and the factors that influence those sessions are presented. Informal storytelling sessions are storytelling sessions that are not being organized by the department. Since informal storytelling happens without respite in organizations, the case study covers only the issues that affect informal storytelling. For the sake of being consistent, this case study will use the term KM storytelling to signify stories used for the purpose of knowledge transfer.

BACKGROUND

UTP is a university that was established in January 1997 by PETRONAS, a Malaysia oil and gas company, at the invitation of the Malaysian government to set up a university. It has always been the objective of the university to produce well-rounded graduates who are not only technically competent but also possess other traits such as lifetime learning capacity, critical thinking, communication and behavioral skills, business acumen, practical attitude and solution synthesis ability.

The courses offered at UTP are geared towards engineering and technology at the undergraduate and postgraduate levels with high emphasis on research and development. Currently, the university offers five engineering programs namely Chemical, Civil, Electrical & Electronics, Mechanical and Petroleum, and two technological programs namely Information & Communication Technology and Business Information Systems at Bachelors undergraduate and post graduate degree.

This case study unit of analysis is the Information Technology and Media Services (ITMS), the information communication and technology (ICT) department of the university. The reason why ITMS is chosen is because the subjects are conveniently accessible to the researchers. Furthermore, the staffs in ITMS rely heavily on their technical and non-technical knowledge in completing their task. The ITMS department is responsible for the ICT facilities and services in the university. It comprises four units: Laboratory, Support, Network and Application Unit. The Laboratory unit consists of lab technicians assigned to computer labs throughout the campus. The main responsibility of the Lab unit is to ensure that the lab is managed effectively and the equipment in the lab is properly maintained. There are also lab technicians who assist the lecturers in conducting lab sessions. For example, the technician of the networking lab assists the lecturers by demonstrating the workability of network equipments. The Support unit provides support services to users in the campus and maintain IT facilities such as ID servers, email servers and network servers. This unit is the frontline between users and ITMS as all IT request by the users are channeled through the Support unit. The Network unit is responsible for the planning, design and maintenance of the university's network infrastructure. The Application unit is responsible for the development of IT applications that aid academic management staffs in their job. ITMS is also responsible for any usage of media services such as the video production of university events.

SETTING THE STAGE

Knowledge Management in UTP

The knowledge in the university is retrieved through intranet and groupware application. UTPnet portal, with its tagline One-stop center for UTP community, is a intranet web-based portal that provides information about the university. UTPnet is a simple portal. The features include latest news and events, calendar and links to other department's website. The ITMS is responsible on the design of the portal. The contents are obtained from the departments who wanted to share their news or events. Knowledge is also shared using information disseminating tools such as email, instant messaging applications and social networking websites. The university does not have a proper document management system. Nonetheless, documents are shared through emails as attachments or stored into a shared drive. If a person decides to share documents into a drive, an email will be sent to intended recipient on the location of the document.

In 2009, the university formally engaged in a university-wide knowledge management initiative. The initiatives start with awareness programs. Thus far, three series of personal knowledge programs have been conducted to inculcate the knowledge sharing culture in the university. The first program was conducted in March 2009 to all UTP's head of departments. The second program was conducted in July 2009 to KM facilitators. Representative for each department in the university were appointed as facilitators to instill KM culture, promote the importance of KM for individual and business success, provide effective learning through formal and informal engagement activities at all levels and promote knowledge sharing in UTP. A 3-day program called Personal Knowledge Management: Train of Trainers was conducted to train the facilitators on how KM culture can be developed in their respective departments. Among the objectives of the program are to sharpen the skill of the participants as effective and competent KM facilitators for UTP and prepare them to train, facilitate and demonstrate KM skills to UTP community. In August 2009, the third personal knowledge management awareness commenced. This program marks the first KM program that involves UTP staffs. The departments involved were Information Resource Centre and Finance & Fund Management comprising 13 Executives and 19 Non-executives. Among the learning objectives of the program were to explain the relationship between Personal KM and Education Division and UTP business requirements, understand the KM benefits and challenges to PETRONAS, Education Division and UTP and KM tools in work processes to improve business operation and efficiency. In addition to learning and exploring KM, the program was also meant to be a sharing platform on how the organization could leverage and exploit human capital for business excellence.

KM Storytelling in ITMS Department

Before the inaugural launching of the UTP KM initiative, storytelling has been used as one of the mediums to transfer knowledge among the staff. In ITMS, stories are used orally in communicating ideas, beliefs, personal stories and life-lessons (Swap et al., 2001). In the context of KM, storytelling is used to describe complex issues, explain events, understand difficult changes, present other perspectives, make connections and communicate experience. For this study, storytelling is referred to as KM storytelling which is defined as the practice of telling stories for the purpose of facilitating knowledge transfer. Generally, stories are told informally within a social environment in an organization but there are also storytelling sessions that are organized by the organization. An example is the dedicated storytelling session in The National Aeronautics and Space Administration (NASA) (Bailey, 2005; Post, 2002). Teresa Bailey (2003) describes her experience as a librarian in Jet Propulsion Laboratory (JPL), in implementing a storytelling project JPL. Todd Post (2002), of a consulting firm called Edutech, describes how storytelling initiatives have increased the level of knowledge exchange in Edutech and NASA. Edutech produces a storytelling magazine for NASA called Academy Sharing Knowledge (ASK).

Similarly to NASA and JPL, ITMS department also has its own dedicated storytelling session. The ITMS department of UTP has two storytelling sessions and one program that stimulate storytelling. The storytelling sessions are ITMS sharing session and Toolbox session. The program that stimulates the use of stories to share knowledge is called the Buddy program. ITMS sharing session involves the whole ITMS department. On the other hand, the Toolbox session and the Buddy program only involve certain units in the ITMS.

The Buddy Program

The Buddy program in ITMS is a seasonal program that was design with the purpose of familiarizing staff with tasks other than their job description. Two staffs from different units are paired together to work as a one single unit in a pre-determined unit. In this buddy system, the less-experienced are paired with a more experienced colleague. The less-experienced colleague is expected to learn as much as possible about the tasks of the more-experienced colleague from a different unit. The Buddy system is crucial to the ITMS personnel as it makes the employees more flexible in their job scope. Furthermore, in a given period of time, the department performs a job reshuffling exercise. Due to the shortage of manpower and differences in job scope among the units in ITMS, the Buddy system is implemented during the semester breaks. The implementation of the Buddy system is not feasible during operational time because of downtime and service availability issues.

The Buddy system is also practiced in the academic environment in which junior academics are paired with a senior academic. The senior academic acts as a mentor to the junior academics. In this arrangement, the junior academics are allowed to choose their own mentors. Every 3 months, the junior academics need to submit a report signed by their mentors to the Human Resource department. The report contains information such as number of meetings, research activities and others.

ITMS Sharing Session

ITMS Sharing session was a session that promotes the sharing of knowledge among ITMS personnel. The 1-hour session was conducted once a week. In this session, a speaker that represents a unit will share stories about anything. This include the unit's experience in handling a problem, project management or even stories that are not related to work such as time management. However, the session lasted for only a month due to lack of participations from the ITMS staffs. There are intentions to reintroduce the sharing session but as yet, there are no plans materialized for this session.

Toolbox Session

The Toolbox session is organized by the Support unit of ITMS. The session is conducted three times a week every morning and gathers Support unit personnel to share stories on the problems that were encountered and the solutions to solve the problems. This session is reminiscent of the work by (Orr, 1997) about Xerox technician who shares stories about their work. This session promotes interaction among the support personnel in which members contribute in building up the story. The stories told in this session also create a shared meaning and understanding among the Support personnel.

CASE DESCRIPTION

Participant Summary

Respondents were selected randomly from ITMS department. Table 1 shows the summary of each respondent's position, job scope and knowledge used.

The Usage of Stories in the ITMS

Within the context of knowledge transfer, there are two type of stories used in ITMS. First, the stories itself are the intended knowledge to be transferred. Stories of this type are used as a visualizing tool. Examples of this type of stories are lessons learnt, how-to and experiences. The second type concerns with the role of the stories in supporting the transfer of knowledge. Rather than using the stories as the knowledge to be transferred, the stories are used to facilitate the transfer of knowledge. From the findings, stories of this type are used to prompt responses from other staffs, to simplify concepts and to obtain shared understanding with other staffs. Figure 1 shows the model that depicts the use of stories as the main knowledge transferred.

Story Role Type 1: Knowledge Visualization

For this type of usage, stories are used to visualize the intended knowledge to be transferred. It is being thought as another way of representing the knowledge. This type of story is in three forms: experience, lessons learnt and how-to.

In knowledge transfer literature, storytelling is used mainly for the purpose of transferring tacit knowledge (Groff & Jones, 2003; Sole, 2002; Wetlauffer-Adcock, 2004). Respondents share stories to transfer tacit knowledge in a form of experience. The following responses support the use of stories to transfer experiential knowledge,

I told stories based on my own initiative. I remember when I was in Singapore as a consultant for an HR system project. I shared the story about my experience in Singapore with my colleagues

Stories help me by learning what other people have experienced. These stories come from informal conversation. This informal conversion has helped me in doing my work.

UTP staffs normally see a lesson learnt document in their e-mail sent by the Health and Safety Environment (HSE) department. Generally, the

Table 1. Respondent summary

Resp. No.	Position (Years of Experience)	Job Scope	Knowledge Used in Job
1	IT Executive (3 years)	IT Support Executive responsible with supporting users with IT facilities such as computers and maintaining server for ID for email and network.	Technical-Explicit and Tacit knowledge
2	IT Executive (10 years)	Attached to the Application Unit. Manages a IT project in the organization	Explicit and Tacit knowledge
3	IT Executive (6 years)	Attached to the Network Unit. Responsible in the design and managing the university's network.	Tacit knowledge
4	IT Technician (8 years)	Attached to the Network Lab. Responsible in managing the lab and ensure equipments are in order.	Tacit knowledge
5	IT Technician (10 years)	Attached to the Support Unit. Responsible in supporting end user server.	Tacit knowledge

Figure 1. The use of stories in knowledge transfer in ITMS department

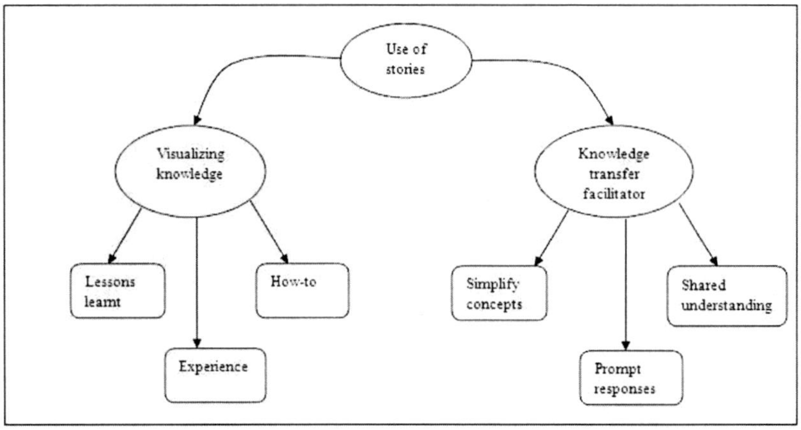

purpose of this document is to make the staffs aware of what have happened and what they could do to avoid such incident from happening again. In ITMS, respondents listen to stories from their predecessors or experienced colleagues on the dos and don'ts in their daily jobs. The responses which suggest lessons learned type of stories are as follows:

Yes, for example how to handle events. From their stories we will know what is needed to execute the event. So, from the experience we know the dos and don'ts

The previous coordinator told me that is what her predecessor did and this is what I learned out of it. I learn something because there are problem faced by them. Based on their experience, I know the problems that they faced and the solutions. With that, I am able to avoid undesired situations.

Respondents also transfer how-to using stories. According to Merriam-Webster online dictionary, how-to documents are documents that give practical instruction and advice. In this context, a how-to does not have to be in a form of documents but also it can be verbal. From the responses, the terms *story* and *how* are noted to see the relationship between story and how-to. Examples of the responses are as follows

Just look at our voice network systems. Stories can be used in describing these technical works. If no explains on how to do all these technical work, we would do things not according to specifications.

We start with lots of How-to documents. So, if anybody thinks that they have done something, they will write a how-to document and stored it in the shared drive. Other people can refer the how-to for any guides.

Story Role Type 2: Knowledge transfer facilitator

For this type of usage, stories are also used to facilitate knowledge transfer. This type of stories complements the intended knowledge to be transferred. From the findings, the stories are used to simplify concepts, to encourage prompt responses and to obtain shared understanding. In UTP, stories are used by a user support personnel to explain technical concepts involved in solving a problem.

The thing some people cannot understand just reading the materials on Google. So, whoever

understands material will explain it verbally to the other person on the material in a simple manner.

I think we could do that because nowadays when we tell convince a message or something we use metaphors to relate technical stuff which allows people to understand much more easily. I think storytelling can be used to simplify complex concepts.

People share stories with each other when they are in a social setting. Respondents alluded that when stories are shared in an informal session, they will get prompt responses from other colleagues in that session. A respondent mentioned a number of times about how he prefer sharing stories during breakfast with friends due to the opportunity of being able to ask questions and get responses. The respondent stated the following,

To me, stories are told during informal sessions. During these sessions, colleagues share knowledge and give suggestions. During these sessions also, I will ask a lot of questions and get responses from them on the spot.

Stories are also used to obtain shared understanding among the colleagues. The respondent who mentioned about the toolbox session demonstrates this use of stories.

Every morning, we will get together and discuss about what happened when doing user support. When doing so in a group, we can get interaction and obtain immediate response on the solution.

The toolbox session allows all the support personnel to get better ideas on some of the problems solved when supporting users. When a colleague tells a story in the toolbox session, everyone will have similar level of understanding on the context, problems and solutions.

Knowledge Management Storytelling Factors

Figure 2 depicts the factors that influence storytelling practice in ITMS department. The findings show the emergence of four themes: trust, comfortability and social environment.

Trust

According to Merriam-Webster online dictionary, trust is defined as assured reliance on the character, ability, strength, or truth of someone or something. Respondents mentioned that they share stories with friends or experienced colleagues which hinted on the existence of trust. This can be seen from statements by the respondents as follows:

Figure 2. Factors the influence KM storytelling in ITMS

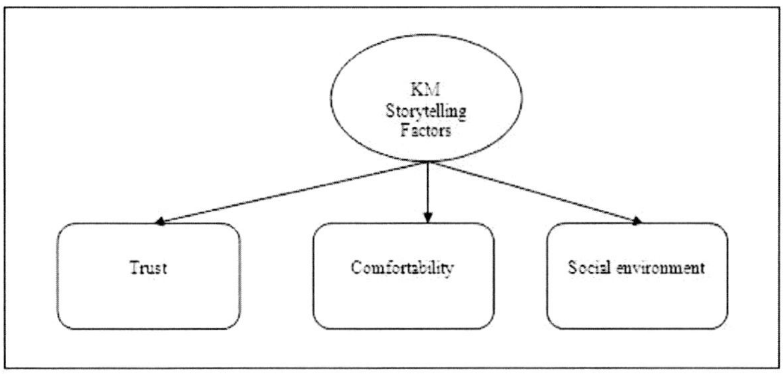

Knowledge is transferred when having breakfast with breakfast. They will tell their stories at that time...giving suggestions on problems.

The sharing of stories in this department is not yet widely accepted. And if they share they share among the colleagues that they are close to.

For some people, if they know something, they will tell their friends.

Words such as *friends*, *colleagues* and *colleagues that they are close* to are one of the indicators of the trust theme.

Comfortability

In order to promote storytelling, the environment in which storytelling happens is crucial. Respondents indicate comfortability as one of the factors that could influence the practice of telling stories. Comfortability in this context refers to the comfortability of the person towards the other people that he or she is engaging in storytelling with and the environment of which storytelling happens. Comfortability with another person relates to trust. In knowledge transfer literature, trust is cited as one of the factors that influence knowledge transfer (Bechina & Bommen, 2006; Fengjie, Fei, & Xin, 2004). This is reflected in the statements,

I might share with people that don't know yet. So, I'd rather share with someone that I'm close with and definitely know that the person don't know about it yet. I'm quite selective in sharing.

I believe people in this department work in silos. And if they share they share among the colleagues that they are close to....

Social Environment

The term *informal* is being used widely by participants to illustrate the suitable type of storytelling settings. The term *informal* used by the respondents is referring to social gathering sessions. Examples of social settings are stories exchanged during breakfast and lunch or other social platforms such as department gatherings or tea talk. Social gathering sessions such as lunch hour or gathering at the pantry are cited as an influential factor in using stories to share knowledge. Some of the responses to support social environment are,

The thing was, I bumped into her at the pantry during lunch and it was just a casual discussion. What amazed me was the rate that the ideas flowed during the informal discussion was way faster than in the meeting room. I further observed that some pretty lively discussions about work actually happened in informal settings i.e. the pantry. I did some very surface research (very, very, very surface) and found that brilliant ideas flow better when people are in relaxed mode.

I think it (storytelling session) must be informal... maybe with food <laugh>people may be there. Food brings people in. I think the session should short and concise.

If I were to share my story with colleagues, I have to look at what story to tell. When I went back from an IT management course last time, I shared my story informally. I don't want to invite people like inviting them to a meeting. I don't want it to look like a meeting. If it were like a meeting, people tend to get scared.

CURRENT CHALLENGES

Issues Regarding KM Storytelling Platforms in ITMS Department

When discussing with the respondents about the issues regarding the KM storytelling platforms in UTP, the respondents were in favor of using ITMS sharing session as the basis for their comments. This happens for two reasons. Firstly, the session involves the whole ITMS department as compared to the Toolbox session and Buddy program. Secondly, the session was cancelled due to lack of participation from the staffs.

Time Constraint

ITMS sharing session is a formal sharing session that utilizes the storytelling approach in disseminating knowledge and supporting knowledge transfer. However, the session has been cancelled due to a number of reasons. The terms *lack of time* or *busy* are mentioned by all respondents to signify the failure of the session which suggest time constraint as an influential factor that leads to the discontinuation of the session.

We do have last time what we call sharing session. We do it every Friday we call everybody and sharing stories with us by unit. We started last year but unfortunately when we start the new year, we never implement it yet back because everybody is busy.. VERY busy.

We had to rely on volunteers to speak at the session. Then we had to find a suitable slot but since as you know different people had things to do at different times it was almost impossible to find a time slot where the speaker was free and a sizeable audience was free as well.

Topic and Content

Topic and content of the stories are two issues that attributed to the low attendance of the session which subsequently leads to its discontinuation. The respondents cited that the *topic* and *content* of the story should be made interesting and relevant to the audience. The topic and content should be made interesting and related to the audience. Respondents stated that in order for them to come to the formal storytelling session, the topic must catch their interest. The term *interesting* is widely used to describe the kind of topic to be presented in the platform. This is supported by the following responses

First of all, the topic needs to be interesting. If not, people would not come.

I think the topic needs to be interesting. It should addresses to a certain group. Your topic depends on who your target audience are. The topic must be applicable to your audience.

Topic must be interesting. If the content of the story is something that people have not heard, people will come to the session.

The term *interesting* is fairly broad. When the respondents were asked about the content, the meaning of the term *interesting* started to emerge. This gave more insights on what the respondents meant by *interesting*.

Stories about my experience, what is so interesting working here. Because the most popular topic here is firstly, my experience, secondly what happened in my previous employment and thirdly, soft skills.

Content is important because the content must relate to work and it is fresh. The content should be like what other people don't know but you know

Sometimes, if it is just talk without facts, it is boring. If you make it too fictionalized, people might not be able to relate to their lives.

Generally, for a story to be deemed as interesting, the story should be relevant to the audience. A respondent stated that *if the story is related to my work that will affect my performance, I will definitely come to the session.*

This respondent will come if the topic and content will improve her job and with the improvement in her job, will improve her work performance. A KM story is also deemed to be interesting if the story is about someone's experience.

CONCLUSION AND DISCUSSION

The respondents of this study acknowledged storytelling as an acceptable approach in knowledge transfer. In UTP the stories are used either as the knowledge itself or as the facilitator of knowledge transfer. Stories as knowledge are transferred in the form of lessons learnt, how-to and experiential knowledge. On the other hand, there are stories that are used to facilitate knowledge transfer. The stories are used mainly to support the knowledge transfer process. The findings indicate that stories are used to simplify complex, prompt responses and obtain shared understanding.

In this case study, we presented the factors that influence informal KM storytelling. The factors of informal KM storytelling are an extension of the existing knowledge transfer factors mentioned in other literatures. We believe that these knowledge transfer factors do influence KM storytelling. Nevertheless, the factors that emerged from this study are more specific consistent with the context of storytelling. The findings suggest that the factors that influence KM storytelling are trust, comfortability and social environment.

With regards to the formal KM storytelling platform, the ITMS department has initiated two storytelling sessions and one program that stimulate storytelling namely ITMS sharing session, Toolbox session and the Buddy program respectively. ITMS sharing session has been discontinued while the Toolbox session and the Buddy program are still ongoing. Respondents prefer to discuss about the ITMS sharing session as the basis for their responses with regards to issues in formal storytelling platform. This is because ITMS sharing session is department-wide session and it was deemed a failure. Two issues emerged from the data: *time constraint* and *topic and content*. The ITMS sharing session suffers lack of participation due the time constraint faced by the staffs and also the uninteresting topic and content of the story.

REFERENCES

Appan, P., Sundaram, H., & Birchfield, D. (1999). Communicating Everyday Experiences. *Proceedings of the 1st ACM workshop on Story representation, mechanism and context,* New York. 17-24.

Bailey, T. (2005). *The story of JPL stories.* Retrieved September 12, 2009, from http://askmagazine.nasa.gov/issues/11/special/index.html

Bechina, A. A., & Bommen, T. (2006). Knowledge sharing practices: Analysis of a global scandinavian consulting company. *The Electronic Journal of Knowledge Management,* 4(2), 10 November 2009.

Bradner, E., Kellogg, W. A., & Erickson, T. (1999). The adoption and use of BABBLE: a field study of chat in the workplace. *Proceedings of the Sixth European Conference on Computer Supported Cooperative Work,* Copenhagen, Denmark. 139-158.

Fengjie, A., Fei, Q., & Xin, C. (2004). Knowledge Sharing and Web-Based Knowledge-Sharing Platform. *CEC-EAST '04: Proceedings of the E-Commerce Technology for Dynamic E-Business, IEEE International Conference,* 278-281.

Gabriel, Y. (2000). *Storytelling in organizations: Facts, fictions, and fantasies.* London: Oxford University Press.

Groce, R. (2004). An experiential study of elementary teachers with the storytelling process: Interdisciplinary benefits associated with teacher training and classroom integration. *Reading Improvement,* 41.

Groff, T., & Jones, T. (2003). *Introduction to knowledge management: KM in business*. Oxford: Butterworth-Heinemann.

Joshi, K. D., Sarker, S., & Sarker, S. (2007). Knowledge transfer within information systems development teams: Examining the role of knowledge source attributes. *Decision Support Systems*, *43*(2), 322–335. doi:10.1016/j.dss.2006.10.003

Katzeff, C., & Ware, V. (2006). Video storytelling as mediation of organizational learning. *NordiCHI '06: Proceedings of the 4th Nordic conference on Human-computer interaction,* Oslo, Norway. 311-320.

Khamseh, H. M., & Jolly, D. R. (2008). Knowledge transfer in alliances: Determinant factors. *Journal of Knowledge Management*, *12*(1), 37–50. doi:10.1108/13673270810852377

LeBlanc, S. M., & Hogg, J. (2006). *Storytelling in knowledge management: An effective tool for uncovering tacit knowledge*. Retrieved January 23, 2007, from http://www.stcatlanta.org/currents06/proceedings/leblanc.pdf

Linde, C. (2001). Narrative and social tacit knowledge. *Journal of Knowledge Management*, *5*(2), 160–170. doi:10.1108/13673270110393202

MacLeod, M., & Davidson, E. (2007). Organizational storytelling and technology innovation. *Hawaii International Conference on System Sciences,* 0, 248.

National Storytelling Association. (1997). *What is storytelling*? 2008, from http://www.eldrbarry.net/roos/st_defn.htm

Nielsen, L., & Madsen, S. (2006). Storytelling as Method for Sharing Knowledge across IT Projects. *Proceedings of the 39th Annual Hawaii International Conference on System Sciences (HICSS'06)* Kauai, Hawaii., 8 191a.

Orr, J. E. (1997). *Talking about machines: An ethnography of a modern job*. Ithaca, NY: Cornell University Press.

Post, T. (2002). The impact of storytelling on NASA and EDUTECH. *Knowledge Management Review*, *5*(1), 26–29.

Santoro, F. M., & Brezillon, P. (2005). Group Storytelling Approach to Collect Contextualized Shared Knowledge. *Proceedings. Sixteenth International Workshop on Database and Expert Systems Applications, 2005,* Copenhagen, Denmark. 388-392.

Sole, D. (2002). *Sharing knowledge through storytelling*. Retrieved September 14, 2007, from http://www.providersedge.com/docs/km_articles/Sharing_Knowledge_Through_Storytelling.pdf

Swap, W., Leonard, D., Shields, M., & Abrams, L. (2001). Using mentoring and storytelling to transfer knowledge in the workplace. *Journal of Management Information Systems*, *18*(1), 95–114.

Wetlauffer-Adcock, C. (2004). *The Storyteller as a Knowledge Transfer Agent: Facilitating Transformational Learning in the Adult Classroom*. PhD, Walden University.

Zhihong, L., & Tao, Z. (2009). Empirical study on the influence of social capital to informal knowledge transfer among individuals. *Journal of Software*, *4*(4), 291–298.

ADDITIONAL READING

Merriam-Webster. (n.d.). Retrieved from http://www.merriam-webster.com/

Universiti Teknologi Petronas. (n.d.). Retrieved from http://www.utp.edu.my

KEY TERMS AND DEFINITIONS

Storytelling: The use of oral or written medium to pass on complex and multidimensional information and ideas through narrative.

Knowledge Transfer: The process of transferring knowledge from one entity to another.

KM Storytelling: The practice of telling stories for the purpose of communicating knowledge. KM stories can be in the form of cases, anecdotes, examples, histories and experience.

Knowledge Visualization (Story): Stories are used to visually representing the knowledge to be transferred in the form of experience, lessons learned and how-to.

Knowledge Transfer Facilitation (Story): Stories are used to facilitate knowledge transfer process such as simplifying complex concept, prompting immediate responses and obtaining shared understanding.

Informal Storytelling Sessions: Sessions in which stories are exchanged in a social environment. An example would be stories exchanged during lunch hour.

Formal Storytelling Sessions: Sessions organized by the department or organizaton in which stories are used in knowledge transfer

Chapter 12
Knowledge Management Enablers and Knowledge Sharing Process:
A Case Study of Public Sector Accounting Organization in Malaysia

Kalsom Salleh
Universiti Teknologi MARA, Malaysia

Syed Omar Sharifuddin Syed Ikhsan
National Institute of Public Administration of Malaysia (INTAN), Malaysia

Syed Noh Syed Ahmad
Universiti Teknologi MARA, Malaysia

EXECUTIVE SUMMARY

In the knowledge based economy era, the success of an organization lies more in its intellectual capital and system capabilities than its physical assets and financial capital. Therefore, professional intellects such as lawyers, doctors, systems analysts, accountants are an important source of intelligence for most organizations. Knowledge Management (KM) in a public sector organization can take advantage the transfer and sharing process of experiences, ideas and expertise of the organization's professional intellects for the organization's knowledge assets with the support of information technology. Therefore, this chapter presents the knowledge sharing model that interconnects KM enablers and knowledge sharing process for KM implementation success in the public sector accounting organization in Malaysia. Survey questionnaires were used as the main instrument to collect data from all public sector accountants working in the Accountant General's Department (AGD) which is an accounting specialist department of the Federal Government in Malaysia. This selected knowledge intensive accounting organization has a large pool of public sector accountants and the core competencies of the organization are to provide timely accounting information and quality accounting services to all agencies of the Federal Government. Therefore, knowledge management (KM) in this case study organization is concerned with the

DOI: 10.4018/978-1-61692-886-5.ch012

flow of accounting knowledge during the sharing process of public sector accountants and getting value through the knowledge flow for the organizational performance. Since the main task of the AGD is heavily related to accounting process and technology for its financial data integrity, the statistical results of this case study research provide evidence that knowledge sharing technology has a high significant impact on the knowledge sharing performance as compared to other moderate significant factors such as leadership roles to create knowledge sharing culture, ICT skills and know-how of accountants and the ICT infrastructure and software for the accounting information system.

ORGANIZATION BACKGROUND

In Malaysia, the public sector accounting agencies comprise of Treasury and the Accountant General's Department (AGD). Both of them are under the direct control of Ministry of Finance.

The AGD is entrusted with the responsibilities of identifying and formulating all accounting standards, principles and procedures and ensuring their compliance by all agencies of Federal and State Government. The AGD is also the main repository of all financial and accounting data pertaining to revenue receipts, trust monies, loans and investments and all disbursements of the Federal Government. In addition, the AGD provides training, consultancy and advisory services on all accounting matters to both the Federal and the State agencies. The vision of the AGD is to be an organization of global repute in public sector accounting and financial services while its mission is to strive for improving public sector accountability through excellent accounting services and management information system.

As a knowledge nexus of public sector accounting knowledge and practices, the AGD will continue to play its role as an advisor, initiator, developer and implementer of accounting systems for the public services sector in Malaysia apart from its primary objective of maintaining and preparing the accounts and financial reports of Federal Government. A high quality of government accounting personnel is crucial on perfecting the administrative and accounting systems in the AGD besides having a computerised system with latest technology.

KM is one of the AGD's information and communication technologies (ICT) plan to leverage on the wealth of data and information available within the organization into corporate knowledge depository to support the informed decision making and its future adoption for accrual accounting system. The current AGD's leadership is also focussed on the technological development (ICT infrastructure and KM technologies) and human resource management programs (job training, learning opportunities, performance appraisal and incentives, job rotation and online mentor mentee).

SETTING THE STAGE

AGD has 10 departments to carry out accounting operations and services at the headquarter office as well as 35 branch accounting offices and 10 self-accounting offices located throughout the country in order to provide efficient and timely services to its customers i.e. the ministries, departments and agencies of Federal Government. The two main accounting services provided by the AGD are the preparation of public accounts of Federal Government and the maintenance of accounting system applications.

The use of computer technology in government is not only restricted to data processing areas but was also extended to several new areas such as communication networks and access to information services. In 2006, AGD took a big step to implement its new holistic approach in accounting system applications by introducing the Government Financial Management Accounting System (GFMAS) to overcome the standalone computerised systems and mainframe based approach for its legacy accounting systems such as Branch Accounting System, Payroll System and Central Integrated System.

GFMAS which is based on SAP solution will provide real time knowledge base, flexible reporting structure and integration capabilities which can provide full data integration between head office and other branch accounting offices. The GFMAS application with its integration capabilities will help enabling standardized and accurate data creation and dissemination between the head office and all accounting offices.

With the introduction of Electronic Government (e-Government) paradigm, the AGD is also currently involved in the integration of its legacy systems with the e-Government application systems such as electronic procurement, human resource management information system, project monitoring system and electronic services. All of the e-Government's applications, which involve monetary transactions and accounting system, need to be connected to the AGD's legacy system.

KM is one of the AGD's ICT plateau plans to help in building the full service capability for decision support and the future implementation of full accrual accounting system. The implementation strategy of KM in the AGD is to improve data integrity and data transformation into organizational knowledge or corporate memory to support the informed decision making process within and across the organization.

In terms of learning resources and communication networks, all accounting staff in the AGD can have access to AGD's websites for internal information and collaboration platform (online information sources, online mentor-mentee program, knowledge portal, staff portal, economic intelligence unit, professional web based forum, groupware) as well as for external information through hyperlink functions. A helpdesk facility in the AGD is used for call management, call tracking, KM, problem resolution, remote control and self-help capability. An accounting process that currently used this facility is the electronic budgetary planning and control system.

The existing human resource programs in the AGD were enhanced to become more strategic with the policies to leverage employees' knowledge by giving them more exposure and avenues to gain new experience and knowledge. Internal training programs for government accounting personnel are conducted by the National Institute of Accounting of the AGD. The government accounting personnel are trained to have adequate and up-to-date accounting knowledge as well as upgrading skill training in other related fields such as ICT skills and know-how. A job rotation program in which accounting staffs are transferred to a different accounting department every three years is essential in broadening their horizons for the growth of their career path. In addition, the emphasis on the mentor-mentee program by the AGD is another milestone to help accounting staff to become more socially adaptable and bring out the hidden potential in people. The actual efforts made by the AGD's leadership for KM organizational mechanisms such as training programs, mentoring practices, job rotation, access to the AGD's website, internet and intranet services, electronic databases and services (ICT and KM tools) revealed that AGD has attained some indicators of KM practices in the public sectors as recommended by the Organization for Economic Cooperation and Development (OECD, 2003).

Nevertheless, KM implementation concept is never being made prioritized in the AGD's routine activities and performance targets because the traditional mode of practices such as meeting the datelines, adherence to client charters and internal circulars, knowledge sharing in a formalized method via face to face meetings are currently more important. Even though most leaders in the AGD had attended KM related seminars and workshops but the KM concept in managing accounting knowledge is not fully understood by the AGD's staff.

CASE DESCRIPTION

The description for this case study chapter covers the following sections: 1) literature review for the conceptual framework, 2) research objective and 3) research findings from the survey questionnaire conducted.

1) Literature Review, Research Objective and Conceptual Framework

KM is the systematic process of creating, maintaining and nurturing an organization to make the best use of its individual and collective knowledge to achieve the corporate mission. The mission is generally viewed as achieving high performance for public sector or maintains competitive advantage in the private sector (Bennet and Bennet, 2003). Therefore, technology, process, people and the organizational structure and organizational culture are the key enablers of the KM. Apart from having effective KM strategies and adequate technology infrastructure, Ruggles (1998) and Taylor & Wright (2004) highlighted that the main barriers to implementing KM were people related issues such as poor understanding of the KM benefits, a lack of top management leadership and a culture that inhibited knowledge sharing. Other barriers in implementing KM would be the risk of losing the crucial information and the improper sanitization of organizational knowledge.

Theoretically, KM is multi-disciplinary in nature and therefore, there are many different definitions of KM and KM process being proposed by various KM researchers from different discipline (Gupta *et al.*, 2000; Yahya and Goh, 2002, Edwards *et al.*, 2005). Since this case study investigates the overall perceptions of accountants in identifying the critical success factors or KM enablers that could contribute to the knowledge sharing model in a public sector accounting organization, the following definitions provide the scope of KM and knowledge sharing process which are being considered for the present study.

In this study, KM is defined a process of "leveraging and articulating skills and expertise of employees with the support of information technology" (Chong *et al.*, 2000, p. 369). KM is further defined by Salleh (2008) as the process of managing, leveraging and articulating knowledge, skills and expertise of professional intellects to gain the value of knowledge and KM investment through knowledge transfer and sharing process with the support of information technology.

The basic goal of KM is concerned with how to manage organizational knowledge that is the tacit and the explicit knowledge so that it can be utilized across organization effectively and efficiently (Gupta *et al.*, 2000). Therefore, effective transfer and sharing of both tacit and explicit knowledge are clearly an important KM process in enhancing organizational performance and innovativeness (Becerra-Fernandez *et al.*, 2004).

The main objective of this case study research is to investigate how public sector accountants working in the AGD perceived the importance of KM enablers in their knowledge sharing process within and across the organization.

The Accountant General's Department (AGD) of Malaysia was selected for an in-depth case study of KM in the public sector accounting organization. The AGD is a knowledge intensive organization and it has a large pool of public sector accountants. KM implementation in the AGD can take advantage of the transfer and sharing process of public sector accountants for the organization's knowledge assets and embedded intellectual capital. Further, the AGD has the key KM criteria such as the leadership focus on technological development and human resources development which is ideal to be empirically investigated to support its effective implementation of KM strategy. Therefore, KM in this case study is concerned with the flow of accounting knowledge during the sharing process of public sector accountants and getting value through the knowledge flow for sharing performance of the organization.

In this case study of knowledge sharing process among public sector accountants in the AGD, the researchers have included the four pillars of Stankosky's KM theoretical framework and previous empirical studies of KM in the public sector organization carried out by Syed-Ikhsan (2005) and by Taylor (2004) into the conceptual framework of this study.

Stankosky (2005) suggested that a successful implementation of KM requires the integration and balancing of the four KM pillars: 1) leadership, 2) learning, 3) organizational structure and culture, and 4) technology in an organizational wide setting. KM activities must have the visible support and follow-through by the leadership and the organization must nurture the environment of open knowledge sharing, collaboration and learning, enabled by the power of leading-edge technology (Stankosky, 2005). Leadership is responsible for practicing strategic planning in making the best use of resources and fostering a knowledge sharing and a learning culture. Organizational structure and culture should facilitate personal interactions and support social interactions to capture the tacit and explicit knowledge within an organization. Technology infrastructures should promote the efficient and effective capture of both tacit and explicit knowledge and to support knowledge sharing in the organization. Learning in KM is described as the acquisition of knowledge or skill through study, experience or instruction and social interactions. The role of learning in leveraging knowledge is to manage information to build organizational knowledge and use it to promote organizational learning and performance. Some examples of learning mechanisms are job training, learning by observation, learning by doing, learning from failures or sharing good practices, apprenticeships, employee rotation across departments, communities of practices and traditional hierarchical relationships (Becerra-Fernandez *et al.*, 2004). Professionals, like accountants, also need to be evaluated, to compete, to know they have excelled against their peers (Quinn *et al.*, 1996).

Effective transfer and sharing of both tacit and explicit knowledge is clearly an important KM process in enhancing organizational performance and innovativeness (Becerra-Fernandez et al., 2004). For the purpose of this study, the speed, reliability and accuracy of knowledge transfer are used to measure the transfer performance of explicit knowledge (Syed-Ikhsan, 2005, Syed-Ikhsan & Rowland, 2004a; 2004b). Explicit accounting knowledge is transferred through documents, databases, administrative process and computerized accounting system. Hernandez (2000) defined tacit knowledge transfer as "the process by which tacit information is moved and shared throughout the organization through individuals and groups across various functions through personal, mechanical and electronic means" (p. 118). To measure the effective sharing process of personal experience, ideas, know how, insight, expertise, talents and intelligence, such tacit knowledge must be shared and made explicit (formalized) in order to have a significant value to an organization (Beckman, 1999). Only formalized tacit knowledge can be represented electronically and be stored, shared and effectively applied (Beckman, 1999).

Hence the knowledge sharing model for this case study adopts the conceptual relationship of four KM enablers i.e. learning, leadership, technology and culture to influence the knowledge sharing process of public sector accountants in the AGD. There are ten (10) KM enablers being identified to be used in leveraging professional intellects' learning experiences, knowledge and expertise for the effective sharing performance of both explicit and tacit knowledge. The identified KM enablers for this case study include 1)

ICT know-how and skill, 2) job training, 3) job rotation, 4) performance evaluation 5) learning opportunities, 6) information sourcing opportunities, 7) leadership support, 8) knowledge sharing culture and 9) ICT infrastructure and software and 10) KM technologies (integrative and interactive technologies for knowledge sharing process). If these identified KM enablers can be managed efficiently and effectively, the explicit accounting knowledge can be effectively transferred and the tacit accounting knowledge can easily be shared and electronically codified to improve the organizational performance of the AGD.

Based on the extensive review of literature, Figure 1 depicts the effects of KM enablers on the knowledge sharing performance.

2) Research Method and Findings

A self-administered survey questionnaire was used as the main research instrument to collect data from all public sector accountants employed by the AGD. All accountants employed by the AGD were required to complete the survey questionnaires as they are primarily responsible for the accounting knowledge, accounting processes, computerised accounting systems and preparation of financial reports. They are also engaged in the operational and strategic decision making processes that affect the future success of KM implementation. Thus, their opinions to the issues raised in the questionnaires have credibility.

The questionnaire was designed and developed based on the literature review, exploratory interviews and previously tested and validated measurable variables from previous empirical studies. The first section of the questionnaire searched for profile information about the respondent. The second section of the questionnaire focused on the hypothesis testing of all the measured variables involved in the conceptual framework of this study. Each question consisted of multiple question items where respondents were asked to indicate their perceptions over the conceptual variables involved in the hypothesis testing of KM implementation factors in the AGD. A seven

Figure 1. Conceptual framework

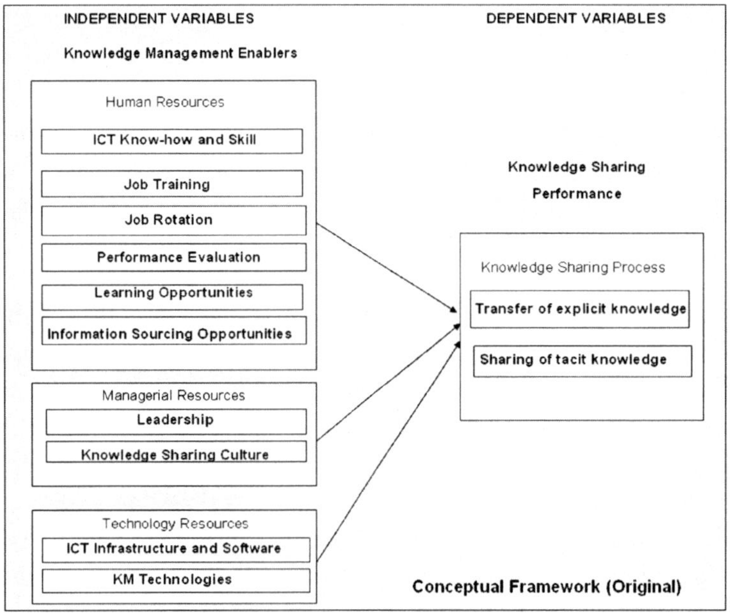

point Likert like scale were used and their level of agreement were measured from the scale of 1 to represent 'strongly disagree', 2 to represent 'disagree', 3 to represent 'slightly disagree', 4 to represent 'neutral', 5 to represent 'slightly agree', 6 to represent 'agree', and 7 to represent 'strongly agree'. The definition of knowledge, knowledge management and 68 question items to test the conceptual framework were included in the questionnaire. Pre-testing and pilot testing of the questionnaires was undertaken and some revisions were made prior to the full administration of data collection.

Out of the 365 questionnaires distributed, 203 respondents (56% response rate) returned the completed questionnaires. About 60% of the respondents were between the ages of 24 and 37 years and the rest 40% were between 38 and 55 years. About 57% of the respondents were female and 43% were male accountants. On average, respondents had been in the AGD for ten years. The majorities (56%) of the respondents are junior accountants and the remaining respondents (44%) are senior accountants (30%) and top management level of accountants (14%). This study emphasizes the managerial position and accumulated work experiences of the respondent in answering the questionnaire.

Factor Analysis was used to confirm that only eight (8) out of ten (10) KM enablers are valid and are in conformity to the literature review and prior empirical evidences. However, two factors (learning opportunities and knowledge sharing culture) were dropped for hypothesis testing due to low reliability results. Based on the statistical results of Factor Analysis, Figure 2 depicts the revised conceptual framework with the effects of KM enablers on the knowledge sharing performance.

On the basis of the factor loadings, all the extracted factors were identified and named accordingly in conformity to the survey literature. Pearson correlation tests and multiple regression models were then used to explore the positive significant relationships between KM enablers (independent variables) and KM performance (dependent variables) as well as to identify the significant KM enablers that could contribute to

Figure 2. Revised conceptual framework

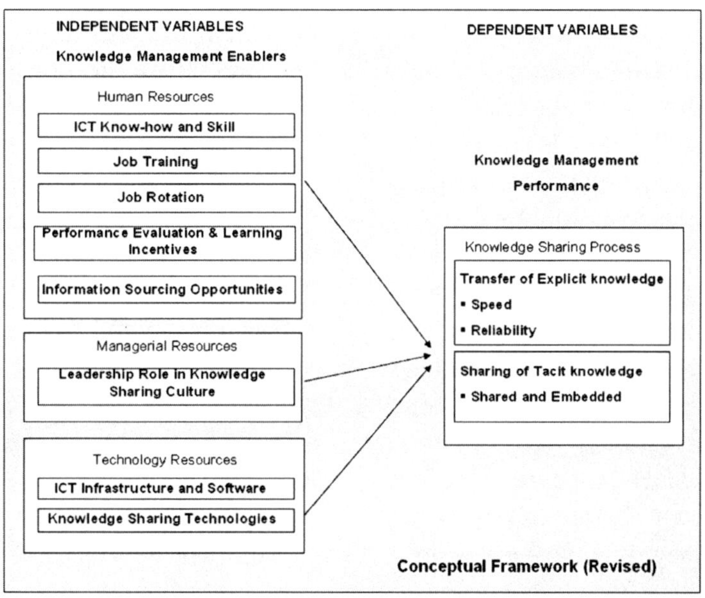

the knowledge sharing performance and organizational performance.

The regression model for finding the relationship between KM enablers and knowledge sharing performance is shown in Table 1. In this standard multiple regression analysis, knowledge sharing performance is considered as an aggregated variable when the transfer process of explicit knowledge and sharing process of tacit knowledge were grouped together. The regression result in Table 1 shows the value of R is 0.716 and the R-Square is 0.513. This indicates that the regression model explains 51.30% of variances in the knowledge sharing performance. Knowledge sharing technologies is one of KM enablers that emerged as a highly significant factor in explaining the overall performance of knowledge sharing process of both tacit and explicit knowledge in the AGD.

In Table 1, knowledge sharing technologies (p-value of 0.001) comprised professional discussion groups and collaborative office environment with KM technologies (web-based forum, online communities of practices, video conferencing, groupware, etc.) is highly and significantly related to the dependent variable at 1% level of significant test. The ICT infrastructure and software (p-value of 0.036), ICT skills and know-how (p-value of 0.038) and leadership role in knowledge sharing culture (p-value of 0.032) are moderately significant and positively associated with the dependent variable at 5% level of significant test. The regression result implies that knowledge sharing model among professional intellects i.e. accountants in a public sector accounting organization in Malaysia is significantly related with KM enablers such as ICT skills and know-how (learning factor), leadership roles in nurturing knowledge sharing culture (leadership factor) and ICT infrastructure and software plus knowledge sharing technologies (technology factor).

The regression model as shown in Table 1 is an appropriate knowledge sharing model for the AGD's effective knowledge sharing process. Since the AGD's core business is primarily associated with accounting process and timely preparation of financial reports, the ICT infrastructure and computerized accounting system are basically important before the AGD can fully create KM environment and deploying KM technologies in the knowledge sharing process of explicit knowledge and tacit knowledge. The finding of this study shows the importance of technology resources for the AGD's KM environment in using advanced ICT infrastructure and KM technologies to systemize, facilitate and expedite the organization wide KM project through effective knowledge sharing process and the successful transformation of individual/group knowledge into embedded organizational knowledge / intellectual capital. Thus, this implies that knowledge sharing processes with the support of knowledge sharing technologies together with the core and advanced ICT infrastructure and software are the most crucial factor for the successful KM implementation in the AGD.

The moderately significant learning variable which seems critical to knowledge sharing performance in the AGD is the ICT skills and know-how of professional intellects. Public sector accountants in the AGD must be trained and educated in using the KM tools and other ICT tools for managing accounting knowledge and accounting services. This can help to ensure that they can utilize the full potential and capabilities offered by the acquired technologies. They must have adequate training internally on how to use computer and the computerized accounting system in order to manage knowledge resource within their organization (Syed - Ikhsan, 2005; Syed-Ikhsan and Rowland, 2004a; Salleh, 2008; Salleh *et al.*, 2008). To enhance job performance of professional intellects, the technology and system know - how in the AGD must be easily transferable and more opportunities are provided for accountants to attend advanced ICT training internally and externally.

Top management should institute non-monetary rewards to encourage employees to share and apply knowledge for effective KM practice. Promotional and reward systems should also be based on their ability to share personal experiences and innovative ideas with others (Liebowitz & Chen, 2003). It is also worth noting that knowledge gained by professional staff through internal and external training courses, life long learning opportunities as well as people networking can enable them to translate their knowledge into the organization's routines, core competencies, business process, strategies and culture (Holsapple & Singh, 2003).

Leadership support is another moderately significant KM enabler in building and maintaining an organizational culture of knowledge sharing and learning process in the AGD. Leadership in a knowledge based organization is also of great importance when dealing with professional intellects with specialized knowledge, particularly, leaders in the AGD. Therefore, leaders should share the burden of decision making with their subordinates i.e. knowledge workers by becoming coaches and mentors that encourage, motivate, asking the right questions and help knowledge workers to learn and get involved (Bukowitz & Williams, 1999). To succeed in KM, leaders in the AGD should change their approach to focus on those KM activities such as feedback on performance evaluation and use KM technologies & tools for communication channels which can help to better

Table 1. Summary of Regression Results on the relationship between KM Enablers and Knowledge Sharing Performance

Relationship between KM Enablers and Knowledge Sharing Performance	**Knowledge Sharing Model** (Knowledge Sharing Performance)
Standard Multiple Regression Result	R = 0.716 **R^2 = 0.513** F = 25.508 Sig. = 0.000***
KM Enablers	**B = Beta Coefficient** **P = P-Value**
HUMAN RESOURCES – LEARNING	
ICT know-how and skills	B = **0.140** p = **0. 038****
Job Training	B = 0.001 p = 0.983
Job Rotation	B = 0.009 p = 0.864
Information Sourcing Opportunities	B = 0.060 p = 0. 468
Feedback on Performance Evaluation and Learning Opportunities	B = 0.104 p = 0.168
MANAGERIAL RESOURCES – LEADERSHIP SUPPORT	
Leadership roles in knowledge sharing culture	**B = 0.184** **p = 0. 032 ****
TECHNOLOGY RESOURCES - ICT TOOLS AND KM TECHNOLOGIES	
ICT infrastructure & software	**B = 0.145** **p = 0. 036****
Knowledge Sharing technologies	**B = 0. 253** **p = 0. 001 *****
Note: Supported hypotheses in boldface type: ***p < 0.01, **p < 0.05, * p <0.1	

handle and manage knowledge and professional intellects in the AGD. Leaders should use their leadership mechanisms such as recruitment and promotional procedure, motivational tools, reward system, communication styles in implementing and sustaining a knowledge sharing culture to facilitate KM (Ribiere & Sitar, 2003).

CURRENT CHALLENGES FACING THE ORGANIZATON

This research finding confirms that knowledge sharing processes and technology resources are important KM enablers in the AGD to enhance KM performance since the core business of the AGD is heavily related to accounting process and information technology. KM process formulation across the organization is the first step to achieve the KM goals. The complete knowledge flow within the organization completely depends upon how smooth the process flows are and what is the impact of knowledge created out of these processes flows on the overall growth of the organization. By adopting knowledge sharing model derived from this case study, it can help the AGD to take the full advantage of the human knowledge and technology innovation for the growth of the organization's embedded intellectual capital.

Therefore, the results of this case study research can enhance the importance of those identified knowledge sharing mechanisms and key learning factors of accountants in the knowledge sharing process in the AGD. AGD has to give more emphasis on those significant and non-significant KM enablers as being identified by this case study in drawing up its future knowledge sharing model.

For KM implementation success, leaders in the AGD should give more emphasis on how to overcome KM challenges i.e. those non-significant KM enablers such as job training, job rotation, information sourcing opportunities and feedback on performance evaluation and learning opportunities in drawing up its knowledge sharing model.

All employees must be trained and educated in using the KM tools to ensure they can utilize the full potential and capabilities offered by these ICT tools. Training programs provided to public sector accountants must reflect the increased importance of managing the knowledge resource within an organization. The job training needs to cover not just the technology based approaches to KM but also people and process based approaches to KM. Equally important to provide accountants in the AGD with the opportunity for regular access to internal and external information sources and networking facilities as well as learning resources for career opportunities.

Frequent feedback on formal performance evaluation, formal discussion of career opportunities and introducing reward system or official recognition to promote learning and knowledge sharing culture may help the AGD in managing its professional intellects and organizational knowledge.

Top management leadership in the AGD should encourage multiple champions to be coaches, mentors and facilitators of learning. Leadership skills are also important to the middle level managers as they are the one who leads the change at the lower levels as well as to maintain employees' morale during the difficult change period.

In essence, for the KM efforts to succeed in the AGD, the power of knowledge sharing process and connectivity concept of KM technologies must be fully exploited. Leaders in the AGD should promote KM awareness and encourage collaborative office environment with KM technologies. Emphasis must also be made on the leadership role in embedding cultural values that knowledge sharing creates more power as knowledge shared becomes knowledge doubled. Also, while enabling knowledge sharing, the critical knowledge should be protected by providing the role based access to all the employees.

SOLUTIONS AND RECOMMENDATIONS

This case study research has attempted to fulfil its conceptual objective by providing a comprehensive knowledge sharing model for KM implementation success by integrating the KM enablers in managing human, technological and managerial resources.

To maximize the benefit of KM investment, the AGD should establish a positive knowledge sharing culture in a more collaborative environment by encouraging employees to use KM tools more often and by providing more training programs and reward system to encourage employees to share and apply knowledge for effective KM.

The following are some suggestions to be considered by the AGD for effective flow of knowledge within and across the organization through KM enablers and knowledge sharing process:

1. Human and social interaction in the business context is important for knowledge creation activity though sharing and transferring knowledge.
2. It is important to create and foster the right culture. KM programs cannot be successfully implemented if the organization lacks the requisite culture i.e. knowledge sharing culture and learning culture should be part of the organizational value system.
3. It is also critically important to conduct knowledge audit i.e. knowing what the organization knows and the expertise it has so that to use the knowledge gap to maximise value creation.
4. Knowing what the organization needs to know to meet its high performance or desired competitive position by recognizing and analyzing the gaps in the organizational knowledge resources and blocks in knowledge flows (gap analysis).
5. Adopting the appropriate knowledge strategies to enable leveraging existing knowledge, creation of new knowledge and the acquisition of the requisite knowledge for high performance or competitive positioning.
6. Operationalizing the knowledge strategies through 1) creating processes to enable identification and dissemination of best practices (a practice that is repeatedly applied and perfected), 2) sharing and creating knowledge through communities of practice such as yellow pages or web based forums (knowledge or social networks that evolve between employees or expertise around a common area of knowledge) and 3) other supporting knowledge creation processes through the knowledge base or IT infrastructure (intranet, knowledge portal, groupware, computerized accounting system)
7. Monitoring, reviewing and tracking results of KM initiatives to measure the effect of KM on performance and productivity. Hence, this requires the development of KM performance measurement index.

REFERENCES

Becerra-Fernandez, I., Gonzalez, A., & Sabherwal, R. (2004). *Knowledge Management -Challenges, Solutions and Technologies*. Upper Saddle River, NJ: Pearson Prentice Hall.

Beckman, T. J. (1999). The Current State of Knowledge Management. In J. Liebowitz (Ed.), *Knowledge Management Handbook* (pp. 1-1 to 1-22). New York: CRC Press.

Bennet, A., & Bennet, D. (2003). The Partnership between Organizational Learning and Knowledge Management. In Holsapple, C. W. (Ed.), *Handbook on Knowledge Management 1: Knowledge Matters* (pp. 439–455). Berlin: Springer-Verlag.

Bukowitz, W. R., & Williams, R. L. (1999). *The Knowledge Management Fieldbook*. London: Prentice Hall.

Chong, C. W., Holden, T., Wilhelmij, P., & Schmidt, R. A. (2000). Where does knowledge management add value? *Journal of Intellectual Capital*, *1*(4), 366–380. doi:10.1108/14691930010359261

Edwards, J. S., Collier, P. M., & Shaw, D. (2005). Knowledge Management and Its Impact on the Management Accountant [Electronic Version]. *The Chartered Institute of Management Accountants (CIMA), London, U.K.*, 1 - 33. Retrieved June 20, 2006 from www.cimaglobal.com.

Gupta, B., Iyer, L. S., & Aronson, J. E. (2000). Knowledge Management: Practices and Challenges. *Industrial Management & Data Systems*, *100*(1), 17–21. doi:10.1108/02635570010273018

Hernandez, M. (2000). *The impact of the dimensions of the Learning Organization on the transfer of tacit knowledge process and performance improvement within private manufacturing firms in Colombia.* Doctoral Dissertation. University of Georgia Athens, Georgia. ProQuest Digital Dissertation, UMI No:9994101, Retrieved on May 20, 2007 from http://wwwlib.umi.com/dissertations/fullcit9994101.

Holsapple, C. W., & Singh, M. (2003). The Knowledge Chain Model: Activities for Competitiveness. In Holsapple, C. W. (Ed.), *Handbook on Knowledge Management 2: Knowledge Directions* (*Vol. 215-251*). Heidelberg, Germany: Springer-Verag Berlin.

Liebowitz, J., & Chen, Y. (2003). Knowledge Sharing Proficiencies: The Key to Knowledge Management. In Holsapple, C. W. (Ed.), *Handbook on Knowledge Management 1: Knowledge Matters* (*Vol. 409-424*). Berlin, Germany: Springer-Verlag, Berlin, Germany.

OECD, (2003).*Conclusions from the result of the survey of Knowledge Management practices for Ministries/Departments/agencies of Central Government in OECD Member Countries, February 3-4, 2003.* Retrieved on December 20, 2005 from GOV/PUMA/HRM (2003)2

Quinn, J. B., Anderson, P., & Finkelstein, S. (1996). Managing professional intellect: Making the most of the best. *Harvard Business Review*, (March - April): 71–80.

Ribiere, V. M., & Sitar, A. S. (2003). Critical role of leadership in nurturing a knowledge - supporting culture. *KnowledgeManagement Research & Practice*, *1*, 39–48. doi:10.1057/palgrave.kmrp.8500004

Ruggles, R. (1998). The State of the Notion: Knowledge Management in Practice. *California Management Review*, *40*(3), 80–90.

Salleh, K. (2008). *Knowledge Management and Public Sector Accountants: Knowledge Sharing Process in the Accountant General's Department of Malaysia.* Doctoral Dissertation, Universiti Teknologi MARA, Shah Alam, Malaysia.

Salleh, K., Ahmad, S. N. S., & Syed-Ikhsan, S. O. S. (2008). *Knowledge Management in a Malaysia Public Sector Accounting Organization: An Integrated KM Framework.* Paper presented at the 9th European Conference in Knowledge Management (ECKM 2008), Southampton Solent University, Southampton, United Kingdom,Publishing Limited, Reading, UK, pp.763 – 774

Stankosky, M. A. (2005). Advances in Knowledge Management: University Research Toward an Academic Discipline. In M. Stankosky (ed.), *Creating The Discipline of Knowledge Management - The Latest in University Research.(pp.1-14)*. Washington, DC: Elsevier-Butterworth_Heninemann.

Syed-Ikhsan, S. O. S. (2005). *Knowledge Management in a public organization: A study on the relationship between organizational elements and the performance of knowledge transfer.* Doctoral Dissertation, Loughborough University, United Kingdom, Loughborough, UK.

Syed-Ikhsan, S. O. S., & Rowland, F. (2004a). Benchmarking Knowledge Management in a Public Organization in Malaysia. *Benchmarking: An International Journal*, *11*(3), 238–266. doi:10.1108/14635770410538745

Syed-Ikhsan, S. O. S., & Rowland, F. (2004b). Knowledge Management in a public organization: A study on the relationship between organizational elements and the performance of knowledge transfer. *Journal of Knowledge Management*, *8*(2), 95–111. doi:10.1108/13673270410529145

Taylor, D. W. (2004). Knowledge Activism by CPAs in Public Sector Organizations. *Malaysian Accounting Review*, *3*, 171–192.

Taylor, W. A., & Wright, G. H. (2004). Organizational readiness for Successful Knowledge Sharing: Challenges for Public Sector Managers. *Information Resources Management Journal*, *17*(2), 22–37.

Yahya, S., & Goh, W. K. (2002). Managing human resources toward achieving knowledge. *Journal of Knowledge Management*, *6*(5), 457–468. doi:10.1108/13673270210450414

Section 3
Knowledge Management Practices

Chapter 13
Knowledge Management Practices in Brazilian Software Organizations: The Case of SERPRO

Tatiana de Almeida Furquim
SERPRO and University of Brasília, Brazil

Sueli Angelica do Amaral
University of Brasília, Brazil

EXECUTIVE SUMMARY

This chapter explores knowledge management practices in a software organization. It argues that software companies are knowledge intensive organizations and therefore they must properly address the matter of knowledge management. This case study highlights the importance of understanding the practices of knowledge management and describes knowledge acquisition, protection, transfer, and application practices in the context of a Brazilian software organization. The authors hope that this chapter increases understanding of existing knowledge practices in software organizations.

INTRODUCTION

Software development and maintenance are knowledge intensive activities. Fast technology development and reduced time to market make software business thoroughly dynamic. Software organizations in general must rely upon the ability of its software engineers to adapt to a challenging environment and, in consequence, their results can be highly influenced by its intellectual capital. In this context, knowledge management can be one influential factor affecting software organizations' performance and therefore must be properly addressed.

Specifically regarding the software engineering context of a Brazilian software organization, results of a recent survey (Furquim, Oliveira, and Amaral, 2007) show that knowledge repositories were not used by software practitioners, who preferred to use the organization's software process portal as a knowledge base. These results motivated the development of this case study

DOI: 10.4018/978-1-61692-886-5.ch013

to understand which knowledge management practices are developed at the referred Brazilian software organization.

Knowledge Management in Software Organizations

According to von Krogh, Ichijo, and Nonaka (2001, p. 38) it is much easier to talk about knowledge management than to act accordingly. To the authors, as it has been practiced in organizations, knowledge management is a disturbing paradigm instead of a reforming one. Alavi and Leidner (2001) recommend a conceptual structure to the analysis of organizational knowledge management based on the view of the organization as a knowledge system. In accord with this structure, organizations consist of four groups of socially enacted knowledge processes: a) creation, b) storage and retrieval, c) transfer and d) application.

From the point of view of Grover and Davenport (2001) knowledge processes can be resumed in knowledge generation, knowledge codification and knowledge transfer/realization. Lee and Choi (2003) believe that knowledge management processes can be resumed in create, store, share and use knowledge. Gupta and Govindarajan (2000) claim that knowledge management comprises two large tasks: accumulating and mobilizing knowledge. The task of accumulating knowledge concerns knowledge creation (learning by doing), acquisition (internalizing external knowledge) and retention (minimizing the loss of proprietary knowledge). The task of mobilize knowledge can be disaggregated into a set of sub-tasks: knowledge identification (uncovering opportunities for knowledge sharing), knowledge outflow (motivating potential senders of knowledge to share it), knowledge transmission (building effective and efficient channels for knowledge transfer) and knowledge inflow (motivating potential receivers to accept and use the incoming knowledge).

Lai and Chu (2000) reviewed different knowledge management theoretical models, and grouped the different knowledge management tasks they found in the literature in a set of six activities: a) initiation, b) generation, c) modeling, d) repository, e) distribution and transfer and f) use and retrospect. To Gold, Malhotra, and Segars (2001) successful knowledge management can be understood as an organizational capability to acquire, protect, convert and apply knowledge combined with a knowledge infrastructure consisting of technology, structure and culture.

To Guptara (1999) the main reason of failure of knowledge management initiatives lies on organizational culture. Figallo and Rhine (2002, p. 119) affirm that real incentives and demonstrations of faith in the new cultural direction can foster a knowledge sharing culture in organizations. Besides culture, there are several other factors that influence the success of a knowledge management initiative. Jennex and Olfman (2005) summarized 12 critical factors that impact knowledge management initiatives and knowledge management systems, which are: integrated technical infrastructure; knowledge strategy; common enterprise wide knowledge structure; motivation and commitment of users; organizational culture; senior management support; establishment of measures to assess the impact and effectiveness of the knowledge management system; existence of a clear goal and purpose for the knowledge management system; learning organization; easy knowledge use supported by the knowledge management system functions; work processes and knowledge protection. Yeh, Lai, and Ho (2006) summarized the knowledge management enablers and success factors found on the literature in four main categories: strategy and leadership; culture; people and information technology.

To understand knowledge management in software organizations it is necessary understand these organizations and their environment. According to Rus, Lindvall, and Sinha (2001), software engineering is a complex business that involves many people working in different phases and activities. Every day new problems are solved and knowledge is created. The knowledge in software engineering is diverse and of enormous proportions. Software organizations, in general, have problems to define this knowledge, to determine where it is and those who have it. The authors state that there are two types of specific knowledge regarding software development and maintenance: knowledge that is embedded in software artifacts and knowledge related to software products and processes. They assert that software organizations have specific needs to be attended by knowledge management, such as: capturing and sharing knowledge about products and processes; domain knowledge; acquiring knowledge about new technologies; sharing knowledge about local policies; knowing who knows what; distance collaboration.

Because of this complex context, in general, research on knowledge management in software engineering focuses on construction and usage of knowledge repositories; knowledge mapping tools; knowledge management and software process; knowledge network and strategies; knowledge enablers and organizational learning (Bjørnson & Dingsøyr, 2008). Some studies address the problem of understanding the relationship of knowledge management practices and software process improvement, like those of Corbin, Dunbar, and Zhu (2007), Falbo, Borges, and Valente (2004), Kukko, Helander, and Virtanen (2008) and Ward and Aurum (2004).

Bjørnson and Dingsøyr (2008) affirm that having an established development process can improve both communication and learning, but they also see that it is important to focus on sharing tacit knowledge. In fact, some authors suggest that software process improvement practices (SPI) can foster knowledge creation and other knowledge management capabilities in software organizations (Arent, Nørbjerg & Pederson, 2000; Kneuper, 2002).

Arent *et al.* (2000) state that software process improvement can promote knowledge creation in software organizations by claiming that to improve software practices it is therefore necessary to create and institutionalize new and better knowledge about how software is developed. Mathiassen and Pourkomeylian (2003) affirm that SPI is a particular form of knowledge creation, sharing and management. Kneuper (2002) states that the main support that SPI can offer to software developers is providing the knowledge they need to perform their tasks. Nonetheless, some scholars believe that researches following this line of investigation are insufficient to initiate the outline of a theory (Ravichandran & Rai, 2003; Schönström & Carlsson, 2003).

ORGANIZATION BACKGROUND

SERPRO (an abbreviated form of "Serviço Federal de Processamento de Dados", a Brazilian Portuguese name that, literally translated to English, is "Federal Data Processing Service") is one of the biggest information technology companies in Latin America. It is a Brazilian publicly owned company, created in 1964 which is geographically distributed throughout Brazil. It has 10 major branches in Brazilian capitals and smaller offices and representations in other 330 municipalities. It has around 10,000 employees among which it is estimated that 3,000 are directly involved in software engineering activities.

SERPRO was created to modernize and expedite strategic sectors of Brazilian public administration. Its main business is to provide information technology services and solutions to the government sector, among which are software and systems development and maintenance, network and connectivity solutions, data center services, etc. It develops technological solutions in several domains, like: human resources, governmental expenditure and revenue, international trade, social control, among others. Amid its main clients there are: the Ministry of Finance, the Ministry of Planning and Administration, external control agencies, among others. Its main market is public finance, corresponding to 82.5% of SERPRO's annual sales, according to the company 2009 annual report.

In addition to its president and vice-president, its board of directors has five members, each of which responsible for one strategic area. The software development business is under the superintendence of development, which is supervised by a software development director. The superintendence of development has software development pools in 10 major Brazilian capitals and some software teams groups in other offices. Since 1999 SERPRO has been investing in knowledge management and software process improvement programs. The company also stimulates the adoption of project management best practices like those of Project Management Body of Knowledge (PMBOK). In 2002 it received its first Capability Maturity Model (CMM) certification. It has a well-defined standard software process that comprises with Capability Maturity Model Integration (CMMI) model and it is now working to include agile methods in its software life-cycles. Following Brazilian policies, the organization is investing heavily in the adoption and usage of open source software and fostering the public software philosophy.

SETTING THE STAGE

According to organizational records there have been knowledge management initiatives at SERPRO since 1999. The company invested in the education of its employees on the subject of knowledge management by supporting post-graduation level courses. Initially, SERPRO's knowledge management initiative was conducted by a committee with representatives in all its units. Later, the company established its knowledge management program, which has coordinated the work of the knowledge management specialists. The specialists were distributed among SERPRO's units and were not subordinated directly to the manager of the knowledge management program. Among their responsibilities was the development of a knowledge inventory, knowledge yellow pages and an organization-wide knowledge repository.

The knowledge yellow pages were implemented in the form of a web information system in which each employee should entry his/her specialties, knowledge and abilities as well as his/her perceived level of mastery on each subject. The knowledge inventory resulted in a knowledge mapping structure in the form of a knowledge tree, which were incorporated in knowledge repositories. Nowadays there is no corporative program specifically regarding knowledge management, although there is a knowledge management project within the company structure. One observable effect of this change is that the usages of the knowledge yellow pages and the knowledge repository have been diminishing in the organization as a whole.

Since 1999 SERPRO has been working on software process improvement (SPI). One of its main results was the creation of the SERPRO's Solution Development Process or simply PSDS (PSDS is an acronym for the Brazilian Portuguese expression *Processo SERPRO de Desenvolvimento de Soluções*), which was released in 2001. The SPI initiative has been supported by high management levels by means of establishing organizational policies, providing resources and monitoring and controlling the whole SPI program. The process of creating and maintaining a well defined software process began with one group of people. Later, specialized groups (GE) that are responsible for each software discipline (i.e. requirements, design, configuration management, etc) were created. Concurrently, software quality assurance (SQA) groups were created, followed by the creation of software engineering process groups (SEPG), whose job was to train and support practitioners in the tailoring and usage of the newly created organization standard software process. Nowadays there are corporative SEPG and SQA groups as well as local SEPG and SQA groups in each software development pool. In general, the groups are mainly constituted of software practitioners, although there are some people that work only with process improvement and do not participate in software projects.

It was necessary to adopt an enterprise-wide technological infrastructure to support the SPI program. The organizational process asset library was implemented in the form of an Intranet portal which comprises the organization standard software process (PSDS), as well as tailoring guidelines, models, templates, technical guidelines, samples, life-cycle descriptions, training material, checklists, etc. Several commercial tools were adopted to support requirements management, configuration management and project management. Some other tools and systems were developed internally, like systems to support SQA reviews and reports, metrics analysis and managerial information systems and tools for managing process changes.

PSDS is widely used within the organization. Initially it was conceived to comply with the CMM requirements, among other reasons. Afterwards, it has been adapted to comply with the CMMI requirements. PSDS's main references are Barry Boehm's spiral model, SERPRO's previous software development methodology which was called MUSE and the Unified Software Development Process (USDP).

PSDS is a well-defined process, in terms of SPI terminology. It means that it is a defined process that has inputs, input criteria, validations, tasks, outputs and output criteria well documented, consistent and complete. PSDS is structured as a group of macro-activities. Each macro-activity has a set of activities that are described in terms of workflow, responsibilities and artifacts. A macro-activity can be related to software engineering, managerial or complementary activities.

In this context, activities are work units that must be executed by an individual who is playing a role within the software process. Activities have a clear purpose, usually expressed in terms of creation or update of an artifact. Each activity is attributed to a role. The granularity of an activity can vary from a few hours to a few days. If an activity is considered complex, it can be decomposed in sub-activities. A role is the definition of an individuals' behavior and responsibilities. However, roles are not individuals. One person can perform different roles in the same software project. Artifacts are work products that are produced during software development or maintenance. Models, diagrams, plans, reports are examples of artifacts. Activities use artifacts as inputs and produce artifacts as outputs. A project can be defined as a group work effort with a defined goal, within an established term and limited resources. Usually projects are originated from SERPRO's clients' needs.

CASE DESCRIPTION

This case study is part of an ongoing research project. The main objective of this case study was to identify and describe knowledge management practices in a Brazilian software organization. To accomplish this purpose, SERPRO was chosen for its distinguished position among Brazilian software organizations. In addition, recent research conducted at SERPRO indicated that SERPRO's software developers preferred knowledge base was the software process portal, even if other knowledge bases were available in the organization (Furquim *et al.*, 2007). These results motivated this case study to understand how knowledge is acquired, protected, transferred and applied within the software unit.

This case study was conducted at one of SERPRO's software pools. The software pool in question had approximately 186 people working directly with software development and maintenance. These people were distributed in 18 divisions. The software pool had a team of approximately 11 SQA consultants and 10 SEPG members. It had, on average, 394 software projects per month. Most of the software projects were conducted according to the organizational standard software process (PSDS), and regular SQA audits were performed within different phases of the projects.

To accomplish this study we conducted semi-structured interviews and observational visits. We interviewed 10 software practitioners and the software pool manager. We elaborated two main interview scripts, one directed to the software practitioners and one directed to the pool manager. Each interview script contained approximately 50 questions. Among the software practitioners interviewees there were project leaders, requirement analysts, configuration managers, metrics analysts, programmers, software architects, software testers, software quality auditors and SEPG consultants. We also realized eight observational visits.

The interviews questions focused primarily on software practitioners and managers' practices and perceptions of knowledge acquisition, protection, transfer and application within software project teams, software maintenance teams as well as in the context of their pool and the organization as a whole. The focus of our case study was on organizational knowledge in general. We made no distinction among knowledge forms, types or taxonomies.

Each interviewee described his/her knowledge acquisition and transfer behavior, as well as his/her practices and perceptions on how knowledge is protected and applied in a personal, group and organizational context. Each interviewee described its usual office routine, their roles within software development and maintenance teams as well as their perception about the effectiveness of knowledge management practices and the organization climate regarding knowledge sharing. Our interviews were recorded verbatim digitally.

The observational visits main purpose was to collect qualitative data to complement and triangulate the results from the interviews. We observed six different software teams during eight days. During our visits we could observe software practitioners and managers' daily routines as well as meetings, decision making contexts, informal talks, presentations, group work, work processes, usage of tools and artifacts and newcomers' reception processes.

We used the software NVivo to perform our data analysis. First we imported our interview transcripts and observational notes. Second we manually coded each document and after reviewing the code we performed text, node and proximity searches to identify possibly missed connections. Eventually we built co-occurrence matrixes for each main construct (knowledge acquisition, protection, transfer and application) relating each main concept to its means and channels.

Before presenting our results we will briefly detail the main constructs of this case study.

KNOWLEDGE ACQUISITION, PROTECTION, TRANSFER AND APPLICATION AT SERPRO

There is little, if any, consensus among knowledge management researchers regarding the terminology of knowledge processes. To our work we chose an approach that is mainly based on Alavi and Leidner (2001) and Gold *et al.* (2001) definitions. We adapted their terminologies and resumed knowledge management capability in four main disciplines: knowledge acquisition, protection, transfer and application (APTA).

Knowledge acquisition has to do with obtaining and creating knowledge. It involves knowledge creation as defined by Nonaka and Takeuchi (1995) SECI model, as well as innovation and obtainment of knowledge that is external to the organization. Knowledge protection encompasses the establishment and maintenance of an organizational memory. This memory can contain knowledge bases and repositories, knowledge trees, knowledge maps, taxonomies, processes, routines, procedures, artifacts, decision contexts, among others. Knowledge codification and modeling is another aspect of knowledge protection.

Knowledge protection also comprises information and knowledge security, access control, security policies and traceability. Knowledge transfer comprehends aspects of knowledge sharing. It has to do with the transfer of knowledge from one point to another within the organization. It includes what Gupta and Govindarajan (2000) denominate mobilizing knowledge and Lai and Chu (2000) call knowledge distribution. Knowledge application is the least explored of all knowledge capabilities in the knowledge management literature. It can be understood as the usage of knowledge to accomplish tasks that are relevant to the organization. The knowledge management practices we found within the software pool are described below.

Knowledge Acquisition

We found the following knowledge acquisition practices within the software pool: training, interpersonal interactions, study of the organizational standard defined process, participation in forums, Internet search and self-study, as illustrated in Table 1.The interviewees in general emphasized that, although knowledge can be acquired through study, human interaction and training, learning depends heavily in practical use of acquired knowledge. According to the interviewees, learning usually occurred when they played their roles in the software process, i.e. performing requirements analysis, managing software configuration, programming, etc.

Newcomers to the software pool were stimulated to browse the PSDS portal. In most of the cases newcomers received general training regarding PSDS's main concepts and disciplines and specific trainings concerning their specific roles within the software process and the tools they were going to use. Examples of typical software development roles were: requirements analyst, project leader, programmer, software designer, and software architect. Some interviewees declared that once they began to play a role in a software project, that information in the form of routines, procedures, technical guides, etc that they found at PSDS became new knowledge to them. In that sense, PSDS can be understood as a knowledge integration mechanism, as defined by Fernandes, Raja and Austin (2005).

Usage of forums is also common among the software developers: 5 interviewees declared they use forums as a source of external knowledge on a regular basis. All interviewees declared that eventually they resorted to Internet search engines to find specialized websites in which they could obtain new knowledge. A few interviewees (n=2) declared that they recoursed to self-study in order to obtain knowledge. In general, when forums and self-study were not effective they required external training.

Table 1. Knowledge acquisition practices

Practice	Means	Channels
Training	Expositive classes	Slides, books, student notes
	Distance learning	Intranet and Internet distance learning environments
Interpersonal and group interactions	Socialization	Meetings, informal gatherings, telephone calls, e-mail, personal network, video conferences, mentoring
Studying PSDS	Reading	PSDS portal
Participation in forums	Anonymous interpersonal and group interaction	Intranet e Internet forums
Internet search	Reading	Internet search engines, specialized websites
Self-study	Reading	Books, periodicals

We observed that prior to initiate external search for knowledge people usually consulted their immediate superior and closest colleagues or other people in their personal network, browsed the PSDS portal and asked SQA and SEPG groups for support. Once they believed that external knowledge should be acquired, they started searching for specialized websites or forums. If specialized websites and forums were not effective, they would consider engaging in training or contracting consultancy services.

Knowledge Protection

The following knowledge protection practices were identified within the software pool: establishment and maintenance of PSDS, establishment and maintenance of PSDS's solutions base, knowledge modeling in terms of process definition and improvement; knowledge mapping sheets; version control, configuration control and access control. The means and channels of each practice are listed in Table 2. The practice of knowledge modeling was mainly related to knowledge externalization (the conversion of experts' tacit knowledge into explicit knowledge according to the SECI model) and combination (the conversion of separated explicit knowledge into systemic explicit knowledge). The storage and protection practices were more related to the technological aspects of knowledge protection.

Access to PSDS and PSDS's solutions base was allowed to anyone connected to SERPRO's intranet. PSDS's solutions base was a knowledge repository that contained outstanding examples of good software engineering practices. The examples usually were previous projects experiences, artifacts, decision contexts or anything that could be considered potentially reusable knowledge. The final decision about whether an experience was appropriate or not to be included in the solutions base was taken by the corporative SEPG. The PSDS solutions base was released in 2008 as a result of seven years of process improvement.

The knowledge yellow pages were used to identify who knows what within the software pool. Each division manager would periodically map its team member's knowledge according to a predefined set of selected subjects. Each team member was supposed to indicate the level of knowledge they had regarding each subject.

We observed that certain CMMI practices, especially those related to process areas like organizational process definition and organizational process focus strongly influenced the software pool knowledge protection practices, in a similar way of that described by Falbo *et al.* (2004).

Knowledge Transfer

As a rule, the primary approach to knowledge transfer in the software pool was socialization. Developers and managers declared that the first thing they did when they needed knowledge was to talk to someone who, according to their judgment, should possess that sort of knowledge. This behavior was observed in all observational visits. As a rule, communication used to flow naturally within the software pool, and it was observed and declared by interviewees that the software pool

Table 2. Knowledge protection practices

Practice	Means	Channels
Modeling	Process modeling, lessons learned and good practices	Specialized groups (GE), Software Engineering Process Groups (SEPG), PSDS solutions base
Storage	Version control, configuration control	Configuration control tools, PSDS portal
Protection	Access control, backup procedures, knowledge yellow pages	Security policies and tools, knowledge mapping sheets

as well as SERPRO as a whole had a knowledge sharing culture.

In addition to interpersonal and group interactions, other evidences of knowledge transfer observed in the software pool were: internal training, usage of PSDS portal and PSDS solution base, mentoring, reviews and participation in special groups and committees. Internal trainings are trainings ministered by SERPRO's employees, as detailed in Table 3. It was observed daily accesses to the PSDS portal but not frequent accesses to the PSDS solution base. The practitioners declared that there were few good practices and experiences in the solution base because it was in its earlier stages.

Mentoring was the favorite knowledge transfer practice among the software developers. According to them, mentoring had a practical dimension that made it preferable to trainings, which could be more time consuming and less effective. According to Dreher and Ash (1990), the mentoring system has another potential advantage: it grants access to relevant social networks. In general, these social networks are repositories of valuable information that is usually unavailable through formal communication ways.

Review activities were also mentioned by interviewees as a fruitful knowledge sharing practice. We observed several evidences of knowledge sharing during review activities. There were different types of review in the software pool: SQA reviews, peer reviews, post mortem reviews and senior management reviews.

SQA reviews were conducted by a software quality consultant in different phases of software projects. The main purpose of SQA reviews was to ensure that the organization standard software process is being used within the software projects. The SQA consultant reviewed the works of the project leader and the project team based on procedures described at PSDS. After reviewing the artifacts the SQA consultant used to hold meetings with the project leaders and some team members, to discuss the findings. According to the interviewees, several doubts were eliminated in this process.

Peer reviews were reviews in which a work product was examined by its author and one or more colleagues within the same area of expertise. Peer reviews were usually based in checklists and procedures, both described at PSDS. Post mortem reviews were conducted at the end of software projects. One of its main purposes was to register lessons learned during the project. These lessons learned would then become part of the organizational memory and could be utilized in other contexts. Our findings reinforce Dingsøyr (2005) affirmation that learning is inherent to reviewing activities. In addition, reviews can represent a structured alternative to the problem identified

Table 3. Knowledge transfer practices

Practice	Means	Channels
Internal training	Expositive classes	Slides, books, student notes
Interpersonal and group interactions	Socialization	Meetings, informal gatherings, telephone calls, e-mail, personal network, video conferences
Browsing PSDS and PSDS solutions base	Internalization	PSDS portal
Mentoring	Coaching and hands-on	Structured meetings
Reviews	SQA reviews, peer reviews, post mortem reviews	Structured meetings
Participation in special groups and committees	Socialization	Structured meetings

for Desouza (2003), when he states that the main obstacle to manage knowledge in software organizations is the difficulty to put software people together to talk and share knowledge. Some interviewees reported that to participate in special groups facilitates knowledge transfer especially between different departments and branches.

SEPG and SQA groups can be considered communities of practice that were formed at SERPRO as an effect of its SPI program. We observed that knowledge transfer was also supported by other communities of practice that were formed within the software pool as well as across the organization, usually grouped by software disciplines like Requirements, Configuration Management and so on.

Knowledge Application

As a rule, when asked about how they applied their knowledge, software practitioners responded by describing their roles within the software process. At first, they usually answered in a general way like "all my activities require the application of my knowledge". When they were required to detail these activities, usually they described the same activities defined at PSDS. As detailed in Table 4, other evidences of knowledge application capability were decision making processes and problem solving abilities, which could be observed individually and in a group level.

Some people were specific when describing evidences of knowledge application, i.e.: "I write and manage software requirements. I produce use case specifications, I write business rules, glossaries and non-functional requirements." Other people simply cited their most common role as an example of knowledge application, while some other people declared: "I cannot prescind from any knowledge I have in order to do my job as a project leader". Although the means and channels of knowledge application were more elusive than the other knowledge management activities, it seemed that having a well-defined process helped to clarify the knowledge application dimension. Notwithstanding, we observed some situations in which some activities were performed in an automatic way, not being clear whether there were real knowledge application. Regardless, one can argue that such automation is only achieved after a long period of practice and experience and can be considered a different manifestation of knowledge application.

To Rus *et al.* (2001) software development is a process in which all people involved have to take a considerable number of decisions, each of which usually has more than one alternative. For example, a project leader must select the project team, and plan the project, which implies the selection of a process and a set of methods and techniques; a designer must select an efficient algorithm and a programmer must decide on which functions or type of variables to use. We noticed that having a well-defined software process is usually helpful in some decision making contexts. Nevertheless, we observed that problem solving and decisions in general were made based in personal knowledge, experience and intuition.

Table 4. Knowledge application practices

Practice	Means	Channels
Playing a role within the software process	Accomplishment of daily tasks and responsibilities	Well defined roles, activities and artifacts
Decision making and problem solving	Usage of personal knowledge, experience and intuition	Solitary reflection

DISCUSSION AND FURTHER RESEARCH

At SERPRO, the SEPG group was responsible for the continuous improvement of the software process by analyzing or proposing improvement changes to the defined process. The SEPG job, in a knowledge management perspective, ultimately was to make tacit knowledge explicit in the form of procedures, artifacts templates and models, workflows, role definitions, etc as well as to combine different pieces of explicit knowledge into systemic explicit knowledge. The SEPG group played a role mobilizing knowledge within the organization, by deploying new versions of the process and supporting it with trainings, mentoring, etc.

We verified that the SQA group also contributed to the task of transferring and mobilizing knowledge within the organization. By continuously counseling software practitioners and examining their work, SQA consultants kept the process adoption alive. In addition, they provided feedback to process improvement groups regarding developers' perceptions about the process and the degree of process institutionalization. This feedback helped to make process improvement changes more effective.

Among the results of our study, one could verify that the software pool presented a capability for knowledge acquisition, protection, transfer and application. Based on our observational visits and in some of our interviews, it seemed to us that knowledge management practices within the software pool were strongly influenced by certain generic SPI practices, such as: train people, assign responsibility, objectively evaluate adherence, establish a defined process, collect improvement information and ensure continuous process improvement.

It is also possible that other generic and specific CMMI practices were related to how knowledge was acquired, protected, transferred and applied in the software pool, although further research is necessary to examine the relationship among CMMI practices and knowledge management activities in the software pool.

It seemed that some SPI practices helped to strengthen the main bases that sustained a knowledge sharing environment at SERPRO. It looked like some SPI practices and results facilitated the creation of shared context and the building of social and intellectual capital. For future research it is recommended to explore the social side of knowledge management situating Communities of Practice (CoP's) Value Network Approach (VPN) and Complexity Theory in KM within the theoretical framework and results, as well as to investigate the role of adopted CMMI practices on enabling knowledge management capability within the software pool.

CURRENT CHALLENGES FACING THE ORGANIZATION

After the completion of our study we could identify a few challenges that SERPRO had yet to face. In our point of view, although software process improvement and knowledge management practices at SERPRO were clearly intertwined, SPI had not been conducted within a knowledge management perspective in the organization. On the other hand, albeit knowledge management initiatives conducted by SERPRO's corporative knowledge management program had been recognized by software practitioners as efficient in the sense of establishing a knowledge sharing culture at SERPRO, they were not totally effective within the software context.

In terms of software organizations, we would consider a knowledge management program to be efficient if it satisfies the basic needs highlighted by Rus *et al.* (2001): capturing and sharing knowledge about products and processes; domain knowledge; acquiring knowledge about new technologies; sharing knowledge about local policies; knowing who knows what and distance collaboration.

Moreover, SERPRO's software departments had specific knowledge needs that could be satisfied by an effective integration of a knowledge management program with the SPI program. Amid these needs one would highlight the incorporation of different ways of work with reference to open source and public software technologies, processes, tools and agile methods as well as establishing software product lines.

REFERENCES

Alavi, M., & Leidner, D. (2001). Review: knowledge management and knowledge management systems: conceptual foundations and research issues. *Management Information Systems Quarterly*, *25*(1), 107–136. doi:10.2307/3250961

Arent, J., Nørbjerg, J., & Pederson, M. (2000). Creating organizational knowledge in software process improvement. In K. Althoff, W. Müller (Eds.), *Proceedings of the second International Workshop on Learning Software Organizations*. (pp. 81-92) Finland.

Bjørnson, F., & Dingsøyr, T. (2008). Knowledge management in software engineering: a systematic review of studied concepts, findings and research methods. *Information and Software Technology*, *50*(11), 1055–1068. doi:10.1016/j.infsof.2008.03.006

Corbin, R., Dunbar, C., & Zhu, Q. (2007). A three-tier knowledge management scheme for software engineering support and innovation. *Journal of Systems and Software*, *80*(9), 1494–1505. doi:10.1016/j.jss.2007.01.013

Desouza, K. (2003). Facilitating tacit knowledge exchange. *Communications of the ACM*, *46*(6), 85–88. doi:10.1145/777313.777317

Dingsøyr, T. (2005). Postmortem reviews: purpose and approaches in software engineering. *Information and Software Technology*, *47*(5), 293–303. doi:10.1016/j.infsof.2004.08.008

Dreher, G., & Ash, R. (1990). A comparative study of mentoring among men and women in managerial, professional, and technical positions. *The Journal of Applied Psychology*, *75*(5), 539–546. doi:10.1037/0021-9010.75.5.539

Falbo, R., Borges, L., & Valente, F. (2004). Using knowledge management to improve software process performance in a CMM level 3 organization. In *Proceedings of the fourth International Conference on Quality Software*(pp. 162-169). Washington, DC: IEEE Computer Society.

Fernandes, K., Raja, V., & Austin, S. (2005). Portals as a knowledge repository and transfer tool – VIZCon case study. *Technovation*, *25*(11), 1281–1289. doi:10.1016/j.technovation.2004.01.005

Figallo, C., & Rhine, N. (2002). *Building the knowledge management network: best practices, tools and techniques for putting conversation to work*. New York: John Wiley & Sons.

Furquim, T., Oliveira, M., & Amaral, S. (2007). Enabling knowledge creation and sharing in software organizations: the case of the Brazilian federal software agency. In C. Stary, F. Barachini, & Hawamdeh, S. (Eds.), *Knowledge management: innovation, technology and cultures: Proceedings of the fourth International Conference on Knowledge Management* (pp. 101-109). New Jersey: World Scientific.

Gold, A., Malhotra, A., & Segars, A. (2001). Knowledge management: an organizational capabilities perspective. *Journal of Management Information Systems*, *18*(1), 185–214.

Grover, V., & Davenport, T. (2001). General perspectives on knowledge management: fostering a research agenda. *Journal of Management Information Systems*, *18*(1), 5–21.

Gupta, A., & Govindarajan, V. (2000). Knowledge management's social dimension: lessons from Nucor Steel. *MIT Sloan Management Review*, *42*(1), 71–81.

Guptara, P. (1999). Why knowledge management fails. *Knowledge Management Review*, *9*(1), 26–29.

Jennex, M., & Olfman, L. (2005). Assessing knowledge management success. *International Journal of Knowledge Management*, *1*(2), 33–49.

Kneuper, R. (2002). Supporting software processes using knowledge management. In Chang, S. K. (Ed.), *Handbook of software engineering and knowledge engineering* (*Vol. II*, pp. 579–608). New Jersey: World Scientific Publishing.

Kukko, M., Helander, N., & Virtanen, P. (2008). Knowledge management in renewing software development processes. In: *Proceedings of the 41st Annual Hawaii International Conference on System Sciences* (pp. 332-342). Washington, DC: IEEE Computer Society.

Lai, H., & Chu, T. (2000). Knowledge management: a review of theoretical frameworks and industrial cases. In R. Sprague Jr (Ed.) *Proceedings of the 33rd Hawaii International Conference on System Sciences* (pp. 1-10) Washington, DC: IEEE Computer Society.

Lee, H., & Choi, B. (2003). Knowledge management enablers, processes and organizational performance: an integrative view and empirical examination. *Journal of Management Information Systems*, *20*(1), 179–228.

Mathiassen, L., & Pourkomeylian, P. (2003). Managing knowledge in a software organization. *Journal of Knowledge Management*, *7*(2), 63–80. doi:10.1108/13673270310477298

Nonaka, I., & Takeuchi, H. (1995). *The Knowledge-creating company: how Japanese companies create the dynamics of innovation*. Oxford, UK: Oxford University Press.

Ravichandran, T., & Rai, A. (2003). Structural analysis of knowledge creation and knowledge embedding on software process capability. *IEEE Transactions on Engineering Management*, *50*(3), 270–284. doi:10.1109/TEM.2003.817278

Rus, I., Lindvall, M., & Sinha, S. (2001). *Knowledge management in software engineering: a state-of-the-art report*. Research Report, Fraunhofer Center for Experimental Software Engineering Maryland and the University of Maryland.

Schönström, M., & Carlsson, S. (2003). Methods as knowledge enablers in software development organizations. In C. Ciborra *et al.* (Eds.) *Proceedings of the 11th European Conference on Information Systems*. Naples: CUEN.

von Krogh, G., Ichijo, K., & Nonaka, I. (2001). *Facilitando a criação de conhecimento: reinventando a empresa com o poder da inovação contínua*. Rio de Janeiro: Campus.

Ward, J., & Aurum, A. (2004). Knowledge management in software engineering: describing the process. In *Proceedings of the 18th Australian Software Engineering Conference* (pp. 137-146), Washington, DC: IEEE Computer Society.

Yeh, Y., Lai, S., & Ho, C. (2006). Knowledge management enablers: a case study. *Industrial Management & Data Systems*, *106*(6), 793–810. doi:10.1108/02635570610671489

KEY TERMS AND DEFINITIONS

Knowledge APTA: The organizational capability to acquire, protect, transfer and apply knowledge.

Knowledge Acquisition: The process of obtainment of external knowledge.

Knowledge Application: The process of usage of organizational knowledge.

Knowledge Protection: The process of modeling and storage of organizational knowledge.

Knowledge Transfer: The process of sharing and mobilizing organizational knowledge.

Organizational Capability: The ability and capacity of an organization to perform a task.

Software Process Improvement: The practice of continuously identify, analyze and improve existing software process to meet organizational goals and objectives.

Chapter 14
Knowledge Management in High-Growth Companies: A Case Study in Serbia

Mladen Čudanov
University of Belgrade, Serbia

Kathrin Kirchner
Friedrich Schiller University Jena, Germany

EXECUTIVE SUMMARY

This case study describes knowledge management practices in Mozzartbet, the major and fastest growing division of Stepanović & Šipka Corporation. Mozzartbet started as sport betting business in 2001. Since then, Mozzartbet's size has increased a few hundred times in just eight years and knowledge increase followed that pace. It can therefore be an example of knowledge management practices in high-growth companies. Technical, organizational and behavioral practices, successes and problems of knowledge management in Mozzartbet are covered, and future plans of top managers in that area are revealed. In this case study the authors describe a relaxed, academic-like organizational culture, organizational structure, strategy of development, industry and regional environment, and other internal and external factors important for knowledge management. Afterwards they present conclusions about knowledge management in Mozzartbet and future directions, according to contingency approach.

ORGANISATION BACKGROUND

Mozzartbet is the largest sport betting business in the Balkans today, the major among several strategic business units of Stepanović & Šipka Corporation and one of the fastest growing Serbian conglomerates. Sport betting is the general activity of predicting sports results by making a wager on the outcome of a sporting event. Betting on events and outcomes in professional sports, especially team sports, has become a high-growth sector with opportunities for sports leagues, their teams, and bookmakers (Forrest & Simons, 2003). In December 2001, Mozzartbet's entrepreneurs Predrag Spasojević and Dušan Petrović started their business with software development in a basement. There is some mythic connection between inadequate space in garages or basements

DOI: 10.4018/978-1-61692-886-5.ch014

and successful entrepreneurship, which is often glorified as a tale from the heroic past of the company. The idea of the two entrepreneurs, formed before Mozzartbet company, was to develop software for ticket processing with a minimal set of functionalities. A ticket is the major document in sport betting, recording the forecast of the player in sport betting for future sport events. If those forecasts are correct, the player gets an amount of money calculated by the software, according to quotas related to the probability of his forecasts. In June 2002 the first version of the software was implemented.

A co-owner of the company Stepanović & Šipka Dejan Čakajac, who is CEO of Mozzartbet today, has started independently a few years before Mozzartbet. His first betting shop was opened in September 2001, a few months before the official start of Mozzartbet in today's form. He developed mathematical and statistical models to calculate quotas for sport matches. Those quotas were sold to the betting companies whose business was dependent on accuracy of those quotas, which proved an invaluable competitive advantage. Hence, quotas were sold with very good conditions to the Serbian betting companies. The starting entrepreneur in that time used his knowledge of statistics to develop one of the two pillars of future success of Mozzartbet company – precise statistically calculated quotas. They were superior in comparison to estimated, especially if considered that during time correct predictability increased with the sample. Quotas developed by his algorithm are considered to be among the best in Europe, which proved invaluable as competitive advantage and source of money inflow in the initial stage of business development.

Quota sales business generated enough money for business development, and more important, more knowledge about emerging legal sport betting sector in Serbia. The future co-owner of Mozzartbet observed strengths and weaknesses, opportunities and threats immanent to the Serbian emerging legal sport betting sector. Thus he found a strategic partner that brought in approximately half of the capital and made a strategic decision to invest in the development of software that could incorporate all aspects of the business process, which presented the second pillar of the future success.

His business was based on the automation of all transactions that were mostly executed manually. Not only the ticket winning amount calculation, acceptance of betting, surveillance and control of the process and other related processes were executed manually, but also they were prone to errors. The possibility of error was increased in crisis occasions, and even slightest mistakes in process, such as being late to stop accepting bets on suspicious sport match could result in major economic losses for the betting company. Won money was paid to players with delay, mostly due to the flaws in calculation and risk management processes. The vision of Dejan Čakajac was to create statistical software, hardware and organizational system that support this process.

SETTING THE STAGE

Serbian economy has started a slow process of revival after political changes in the year 2000. The period of economical growth and stability after Second World War was ended at the beginning of eighties, with division of Socialist Federative Republic of Yugoslavia, in which Serbia was one of six constitutive states. Civil war that followed ravaged economy of the whole region and culminated in conflict with NATO in 1999. After that, regime was changed on 5th of October 2000, and economy hastened process of transitional reforms, that could be tracked in the following table. Also, major sources of income reoriented from agriculture to production and services, causing growth in knowledge-intensive industries.

Although software for supporting betting companies existed in Serbia, a whole system with holistic approach was not provided. Thus an al-

Figure 1. GDP per capita in Serbia since 2000

	2000.	2001.	2002.	2003.	2004.	2005.	2006.
GDP per capita	829	1.375	1.933	2.569	2.938	3.273	3.518

liance was formed between the owners of Stepanović & Šipka represented by Dejan Čakajac and the two software entrepreneurs from Mozzartbet, which formed an initial team of seven people. Their goal was to further develop the initial version of software and to sell a full package of quota, software, hardware and technical support to betting companies for a variable amount of money that was calculated as the percentage of the profit of the betting companies.

The general state in sports betting industry in Serbia was totally different from Mozzartbet initiative. Although highly competitive, industry was based on old paradigms. Most of sport betting companies in Serbia that existed before 2000 are acquired or bankrupt today. Burdened by the legacy of the political and economic crisis period other sport betting companies founded their competitiveness on informal tips, individual knowledge and skills. Instead of organization, information flow, cash flow control and similar formal mechanisms, insider's tips from sport clubs, deeply tacit individual knowledge and skill to transform those tips and other information into acceptable quotas and strong cash reserve were considered to be the path to success in sport betting industry. Connections with organized crime were widely speculated for companies in sport betting industries. But as the whole procedure of sport betting had to be formalized according to legislature changes in Serbia, transactions changed from gray economy to taxable income. Therefore, the ability to manage information and knowledge gained more importance in Mozzartbet. Sport betting chains which have kept strategies that made them successful before 2000 have lost competitive power in an environment that was deeply changed. Unlike those firms, Mozzartbet, that pioneered new values and concepts, and several surviving companies that successfully followed that example, recognized the fact that knowledge and intellectual capital are now the bases of a new source of wealth for organizations and are drivers for economic and social development (Cantu et al. 2009).

After the software was implemented, money was invested in buying several betting shops. Such a betting shop is a place between coffee-shop and bank outpost where players come to place their bets, watch popular matches, analyze match results and socialize. Therewith a direct "playground" for the software was established, where its usage could be directly analyzed and errors could be found quickly. Because betting business is deeply dependent on good software and Serbian government only allowed legal sport betting, the company started an exponential market expansion. The number of betting shops has grown from only one in September 2001 to 16 in September of 2003 and to around 470 (340 in Serbia alone), with more than 1000 acceptance desks in four countries in September 2009. Thus a tremendous growth of the company has occurred in only eight years and three months.

Today, the company has several businesses vertically integrated to the core business:

- **Bookmaking Quotas**: initial business, based on providing quotas to other competing sport betting companies
- **Software Development:** maintaining software and developing new functionalities
- **Betting Shops:** core business that profits from sports betting

- **Internet Betting:** an additional independent project started after realizing the growing demand in internet betting in western countries
- **Betting Machines:** additional source of income, based on casino-type machines located in betting shops
- **Mozzart Bars:** additional source of income by selling beverages to people that place bets, track scores, or just watch popular matches in betting shops
- **SMS Content Providing:** new business area, "question mark" in BCG matrix of the company, aimed to provide mostly advertising content to customers in the future
- **Pre-Paid Mobile Operator Credit Sales:** secondary business, additional source of income using advantage of proximity of possible customers
- **Ticket Service:** secondary new business area, additional source of income that uses advantage of proximity of possible customers, mostly for sport events tickets
- **Physical Security:** Urban Guard Union secures the betting shops, money transfers etc.
- **Betting Machines Production:** product division of the company, that produces computer-based betting machines used both in-house and sold on free market
- **Publishing:** publishes Mozzartbet sport weekly magazine that is 3rd magazine in Serbia with monthly circulation around 200,000
- **Logistics:** supplies company outposts in four countries, covering distances with more than 150,000 km monthly

The company is positioned in four countries: Serbia, Romania, Macedonia and Bosnia and Herzegovina. In Serbia alone, company holds 40% of the sport betting market. Over 2,000 employees work in the company today, including 25 administrators and over 30 software developers and testers. The company has the strategic goal to become the regional leader in betting in Balkans. This growth is represented in the following table, with basic financial data that cover the previous four years of business.

Processes in the company are based on technologies and infrastructure with more than 2000 computers in four countries that are connected with the main system. The whole system runs on three high-performance servers and ten more regional server machines with lower performances. The software is distributed, and analysis and strategic management is supported by a data warehouse.

Fast growth does not only lead to success. A number of studies discuss problems of organizational growth (Garnsey, 1998; Kazanjian, 1988;

Figure 2. Development of Mozzartbet company from 2005-2008

in Thousand EUR	2005	2006	2007	2008
Total assets	1,461	2,700	4,467	5,928
Owners' Equity	1,103	1,891	2,135	2,490
Total revenues	2,226	6,658	40,441	43,095
Total costs	2,049	5,932	39,336	41,570
Profit	71	722	905	1,369

Rico, 2003). It can be regarded as an evolutionary process which involves the accumulation of knowledge unique to the firm. (Penrose 1995, p.13) In Mozzartbet, growth has caused some technical problems, because resource demands have grown exponentially, and large parts of the software had to be completely recoded. Entire organization has changed dramatically, departments were restructured, and it was important to find some invariables in these changes that will keep best features of old organization and define future organization during the period of growth and transformation. Today, Mozzartbet is divided to several functional departments, with betting shops organized as divisional outpost. In the context of knowledge management, the IT sector is of major importance, because it is the source or target of most knowledge in company. It is structured in two major divisions, system administration and software development, which is divided into several organizational units: functional analysis, software architecture, software development, software testing and deployment (Figure 3).

On concrete tasks, matrix organization is applied according to priorities assigned to project coordinators, represented in Figure 4.

Organizational structure has remained flat, with less hierarchical organizational structures and a more active management team which was regarded to increase the dynamics of the knowledge generation process (Cantu et al. 2009). Organizational culture have remained the same in software development sector since the founding. Although the number of employees increased more than 10 times, employees there enjoy and respect benefits of relaxed, academic organizational culture. That is not unusual, e.g. Harrison and Carroll (1991) discussed that rapid organizational growth often assist in establishing cultural stability. Company started with several software engineers, and all of them were connected by personal friendship. Although during rapid growth, human resource politics was to select candidates not only by their formal education and technical skills, but also, sometimes even more important, by their interpersonal skills and team-working abilities. Applicants were rejected in several cases, albeit having sufficient technical knowledge because it was estimated that they will have problems in integrating in the existing culture. Recommendation of existing employee is extremely highly valued in the process of human resources selection, and if the new employee works well, the employee who recommended him or her gets a substantial bonus, even up to 40% of the monthly salary. Such bonuses were awarded since the company was established, and after formal introduction of HR function in the com-

Figure 3. Functional organization of Mozzartbet software development department

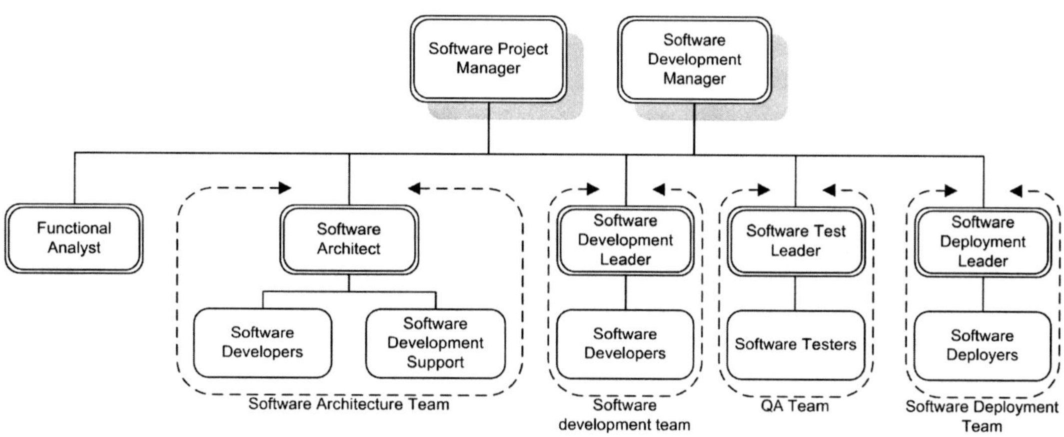

Figure 4. Matrix organization of Mozzartbet software development department

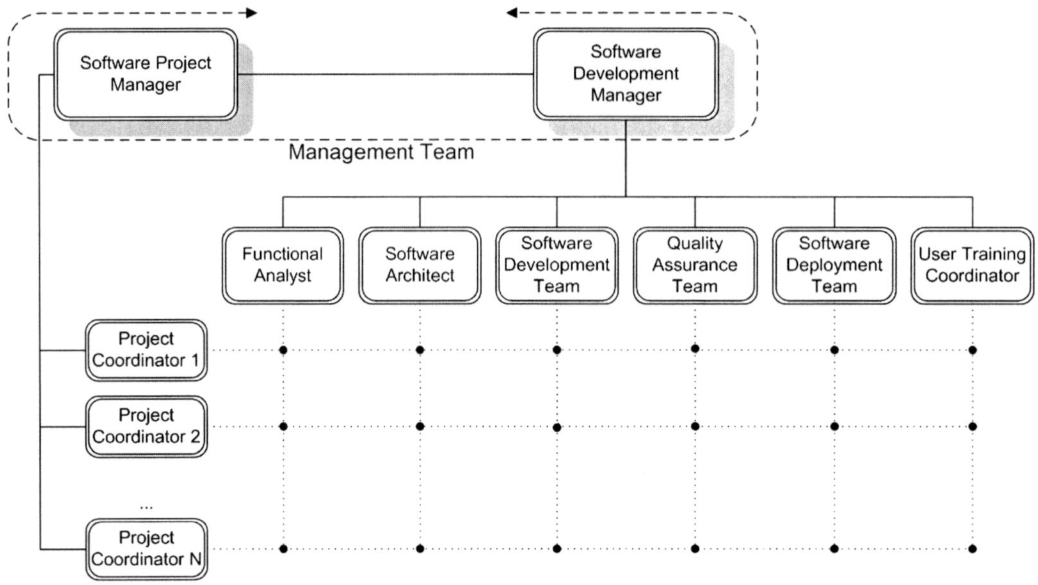

pany that policy was accepted as a guideline for employee motivation.

Especially in the software engineering department, management structure consisted exclusively of senior employees from that sector, mostly people that worked there from the department establishment. Even if they do not hold manager positions, their directives are widely accepted among junior employees, they are regarded as authorities, and respected as "tribal" elders. Therefore, there is a large amount of informal power in the software development department. This in turn strongly influences the knowledge management practices in the company.

The academic culture, specific for software engineering department, was characterized by a relaxed attitude of employees, flexible working time, and a constant development of interpersonal relations among employees. For example, employees are required to work eight hours a day, but have a flexible working schedule. There is no strict control, but all interviewed employees state that there is no misuse of that freedom, even that they work more than they would if they had strict control, and there is common practice of overtime work to keep deadlines in projects. Several times per year, sport events are held, and employees often take part in excursions, extreme sporting events, social events etc. Even in their free time, they mostly choose to socialize among themselves, and weekly sport matches are organized without any interference from the company, and most players are software development department employees and their friends. During working hours, they use a Nintendo® Wii™ console for relaxation, which is sometimes condemned by other departments. All that resulted in a large number of friendships, even family relations among employees, and negative mood among employees, which Jordan *et al.* (2006) described as negatively affecting performance, was on a minimal level. Most of the employees know each other quite well, they have studied together, worked together before or they had another way of interaction outside work.

With employees, the department assesses motivation, and willingness to acquire new knowledge. If some lack of knowledge in technical, but not important, areas is observed, it is often tolerated in the company. The major attitude is that if someone has core software engineering skills the rest will be "learning by doing".

With such means a relaxed environment of friendship and collaboration is established. All interviewed employees were sure that there is no personal conflict in that department. Interestingly, conflicts about technical issues are common, especially in the field how problems should be technically solved. According the report from interviewed employees, those conflicts are never deepened to personal feelings, and remain functional to the goals of the organization. Organizations in which functional conflict is a part of the culture tend to be more creative and responsive to clients and have higher employee job satisfaction (Jordan & Troth 2002), which is observed in this case study.

However, the culture was different in other parts of the company. Without the existence of strong nucleus that could attract similar parts for incorporation in the structure, other departments in the company could not nourish and develop a specific culture, but instead accepted the culture that was introduced from the top management, specific for key people in department who were head-hunted from best companies in the environment. Among database administrators, for example, there is different dominant culture. Since the nature of their work was more operationally defined, and they have a large number of crisis situation where timely reaction is important, their culture is more task oriented. Such culture is still similar to the culture in software engineering department, but more rigid and more formal. The rest of the company has strong culture that originated from top management, and is strongly influenced by entrepreneurial spirit of the founder and co-owner. It is most resembled by Zeus culture observed by Handy (1996) which has strong compliance with personal beliefs of the owner.

The founder of Mozzartbet also recognized a role of top management similar to the perspective of Ndlela and Toit (2001) that states that the role of top managers in implementing knowledge management is to review the internal and external environments of the enterprise in order to understand its strength, weakness, opportunities, and threats in conducting knowledge management activities. Along with CTO and other top managers he established an environment for bottom-up knowledge management initiatives, but he did not take part in concrete initiatives. The only member of top management that was concerned with operative issues regarding knowledge management was CTO. With the support of top management, a wide set of actions connected to different aspects of organization was established, giving holistic approach that Du Plessis (2007) as well as Holsapple & Joshi (2003) recognizes as one of the important factors of knowledge management implementation.

Because the main body of knowledge was in software engineering department and in relation between software engineering department and betting shops, such cultural difference could cause some problems. The knowledge of betting process, including quota calculation, communication with betting shops, algorithms for calculating tickets, algorithms for real time assessment of acceptance change and the maximum possible amount allowed per quota, risk assessment is invaluable for the success of the company. That knowledge is the main topic of our case.

CASE DESCRIPTION

The major body of knowledge was created during the initial development of the software in Mozzartbet. Knowledge of problems in software development resided mostly as wetware, or "the knowledge stored in the wet computer of the human brain, including belief, skills, talent, among others" (Taylor, 1999). Most important, this knowledge was not all tacit in nature. Tacit knowledge is defined as difficult to write down or to formalize (Nonaka, 1991). Polanyi (1966) considers it as human knowledge starting from the fact that we can know more than we can tell. The explicit knowledge is articulated, codified, and communicated in a symbolic form and/or natural language, e.g. a product manual that is given to the customer (Alavi & Leidner, 2001). At the beginning there was no formalized transfer of knowledge between employees in Mozzartbet. Whole communication was relied on informal means.

The transfer of tacit knowledge is difficult, if not impossible, as Huxley (2004) said: "Sensations, feelings, insights, fancies - all these are private and, except through symbols and at second hand, incommunicable. We can pool information about experiences, but never the experiences themselves. From family to nation, every human group is a society of island universes." (p.13) There are cases of transfer of tacit knowledge, and Cavusgil *et al.* (2003) discuss transfer of tacit knowledge between firms, checking following hypothesis: the greater the relationship strength between two firms, the greater the extent of tacit knowledge transfer between the firms; the higher the firm innovation capability; the higher the firm innovation performance. The relationship between firm relationship strength and the extent of tacit knowledge transfer is stronger for firms with greater collaborative experience than firms with less collaborative experience and the relationship between firm relationship strength and the extent of tacit knowledge transfer is stronger for larger firms than for small firms. Similar conclusions could be made for organizational departments inside one company. Greater extent of tacit knowledge sharing in large companies can be explained by too wide definition of tacit knowledge, and it may be solved by Amborsini and Bowman (2001) who suggest several levels of tacit knowledge:

- Deeply ingrained tacit skills
- Tacit skills that can be imperfectly articulated
- Tacit skills that could be articulated
- Explicit skills

This division should not be taken as discrete, because the boundary between the explicit and tacit type of knowledge is both porous and flexible, so there is traffic between the domains (Spender, 1996).

No matter what the type of knowledge in Mozzartbet was, in the first years it relied only on informal means of transfer. Interviewed employees describe this situation with good sense of humor:

"There were at most five of us in the office during the working hours. We were all in the same room. If I wanted to know something, I would slightly bend my head toward a person that we presumed could have the knowledge, and say: "Hey Pete, do you know this?" If we had no idea who could know that, we would say: "Hey guys, does anyone know…" and ask our question. Our biggest obstacle to knowledge management in that time was our habit to listen to the music over headphones during coding process, so sometimes we had to be a bit loud in calling a particular person. Since it could break concentration of other colleagues, the only technology we have used was communication messenger, for the purpose to nudge that person to get his/her attention."

Thus, all explicit knowledge was distributed through informal means, and tacit knowledge transfer was occasionally done through discussing interesting topics during lunch debates. Business was going well, and it seemed that knowledge management practices will remain sufficient until one of the greatest shocks stroke the organization: sudden success. Business indicators grew more than 2900% in just six years, and its departments grew to ten times its initial size. During one year, more than 100 betting shops were opened, six times more than the number of betting shops that were served during the first successful year of that software. Since software scalability may involve numerous issues (Mens & Demeyer, 2001) code had to be rewritten, and development and maintenance of the software included a huge amount of work that is done by around thirty software engineers and twenty-five administrators.

In such coincidences, knowledge management practices had to be changed. Thus, for Mozzartbet to develop competitive advantage (Husted & Michailova, 2002), identifying, capturing, sharing and accumulating knowledge became crucial. Furthermore, with knowledge sharing human capital is developed, employees can improve their job performance and organizational performance with new knowledge (Hsu, 2008). Everyone at Mozzartbet accepted such fact, but still today there are two groups in software department: one that think that knowledge management practices need to be more formal, and another group that thinks that knowledge management procedures on the contrary need to be less formal. Several technologies and practices are used with purpose of Knowledge management:

- JSPWiki explicit knowledge database
- Internal Newsgroups, mailing lists and administrator logs
- Betting shop knowledge support system
- Communication and collaboration tools
- Thursday's lectures (Formal knowledge exchange programs)
- External knowledge acquisition
- Master-apprentice relationship and cross-functional practices
- Informal practices

These technologies and practices support the whole knowledge management process as it is defined by Alavi and Leidner (2001) with knowledge creation, storage, retrieval, transfer and application of knowledge. In Mozzartbet company, technologies and practices are ranked differently by importance in dependence of department of the interviewed employee, but overall impression was that software and system development process is based on JSPWiki and integrated internal newsgroups, mailing lists and administrator logs as knowledge support. Rest of the company sees software and system development in Mozzartbet as black box, and perceives formal knowledge support system for betting shops as only important knowledge management initiative. However, wiki and newsgroup/mail/log are ranked above betting shop knowledge support system because of frequent usage. Communication and collaboration tools are also ranked relatively low because they are based on knowledge sources listed above, and as a tool just communicate that knowledge. Thursday's lectures and master-apprentice relationship still have minimum of formal support, while external knowledge acquisition and informal practices are ad-hoc initiatives so they are ranked relatively low. However, there is great potential for development of those practices that could ensure higher ranking in the future.

In this case technologies and practices require communication means that are used for knowledge transfer. Following table gives description of intensity of communication in Mozzartbet with focus on software development and system administration.

JSPWiki Explicit Knowledge Database

The company is currently using JSPWiki as a base for the development of a knowledge management tool. JSPWiki is open source and uses the standard J2EE components of Java, servlets and JSP. It is primarily used for company intranets and active developer community exists for its support and development. In Mozzartbet's JSPWiki system, knowledge is structured in more than 1100 articles, around 170 MB of refined explicit knowledge of the company.

The usage of JSPWiki as a knowledge management tool has started in 2006. It was actually an incident, as one of the software developers installed JSPWiki to check its features, and a group of senior employees decided that is should be established as a tool for knowledge distribution. It was assessed as a good replacement for face to face discussions about trivial and common issues. Thus the wiki focus was the description

Figure 5. Usage of communication means as knowledge transfer media in Mozzartbet

Communication means as knowledge transfer media	Usage in knowledge transfer (rank 1 – rarely, rank 5 – very often)
Wiki	4 – Often, every problem should be first checked on wiki, but some individuals still restrain from using it.
Mailing list, newsgroups	5 – Very often, it has become part of common practice to use it to communicate all relevant knowledge and information.
Betting shop knowledge support system	2 – Occasionally, from the side of software development and system administration, but it is often used service from the side of betting shops.
Call center, other voice communication and collaboration tools	4 – Often, call center is much more used by betting shops employees, while Jabber as collaboration tool is almost exclusively used by software developers and system administrators
Formal lectures	2 – Occasionally, every Thursday.
Formal face to face / phone external communication	1 – Rarely, only if it is directed by top management.
Informal face to face / phone / messenger external communication	3 – Moderately, especially with former Mozzartbet employees
External public access knowledge domains (specialized forums, blogs, etc.)	4 – Often, depending on the individual preferences, some employees use such sources are even more than internal wiki
Formal internal communication	1 – Rarely, and it often indicates conflict situation of crisis.
Informal face / phone / messenger internal communication	4- Often, informal communication is considered to be among the best

of knowledge concerning most used modules and functionalities of the software, mostly written by senior software developers. There is a strong attitude in the company that the pages concerning description of ticket acceptance and calculation, along with other main processes are best described and most used in wiki.

One of the main problems that were observed for further usage of wiki is the low level of confidence. Maintenance of knowledge in the wiki system is not formally obliged, and therefore people always tend to utilize other ways of acquiring knowledge, preferably direct communication. But two exceptions are information related to:

- important parts of the software system
- parts of the system maintained and developed by senior programmers

These two exceptions are sets with rather large intersections. In practice, it means that if a software developer or an administrator needs knowledge about something he or she knows that is developed by senior developers, or that is very important for the system, his or her first choice is often wiki. In other cases, it remains as an optional tool. Interviewed employees stated that junior employees always tend to ask their peers first about other issues, and if they reply: "Check that on the wiki, there is a good page on that" then they turn to wiki.

Senior developers mostly have a preference towards wiki. Because they often check the work of junior employees, it is usual to hear dialogues such as: "You have done a great job with this functionality, why don't you spend some more time and describe that on wiki, so other guys can see that?" There is no obligation, however, so a major problem occurs with updating. Knowledge sharing practices are affected by cultural expectations such as: what knowledge should be shared within the organization and what should be hoarded by individuals (Zheng *et al.* 2009), but it is not a major problem in Mozzartbet. Updating often depends only on the personality and style of a developer and of his or her working schedule. Culture here does not point employees toward knowledge hoarding, but toward tangible short-term results instead of intangible long term benefits. Small changes in the implementation are therefore often not documented in wiki pages, and that decreases the level of confidence. Therefore people don't use it as a first option. Problem described here, employees solve by asking non-updaters personally, because some tacit knowledge about styles of wiki updating of developers exists among employees, and after a year or two employees know who updates wiki rather regularly and who does not. That tacit knowledge was never tried to be formalized and expressed.

It is, however, wrong to say that awareness of the problem does not exist. Still, major obstructions are priority of tasks and reward system. Developers usually have urgent tasks, and updating the knowledge base is not among them. So such activities are rather low on the level of priorities, and although deemed important, not done if there are concrete tasks in coding. Other, maybe more important problem is that wiki updates are not connected to the reward system. Reward system is dependent upon top management. Input of senior software employees is respected, but only as an advice. It is always concerned when decisions about rewards are made, but not always applied. Also, there is a large time gap between such advice coming from senior staff to management and actual reward. Therefore, junior developers see no direct connection between knowledge management practices advised by their senior colleagues and rewards. Only informal power and respect for senior peers maintain knowledge on wiki created and updated at least at some level. Besides technical knowledge that resides there, organizational aspects are also presented there, and as an example article from wiki we present a description of a Software Project Manager position. Some wording is characteristic for SCRUM methodology (Cohn, 2009), and is leaved as it is. Despite mentioned problems, wiki is still a usable tool for knowledge sharing that eases the process both for senders and receivers of knowledge in Mozzartbet but leaves space for clearing of questionable issues in an old fashioned direct communication that is more compatible with organizational culture.

Internal Newsgroups, Mailing Lists and Administrator Logs

With growth of Mozzartbet, coordination and communication problems emerged and usage of newsgroups was introduced as best solution. Today there are 17 newsgroups, with administrator logs newsgroup being the most important one. Every department has its own newsgroups, and there are several mutual, like "Important notices" and "Entertainment". Every department has access to all newsgroups in order to improve inter-departmental communication. For purpose of sharing knowledge among software engineers and administrators, an administrator log newsgroup was initiated, and it will be described as technical example. That newsgroup has started in April 2005 and since then 2 records are added each day. There are more than 3300 records, 2KB in average in plain text format. An example of a record is given below.

Although newsgroups were designed to provide users with daily information, after several months of usage another value was observed. Newsgroups and mailing lists have proved to be an invaluable source of knowledge that could be obtained by searching, compiling and comparing different expressions. Lots of knowledge about cause and solutions of system administration problems were generated by analysis of newsgroups, often in combination of different logs. There are however no semantic search (Baeza-Yates *et al.*,

Table 1.

Software Project Manager – position description
Software Project Manager is process guardian. His/her task is to lead projects in the software development department. Since the first discussions with clients, up to training of software users Software Project Manager needs to be involved and manage the process. Using different techniques, meetings and documents Software Project Manager ensures that the project is finished in respect to a planned schedule. The task of Software Project Manager is to perceive and eliminate potential problems as soon as possible. Software Project Manager tracks the state of each managed project and reports to his/her superiors. If the department works on a project specific by size or some other features, Software Project Manager can choose a Project Coordinator (among team members assigned to the project), who gets some instructions in managing the project. Project Coordinator is not a formal position. Through cooperation with Software Development Manager, he/she defines priorities of project tasks within the project in order to maximize success of projects. They mutually plan projects taking care of bottlenecks elimination, especially if one employee is involved in different projects. They also mutually make plans for software versions management. In cooperation with Software Development Manager and management of the company he plans and prepares future projects that will be done by Software Development Department.

2008) or data mining procedures, although those are planned in the future. With data mining, e.g. clustering, interesting information can be recommended to a user. At first, the user has to specify which types of information she/he is interested in. The news is clustered by creating groups based on similarities found on the data. After that, only the most interesting groups of information are presented to the user (Paliouras *et al.*, 1999). In Mozzartbet, the problem is that knowledge, when discovered, is not always shared on wiki or stored by other means, but sometimes resides only in the memory of the person that discovered it, so discovering knowledge has not been further developed. Instead of that, some obligations were made for knowledge sharing over newsgroups.

Newsgroups can cover three main aspects of knowledge:

- User administration, aimed to supply basic support for the users, mainly for troubleshooting problems connected to new software version deployment
- Daily log, aimed to give knowledge and information about daily changes in the system, and share common problems that occurred during each day, along with proposed solutions.
- Intervention issues, which serve as a basis of knowledge about all interventions in the system performed by software developers, which generate features that will be further developed

Example of a record in this newsgroup is given in Table 2.

Specific technical terms that are used in the newsgroups are:

- JIRA – software for software development task management
- SAO-4301 etc – identification of JIRA tasks
- CS – central server

Mailing lists are also created for each specific project, and serve as communication means for project participants. Intensity of communication is proportional to project dynamics. A large amount of information is stored in those mailing lists, and knowledge specific for that project can often be searched and created effectively. There are several active mailing lists, for example "Point of sale application", Mozzart club Romania", "Thermal printer Macedonia". Also, mailing lists are used for socializing of the employees, such as "Football matches on Monday evening scheduling".

Table 2.

JIRA
sao-4301 – Application problem on laptop of Dragan Bosnian (machine listing). Software developers informed, check tomorrow
sao-4268 – New SUM installation is finished, needs to be verified.
sao-4304 - Ljig, SUM problems. It seemed like broken power supply or Motherboard (after electricity problems), but the problem is in micro switch that wont start SUM
sao-4310 - Sabac 1, support computer problem – restarted itself. Problem was solved by shuting down for 20 min, check in following few days.
sao-4311 – SUM changed
sao-4309 - Subotica 6 – timer turned back by logging
sao-4306 - Grocka – additional computer has monitor problems, check tomorrow and replace if needed
sao-4305 – Software development task assigned because of connection breaking towards CS, probably will be solved by Mile

Betting Shop Clerk Knowledge Support System

As the battle for market share was intensified on Serbian betting market, new products were introduced for sport betting clients. New betting options, loyalty cards, special conditions for certain occasions, all demanded new software functions for support. Those new functions did not only increase demand for knowledge in the IT related departments of software engineering and system administration, but also for the employees who work in betting shops. First solution was informal, and it included informal orientation and initial training given by older employees or software engineers. But that put too much weight on their working schedules, and other formal solution had to be found.

Betting shop clerks and other related employees are supported by knowledge in their daily work. According to the interviewed employees, their job is today more knowledge demanding than job of the bank clerk. There are almost 700 options for input of a betting ticket, which are result of increased complexity and new betting services. Special situations (robbery, burglary, jack-pot winning, breaking of public order etc.) have to be addressed in a proper way that is defined using accumulated knowledge. Although there is unquestionable part of tacit knowledge, it can be described to some extent in explicit form and compiled on media to support operations in betting shops.

In order to maintain the use of the knowledge support system, formal education process had to be established. After studying a huge amount of training material, an examination has to be passed. It is held by a committee of Mozzartbet employees. There is one employee whose main task is to ensure that process of education of potential betting shop clerks is done according to standard, and he is the head of examination committee in most cases. As an illustration of knowledge scope, we give the contents of the textbook for the preparation of supplemental material (see supplement 1).

If there is some problem that is out of the scope of support, the betting shop clerk can dial the call centre directly. The call center is supported by the same textbook and link to software development and system administration department and their knowledge support. The call center operator can try to solve the problem, and if the problem is successfully solved it is recorded in their internal log. If the operator cannot solve the problem, he creates a task in the JIRA task management system, and addresses it to software development or system support. When the problem is solved later, it is documented in JIRA system. If a problem can be solved using knowledge from the book, there is a formal procedure for disciplining the manager of the betting shop that posted such a problem. To sum up, there are three levels of knowledge support in betting shops:

1. Informal knowledge support, that was originally the only one knowledge support in the company
2. Explicit knowledge support by textbook, that serves as basis for formal training, education and examination process and everyday activities manual
3. Call center as a knowledge broker for atypical knowledge support

Communication and Collaboration Tools

There are two major communication and collaboration tools used in Mozzartbet. The first is classic telephone communication, enforced through call center that is supplied with knowledge support to solve most simple knowledge-related problems, and the second is instant messaging service Jabber. The Jabber service is a safe alternative to other free Instant Messaging services such as Skype, GoogleChat, MSN Messenger etc. All data connections are encrypted so a third party cannot eavesdrop on communication. Jabber also supports creation and joining of chat rooms for group communication. It is very important to note that Jabber is an open standard, where anyone who has a domain name and a suitable Internet connection can run their own server and communicate to users on other servers. The standard server implementations and many clients are also free and open source software, and communication to other mentioned instant messaging systems is also supported.

While talking to the employees, an important gap in the communication between betting acceptance centers and software development department along with database administrators was observed. The knowledge that was needed for usage and trouble-shooting in betting acceptance centers was created in software development, and an efficient way to transfer it to the betting shops. That was not a problem while there were less than 20 betting shops, but as the number grew, number of calls greatly increased, especially calls concerning trivial problems. Therefore, a manual describing usage of the software and common troubleshooting algorithms was prepared in aim to capture all explicit knowledge and help process flow. But a minor percentage of problems could not be solved with that manual, and in order to save time of developers and maximize efficient use of knowledge resources, communication structure represented on Figure 6 was developed.

After major problems that were the result of sudden growth in 2007 the practice of training of workers in betting shops was established, along with corporate training center that issues certificates. But unpredicted problems still needed a procedure for solution. Therefore, knowledge was concentrated in the call center, which resembles knowledge broker. All betting shops can communicate only with call centre, and if problems can't be solved at that level, calls are transferred to employees that have the knowledge to solve that. This type of formalization relieved software developers from simple technical support activities for betting shops.

On the other hand, other departments in the company can communicate freely, and have a network structure connecting them. The exception is security service in organization, which also uses video surveillance as a one-way communication channel. In software development and administration department, Jabber instant messaging service is also used as a mean to enhance communication and support group discussions.

Thursday's Lectures

Among practices for tacit knowledge management most active are Thursday's lectures. With the support of top managers, senior employees and managers in software development and administration have established these lectures as a mean for increasing and sharing knowledge. The practice is applied for almost a year, and generally everyone is satisfied with the initiative. Every Thursday a college-style lecture is held, usually two academic hours long. The topic is suggested by either senior employees and managers or employee that has developed some knowledge that might be interesting. Lecture has to be delivered with power point or similar presentation tools, and all other technical teaching support is welcome.

Presence is not mandatory, but is strongly expected. Because department lacks one large classroom, lectures are held from 14-15h for first group and then again, same lecture is presented to the second group from 15-16h. Topics ranged from formal advancement in software development such as new support frameworks, software engi-

Figure 6. Call center lines of organizational communication

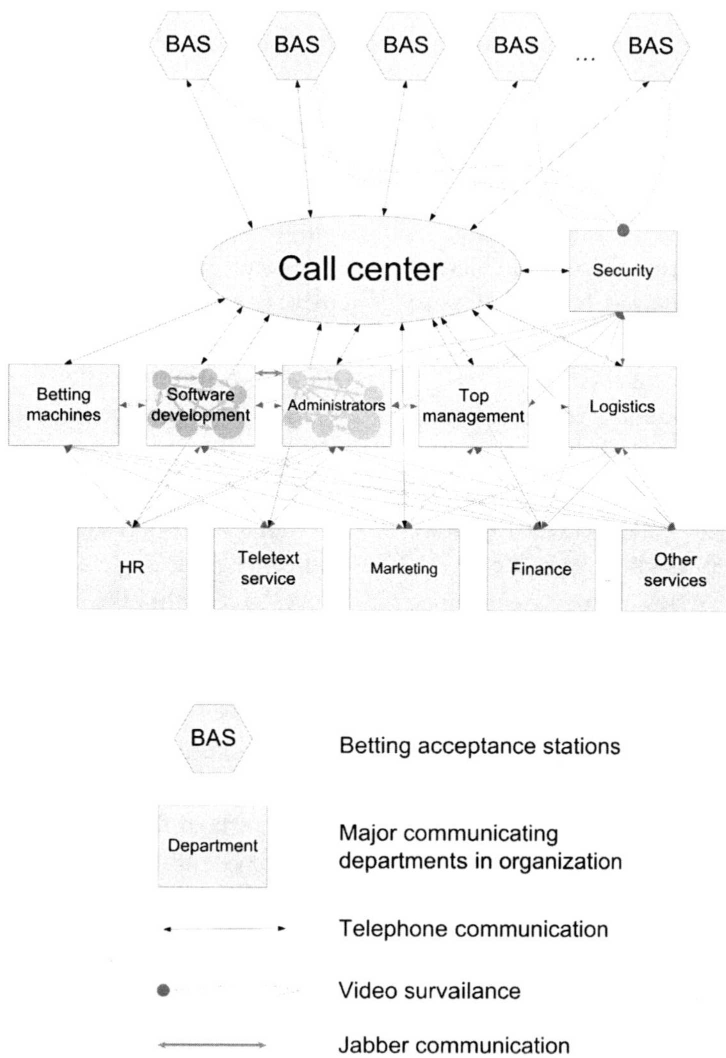

neering management practices such as SCRUM to presentation of someone's work on a part of a system. Although there is no formal reward for holding a successful lecture, it is considered to be an honor, and is accepted among employees. That practice started in order to eliminate "single points of failure" connected to tacit knowledge that is concentrated to single employee who develops some distinct part of the system, and to facilitate acquisition of related technical knowledge that belongs to new cutting edge knowledge.

External Knowledge Acquisition

Sudden growth caused some new problems in the enterprise. New departments of the company which did not even exist as jobs few years ago had to be established and be successful as other departments. Top management realized that there are huge gaps in knowledge available to employees among different departments. In order to fulfill those knowledge gaps, top management did not choose to develop existing employees, but instead due to the lack of time recruit best employees available in Serbia and surrounding countries, practically acquiring their knowledge.

Knowledge gaps in finance, HR, marketing and other newly established departments were therefore filled by recruiting best available staff from leading companies in the environment. That practice was also adopted by software development and administration department. Since employees have a lot of connections, and are motivated by large bonuses (up to 40% of salary) they can receive for recommending a good person, a lot of head hunting practice is performed in areas they find company lacks knowledge in. Mozzartbet is regarded as one of the better companies in Serbia to work for as an IT expert. So, it is not too hard to attract the right person. For example, a database administrator was recruited from a major telecommunication company in Serbia. Some knowledge management and human resource management flaws existed in his former company and his transfer to Mozzartbet resulted in drop of technical performance of his former employer. On the other side, Mozzartbet has earned more money than invested in his transfer just in first few months, since a new, cheaper and more effective solution for providing internet service and communication with betting shops was established. That solution opened new possibilities for the company to act as a content provider. Other forms of external knowledge acquisition are also performed through hiring external consultants in fields where company lacks internal knowledge, and human resources development through certified seminars and symposium attendance.

Master-Apprentice Relationship and Cross-Functional Practices

One of the most important knowledge in Mozzartbet is, as mentioned, knowledge about core processes in betting. Since that knowledge is unique to the company, for software engineers it is acquired in several ways, and one among them is master-apprentice relationship. Each new employee in software engineering and sometimes IT administration tasks is given a mentor, and passes through extensive employee orientation program. On the job training is carried out by senior programmer, and usually lasts for a month.

Each new employee is assigned to one new function of the system and it is his or her duty to develop it within the first month. That functionality is often trivial, even useless, but if possible chosen in that way to be connected with all major parts of the system. That choice is carefully made, because influence should be one-way and that new function should not be able to cause bugs in other parts of the system, nor should other parts of the system depend upon it. During that time, he or she can count on support of his mentor and most important product of that orientation period is knowledge of the entire system that employee acquires.

In 2009, cross functional practices were established for employees from IT related departments to gain knowledge about appliance of their software solutions. Job rotation that is done can include all possible tasks that utilize the software, such as direct betting, technical support etc. Software developers work for some time as system administrators, administrators work as call center clerks, and permanent job rotation, especially to management positions, as a form of advancement in the company. This is practiced and preferred to external recruiting of employees where possible.

Informal Practices

Knowledge is created, shared and acquired in Mozzartbet in number of different practices, mostly connected to the academic culture of the enterprise. For example, one employee shared his knowledge of one specific project with his former peers using internal messenger service for a month after he left the Mozzartbet. It is commonly accepted that if someone leaves the company he or she should share knowledge related to the company if needed. Until now, there were no problems regarding that, and former employees even keep to socialize with their old peers and participate in Monday evening football matches, or even come to their former company and join former colleagues for a lunch break to catch up what they missed. Chief Technical Officer and CEO are more and more aware that such informal practices cannot be relied upon, especially as the company grows, and must be supplemented with formal procedures, keeping the goodwill and existing organizational culture.

Other informal practices include teamwork, which is supported by SCRUM appliance in software project management. Company follows existing culture of academic openness, and knowledge exchange is estimated to be on a very high level among team members that work on the same project. When some issues are observed, they are monitored and management acts according to priorities.

One more informal practice of knowledge sharing is role playing, used to make knowledge sharing more interesting. Participants are given tasks to play roles of other positions in the company, and afterwards discussion is aimed to searching and sharing needed knowledge. As an example, e-mail sent to Product Owners by CTO is given. This concrete role-playing should encourage them to share their knowledge about dealing with users that might be overzealous in assigning tasks and lack technical knowledge.

CURRENT CHALLENGES FACING THE ORGANISATION

Interviewed employees had pointed towards main problems of knowledge management in Mozzartbet:

- Organizational problems
- Cultural problems
- Technology problems

Common attitude is that organizational problems are most important, and that the lack of formal procedures and organization for knowledge management needs to be improved. Those issues were not as important while the company was small, but today as company continues its growth; managers think that it is urgent to establish that before company starts to lose much money. Regarding cultural problems, positive aspects of existing culture in software development and system administration should be used to improve attitude towards knowledge sharing. A knowledge-friendly culture is one of the most important factors of KM success. That includes that employees have a positive attitude towards knowledge sharing (Davenport *et al.*, 1999). This is regarded as important in Mozzartbet, but is very low on priority list, especially when compared with visible results on deadline bounded projects. Technology problems are currently least

Table 3.

Tomorrow on MoneyFlow project presentation, I will not be present as a CTO, but as system user Peter Peterson. Peter manages numerous Betting Shops, in several countries. He is an "expert" in computers, has latest model laptop with 3G modem, Facebook account, wireless mouse, and HE KNOWS HOW TO SEND AN E-MAIL, but not to attach files.   **Peter has problems with money collection form Betting Shops, and thinks that there is too much money locked in the process, i.e. that the sum of money could be lowered without influencing day to day business.**   **Peter is invited for presentation of a complete MoneyFlow module on a copy of a real Mozzartbet system, and he is very happy because he finally hopes to see all his money on one stack. The meeting is scheduled on:**   http://sastanci.mozzartbet.com/day.php?year=2009&month=10&day=02&area=2&room=1 **(link is inaccessible outside Mozzartbet computer network).**   **With outstanding efforts from IT department, Peter was given an explanation that some of issues are still developed, so something MIGHT NOT WORK in a copy of a real system. Peter has accepted, but still insists to see all, even if something does not work.**   **Peter is very fond of new ideas, and he likes to think of new assignments for software developers while he is being presented current solution. He expects that future functionalities will be discussed, too. He must be very satisfied with our work, because he would like to assign much more for us.**   **Guys, good luck with Peter tomorrow.**

perceived, because current technology capabilities mostly cover existing needs of knowledge management. Interviewed employees agree however that the technical capabilities of wiki should be improved, and that new technologies should be timely prepared and tested in informal use, even before formal goals are defined.

In that context, CEO and CTO have developed several guidelines for future development of knowledge management in Mozzartbet:

- It was observed that noise in knowledge sharing is a large problem, so processes need to be formalized.
- Recognizing fields where organization lacks knowledge needs to be accepted as important task, and assigned to one of the senior managers.
- Top management plans to create a data warehouse and perform data mining on several important parts (stored tickets, issues, logs etc.).
- Communication of knowledge, especially among different countries needs to be formalized and improved,
- SCRUM approach to software development projects will be used to improve knowledge sharing.
- Effort has to be invested in solving organizational problems hindering knowledge sharing between users and software developers.

Regarding first guideline, CEO and CTO will focus more on behavioral aspects of the system, because in knowledge transfer such aspects are more important than system architecture (O'Dell & Grayson 1999). Obligations for knowledge sharing need to be formalized and following Pareto principle: Roughly 20% of most important knowledge should be shared in explicit form via existing technologies. That will ease the second task for senior managers. After several positive examples of benefits after external knowledge acquisitions, top managers have agreed to give more priority to find existing knowledge gaps. Also, a formal program for knowledge creation has been envisioned. Huge amount of data that exists in transaction databases could be analyzed and mined for knowledge, using statistical analysis to find relations among parameters. The problem of knowledge sharing among different countries is also recognized, because limits are more difficult then within one country. Each country currently functions as independent knowledge island, and top management plans to exchange senior staff to improve knowledge. That staff will be responsible for knowledge dissemination as one of key tasks in the other country. Personnel rotation can enhance the communication between knowledge holders and recipients, and eliminate the cultural differences between the two. At the same time this will also help to establish a personal network which can enhance their knowledge and technology dissemination (Dong 2009). Already practiced and connected with knowledge management, SCRUM approach will become standard for development to foster better knowledge creation and sharing. An existing problem in knowledge sharing between software developers and software users in different departments are planned to be solved with broadening the role of established call center and using it for active communication of selected knowledge to parties, instead of passively connecting knowledge senders and receivers.

SOLUTIONS AND RECOMMENDATIONS

Fast growth of a company can be the cause of its future demise. The same qualities that caused success can cause an organization to stretch beyond boundaries and capabilities of formal control, leading the organization towards control crisis stage that Greiner (1972) states could be resolved by growth through coordination. Organization has, due to its nature and available resources, already developed principles of information management, and Davenport and Marchand (2000) find that information management comprises a large part of knowledge management. The company should use an existing information management system as a foundation of a future knowledge management system.

Furthermore, it is recommended to keep the focus on preserving existing organizational culture in IT related departments, because it can provide a good background for further knowledge sharing improvements - relaxed organizational culture and decentralized, liberal management styles can be beneficial for ICT application in general (Cudanov *et al.*, 2009). Existing KM tools and practices, like JSPWiki explicit knowledge database, internal newsgroups, mailing lists, administrator logs, communication and collaboration tools, Thursday's lectures practice, external knowledge acquisition, master-apprentice relationships, cross functional practices, the betting shop clerk knowledge support system and informal practices should be preserved as much as possible, but should be subjected to the process of formalization. Cross-border knowledge sharing in different countries covered by business needs to be established as formal practice. It is essential to set guidelines, e.g. that every entry in JSPWiki contains identifying information, assign responsible person to monitor content and delete inappropriate entries (Dearstyne, 2007). A positive example is IBM (2005), where blogging guidelines were defined.

In addition, an organizational reward system related to knowledge creation, external acquisition and sharing should be established (Holsapple & Joshi 2003). With that, employees should be more motivated to take voluntarily part in KM initiatives at Mozzartbet. Although Mozzartbet is way ahead its competitors, and average companies in Serbia by means of knowledge management, top managers should not be relaxed. Organization cannot rely on informal practices, organizational culture and attitude of individuals as key success factors of KM in the future. Top managers are already ahead competition because they have not only recognized KM as useful practice but also deem knowledge of Mozzartbet as one of the most important factors of competitiveness. It might be a great chance to formalize all existing informal strength factors of KM and preserve and enrich knowledge of Mozzartbet as utterly important resource of the organization without constraints from human resources. That could be one more factor ensuring leader's position in the future.

REFERENCES

Alavi, M., & Leidner, D. E. (2001). Review: Knowledge Management and Knowledge Management Systems: Conceptual Foundations and Research Issues. *Management Information Systems Quarterly*, *25*(1), 107–136. doi:10.2307/3250961

Ambrosini, V., & Bowman, C. (2001). Tacit knowledge: Some suggestions for operationalization. *Journal of Management Studies*, *38*, 811–829. doi:10.1111/1467-6486.00260

Baeza-Yaes, R., Ciatamita, M., Mika, P., & Zaragoza, H. (2008). Towards Semantic Search. *Natural Language and Information Systems*, 4-11.

Cantú, F. J., Bustani, A., Molina, A., & Moreira, H. (2009). A knowledge-based development model: the research chair strategy. *Journal of Knowledge Management*, *13*(1), 154–170. doi:10.1108/13673270910931233

Cantu, L. Z., Criado, J. R., & Criado, A. R. (2009). Generation and transfer of knowledge in IT-related SMEs. *Journal of Knowledge Management*, *13*(5), 243–256. doi:10.1108/13673270910988088

Cavusgil, S. T., Calantone, R. J., & Zhao, Y. (2003). Tacit knowledge transfer and firm innovation capability. *Journal of Business and Industrial Marketing*, *18*(1), 6–21. doi:10.1108/08858620310458615

Cohn, M. (2009). *Suceeding with Agile: Software development using Scrum*. Reading, MA: Addison-Wesley Professional.

Cudanov, M., Jasko, O. & Jevtic M. (2009). Influence of Information and Communication Technologies on Decentralization of Organizational Structure. *Computer science and information systems – COMSIS Journal*, 6(1), 93-109.

Davenport, T. H., De Long, D. D., & Beers, M. C. (1999). Successful Knowledge Management Projects. In Cortada, J. W., & Woods, J. A. (Eds.), *The Knowledge Management Yearbook 1999-2000* (pp. 89–107). Washington, DC: Butterworth-Heinemann Ltd.

Davenport, T. H., & Marchand, D. A. (2000). Is KM Just Good Information Management? In Marchand D.A., DAvenport T.H., & Dickson T. (eds.) *Mastering Information Management*, London, UK: Prentice Hall, 165-169.

Dearstyne, B. W. (2007). Blogs, Mashups, & Wikis Oh, My! *The Information Management Journal*, *41* (4), 24-33.

Dong, S. (2009). The Tacit Knowledge Transfer of Transnational Corporation. International. *Journal of Business and Management*, *3*(6), 148–150.

Du Plesis, M. (2007). Knowledge management: what makes complex implementations successful? *Journal of Knowledge Management*, *11*(2), 91–101. doi:10.1108/13673270710738942

Forrest, D., & Simmons, R. (2003). Sport and gambling. *Oxford Review of Economic Policy*, *19*(419), 598–611. doi:10.1093/oxrep/19.4.598

Garnsey, E. (1998). A Theory of the Early Growth of the Firm. *Industrial and Corporate Change*, *7*(3), 523–556.

Greiner, L. (1972). Evolution and revolution as organizations grow. *Harvard Business Review*, *50*(4), 37–46.

Handy, C. (1996). *Gods of Management* (p. 14). New York: Oxford University Press USA.

Harrison, J. R., & Carroll, G. R. (1991). Keeping the Faith: A Model of Cultural Transmission in Formal Organizations. *Administrative Science Quarterly*, *36*, 552–582. doi:10.2307/2393274

Holsapple, C. W., & Joshi, K. D. (2003). A Knowledge Management Ontology. In Holsapple, C. W. (Ed.), *Knowledge matters* (pp. 89–124). Berlin: Springer.

Hsu, I.-C. (2008). Knowledge sharing practices as a facilitating factor for improving organizational performance through human capital: A preliminary test. *Expert Systems with Applications*, *35*(3), 1316–1326. doi:10.1016/j.eswa.2007.08.012

Husted, K., & Michailova, S. (2002). Diagnosing and fighting knowledge-sharing hostility. *Organizational Dynamics*, *31*(1), 60–73. doi:10.1016/S0090-2616(02)00072-4

Huxley, A. (2004). *The Doors of Perception and Heaven and Hell* (p. 13). New York: Harper Perennial Modern Classics.

IBM. (2005). *IBM Blogging policy and guidelines*.Retrieved from http://www.snellspace.com/IBM_Blogging_ Policy_and_Guidelines.pdf

Jordan, P.J., & Lawrence, S.A., Troth, & A.J., Jordan. (2006). The impact of negative mood on team performance. *Journal of Management & Organization*, *12*, 131–145.

Jordan, P. J., & Troth, A. J. (2002). Emotional Intelligence and Conflict Resolution: Implications for Human Resource Development. *Advances in Developing Human Resources*, *4*, 62–79. doi:10.1177/1523422302004001005

Kazanjian, R. K. (1988). Relation of Dominant Problems to Stages of Growth in Technology-Based New Ventures. *Academy of Management Journal*, *31*(2), 257–279. doi:10.2307/256548

Mens, T., & Demeyer, S. (2001). Future trends in software evolution metrics. *Proceedings of the 4th International Workshop on Principles of Software Evolution,* Vienna, Austria. 83-86. New York, NY, USA: ACM Publishing

Ndlela, L. T., & Toit, A. S. A. (2001). Establishing a knowledge management programme for competitive advantage in an enterprise. *International Journal of Information Management*, *21*(2), 151–165. doi:10.1016/S0268-4012(01)00007-X

Nonaka, I. (1991). The knowledge-creating company. *Harvard Business Review*, *69*(6), 96–104.

O'Dell, C., & Grayson, C. J. Jr. (1999). Knowledge tranfer: Discover your value proposition. *Strategy and Leadership*, *27*(2), 10–16. doi:10.1108/eb054630

Paliouras, G., Karkaletsis, V., Papatheodorou, C., & Spyropoulos, C. D. (1999). Exploiting Learning Techniques for the Acquisition of User Stereotypes and Communities. *Proceedings of the Seventh International Conference of User Modeling*, 169-178.

Penrose, E. (1995). *The Theory of the Growth of the Firm*. Oxford, UK: Oxford University Pres. doi:10.1093/0198289774.001.0001

Polanyi, M. (1966). *The Tacit Dimension*. New York: Doubleday.

Rico, Z. (2003). Fast Growth in Control. *Accountability, 203-221*

Spender, J. C. (1996). Making Knowledge the Basis of a Dynamic Theory of the Firm. *Strategic Management Journal*, *17*, 45–62.

Taylor, P. (1999). Thinking about new learning environments (report no. TAY99147), *Australian Association for Research in Education. RetrievedSeptember18, 2009*http://www.aare.edu.au/99pap/tay99147.htm

Zheng, W., Yang, B., & McLean, G. N. (2009). Linking organizational culture, structure, strategy, and organizational effectiveness: Mediating role of knowledge management. *Journal of Business Research*, article in press - http://dx.doi.org/10.1016/j.jbusres.2009.06.005

KEY TERMS AND DEFINITIONS

High-Growth Company: develops very fast; it has a growing customer base, many new business opportunities, and a competitive advantage over other competitors.

Sport Betting: is the general activity of predicting sports results by making a wager on the outcome of a sporting event.

Organizational Culture: describes psychology, attitudes, experiences, belief as well as personal and cultural values of an organization.

Software Engineering Department: is a part of the company, where employees develop, operate and maintain software with a systematic approach.

Tacit Knowledge: is defined as difficult to write down or to formalize and therefore difficult to be transferred to other people.

Explicit Knowledge: can be articulated, codified, and communicated in a symbolic form and/or natural language, e.g. a product manual that is given to the customer.

Informal Knowledge: is acquired through personal experience, outside of formal learning environments such as schools or training courses.

Formal Knowledge: can be found in books, manuals and documents and can be shared in formal learning environments like schools or training courses.

Chapter 15
Cross-Cultural Knowledge Management Practices to Support Offshore Outsourcing

Biswadip Ghosh
Metropolitan State College of Denver, USA

EXECUTIVE SUMMARY

In offshore business process outsourcing (BPO) arrangements between US client firms and Indian vendor firms, a "culture gap" has been reported between the organizations. This gap originates from (1) macro level differences between the two firms in their operating procedures and (2) micro level differences in the cultural proclivity of the individual workers. This gap often hinders the much needed bi-directional knowledge sharing in the outsourcing organization. To close this gap and improve knowledge sharing, effective organizational practices are needed. This chapter studies a case where a set of technical support processes were outsourced by an US based multinational client to an Indian vendor. Data was collected using interviews of client-side, process and business managers to understand what challenges were faced in the organization after BPO and what practices facilitated cross cultural knowledge sharing. A set of organizational enablers and collaboration tools were reported by the interviewees to be successful in bridging the gaps and enhancing bi-directional knowledge sharing. This case demonstrates the benefits of three organizational practices – (1) bridging, (2) bonding and (3) linking, that build social capital in the form of relational norms and connections among workers. These relational practices, together with social capital, complement the use of collaboration tools and technologies to improve knowledge sharing between the client and vendor teams.

BACKGROUND

The concept of transferring the operational ownership and execution of one or more business processes is referred to as Business Process Outsourcing (BPO). Due to the current globalization trends and improvements in computer and telecommunications networks, US firms are pursuing offshore outsourcing to remote countries like India and China. Offshore outsourcing allows the US

DOI: 10.4018/978-1-61692-886-5.ch015

firms to tap into large pools of offshore workers at significantly lower labor costs. Additionally, US firms benefit by getting access to large overseas markets for products and services in the foreign countries and take advantage of time zone difference to staff round the clock operations.

In strategic outsourcing, core processes in the primary value chain of the client (such as customer support or supply chain management) are offshored. Core processes differentiate firms in the marketplace and hence cannot be modularized. Moreover, core processes are usually connected to other process in the client organization and are a source of unique value. Research indicates that, to support the outsourcing of these core processes, complex and customized inter-organizational systems need to be deployed (Carmel and Agarwal, 2002). While modularized tools and procedures can easily support non core processes such as human resources or benefits management (Lacity and Willcocks, 1998), outsourcing of core processes often require differentiated procedures and the integration of tacit knowledge from both organizations to run effectively. Several knowledge management initiatives, such as balanced scorecard metrics, real time dashboards and identifying subject matter experts were identified as part of a list of 20 best practices that help overcome the problems inherent in offshore outsourcing (Rottman and Lacity, 2004). Extensive tacit knowledge needs to be utilized in the case of complex technical support processes such as telecommunications network design, installation and management (as in this case study). This tacit knowledge can be situated among personnel on either the client side or the vendor side. Specifically, product design knowledge may reside on the client side, while local customer and infrastructure knowledge may reside with the vendor. Such knowledge exchanges can be for several situations – (1) addressing unexpected situations when codified explicit knowledge does not exist to handle an issue, (2) learning to understand the complexity and interdependency of various support scenarios – ie, becoming fully mindful of the undocumented "ripple effect" of various technical approaches, and (3) enhancing the client worker's understanding of the offshore infrastructure and customer environment and (4) enhancing the vendor worker's understanding of the technical nuances of each product's features/capabilities. In such cases, KM systems are needed to support bi-directional tacit knowledge sharing to allow the outsourced processes to run efficiently.

The research on interorganizational knowledge management shows that knowledge sharing is particularly difficult between organizations (Burgess, 2005). Factors that impact knowledge sharing at the macro or organizational level include the characteristics of the organizations, their work related practices, how work activities are viewed and how systems are used in addition to the type of knowledge that needs to be transferred (Argote, 1999; Ko, Kirsch and King, 2005; Inkpen and Tsang, 2005). Knowledge sharing can be hindered due to macro level differences in organizational characteristics such as (1) work practices being process oriented or result oriented, (2) employee oriented versus job oriented, (3) parochial versus professional, (4) open systems versus closed communications climate, (5) loose control versus tight control and (6) rule oriented versus customer oriented (Hofstede, et.al., 1990). These differing organizational norms and structural issues lead to differences in how knowledge sharing systems are viewed and utilized by the two organizations. For example, differences in process versus results orientation can lead to how strictly operating procedures/processes are followed and have been reported to cause secondary problems after the initial trouble has been resolved. Additionally there can be differences in the governance, management and control of the work activities and how workers use the available systems. A highly parochial or employee oriented culture can result in less strict enforcement of deadlines and miscommunications due to "clannish" reporting that can lack details to hide situational issues (Pothukuchi, et.al., 2002).

These differences in the organizational orientation can lead to mistrust and hamper the ability of the client and vendor to support the customer holistically, as neither side can communicate or identify with the nuances of the situations faced on each side.

At the micro or worker level, there is often a strong cultural clash between the client and vendor personnel. Indian cultural norms remain very different from US cultural norms, particularly in the dimensions of power distance and individualism (Hofstede, 2008). Power distance is the extent to which the less powerful members of the organization accept and expect that power is distributed unequally. Individualism refers to how weak an employee's loyalty is towards the work group. Indian culture is significantly higher in power distance and significantly lower in individualism compared to US culture. Misalignment in these dimensions hampers the ability of client and vendor firms to interact effectively. Knowledge sharing at the micro level is hindered as underlying cultural differences lead to conflicting practices, misunderstandings and interaction problems. These must be alleviated by addressing the psychological hurdles for collaboration by building relationships or social capital among the workers and providing opportunities for interaction (Willcocks, et.al. 2004; Levina, 2005). By building social capital and common working norms and trust among the client and vendor workers, two additional benefits are incurred – (1) Staff feel safe to explore and share new ideas without fear of failure, leading to more innovation in complex process execution, and (2) shared objectives and vision is developed between client and vendor staff that establishes joint responsibility for delivering "holistic" customer support, rather than piece-meal service events. In a case study of collaborative IS projects, Bhandar, et. al. (2007) found that knowledge integration across organizational boundaries is particularly important for developing cohesion and consensus among the teams.

SETTING THE STAGE

Hansen, Nohria and Tierney (1999) identify two enabling strategies for knowledge sharing - a personalization strategy for sharing tacit knowledge with emphasis on building relationships verses a codification strategy for sharing explicit knowledge with emphasis on IT (Information Technology) solutions. While various IT solutions can facilitate the sharing of explicit knowledge between firms (Hislop, 2002) organizational practices can play a significant role in tacit knowledge sharing by building social capital, a concept from social capital theory (Nahapiet and Ghoshal, 1998).

Social capital theory (SCT) is centrally concerned with the significance of relationships, or social capital on the effectiveness of organizations (Nahapiet and Ghoshal, 1998). SCT argues that effective knowledge exchange and transfer occur when sufficient opportunities for interactions are present between individual knowledge workers. SCT can be applied to explain the behavior of resources embedded in inter-organizational environments such as networks of practice, where norms are formed to facilitate knowledge transfer (Kankanhalli, et. al, 2005). Several research studies have found a high correlation between social capital and knowledge sharing (Yang, et.al., 2009). Simply stated, social capital comprises a set of values, norms and informal sanctions that are shared among members through interactions that permit them to cooperate with each other. Social capital can provide the lubricant for workers to get support and advice well beyond the organizational hierarchy, to enable them "to get things done more effectively." The three dimensions of social capital – (a) trust, (b) common identity and (c) common norms play a large part in the effectiveness/success of the outsourced processes and how much cooperation is achieved between client and vendor personnel and the extent of tacit knowledge sharing (Inkpen and Tsang, 2005). Trust enables the workers to work more cooperatively, limiting the power and positional rivalries. By building a stronger identity, a common goal can be facilitated and common norms enable members to transcend the diversities that are inherent in a multi-cultural organization.

The development of social capital requires the active and willing engagement of citizens within a participative community. Prior research has identified three types of organizational practices for that support social capital creation: (1) Bonding, where individuals are brought together to work collectively, (2) Bridging, when organizational boundaries are crossed and understanding is developed by both sides, and (3) Linking, where connections are established for ongoing activities (Kowch, 2005). An organization is better placed to create social capital through community based interventions that are public and non-intimate (Schmid, 2000). When people's tasks, goals and objectives are more closely related, such that interactions are fostered, it develops bonds of understanding for each other's culture and establishes bridges to cross these differences. This leads to the creation of relational norms and supports the creation of the dimensions of social capital – trust, common identity and norms, which helps the organization to overcome cross cultural issues and improve knowledge sharing (Wasko and Faraj, 2000). In a large empirical study of public organizations in Singapore, components of social capital such as organizational norms and structure were found to moderate the usage of electronic knowledge repositories in managing knowledge (Kankanhalli, et.al., 2005). Strongly tied teams communicate more frequently, maintain more and different kinds of relations, and use more types of media to communicate, thus creating stronger ties. Such stronger ties are more robust as they rest on multiple relations and media and can also influence workers to expand their use of media in exchanging knowledge to support their work activities (Haythornthwaite, 2002). SCT is also found to improve the quality of knowledge contributions to electronic networks in legal practice (Wasko and Faraj, 2005).

CASE DESCRIPTION: XYZ COMPANY

XYZ is an US based multinational company that is an established player in the sales and service of telecommunications products and has a global market presence. XYZ designs and manufactures a variety of communications hardware and software platforms and multiple applications – such as CRM, conferencing solutions and other telephony based applications. XYZ has a large services business and capabilities to support a customer's entire end to end solution. This experience starts with consulting and design services that works with a business to analyze needs. XYZ's offerings continue into integration and implementation of the agreed upon solution. Subsequently XYZ sells maintenance contracts to the customer (large multinational customers such as multinational banks, as well as to smaller regional business customers such as hospitals, universities) on a per port basis. A service contract entitles the customer to extended service beyond the product warranty period. Service entitlements include help desk, break fix and network monitoring, maintenance, administrative and management reporting activities as well. The services business is an important part of XYZ's overall operations. Upwards of 50% of XYZ's revenue (over $2 billion a year) and over 100% of XYZ's profit comes from the Services business.

The business strategy of XYZ has a multinational focus. They have operations in over 50 countries around the world. The firm's off-shored activities range from R&D support tasks to supply chain sourcing to sales and services. Sales and services still constitute over 80% of global activities, however with other corporate functions concentrated in the USA. XYZ has heavily pursued offshore outsourcing of its labor intensive service delivery processes. Initially, software update services were outsourced to an offshore vendor to increase available headcount to staff for Y2K (Year 2000) related software update efforts.

During the last decade, off-shoring was expanded to support more knowledge intensive processes in areas of new product introductions, field trials, network design and installation. Moreover, XYZ's customers have also been riding the outsourcing wave and increasingly deploying IT facilities and back office call centers in offshore locations, such as India. Therefore XYZ's outsourcing has helped reduce operating costs through labor arbitrage and provided in-region internationally located technical personnel. XYZ's off-shoed technical support processes require extensive bi-directional knowledge sharing to build up the vendor's workforce capabilities with product knowledge, and to build up the client's capabilities on offshore infrastructure and customer environments.

Data collection for this case study involved interviewing 4 service delivery process managers (SPM), 2 services delivery systems analysts (SE) and 1 manager of services strategy and business development (BDM) in the client organization at their US location. The BDM was responsible for the operational readiness and the mobilization of new managed service business offers. The SPM handled the details of the operational processes used to deliver service to the end customer including readiness and ensuring process excellence through business process management. The SE provided technology planning and solutions to support these service processes.

The interviewees were identified using business contacts and formed a convenience sample. Semi structured interviews were conducted, each lasting roughly 45 minutes. Total time spent in interviews was roughly 6 hours. The questions posed are listed below:

1. The reasons for pursing offshore BPO and any subsequent organizational challenges they faced.
2. Their observations regarding organizational issues resulting from the BPO.
3. What knowledge management practices and technologies are used to address any of the challenges in the BPO?

CURRENT CHALLENGES FACING THE ORGANIZATION

After outsourcing, XYZ is facing challenges to effectively leverage knowledge management capabilities into the vendor's workforce. Both BDM and SPM managers responded to question #1, stating that increasing off hours coverage, lowering the operational costs and local presence was instrumental in deciding to outsource. The quotes indicate the challenges faced by XYZ, in their goals to transfer knowledge to their outsourcing vendor. A BDM said,

"Our business goals were to improve the service experience of a set of customers, who choose to pay us to manage the headaches of keeping their networking devices on the latest software releases and running the latest security patches, etc. In our dynamic multi-vendor environment with thousands of devices under management, the new developments occurred hourly. It was impossible to get that knowledge to our offshore partner. This was not another 'canned' call center app (sic) that they were well accustomed to, rather our processes involved a lot of dynamic adjustments to their work activities."

When asked about the challenges, the SPM added the following to question #1,

"The costs of developing and trying to get knowledge into their work processes became very expensive. Our service process designers did not understand the vendor's culture. The BPO vendor was not flexible and did not understand our business priorities. Changes often led to more problems and follow on changes."

The BDM mentioned in response to question #2 and indicated the need for intellectual best practices to better handle customer,

"While our plan was to position new services as an add on option for existing customers, the BPO vendor preferred a clear separation between each service event. It became hard to provide a unified view to our customers and support calls were being transferred back and forth with a focus on service events rather than providing a solution to the customer's needs."

The SE mentioned the need to boost bi-directional knowledge sharing between the client and vendor workers,

"Root cause analysis indicated the troubles were due to incorrect installations and inadequate knowledge sharing. Each side was operating without understanding the other. The vendor was attempting to complete billable service events without regard to the overall service requirements, while the client was trying to emphasize the later without understanding the work culture on the vendor side."

Clearly, the managers noted the knowledge sharing difficulties causing poor support from the vendor. In summary, the challenges facing XYZ in two areas:

- Finding better ways to share knowledge between the teams. The client's goal is to develop technical support processes using highly experienced technicians in the US and deploy them to be executed by cheaper offshore workers. The content tools can guide the agent through a resolution flow to get to the right solution in the knowledge base. However for those cases where a new solution must be developed, technicians need to effectively collaborate and share knowledge.
- Higher level service capabilities need to be deployed offshore as XYZ's customers have been moving IT infrastructure and facilities to offshore locations. It has become difficult to support those installations from a US based operation and maintain 24/7 rapid response in a global environment. This is calling for deploying some of the higher tiered support services to the offshore vendor. These services need a holistic approach rather than piece-meal service events.

CLIENT INITIATED KNOWLEDGE MANAGEMENT SOLUTIONS

XYZ has always focused on knowledge management in its services business by deploying tools to support service operations. These tools are complemented thru work practices that encourage interaction and build social capital among the workers in the client and vendor organizations.

The technological solution involves updating their knowledge base with more current knowledge and service scenarios and global knowledge. The organizational component involves training and building the vendor's capability through various programs – mentoring, joint work exercise, collaboration, building cultural bridges and establishing the norms for the relationship.

Technology Components

XYZ was found to have several KM systems and technology components to provide bi-directional transfer of knowledge from the client to the vendor and vice versa (Table 1). XYZ's knowledge management platform (KMS) is based on Web2.0 using Wiki templates, blogs for discussions and social bookmarks for problem research/training (O'Reilly, 2005), Access to these systems have been deployed to vendor personnel as part of the outsourcing strategy. Documents are categorized into several levels – by product, by release, by configuration, by vendor, by customer network and by configured application(s). The company has built up this knowledge base using experience gathered in many years of service delivery. The number of variables that differ in each customer's unique installation can be very large and knowledge must be very judiciously reused from the KMS.

Management and Organizational Components

XYZ realized that to make the knowledge management tools more effective, they had to be complemented with organizational practices. The vendor personnel needed to be mentored and their work practices needed to be made compatible with XYZ's needs. A mentoring process was put in place to increase vendor capabilities and encourage interactions between client and vendor staff on suitable learning projects (Figure 1). The vendors could apply for client mentoring and assistance on an upcoming project. The client established a review board to screen each request for knowledge potential and optimal fit for the goals. If a project is selected, then personnel from the vendor and client work collaboratively on the project over a span of 4-6 weeks. Advantages were seen in the exchange of norms, work practices, common language and standards as well as tacit knowledge that is difficult to codify and contextualize. Social capital is built, as obligations are set and met, resulting in the development of trust, understanding of cultural diversity and establishment of joint ownership for work. The latter was critical, as the BPO goals were broader than cost savings. Project meetings are held frequently and the client personnel served as mentors in the relationship – thus there was an expectation of value to be obtained from the process on the part of the participating vendor staff. Likewise, the client participant was able to build up site specific and international expertise. Both parties, client and vendor, are motivated to participate in the mentoring program to improve network installation quality and minimize downstream customer network issues. Hence, the program established the preconditions necessary for social capital to form (Nahapiet and Ghoshal, 1998).

There are three classes of organizational enablers that support building social capital and that can help improve knowledge sharing. These three classes are:

Table 1. Information technology deployed for knowledge sharing

Knowledge Management System	Objectives and Usage Goals	Average Monthly Usage*
Listservers, Discussion Boards	Capturing threads of discussions on topics raised by team members and their subsequent contributions.	29 unique issues or threads of discussion created per month
Checklists	To guide technical support work from past experience and ensure that adequate data collection and situational analysis is being done. These checklists support building best practices on directed tasks such as configuring phones as well open-ended tasks such as collecting customer information for network designs.	3 New Checklists per month were developed
Lessons Learnt Lists, FAQ	To ensure that new "rules of thumb" and experience is captured and shared for future use	54 lessons learnt bullet points were documented per month
Training Presentations	Training materials are developed by scouring the listservers, discussion boards and FAQ. Presentations include all the listerv threads and their resolutions, the list of lessons learnt and pointers to any checklists or process/product document that is considered a "*must read*".	Monthly training presentations created by the client personnel are imparted to all vendor personnel.

(*Average monthly usage data was collected over an 8-month period.)

Figure 1. Off shore mentoring and knowledge management system

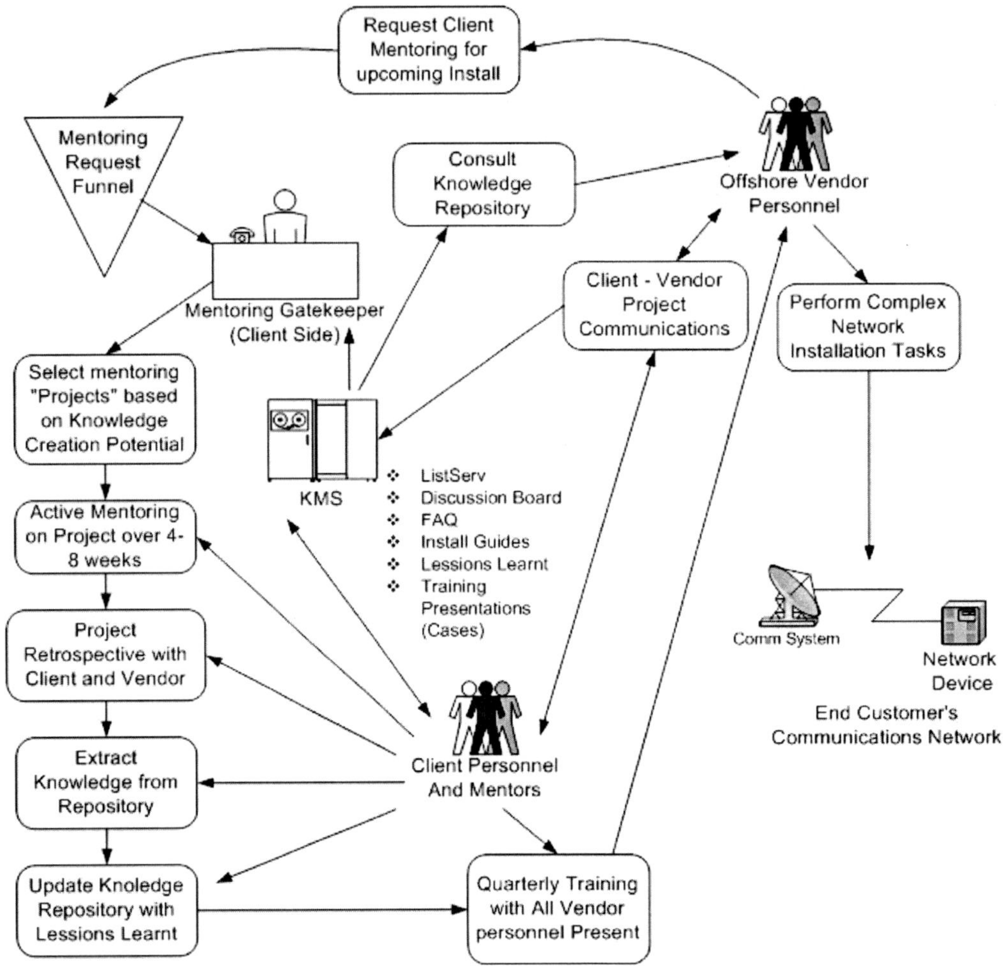

- Bonding activities, where individuals are brought together to build stronger relationships and understanding of each other cultural diversity. Activities such as face to face meetings, video conferences and information communications fall into this category. These activities generate "bonds" and relational capital (Nahapiet and Ghoshal, 1998) between members of the client and vendor teams that allow them to approach each other in the future more comfortably and understand their inherent diversity.
- Bridging activities allow intellectual capital (Nahapiet and Ghoshal, 1998) organizational boundary spanning norms and channels to be developed by both sides to support work activities. Practices such as publishing a join directory of subject matter experts, work status dashboards and other IT tools to support collaboration can create bridges between the two organizations. These "bridges" between the teams allow them to work more efficiently.
- Linking activities allow so that roles on each side are more closely allied and understood better. Activities such as joint training sessions, joint review meetings, workflows that cross organizations are in this category. Rather than keep the teams

and their work as separate as possible, linking activities build "links" to encourage ownership of the end results of a project instead of only the deliverables during the work activities. The benefits of linking activities are seen in more flexibility among team members.

OUTCOMES AND RECOMMENDATIONS

The result of these practices was to develop practices that improved bidirectional knowledge sharing. These helped the client personnel to better define service processes that suited the vendor and helped the vendor personnel to better use the knowledge. An SE mentioned the better use of the KM systems on the vendor side:

"The knowledge was all there in the KM tools – with the mentoring program, we got the vendor to understand the practice and utilize it correctly".

The SPM mentioned the need for regular interaction and communications to develop better work relationships:

"The regular communications and scheduled bimonthly training sessions developed a strong relationship of trust and identification with the offshore people."

Another SPM mentioned:

"They developed confidence and were able to engage us as peers. We saw an increase in the usage of our knowledge bases and tools"

This case demonstrates the importance of knowledge management systems in supporting offshore business process outsourcing. The sharing of knowledge in a cross-cultural environment between two different companies is particularly challenging and problematic. In an outsourcing scenario, the client and vendor teams work independently and have different work practices and cultures. However, since tacit knowledge resides within individuals, the onus is on the individual to share his/her knowledge. The effectiveness of the outsourced processes can be enhanced when this tacit knowledge is leveraged into work situations. The organization can provide tools such as discussion boards to share knowledge, but without effective organizational practices these systems become ineffective, as cultural predispositions inhibit their use among workers.

This case study describes a set of robust business practices that were effective in allowing workers to build social capital by facilitating interactions among workers to promote the development of common norms, identity and trust. The social capital created in the work group helped the sharing of tacit knowledge among workers and resulted in more effective process execution.

Organizations that are pursuing strategic outsourcing by off-shoring core business processes, that cannot be easily severed from their other in house processes, need to build their KMS using a socio-technical approach. The following practical recommendations can be supported from this case study:

1. The KMS must consist of collaboration technologies such as Wikis, discussion boards, FAQ and knowledge bases.
2. Additionally, the technological tools must be complemented with an organizational component involving building social capital through various programs – mentoring, joint work exercise, collaboration, that build cultural bridges and establish working norms among the workers. The results of this study support the notion that benefits of SCT can be seen in downstream knowledge sharing processes.
3. A client needs to invest in ongoing organizational development as a routine part of their

relationship – so that tacit knowledge can flow back and forth. The client organization in this study quantified the net monetary benefits from these practices at 5% of the outsourcing contract at an annual expense of 1% of the outsourcing contract value. Other organizations pursing offshore BPO can also reap benefits by building social capital. The benefits accrue from fewer repeat service requests and rework, better match of installations to the customer's environment and infrastructure, and less downtime for end customers.

This case study shows that if a mutually beneficial mentoring program is instituted that increases interactions among knowledge workers, that can establish working relationships and create shared work practices, social and intellectual capital; this results in greater knowledge sharing. As a result the BPO outcome is improved. Practitioners may note that the investment into building successful working level relationships in a BPO can be worthwhile.

REFERENCES

Argote, L. (1999). *Organizational Learning: Creating, Retaining and Transferring Knowledge.* Boston: Kluwer Academic, Boston

Aron, R., Clemons, E.K. & Reddi, S. (2005), Just Right Outsourcing: Understanding and Managing Risk. *Journal of Management Information Systems* 22(2),.37-55.

Brown, J.S. & Duguid, P. (1991).Organization Learning and Communities of Practice: Toward a unified view of working, learning and innovation. *Organization Science* 2(2), 1991, pp. 40-57.

Burgess, D. (2005).*What* Motivates Employees to Transfer Knowledge outside their work unit? *Journal of Business Communications* 42(4), 2005, pp. 324-348.

Carmel, E., & Agarwal, R. (2002). The maturation of Offshore Sourcing of Information Technology Work. *MIS Quarterly Executive, 1*(2), 65–77.

Hansen, T., Nohria, N., & Tierney, T. (1999). What is your strategy for Managing Knowledge? *Harvard Business Review*, (March-April): 106–116.

Haythornthwaite, C. (2002). Strong, weak and latent ties and the impact of new media. *The Information Society, 18*(5), 385–401. doi:10.1080/01972240290108195

Hislop, D. (2002). Mission Impossible? Communicating and Management Knowledge via Information technology. *Journal of Information Technology*, (17): 165–177. doi:10.1080/02683960210161230

Hofstede, G. (2008) Geert Hofstede Cultural Dimensions.Retrieved from http://www.geert-hofstede.com, Accessed May 15, 2008.

Hofstede, G., Neuijen, B., Ohayv, D. D., & Sanders, G. (1990). Measuring Organizaitonal Cultures: A Qualitative and Quantitative Study Across Twenty Cases. *Administrative Science Quarterly*, (35): 286–316. doi:10.2307/2393392

Inkpen, A.C. & Tsang, E.W.K. (2005).Social Capital, Networks and Knowledge Transfer. *Academy of Management Review* 30(1), 2005, pp. 146-165.

Kanhanhali, A., Tan, B. C. Y., & Wei, K. K. (2005). Contributing Knowledge to Electronic Knowledge Repositories: An Empirical Investigation. *Management Information Systems Quarterly, 29*(1), 113–143.

Ko, D. G., Kirsch, L. J., & King, W. R. (2005). Antecedents of Knowledge Transfer from Consultants to Clients in Enterprise System Implementations. *Management Information Systems Quarterly, 29*(1), 59–85.

Kowch, E. G. (2005). Do we plan the journey or read the compass? An argument for preparing educational technologists to lead organizational change. *Research Journal of Educational Technology, 36*(6), 1067–1070. doi:10.1111/j.1467-8535.2005.00577.x

Lacity, M. C., & Willcocks, L. P. (1998). An empirical Investigation of Information Technology Sourcing Practices: Lessons from Experience. *Management Information Systems Quarterly, 22*(3), 363–408. doi:10.2307/249670

Levina, N. (2005). Collaborating on multiparty information systems development projects: A collective reflection-in-action view. *Information Systems Research, 16*(2), 109–130. doi:10.1287/isre.1050.0055

Nahapiet, J. & Ghoshal, S. (1998) Social Capital, Intellectual Capital and the Organizational Advantage. *Academy of Management Review* 23(2), 1998, 242-268.

Pothukuchi, V., Damanpour, F., Choi, J., Chen, C. C., & Park, S. H. (2002). National and Organizational Culture Differences and International Joint Venture Performance. *Journal of International Business Studies, 33*(2), 243–265. doi:10.1057/palgrave.jibs.8491015

Rottman, J.W. & Lacity, M.C. (2004). *Twenty Practices for Offshore Sourcing. MIS Quarterly Executive* 3(3), 2004, 117-130.

Schmid, A. A. (2000). Affinity as social capital: its role in development. *Journal of Socio-Economics*, (29): 159. doi:10.1016/S1053-5357(00)00059-7

Wasko, M. M., & Faraj, S. (2000). It is what one does: why people participate and help others in electronic communities of practice. *The Journal of Strategic Information Systems, 9*(2-3), 155–173. doi:10.1016/S0963-8687(00)00045-7

Wasko, M. M., & Faraj, S. (2005). Why Should I Share? Examining Social Capital and Knowledge Contribution in Electronic Networks of Practice. *Management Information Systems Quarterly, 29*(1), 35–57.

Willcocks, L. P., Hindle, J., & Feeny, D. Lacity, M. 2004. IT and Business Process Outsourcing: The Knowledge Potential. *Information Systems Management*, Summer, pp. 7-15.

Yang, S., Lee, H., & Kurnia, S. (2009). Social Capital in Information and Communications Technology research: Past, Present and Future. *Communications of the AIS, 25*(23), 183–220.

KEY TERMS AND DEFINITIONS

Knowledge Management Systems: knowledge management systems are specialized applications, procedures and policies that advance the acquisition, conversion, transfer and application of knowledge.

Tacit Knowledge: Tacit knowledge refers to knowledge that is inside a person's head and wholly owned by that person. The sharing of tacit knowledge can be achieved through interactions between the knowledge owner and the requester.

Business Process: A business process is a collection of related activities that produce a product or a service of value to the organization, its business partners, and/or its customers. A process has inputs and outputs, and its activities can be measured.

Business Process Outsourcing: Business process outsourcing refers to transferring the operational, managerial or other responsibilities related to a client firm's business process to another firm, referred to as the vendor.

Social Capital: social capital refers to a set of relational norms that are created inside a workgroup that helps the group function better as a team or unit.

Service Entitlements/Contract: Service entitlements or Service level agreements (SLA) refer to the contractual service delivery obligations that are based upon a service contract between the provider of the contract and the customer.

Culture: A preferred form of operating norms and expectations for a group of people is loosely referred to as culture. Cultural groups can be along geopolitical lines (i.e., by nationalities) as people from the same locale generally have acquired similar norms from other country-mates who have those norms.

Organization/Organizational: An organization is a group of persons with a common objective. For a typical business organization, the objective may be to earn a profit from engaging in commercial activities. Organizations are complex as they have a structure of relationships, power, objectives, roles, activities, communications and other factors that exist as persons work together and interact with each other to fulfill the objectives. Anything pertaining to an organization is referred to as organizational.

Chapter 16
Understanding the Behavior of Knowledge Management Pathways:
The Case of Small Manufacturers of Footwear in Peru and Brazil

Jose Manuel Cárdenas Medina
University of São Paulo, Brazil

Mauro de Mesquita Spinola
University of São Paulo, Brazil

EXECUTIVE SUMMARY

Within small-size firms, it is too difficult to understand how the process of knowledge transfer happens. It is through a close observation of their mind-set and ethos that it is possible to construct a conceptual map of their transfer of knowledge processes. Within the present case; the relations among footwear manufacturers enabled the generation and sharing of knowledge. Thus, this chapter seeks to introduce the behavior of knowledge management and its source by studying the social behavior of micro-entrepreneurs from Peru and Brazil that create relationships, practice communities and other forms of knowledge sharing. Best knowledge-based practices emerged from the adaptation and imitation from others. Three ways were recognized: distance-to-market; distance-to-consumer, and inner circle. And, three ways of social memory creation within the core of the small-firms were observed: memory of product, market report and social memory.

BACKGROUND

In Peru, two sectors were born, around 1990s, after strong pressures from government and the globalization process. In Brazil, on the other hand, from the 1980s there were several clusters of the footwear industry. Small size organizations in both countries have long history. The study of PROEXPANSION (2006) and the article of Schmitz (1995) provided detailed history of the footwear industry. Clusters were formed for the aim of enhancing commercial opportunities

DOI: 10.4018/978-1-61692-886-5.ch016

or taking advantage of geographical location. The development of small business activities next to other related business bring benefits for managerial approaches, production flows, provides guaranty access to local market, and enables micro-businesses to be outsourcers for bigger firms. But, this fact does not guarantee access to technical and technological advances or innovation. The role of the knowledge within these organizations (cluster) would be invisible, because the best practices are conducted by the same people in the organization all the time; and learning from outsiders, e.g., know-how from external producers, is lost.

To understand the micro-firms' behavior it is important to live with the managers in their daily activities. It is through a close observation of their mind-set and ethos that it is possible to construct a conceptual map of their transfer of knowledge processes. It is, the kind of innovation that is a mix of environment and an adaptive process; where it is most important to evolve by relationships and knowledge from others than innovate goods or services. This was shown by a group of footwear small-manufacturers from Peru and Brazil; whose environment pertains to a turbulent ambient of social pressures; including political and competitive factors.

The ethos represent a set of religion, language, habits and other predominant characteristics from the population is intended to shape organizational practices within the firms and furthermore for the understanding of the behavior of entrepreneurs.

Peruvian firms encourage the social integration with other firms looking for market integration or learning about best-practices. In this way, Peruvian firms recruit their workforce for short time (one or two days) only to complete a batch or aiming to help jobless people. On the other hand, firms act as customers as well as vendors; producers are customers of other producers. The result is an empathy with customers and consumers. That is reflected in innovative products and processes to market features and demands.

The Brazilian mind-set, on the other hand, tends to enhance the sales by taking the better practices, and looking to the market. The focus is on the buyer and the fashion. Thus, it is important to understand the own ethos in order to sell in Brazil. The complexity in recruiting the labor is bigger. And the market is globally consolidated. In this way, the small manufacturers need to dislocate themselves in order to attend a market dispersed geographically.

In order to understand the generation and sharing of knowledge in the small footwear firms of the two countries, five independent micro-firms, from each country, were assessed. In order to understand the role of, through their organization forms and their behavior.

The difference is motivation moved small-manufacturers to join, or not to join, others. Like in the biology, microorganisms (or small-firms) form colonies or populate bigger organisms to be representative in their search for survival. All small businesses need to learn, modify, adapt and, finally, innovate, in the search for the next evolution. This process entails a difficult pathway to generate an innovative and creative process through survival and/or taking advantage of opportunities. By this way they are called to compound a bigger system; because the efforts of a big mass of small-firms generate welfare and enable other companies to understand the mind-set of customers and consumers.

In order to integrate culture and entrepreneurial features, it is essential for small businesses to consider social chains. As innovation inside small firms is limited by tightness of funds for research projects, so, producers and vendors shape strategic alliances among themselves.

The objective of this chapter is to examine the cultural and social behavior of knowledge within the independent small-manufacturers in their search for new markets or in explaining the mindset of the costumer and consumer. It seeks to show how the self-organization of an innovation force helps to create alternative process to recruit and modify habits and culture needed to adapt with environmental dynamics without modifying their production setup. The knowledge management, in this context, shows that the innovative behavior arises from the overlapping between culture and imitation. And, that the learned lesson from many small-firms, interacting with each other and acting as vendors and consumers at the same time, creates a great repository of knowledge and organizational memories.

SETTING THE STAGE

Small-firms have the tendency of growth in both countries. The big picture is like a great sea of small-firms and some islands that represent big firms (Villaran, 2001). Each firm could be very different from others, for instance, a small-firm within a regional cluster is very different than one that is a spin-off. This is important because, culture also encloses different features in each country. Markets are not only local but regional. And, producers tend to achieve these markets by understand a set of multi-cultural behaviors, beliefs and desires. In most cases, culture drives the internal politics of firms despite their stage of maturity. Organizational culture is also different in each case but guided by a set of features (or culture) as the biggest force. Thus, there is a sum of conditions that create differences among small firms.

The social behavior of the small-manufacturers starts with a little knowledge of the environment. The administrative processes are empirically created and best practices result from years of work and interaction with other similar firms.

Case Description

Six micro-firms were studied from the footwear sector in Arequipa city – Peru and five micro-firms in the São Paulo city- Brazil. The methodology used was a case research method. The objective was to extract all possible information about production systems, market characteristics and cultural features, and to understand how those factors could generate changes across these organizations.

All members of the family firms, *in loco*, were interviewed with the aim of making them feel comfortable. In Brazil; footwear micro-firms represent specific conglomerates in specific cities. Thus, it was difficult to find micro firms in the footwear sector and that do not belong to a productive arrangement. Characteristics of the micro-firms in each country are introduced in the Table 1 and in the Table 2. (FE = familial entrepreneur; PE = private entrepreneur)

In order to extract information from all areas in each firm, it was decided to study logistics, market, administration and human resources. The most important factor was to understand the behavior of people around the entrepreneurs because customers and consumers are important agents in the development of the innovative processes and products. Besides, specific features of the entrepreneurs were also considered, i.e. folklore, music, ethos, etc.

Arequipa city is one of the three Peruvian cities that was representative of the footwear, tanner and saddles production and São Paulo is a cosmopolitan city that has some footwear clusters. It is important to indicate that each interviewed micro-entrepreneur belongs to independent firms and has a relatively long experience (about twenty or more years) in the market and/or production.

Management and Organizational Concerns: Post-Fordist Entrepreneurs

The *Fordist* (Ford's assembly line production) model considers a human like an appendix to a machine. According to Zarifian (1995), the "productivity of the work" in this vision would not be another thing but an organization of the speed of work and of the production flow that results

Table 1. Characteristics of the Peruvian micro-firms

Micro firm	Lifetime (years)	Charact. of the owners	Principal good	Labor (#)	Informal Labor (#)	Sales	Professional workers
AP	21	FE: mother & 3 sons	Tennis	3	From 6 up to 12	Wholesaler & retailer	--
BP	17	EF: 3 brothers	Tennis	4	From 4 up to 12	Wholesaler & retailer	1
CP	18	FE: a couple(husband and wife).	Sport shoes, elegance shoes	6	From 12 up to 15	Retailer	2
DP	30	FE: owner and his nephew	Shoes (scholar/ men) & Renewal	3	—	Wholesaler	--
EP	25	PE: owner.	Femme	1	8	Wholesaler	1
FP	30	FE: two brothers.	Tennis/scholar tennis	4	From 5 up to12	Wholesaler & retailer	1

from diminishing the working time in each unit of work. According to Zarifian (1995), the work would undergo some changes in its organizational structure, especially in three aspects: a) the separation of two objects: *work*, as a set of elementary operations that can be rationalized and be placed in the factory and, *labor*, that would be the set of capacities bought in the work-market; b) the flow, as the central criterion of the industrial production is restored; and c) the workers, are immobilized *in the space*, so that, no longer they can move freely like when they were craftsmen or farmers. They are immobilized *in time*, because they must respect the schedules rigorously.

Obviously, the worker participates with his/her brain and therefore, he/she accompanies the sequence of the learning curve. So that, he/she would not be just working, but also obtaining abilities that could be increased with time. Perhaps this was the understanding of the framers of the *post-Fordist* school (e.g the *socio-technical school* and the *Toyotaism*). Following this idea, it is possible for a person who owns abilities obtained during long time to carry out same task generates the concept of "emergent entrepreneurs" (Pinilla,

Table 2. Characteristics of the Brazilian micro-firms

Micro firm	Lifetime (years)	Charact. of the owners	Principal good	Labor (#)	Informal Labor (#)	Sales	Professional workers
AB	20	FE: father& 2 sons	Baby shoes	6	From 12 up to 25	Wholesaler & retailer	6
BB	17	FE: 2 brothers	Casual shoes	3	From 20 up to 30	Wholesaler & retailer	3
CB	18	FE: a couple (husband and wife).	Letter purse	2	From 2 up to 15	Retailer	--
DB	14	FE: a couple (husband & wife)	Slices re-seller	2	From 3 up to 8	Wholesaler	—
EB	11	PE: couple with two sons	Tennis	2	From 3 up to 8	Retailer	—

2004, p. 15), people who can produce goods that need limited technological capacity. And, that soon could join similar producers, in order to shape conglomerates of micro-firms (clusters, and other forms of productive networks).

Thus, we understand *post-Fordists entrepreneurs* as those people who inherited abilities via familial transference or that work along a period in factories (regardless of the form of organization) and afterwards dislocate themselves to create their own companies of goods derived from the work that they have manufactured before. Entrepreneurs, that need improvement to become suppliers of small parts of machineries or for more complex goods, restore the hierarchical and nonhierarchical forms within their scheme of familial or nuclear work. They also develop forms of self-organization with similar companies, suppliers and market, once they need others. This understanding leads to comprising the knowledge generated, shared, and contained within a social behavior.

Management and Organizational Concerns: Knowledge Management within the Micro-Firms

The knowledge of micro firms is generated from the practices developed by the day-to-day activities. In this sense, the learning of the entrepreneurs is formed by family and business environment that surrounds them. Best practices emerge from the adaptation and imitation of others. So, three ways are recognized and introduced: distance-to-market; distance-to-consumer and inner circle (Figure 1).

Distance-to-Market (D2M)

The time-to-market refers to the space of time needed to put a new (or relatively new) good or service in the market. In this sense, the *start-up* companies are those that diminish this time as maximum as possible. The distance that exists between the producer and the market has a dyadic form: a) is understood like the time consumed by learning the customer desires and, b) is understood like the physic distance from the facilities to the geographic position of the vendor. The relative importance of this distance lies in the need of the producer to enhance both characteristics in the good or service, usability and price. Because, the cost to produce an item is an important constraint and the micro-entrepreneur is not able (economically) to make experiments.

Small-firms reduce the risk of not responding to the demands by imitating the most important brands for their goods, but they cannot get the quality. Then, their focus shifts to how-to-do or how to reproduce the top goods. This process creates some relationships, e.g., bargain with suppliers and with other producers that have facilities to let them produce certain items with certain technology. It

Figure 1. Behavior of the knowledge flows within the small-firms

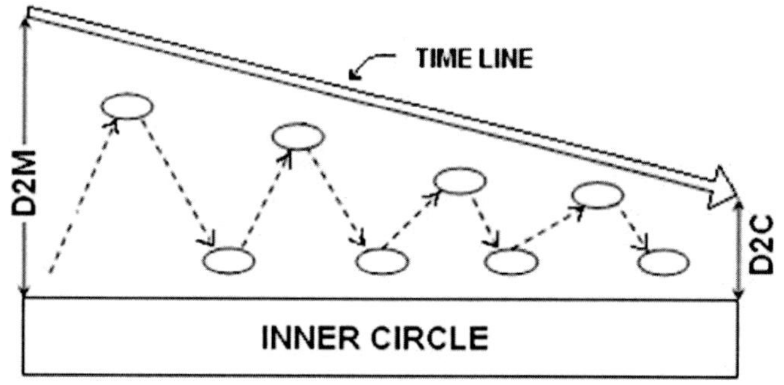

enables knowledge flow among competitors by linking them with other agents.

Distance-to-Consumer (D2C)

The main difference between customers and consumers is that the former could be buying a good for a third party; for instance, a mom buying diapers for a baby. So, the distance or time that the manufacturer needs to understand the desire of the final consumer is through interactions (conversation, bargain, etc) and experience in selling his/her own products. However, the knowledge flow created between the vendor and the client is subject to regional characteristics. A small manufacturer, usually, brings a product for the whole country and, sometimes, he/she crosses the national frontiers in order to sell it. The demands in each region visited by the producer vary according to raw material needed, colors, prices, and the final use or the recycled uses for the good.

The sales process introduces a second way of knowledge transfer. By interacting with the clients from different regions; the vendor understands new requirements for the good and brings new point of view to the core of his/her firm and disseminate it. Hence, the firm, as a whole, will be able to customize items with the aim of attending to differentiated orders.

Inner Circle

The inner circle refers to the need to create a hierarchy in order to establish who makes what. The hierarchy in the familial environment is given by the capability of each member; once the core of the firm is formed by members or friends, the skills of each one are recognized by the rest. The person that feels better in doing "some routine" is assigned to do it, democratically. But, there is a gap in this social behavior; the knowledge is not shared as much as possible. The knowledge embedded into a function is too sticky (Sulansky, 1996) and the flow of the knowledge is not enhanced as is possible. Thus, specific capabilities shape the personality of each member and each member develops its own capabilities. But when he/she needs or wants to be independent of the family ties, within the knowledge transfer process, it is necessary to develop a new element in his/her position. This creates a barrier to knowledge flows as a whole, but presents an advantage as the firms develop mechanisms in order to underpin troubles.

Management and Organizational Concerns: A Social Memory of the Post-Fordist Entrepreneurs

Within the text of Walsh and Ungson (1991), organizational memory is understood as repositories of knowledge: a) people, b) culture c) procedures, d) the organizational structure and e) the physical environment of the place work. In other work, Argote and Ingram (2000) identified three elements: a) individuals, b) software tools and hardware, c) the tasks; and in turn, the authors believe that the interrelations between the three elements are also in the form of knowledge repositories. In the interrelations of person to person, all of which constitute a social network to store knowledge and the interrelationship of tasks that generate a routine process is also a repository.

The construction of memory, on the other hand, refers to the process of storing information based on the history of the company, which may be, therefore, recovered and assisted in the decision making process. The information is stored and successful experiences and situations of failure must be easily retrieved and made available to persons (Hackbarth; Groover, 1999).

In the case of small firms, the construction of an organizational memory is related to the idea of a "social memory". It implies cultural aspects that form part of the folklore of a specific population and suggests a relationship of common characteristics accepted by the vast majority of the components. It is often called, in general terms, culture. The most important features of culture are:

language, religion, power distance and machismo (Hofstede, 1991). So, these features remain thereafter and are universally accepted. Even when they are formed into the inner of small-sized organizations, they are fraction of memory that shapes the organizational culture. Thus, it is possible to say that this fraction is a "social memory".

The daily life of people or specific agents will be accumulated in their life so that after a while, these experiences shape part of everyday life and allow the behavior of groups of people to go through problems and will accept a certain "leadership" (Hallbachs, 1990). This social memory does not resemble the organizational memory; the former refers to organizational matters accepted within a productive cell whereas the second refers to patterns of process accepted on a relationship of ruler and ruled, as in the concept of governance of Weber (2000).

In the field of micro and small enterprises, construction of an organizational memory is not different from that of larger firms. However, the social networks that are formed within organizations are formed by small-entrepreneurs, create other relationships of interest and create effective collaboration over the processes of development of new products and over the formation of new social networks with business.

Entrepreneurs often face difficulties in their day-to-day business activities, thus generate knowledge from these problems (Villaran, 1998). Knowledge management becomes visible when these companies are parts of clusters (Schmitz, 1995; Rabellotti, 1993). When it comes to individual companies, the knowledge management model can be a little different. Once these companies are organized differently, they could generate an organizational memory of their activities differently. It could be said that social memory is more linked to clusters with the governance exercised within the conglomerate and the social memory of isolated small-firms (or post-Fordist entrepreneurs) is more linked to social networks, products and markets.

Solutions Developed in Order to Attend to the Knowledge Management Requirements

According to the observation of daily practices of employers, the generation of knowledge in shared cycles (Figure 2), was based on flexible specialization and relationships between producers, traders, non-producers (universities, government and non-governmental organizations), consumers and users. These cycles are composed of the following elements: a) learning, b) needs and, c) opportunism.

Learning

Learning is part of the socialization process (inherent knowledge or knowledge that is transferred from parents to sons or acquired in the school). Knowledge develops like a spiral, in similar way that was proposed by Nonaka and Takeuchi (1998), but extends to include all stakeholders. Piore and Ruiz (1998) presented a case from the Mexican furniture industry: one day, one of the carpenters began to sell furniture concerning to Louis XVI model, for a price much more attractive than the market price. Shortly afterwards, almost all the furniture had the same cabinet with very similar characteristics and belonging to the same catalog. The perception of was that "knowledge is in the air" to be picked up by employers for the benefit of streamlining daily procedures involving constant innovation. The sources of this learning were the ones that appeared through the external and internal agents like recruitment, subcontracting, conversations, and flexibility in production and modification of products.

Needs

"Needs" are considered as the "distance-to-the-buyer" (time-to-customer). That is, the ability of the employer to understand the needs of users and/or consumers as they are psychologically and geographically closer to them. If the distance is short, the entrepreneur can react more quickly to changes of business environment and to have greater flexibility in the supply of goods and services. The means used in order to understand and react to unexpected demands that arise in the external environment are through the following: a) the supply chain; b) the "distance to the buyer", c) the unexpected demands and d) the possibility of technology transfer.

Opportunism

It may be argued that opportunism is a natural characteristic of human beings. In pursuit of special interests, human beings seek to meet expectations based on the spontaneity of resources available in the environment. For small businesses, opportunities imply lower costs, new markets, high seasonal demands, technology transfer, and others. As expressed by the entrepreneurs, the following actors are considered as "creators" of business opportunities or chances (in no particular order): a) government; b) large companies; c) University; d) NGO's; e) other employers from similar business and with the feature of be small-firm entrepreneurs. Although not always the opportunities created are transparent, once the opportunities for small-firms pass through a funnel, the greatest needs or opportunities of the agents are passed for these entrepreneurs.

On the other hand, the knowledge management, within the micro-firms is better understood when concerned with the knowledge generation through internal or external memories. This approach concerns with the learning pathways of the small-firms owners. By living together with people who have the same culture, the entrepreneurs acquire and bring out knowledge or new ways to do the same operation. Here are introduced three ways on social memory creation within the core of the small-firms: memory of product, market report and social memory.

Memory of Product

Within the small-firms environment it was noted that workers did not have a specific time contract job. So, they could work even just for a day in one firm and the next day in another. In this way, all that they can learn within a firm wakes away with them although entrepreneurs know that this leads to learning "company secrets". Entrepreneurs also learn other secrets by recruiting people who

Figure 2. Knowledge creating and sharing cycle

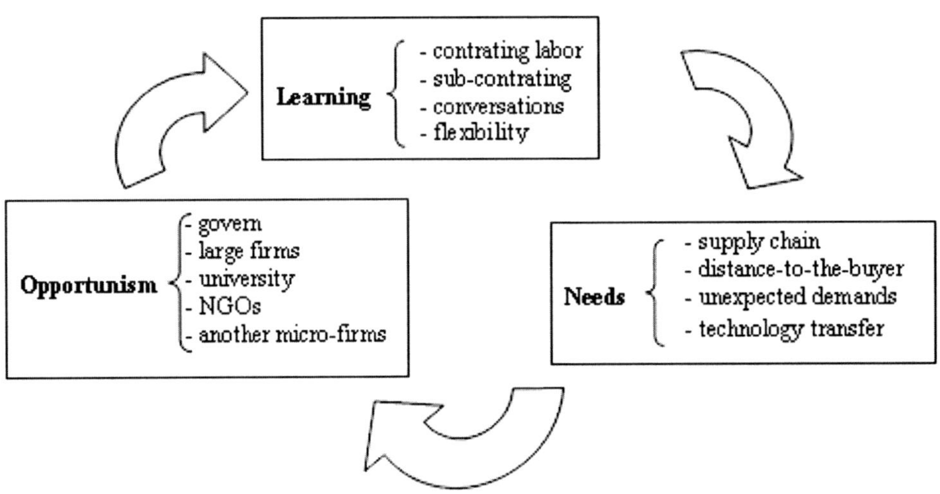

have worked closely with a competitor, without thinking that this could be an industrial espionage.

Market Report

The small-firms market usually deals with one type of customers whose loyalty and knowledge contributes to continued success and profitability of these firms. Organizational knowledge of these firms is comprised by the cultural characteristics of customers. Prices and features that the client is willing to accept and pay are determined in advance. The market report also helps in finding new customers. This implies a dislocation to different geographical markets, but aiming at people with a mind-set similar to the archetype of the loyal customers. That information is stored in the individual memory of the person who deals with new markets, but he/she may subsequently train another member, with the aim of knowledge transfer.

Social Memory

It was noted that the learning process of entrepreneurs cannot be established at the beginning as family members constantly pass through different parts of the production and sales processes. All family members know each stage of the product life cycle and are forced to become involved with each other. Work experience is the essence of organizational memory within these organizations.

It was also observed that most of the knowledge that the "post-Fordist entrepreneurs" apply in their processes and innovations were coming from the exchange of ideas with the clients and their perceptions of the culture and habits.

CONCLUSION

This chapter introduces a case study that seeks to explore the entrepreneurial behavior of small-firms from Peru and Brazil and then to examine the social forces that move them to manage their knowledge, in order to innovate and increase their organizational memory.

The small-firms were founded by 'post-Fordist' entrepreneurs that improve their practices through direct examination of customer and consumer behavior. In this sense, they seek to understand market requirements and to search for new materials or other features in their products with the goal of reducing prices or giving some differential look to their goods/services.

The knowledge embedded in the best practices is inherent and continuously improved. It was suggested also that the knowledge into the environment of the small firms is more linked with the social activities, and the work practice and market knowledge; including sales, production and other activities related to the daily activities of the small firms. The pathways concerned with the knowledge creation are more linked with the recruitment process and the best form of learning the workers are involved in during the production process.

The knowledge management approach is peculiar to each firm and thus it is too difficult to generalize. But the pathways of learning is explained and used, showed similarities in knowledge generation and differences in its sharing.

Finally, it is possible to conclude that the knowledge embedded in small business firms starts with memory creation that, in turn, starts with the interaction with the social environment. So, the larger part of learning in the small-firms comes from customers and consumers.

REFERENCES

Argote, L., & Ingram, P. (2000). Knowledge Transfer: A Basis for Competitive Advantage in Firms. *Organizational Behavior and Human Decision Processes*, *82*(1), 150–169. doi:10.1006/obhd.2000.2893

Hackbarth, G., & Grover, V. (1999). The Knowledge Repository: Organizational Memory Information *Systems. IS Management, 16*(3), 21–30.

Halbwachs, M. (1990). *A memória colectiva*. São Paulo: L.L. Shaffter.

Hofstede, G. (1991). *Cultures and organizations: Software of the mind*. London: McGraw-Hill.

Leonard-Barton, D. (1990). A dual methodology for case studies: synergistic use of a longitudinal single site with replicated multiple sites. *Organization Science, 1*(1), 248–266. doi:10.1287/orsc.1.3.248

Nonaka, I., & Takeuchi, H. (1997). *Criação de conhecimento nas empresas*. Rio de Janeiro: Campus.

Pinilla-Cisneros, S. I. (2004). *Condiciones de éxito de los emprendedores emergentes de Lima en el contexto de globalización. Magister disertation*. Perú: Universidad de San Marcos.

Piore, M., & Ruiz, C. (1998). Innovation cases into the mexican maquila. In Kagami, M., Humphrey, J., & Piore, M. (Eds.), *Learning, Liberalization and Economic Adjustment*. Tokyo: Institute of Developing Economies.

Piore, M., & Sabel, Ch. (1990). *La segunda ruptura industrial*. Madrid: Alianza.

PROEXPANSION. (2006). *Estudio sobre cluster y asociatividad*. Document prepared for PROMPyme. Lima. Retrieved February 02, 2010, from http://www.programa-area.org/img/documentos/309-estudio-sobre-clusters-y-asociatividad-peru-pdf.pdf

Rabellotti, R., (1993). Industrial districts in Mexico: The case of the footwear industry. *Small Enterprise Development*, 4(3).

Schmitz, H. (1995). Small shoemakers and fordist giants: tale of a supercluster. *World Development, 23*(1), 9–28. doi:10.1016/0305-750X(94)00110-K

Szulanski, G. (1996). Exploring internal stickiness: Impediments to the transfer of best practice within the firm. *Strategic Management Journal, 17*, 27–43.

Villaran, F. (1998). *Riqueza Popular, pasión y gloria de la pequeña empresa*. Peru/Lima. Congreso del Perú.

Walsh, J., & Ungson, G. R. (1991). Organizational memory. *Academy of Management Review, 16*(1), 57–91. doi:10.2307/258607

Weber, M. (2003). *Ciência e política: duas vocações*. Brazil, São Paulo: Martin Claret.

Zarifian, P. (2001). *Objetivo competência por uma nova lógica*. Tradução [Trylinski. Brazil/São Paulo. Atlas. Key Terms and Definitions]. *Maria Helena, C*, V.

KEY TERMS AND DEFINITIONS

Small and Micro Firms: There is no agreement on a definition of small and micro enterprises within the context of these both countries. Some definitions are linked to the number of people involved into the firm, and other specifications are concerned with dividends that these firms accruing annually. But it is clear that it varies by the kind of business in turn. Thus small and micro-firms, within the context of either country, could be defined like self-organized little groups, too likely with familial ties. And, that create their own hierarchy with similar firms and their own governance into the core of each one. They are small business with great flexibility to align quickly their scope to the demands in turn.

Social Memory: It is the construction of shared symbols that a group of people accepts like facts; within a specific population it is formed by habits or folklore included into the shared knowledge about popular arts, urban legends and other forms of popular language creations. Is important to understand the social memory about a population because it answers some questions about specific behaviors or believes.

Entrepreneur: Schumpeter understood this kind of people as a superhero that had the competence to do something to enhance welfare state. Following this concept, an entrepreneur is involved into a group of people whose think is about how to change rational processes or mechanism associated with sales, production or ways to do. He accomplishes with a social memory, but it appears that he has an additional understanding of the specific situations and is next to solve problems or understand generated demands.

Organizational Memory: It is understand, in the scope of this case, like constructs created into the inner of one family or firm. There is a cumulative learning through defining processes or activities, bargaining with other families or firms, etc. Such procedures owe modifications about the environmental perspectives, this is, that knowledge about and environmental social memory generates a specific *organization culture* and different approaches about the way of produce and in the way of sell. Within micro-firms, most of the knowledge is stored into the owner's mind. Likely, because he is an entrepreneur or because the origins of the firm; most of the time micro-firms originate themselves over a family house. Then, the infra-structure, relative to the firm, is not adequate to productive purposes. Thus, structures and infra-structures have an important role in the organizational memory from little firms.

Knowledge Management: Knowledge management is a science area concerned with the generation, sharing and diffusion of knowledge. In this sense, it has many ways to understand this concept. However, it must understand, in this arena, like the capability of the stakeholders to share skills and learn about techniques what are visible to test. For instance, manage the people unaware to manufacture shoes. At begin, an experience observer needs around ten minute to know if the beginner has potential to this job. With that information the time elapsed, between to hire a new worker and get an experienced one, diminish. This is because knowledge management is most understand, in this arena, like the capability of embed skills and reproduce it.

Post-Fordism: The name given to the period that comes after the Fordist period. This period was characterized by a linear production where the worker was considered as a machine appendix. Henry Ford, implants a work characterized by production in large scale and little batches. It had no structural micro-firms in this period, because the workers were most interested into work earning a good salary and less about to generate little entrepreneurships. Belong the time, big firms needs to buy supply at low cost. So, buy form little producers passed to be a better way to enhance this commerce.

Self-Organization: Insects and other organisms have a specific behavior about how to work benefiting the colony as a whole. This behavior is understood as "stigmergy"; it refers to the communication among two or more agents form the same origin in absence of a language. It is possible because each one has the capability of understand the other requirements. This biological approach is repeatedly observed form clusters of firms. And, observed on micro-firms articulation.

Section 4
Knowledge Management Issues

Chapter 17
HRM Adaptation to Knowledge Management Initiatives:
Three Mexican Cases

Laura Zapata-Cantú
Tecnologico de Monterrey, México

Jacobo Ramírez
Tecnologico de Monterrey, México

José Luis Pineda
Tecnologico de Monterrey, México

EXECUTIVE SUMMARY

Organizational knowledge and human capital are increasingly regarded as key levers of competitive advantage in today's global, dynamic and complex business environment. People are the ultimate knowledge creators and bearers; although organizations may have memory systems, they do not think by themselves. Human resource management (HRM) has a strategic role facilitating knowledge management (KM) initiatives, specifically in activities such as: recruitment, training and compensation. Although many studies raise the question of how HRM shapes KM, limited research in Latin America has explored this matter. This chapter aims to explore how HRM policies and practices have been designed in three organizations located in Mexico, in order to support KM initiatives. The findings suggest that the strategic role of HRM in supporting KM initiatives must be expanded. HRM policies, practices, and compensation systems, such as incentives and bonus packages to motivate employees to create and share knowledge, need to be redefined. Recruitment, selection and, training and development must be tailored to obtain a successful implementation of the KM program.

BACKGROUND

Nowadays, firms are facing a more dynamic environment, where uncertainty and technological changes dominate. As a result, organizations have to base their strategy and competitive advantage on intangible resources and capabilities. Managers should know how to administer the challenges and advantages that the environment presents, particularly in cases where the time factor is vital to develop, implement and transfer innovation

DOI: 10.4018/978-1-61692-886-5.ch017

systems for their survival. Furthermore, firms must develop flexible, administrative capacity to effectively coordinate their internal competence. The more dynamic the environment is, the more firms focus their strategy on resources and capabilities (Grant, 1996). In this context, one must take into account that while some organizational resources are easy to transfer, such as processes and equipment, others, such as knowledge, an intangible asset that is usually tacit and organization-specific, are less mobile. Knowledge is a valuable, generally essential, rare and inimitable resource (Barney, 1991:112.), and it constitutes a source of competitive advantage (Grant, 1996:376). The key resource is knowledge, which is intrinsic to human capital. People are the critical creators and bearers of knowledge. Organizations do not think by themselves; they need employees to leverage their knowledge. In this particular framework, it is relevant to explore how HRM policies and practices have been designed to manage this intangible resource.

Lado and Wilson (1994, p.699) in Minbaeva (2005:126) suggest that HRM practices "can contribute to sustained competitive advantage through facilitating the development of competencies that are firm specific, produce complex social relationships, ... and generate organizational knowledge." Managing HR to achieve better knowledge-related outcomes means "retaining personnel, building their expertise into the organizational routines through learning processes, and establishing mechanisms for the distribution of benefits arising from the utilization of this expertise" (Minbaeva, 2005:126).

As emphasized by Huselid (1995), HRM practices influence employees' skills and competencies through the acquisition and development of a firm's human capital. The competitive advantage of a firm is dependent on the existence of HR with relevant competence profiles. An analysis of the competencies needed for different positions, along with an analysis of a firm's current pool of employee competencies, helps the organization hire people with the desired skills and knowledge (p. 590).

The emergence of new forms of working practices, such as flexible working practices, reflects that need. Most traditional KM systems rely on the assumption that knowledge can be conceptualized as an object that can be identified, separated from its initial context, and handled in information systems (e.g. Nabeth *et al*, 2002). Thus, managers need to understand the meaning of knowledge in order to plan knowledge handling processes and to enable them to judge how knowledge affects people, organizational culture, management activities and other processes within the organization. Managing knowledge is not the same as HRM. KM involves managing intellectual property; that is, managing the development and transfer of organizational know how. It is more multifaceted than simply managing people (Teece, 2000:147). Organizational structures and management practices are changing to facilitate the implementation of KM in organizations.

The Study and its Methodology

In KM literature, there is relatively little consideration of cultural and human issues, even when the key obstacles exposed as knowledge transfer inhibitors are those related to the "human" side of KM. The main purpose of the present study is to determine the HRM practices and policies that 1) enhance the generation of knowledge and 2) support knowledge transfer in an actual business environment in Mexico. The present chapter explores the cultural and social embeddedness of knowledge from the evidence obtained through exploratory case study research from three Mexican firms, one in manufacturing industry and two in the service sector.

Exploratory qualitative research has been designed to investigate how HRM has had to adapt to KM initiatives in three Mexican enterprises. The case study method was used given that it fits well with the maturity stage of the theory involved and with the relative emergence of the phenomenon under study. This allowed us to establish detailed evidence and interpretation (Yin, 2009).

The three case studies were not chosen randomly; companies were selected on the basis of having innovative processes that contribute to their competitive advantage. Our interview protocol was designed around these issues: (1) knowledge generating activities, (2) knowledge transfer mechanisms, and (3) HRM practices that support KM initiatives.

The interview information was complemented with internal documentation provided by the firms. Face-to-face semi-structured interviews were held with HR managers and employees at different levels of the organization. The interviews lasted 60 and 90 minutes each and were audio-recorded. The interviews were transcribed verbatim and were circulated among all the authors in order to construct the full write-ups of each case. The interviews took place in 2008. In the following section, each case is analyzed separately.

SETTING THE STAGE

Information and Knowledge

Most traditional KM systems rely on the assumption that knowledge can be assimilated to objects that can be identified, separated from their initial context, and handled in information systems (Nabeth et al., 2002). Thus, organizations, and in particular managers, need to understand the meaning of knowledge, in order to plan knowledge handling processes and to enable them to judge how knowledge affects people, organizational culture, management activities, and other processes within the organization.

Knowledge as a business resource can be a source of sustainable competitive advantage, if it has two important properties: (1) it derives from accumulated experience, rather than specific tasks, meaning that it largely consists of tacit knowledge, which makes it difficult to transfer from one firm to another or for competitors to imitate it; and (2) it is not only connected to individuals but also to the way these individuals interact (Nonaka & Takeuchi, 1995).

Most people think of knowledge as a recipe - a predefined procedure - for dealing with a concrete, routine situation. However, in a dynamic environment few situations are repeated, which means that most situations are novel, particularly in their details. Knowledge must provide people with the ability to envision ways of handling different situations and to anticipate their future implications. Knowledge - in the form of mental models, scripts, and schemata - enables people to work in novel situations. They are thus able to assimilate not only concepts and predefined methods and judgments, but also to make connections between these and other complex concepts, meta-concepts, and mental models (Wiig, 2000).

Personal knowledge cannot be shared directly. This explains why people can carry out the same task differently from their colleagues even when a standardized process exists. The answer lies in personal knowledge, experiences, attitudes and context (Polanyi, 1962). Everyone has his or her own way of internalizing and making sense of the information received in order to interpret it and turn it into new knowledge. Knowledge is built through a complex learning process and results in highly individual mental models and associations. However, even under the best circumstances, only a small fraction of an individual's applicable expertise can be articulated and shared. Frequently, only concrete, operational or routine knowledge can be communicated. Deep, broad insights are generally not available and may not even exist, except as the abstract ability to reason. Significantly, when experts disseminate knowledge openly and widely, they tend to be considered important by their peers and gain status and recognition (Szulanski, 1996).

In sum, information needs to be interpreted, analyzed and transformed by people–knowledge workers. There should be an understanding of what knowledge workers do and in what ways they use knowledge and information to create value in organizations (Davenport, 2003). Most debates on KM depart from the premise that organizations do not recognize the full value of the knowledge "created" by their employees. The following section presents our definition and view on knowledge workers.

The New Work Force: Knowledge Workers

In 1969, the definition of knowledge worker proposed by Peter Drucker sought to describe the successors of factory workers: "*The most valuable asset of a 20*th *century company was its production equipment. The most valuable assets of a 21st century institution (whether business or non-business) will be its knowledge workers and their productivity*". However, knowledge workers can be defined as participants in the knowledge-based economy. This definition has the advantage of being cross-sectoral but the disadvantage that it is far from specific (Brinkley, 2006:16). Brinkley (2006) suggests three ways to define knowledge workers:

- All those who work in the top three standard occupational classifications (managers, professionals, associate professionals).
- All those with high levels skills, indicated by degree or equivalent qualifications (professional diplomas and/or higher education).
- All those who perform tasks that require expert thinking and complex communication skills with the assistance of computers.

On the other hand, Rüdiger and McVerry (2007) explain that neither definition perfectly captures what knowledge workers really are about, how they differentiate themselves from non-knowledge workers or how they perform their jobs.

Today, knowledge workers comprise the majority of the work force. While at the beginning of the 20th century, unskilled labor accounted for about 90% of the work force, today that figure is closer to 20% (Spira, 2005). Taylor showed that manual work consists only of simple, repetitive activities. Knowledge, that is, the way in which these simple, unskilled motions are put together, organized and executed, can make these activities more productive.

In the context of emerging economies, the example of Mexico is presented. In 2004, 17% of the total Mexican workforce consisted of knowledge workers (Brinkley, 2006). This may sound rather high, but if we compare it with the Netherlands (48%), Canada (39%) and Spain (31%), it is not high enough. This means that in Mexico, the majority of production results from manual work rather than knowledge work.

The shift from factory to knowledge workers represents a significant challenge to managers, who manage workers in more traditional roles. The basic difference is that a greater stock of knowledge supports a higher level of productivity (Grant, 2000). Different workers use different types of knowledge; this knowledge takes different lengths of time to acquire, and without knowledge, the employee is completely unproductive.

Besides, it is difficult to compare the knowledge intensity of different occupations. For survival in competitive environments, all employees, especially HR managers, have always worked to build the best possible knowledge base within their area of responsibility and to ensure that it can be easily disseminated to all organizational levels (Wiig, 2000:3). The HRM department has to provide a kind of storage space so that all workers can access most of the firm's knowledge. This storage space may be located within the firm's Intranet or a database containing documents that describe important processes and activities.

In Mexico, it is mostly transnational companies that use the Intranet as a space where employees can access information about the enterprise's main activities, including information about who has expertise in a specific area. For employees, the Intranet is a convenient means to establish virtual contact with other members of the organization and to keep up to date with the organization's activities. Effective KM depends on an explicit understanding of the complex ways in which people use their knowledge – their thought processes - to conduct their work. It means understanding what people have to know and how they access this information so that they can act effectively in different situations. The following section will elaborate on the knowledge management process.

Knowledge Management Process

Knowledge-based economies, which are directly based on the production, distribution and use of knowledge and information (OECD, 1996), challenge the basic processes of KM: how to create, share, transfer and store knowledge. These processes may pose a threat to traditional productive processes, and this is in part what KM research studies.

Organizations have always managed knowledge, even if they have not always talked about it in those terms (OECD, 2004). Nevertheless, there is an urgent need for KM to be seen as a corporate strategy and as a key precondition for increased productivity in the knowledge economy more broadly (Brinkley, 2006).

Early KM initiatives approached the challenge of improving knowledge workers' productivity from an organizational perspective by treating knowledge as an object. Reviewing the studies that have dealt with the theme of KM and its contribution to a firm's competitive advantage, three main aspects can be identified. First, at the individual level, it is the individual employee, who is responsible for knowledge creation within the firm (Nonaka & Takeuchi, 1995; Grant, 1996). Second, at the organizational level, the knowledge that has been created within the organization by some of its members should be transferred to other individuals so that it may be known to all. Third, once transferred and received, the knowledge generated should be integrated into the existing organizational knowledge base (Zárraga & García-Falcón, 2003:81).

Knowledge processes, such as generation, transfer and utilization, are interrelated and often cannot be separated. In the following sections, each phase of the KM process- generation, transfer and utilization- is described.

Knowledge Generation and Utilization

The knowledge generation and utilization processes are presented together in this section, given that when an organization utilizes knowledge, it can also generate it, and vice versa. Knowledge generation is mainly an institutionalized activity, so each organization must be able to establish its own creative routines and human intervention to make this process possible. The challenge is building systems that collect the learning processes acquired during projects and ongoing activities, capturing that knowledge in a database or document, and then spreading it throughout the entire organization (Grant, 2000).

Enterprises mainly create and utilize knowledge through social interaction. Socialization, as Nonaka and Takeuchi (1995) have explained, is a process of converting new, tacit knowledge into tacit knowledge through joint formal and informal activities, such as spending time together, working in the same office and chatting. In the knowledge generation process, organizational culture, which results from experience based on tacitly shared norms of coordination and collaboration, can influence the way in which individuals interact within an organization. In fact, a culture that promotes intensive communication, accepts new ideas, and is prepared to explore new processes and activities favors the generation of knowledge (Zapata, Rialp and Rialp, 2009).

This knowledge generation process leads to increased firm competitiveness. It may be stimulated through non-hierarchical organizational structures, an active general management, and by motivating employees to innovate and learn lessons that allow them to obtain new and better knowledge (Nonaka *et al.*, 2000).

The dynamic environment, which firms are currently facing, not only provides employees with the motivation to generate new knowledge, it also gives them the opportunity to learn. At the same time, the experience that the employees can obtain and the possibility of applying their knowledge in diverse activities within the organization, motivates them to learn new tools and create new processes or ways of doing things. As it can be seen, the creation and utilization of knowledge requires human intervention. Few organizations pay much attention to the creation of knowledge, but many emphasize its utilization through information technologies.

Knowledge Transfer

For Gooderham (2007:36), knowledge transfer can be referred to as the accumulation or assimilation of new knowledge in the receiver unit. However, researchers like Minbaeva et al. (2003:587) argue that pure transmission of knowledge from the source to the recipient has no useful value if the recipient does not see the potential of the new knowledge and hence, does not utilize it in his or her own activities.

Besides reaffirming the importance of communication, this suggests that there is another important aspect that influences which tools knowledge workers choose in order to support their work; they prefer to use *personal* tools, i.e. tools that they can control and customize to their own needs. Cognitive tools, such as mind mapping software, other visualization tools or personal information management tools are designed for the solitary worker and do not address the social context of knowledge creation (Efimova, 2004). That said, one tool has become the dominant support for knowledge work–email. Email is successful because it is *personal* and *social* at the same time.

Informal communication mechanisms, such as face-to-face interaction, corridor chats and IT systems, facilitate interaction among colleagues and improve the knowledge transfer process. Meetings are another appropriate way to transmit complete messages.

On the other hand, technology alone is clearly not enough to induce a person with knowledge to share it with others. Technology can improve access and help bring the right knowledge to the right person at the right time, but this is not sufficient. Document management systems are only helpful in locating information. Using information, sharing it, making sense of it and discussing it are all done elsewhere.

There are also several important contextual dimensions of knowledge transfer. First, firms must seek to operate with an open culture that facilitates knowledge transfer. What is needed is an organizational culture that motivates the members of the organization to search for new ways of doing things and where these efforts are facilitated by flexibility and interaction among colleagues. Such a culture should be open in order to encourage employees to share their knowledge with colleagues, even in contexts where time is at a premium. Time is vital for firms that operate in dynamic environments; hence the knowledge to be transferred must be assigned a high priority within the organization. That is, its transfer should be planned just like all the firm's other important activities (Zapata, Rialp and Rialp, 2009).

Other significant contextual elements within the organization include management support and close proximity among employees. As Cummings (2004) comments, great physical distances make communication among members of a group difficult since opportunities for informal contact are reduced.

Knowledge as part of intellectual capital is the key to a competitive advantage in the knowledge economy; hence HRM should seek to become an integral part of corporate strategy, enabling knowledge to be managed in the interests of organizational performance (Thite, 2004).

Knowledge Management and Human Resource Management

The importance of KM lies in the knowledge possessed by individual employees, even if this is an inherent social construct. The increased reliance on knowledge workers is changing the role of HRM (Beatty *et al.,* 2003). Managing knowledge is not the same thing as just managing human resources. KM involves managing intellectual property; that is, managing the development and transfer of organizational know-how. It is more multifaceted than simply managing people (Teece, 2000: 147).

In order to begin discussing the different approaches to HRM adaptation to KM projects, it is important to give an account of HRM's principal activities. HRM functions have emerged parallel to the economic shift from agrarian to manufacturing to services and now, to information and knowledge (Beatty *et al.,* 2003: 107). One factor in the changing role of HRM is the increased reliance on knowledge workers (ibid). Under this perspective, HRM involves a series of organized activities conducted within a specified time and designed to produce behavioral change. Within HRM, the most common activities are recruitment and selection, training (learning for the present job), compensation and incentives, and performance evaluation, as well as staffing, career development and internal communication and compensation (e.g. Nadler, 1994).

Organizational structures and management practices are changing to facilitate the implementation of KM in organizations. The following section presents the different configuration to recruitment and selection, training and compensation in the context to support KM initiatives.

Recruitment and Selection

Employee selection is a fundamental aspect of HRM and assumes that individual differences make a meaningful difference in job performance. Therefore, what is needed to maximize performance is a good match between personal characteristics and job requirements (Huo *et al.,* 2002). The following sections present an analysis that shows how recruitment and selection processes have changed.

Employee Profile

According to academic literature, the employee profile in firms where knowledge is relevant has been modified. Intangible skills refer to a workforce with tacit or difficult to measure capabilities, such as the ability to work in a team-based system, innovation, flexibility, problem solving capability and good interpersonal relations (Ramirez and Fornerino, 2007).

Recruitment and Selection Tools

Methods of recruiting employees can be a good indicator of management style and, at the very least, can tell us something about the formality of the employment relationship (Cully *et al.*, 1999: 60). Therefore, in terms of recruitment tools, the Internet and assessment centers were the primary means analyzed. The Internet was chosen because it has become a leading recruitment tool for potential employees and employers searching for positions with a highly intensive use of knowledge (e.g. Buckley *et al.,* 2004).

Training

It is argued that training shapes the skills and knowledge of the workforce. Knowledge management projects tend to be proactive for training purposes; they try to anticipate the training needs for their workforce (Cascio, 1990). Training programs in KM tend to focus on problem solving, communication, technical skills, job rotation and mentoring relationships (Lepak and Snell, 1999; Ramirez and Fornerino, 2007).

Compensation

In a KM environment, one important policy and practice to review is the compensation system, as a function and as a motivator to knowledge sharing, especially in the Latin American context, where people are afraid to share their knowledge and, in some cases, employees may mistrust the source (Davila and Elvira, 2007). It is argued that incentives should be introduced to promote knowledge generation, effective knowledge sharing, learning and the application of the best knowledge for work purposes (e.g. Tayeb, 2005).

Other important factors include the employees' perception of freedom and their willingness to share knowledge with others inside and outside the organization. This dynamic reflects the quality of management of the enterprise as a whole (Barrett, *et al;* 2004). As Hislop (2003:185) point outs, there is an inherent tension between workers and the organizations for which they work over who owns and controls their knowledge.

MULTIPLE CASE STUDIES

In order to explore the phenomenon under study—how HRM is adapting its activities to support KM initiatives—three Mexican cases were selected. The selection was based on their innovative processes, even if they do not have a specific KM project. The first case, Systems, is a medium enterprise whose main activities are related to the implementation of telecommunication solutions and applications. The second case, Snacks, is a large firm that manufactures food and snacks. Finally, the third case, Satellital, is a medium enterprise that provides solutions and consulting services for the optimization of the supply chain.

CASE SYSTEMS

Systems enterprise was founded in 1991 by a visionary Mexican, who focused on the way in which telecommunication tools facilitated and optimized processes that, at the same time, had been successfully implemented by firms. It began its operations in western Mexico with six employees and is currently a medium-sized firm (140 employees) with coverage across Mexico. *Systems* provides consulting services for projects related to the design and implementation of solutions and to advanced applications in telecommunication. Additionally, it is strongly oriented to business process optimization.

Among its services, *Systems* offers a range of solutions, from convergent network applications to complete solutions for contact centers and CRM (customer relationship management) applications. In the field of convergent networks, the main products are voice and data transfer applications in local and wide area networks, IP telephony and unified messenger. Other services are the contact centers, which range from conventional call center solutions that consist of the intelligent distribution of phone calls, to sophisticated multi-channel contact centers, where interactions via telephone, email and instant messenger can coexist and integrate into different CRM systems to provide customers with a consistent response no matter what contact channel they use. Regarding CRM business applications, *Systems*' approach is a complete integration with the contact center to automate and improve customer service processes and increase customer retention. These services are used by universities, hospitals, public administration offices, financial institutions and manufacturing enterprises.

Systems' IP telephony solutions are used today by its customers as a competitive edge and provide appropriate flexibility and scalability across the firm.

Knowledge Generation

Organizational culture in *Systems* has played a key role with respect to knowledge generation. As the operations manager pointed out: *"employees work in a place where they feel free to innovate, to change the way they do things."* As it is widely recognized, more intensive communication along with an accepting culture opens up to new ideas, which is also prepared to explore new processes and activities, favors the generation of knowledge.

This dynamic knowledge generation process, which will improve the firm's competitiveness, is favored by a non-hierarchical organizational structure and makes visible an active role played by general management. *Systems* has four levels in its organizational structure: 1) the Board of Directors – headed by the founder, 2) the Top management team –comprising four functional areas, 3) the Middle management team – comprising the operative supervisors, and 4) the Operative team.

Systems employees' motivation to innovate and learn lessons allows them to obtain new and better knowledge. According to the interviews, *Systems'* employees describe the firm as an organization that makes it is possible to learn continually. The dynamic environment, which *Systems* has created, provides its employees not only with the motivation to create new knowledge but also with the opportunity to learn. Employees consider that they can learn a great deal about the work they carry out and from the experience achieved by applying their knowledge. This knowledge arises from the fact that they are informed about the areas in which services should be improved both for internal and external clients.

Systems has realized that its clients are the main focus of its business—its reason for being in the market. Consequently, it has initiated a strategic drive to highlight their customers' knowledge.

Knowledge Transfer

Systems' employees use different communications channels to interact and share their knowledge. One of them is a knowledge base, which makes the response time to customers (internal and external) more efficient, specifically in the area of maintenance. Employees working on a customer report first need to access the knowledge database in order to find out how the same problem has been solved in the past. The knowledge database works as a decision tree. All maintenance reports are entered into this database.

Another current communication method involves distribution lists that enable employees to interact with all members of the organization. This communication channel is frequently used to share experiences or seek assistance from another member of the firm in order to solve a problem. At the same time, *Systems*' employees have other means of sharing their knowledge, such as the telephone, messenger, face-to-face chats and formal and informal meetings. The use of formal means of communication, which facilitates interaction among employees, in addition to the informal mechanisms of physically present communication, such as face-to-face and hallway chats, can also improve the knowledge transfer process.

Effective knowledge transfer requires an organizational culture that motivates the members of the organization to search for new ways of doing things. The *Systems* case shows how culture is a relevant factor in the knowledge transfer process, especially in creating a sharing atmosphere. The employees are committed to sharing their *know-how* with their co-workers. At the same time, the top management has committed itself to acquiring the media that makes knowledge transfer possible.

As regards the contextual factors affecting knowledge transfer, the maintenance manager pointed out,

"Here in Monterrey we do not have a problem with sharing our know-how. Our challenge has been

deciding when we should share it with coworkers in other cities. Before the creation of knowledge database, we had to phone the expert to try to resolve a specific situation. IT introduction has diminished geographical distance limitations; in this case, through the knowledge database."

Another important aspect to consider is that people are often afraid to share their knowledge. They believe that they will lose the advantage that their expertise gives them among their peers and within the organization. In *Systems*, top management is attempting to introduce a commitment letter. The employee who has been given the opportunity to attend a training course outside the firm has to disseminate this new knowledge throughout the firm. The firm is clearly actively thinking about how its employees can become trainers. In this way, HRM plays an important role in knowledge transfer.

KNOWLEDGE MANAGEMENT AND HUMAN RESOURCE MANAGEMENT

Recruitment and Selection

The industrial sector in which *Systems* operates depends on the creation of specific knowledge needed for quality work in the telecommunications sector. *Systems* recruits highly qualified engineers, certified by its business partners, and graduates of Mexico's most prestigious universities. *Systems* does not have a special method for recruiting employees but believes that employee retention is related to commitment and trust. Nevertheless, in some cases, employees have left the company for a higher salary.

Training- *Systems* invests in more than 60 hours of training per employee per year. Training is primarily focused on state-of-the-art technologies developed around the world. Knowledge itself plays a tacit role, and even when it cannot be codified in the form of a written document, it can be taught by means of *mentoring* and learning. In this sense, organizations have to consider specific ways to ease the training process, for example, formal training of workers in the workplace, e.g. by their coworkers, as in the case of *Systems*.

Compensation- A high pay package with performance incentives motivates personal knowledge creation, sharing and use, and it also helps to attract and retain employees. In this sense, *Systems* is thinking of developing incentives for employees to share their knowledge among coworkers. Even though employees are committed to transferring the knowledge obtained in training courses, they will receive a bonus for this activity in the future. The employee will benefit doubly: first, by receiving the opportunity to take a course outside the firm; and second, by receiving payment just for transferring this new knowledge to coworkers.

CASE SNACKS

In 1921, *Snacks* was founded by three brothers in the northern Mexican city of Monterrey. In 1991, the company was taken over by an American multinational, turning *Snacks* into a global leader in the cookie market and into Mexico's largest cookie manufacturer. The company offers its consumers a wide variety of high-quality products for every lifestyle, producing pastries, oats, cereals and other related products. It has production facilities in five states across Mexico.

Snacks has achieved sustained growth to become the cookie market leader in Mexico, as well as consolidating its leadership in the US Hispanic market. It has also reached this market position by being fully customer oriented, constantly offering innovative, high value-added products and anticipating its customers' needs.

Its accomplishments include one of its greatest sources of pride: its people. *Snacks* has always taken an interest in its employees' well-being. Consequently, the company obtained the highest level of employee satisfaction of all the companies that comprise the international corporation to which it belongs.

In March 2005, *Snacks* was certified as one of "The Top Ten Places to Work in Mexico," awarded by the Great Place to Work Institute México.

Knowledge Generation

For *Snacks,* external professional consulting is a highly important source of knowledge generation. It resorts to these services in two very specific situations: to deal with a situation that is unusual for the organization and to anticipate changes in the environment. An example of the former situation is described by the human resources coordinator,

"From the bird flu to the current influenza crises, we have had to have a contingency plan in place. Consultants shared their knowledge on how the viruses are transmitted, the medical side, the consequences of the virus, how long it takes to transmit the virus, the different stages of an epidemic and a pandemic, the WHO website. All of this information has been useful for organizing informative workshops and defining alternative strategies. In human resources, in particular, we had to decide which positions were indispensable for the organization to continue its operations and how employees could work remotely from home."

A source of internal knowledge creation stems from the corporation's and employees' own initiatives. The company is currently focusing on promoting a culture of innovation. In fact, the organization has named this year the Year of Innovation. *Snacks* has allocated funds specifically to highlight and support this initiative. At the beginning of the year, all the employees, regardless of their organizational level, could disclose any innovation idea they had for the products and/or processes. HRM sends graphs every month on their progress and the contributions to innovation made by each area, and an award ceremony is held every three months. The employees who have contributed the strongest innovation ideas receive the innovation bracelet.

At *Snacks*, the organizational culture and the support of the senior management are fundamental to the generation of knowledge. Moreover, one of the coordinators interviewed commented that the customers themselves are an indispensable source of knowledge generation; *"I believe that since our organization is 100% market-oriented, our only boss is the customer. We always focus on thinking about what the customer is thinking."*

Knowledge Transfer

Even though *Snacks* is a large international company, knowledge transfer is possible with formal, informal and technology-based communication mechanisms. For example, when the company's values were made official, this was shared in a cascade, from one hierarchical level to another, from one boss to another in person. Human Resources trained and empowered directors from all areas, and they, in turn, trained their managers and collaborators in person, face to face. Since the information being transferred was related to values, it was important that it be carried out face to face, so that the boss could share his or her vision of and commitment to said values.

When the new compensation system was designed by the external consulting firm, this new knowledge was transferred in person with the area heads and their work teams. The program was called "Ticket to a Safe Future."

Technology-based communication media is predominantly used when the company wants to share new knowledge that does not involve all of the organization's employees or when it does not entail a radical change in operations.

The commitment shown by the senior management to informing employees about new ways of operating is an organizational component that facilitates knowledge transfer. This commitment is reflected in the attendance by every single employee when specific sessions are held to transfer new knowledge, such as the "Ticket to a Safe Future" program. Moreover, all the relevant information that was shared in person was documented and disseminated through the organization's Intranet.

Knowledge Management and Human Resource Management

Recruitment and Selection

Since the philosophy of permanent innovation to compete in the marketplace and the start of the "Year of Innovation" at *Snacks*, modifying the recruitment and selection practices has been imperative for this organization. As one of the *Snacks* employees mentioned, "It would be illogical for a group of directors, managers and collaborators to have certain initiatives but not the personnel to support them in their implementation because they lack the competencies to do so. Moreover, the creativity assessment that has been included is very important."

Training. In a continually transforming organization, training is a critical topic for *Snacks*. As mentioned in the knowledge transfer section, any new knowledge that arises in the organization, either as the result of consulting or an internal initiative, is shared with employees through training so that they can understand and use the same knowledge. HRM has technological mechanisms as protocols so that training can be face-to-face, in person.

Compensation/performance evaluation. The year of innovation at *Snacks* has resulted in employees, who have participated with initiatives, being rewarded and recognized. There is even a bonus for each idea implemented. Moreover, one of the variable compensation bonuses was redesigned to ensure that almost 10% of an employee's total compensation is related to performance-related innovation.

"*For example, the percentage of the innovative products sold out of the total sales. So, of all the products you sold, how many were innovative products? You can see that we are committed to innovation as far as possible. We even link it to compensation.*"

CASE SATELLITAL

Satellital is a medium enterprise that is part of an international group headquartered in Monterrey, Mexico. It was created from the merger of three strategic business units as a solution integrator for the supply chain. *Satellital* concentrates on services based on satellite telecommunications. It provides services in two main fields: remote asset monitoring and mobile satellite telecommunications. Additionally, *Satellital* develops high value-added business solutions based on the integration of information and wireless telecommunication technologies, which are aimed at optimizing its customers' core business processes in Mexico, North America and Central America.

This enterprise has instituted a philosophy with guiding principles and values that leads its personnel towards achieving clear objectives and fosters daily interaction and the development of each individual. Its mission is to develop high-growth business services based on information and telecommunication technologies.

Its main activities are: (1) Solutions for controlling and optimizing the logistics of land and sea mobile assets through remote monitoring. (2) Reliable solutions for mobile satellite communications (radio, telephony, broadband) for work teams operating in remote locations where conventional means of communication are inadequate. (3) Technological solutions to drive automotive financing through enhanced risk control and better portfolio management. *Satellital* offers concessions to exploit the transmission and reception rights of frequency band signals associated with the foreign satellites MSAT and Inmarsat. Furthermore, it also has the concession to install, operate and exploit a public telecommunications network using those satellites. All the equipment used for the provision of satellite services that it offers under its public network has been approved, which currently, makes it the only company in Mexico that has these authorizations, as granted by the Ministry of Communications and Transport.

Knowledge Generation

External knowledge stems predominantly from the supplier through its equipment and training in the use of the same. Nevertheless, *Satellital* has sometimes discovered faults in the supplier's equipment and, when this occurs, it conducts a diagnosis, documents it and shares it with the rest of the organization. It also shares this new knowledge with the supplier.

The customers offer another form of external knowledge as they too identify faults, and *Satellital* undertakes to share this knowledge with the supplier. As one of the area coordinators said,

"We all learn from mistakes. However, we don't exactly change the world; we leave that to the supplier. So, the supplier has to make changes and then generate new documentation and this documentation is valid and we retrain on the basis of said documentation."

Likewise, customers offer *Satellital* feedback on improvements to their systems and processes. A clear example of the creation of internal knowledge is the development of its own software.

"In terms of software applications, we have developed software that offers a point on the map, so to speak, in very simple terms... that involved numerous functions. There is a whole series of reports stating that it generates value for customers, ranging from those who exclusively request the location of their equipment to those who require more detailed information, such as arrival time estimates, productivity, fixed-asset location and trip start dates and times."

The development of applications implies training all the members of the organization in their use. This ensures that all the members of the company participate in the new knowledge.

An important component in *Satellital* that facilitates the creation of knowledge is the prevailing organizational culture. The creation of knowledge and innovation are encouraged and must be presented in writing for authorization. In this document, designed to request new ways of doing things, employees must notify why they want to do it, how they will do it and whether it can be feasibly done in the company.

Furthermore, a high level of self-directed learning exists. The company has to keep up to date on telecommunications and value-added legislative topics, particularly since customers always want to be on the cutting edge of this field.

Knowledge Transfer

Document processes, or descriptions of equipment technical, are the main formal communication media used at *Satellital* to transfer its knowledge throughout the organization. An area coordinator commented,

"Personally, I have had to create documents that offer a technical description of certain services or pieces of equipment so that we can all have the same vision of what we are offering."

Training is another medium for transferring knowledge. This practice comprises two components that clearly inhibit the transfer process: technical language and people's attitudes. With regard to the language, one of the members of the company commented,

"I have realized that it's very hard for everyone to reach the same level of understanding. I need to bring what I know as a technician down to a common language, where a person from the retail area with no technical knowledge and who

manages customer services can understand the advantages of the technology, in order to assist customers."

In relation to people's attitudes, employees do not feel comfortable asking questions if they have a query. "*Handing out documents is very easy, but you have to have the humility to say I don't understand it and come and ask.*" Gaining people's trust has been very difficult.

Organizational culture, rather than being a component that facilitates knowledge transfer, as mentioned in the literature, limits this process at *Satellital*. With all the changes in the organizational structure, it is not clear what the positions entail and which managers should direct and foment training or knowledge transfers. One of the coordinators said,

"We haven't been able to generate a knowledge transfer and assimilation methodology. It's not that we don't have enough systems, and even though the knowledge is in a computer ... the computer won't transfer it per se. A culture should be created for this."

Knowledge Management and Human Resource Management

Recruitment and Selection

As a result of the changes in organizational structure and the dependence on headquarters in functional support areas, such as human resources and finance, *Satellital* has opportunity areas in human resource practices and, above all, in the management of its tacit knowledge.

Recruitment and selection are key practices for the effective management of knowledge. There is a high staff turnover rate in *Satellital's* customer service area. One of the possible reasons for this is the lack of job descriptions. In fact, even the coordinators do not have job descriptions. "*It's incredible. I provide services for the administrative area, the retail area; I am working on a project we have with the government. I know what I have to do and government participants ask me to do other things, but since they don't say I can't do them... it's not in writing anywhere.*"

Training- Training is of utmost importance for *Satellital*. As mentioned beforehand, in knowledge generation and transfer, all the knowledge, both external and internal, generated, must be shared with the entire organization. Training is provided internally, that is to say, the employees themselves offer courses and assess learning, mainly for technical matters, value-added and returns on investment that could be obtained by the customer. Training is offered in both the retail and the customer service areas. Moreover, the headquarters offers courses, since certain competencies must be fulfilled.

External training is mainly for the telecommunications technicians who have to keep up to date on this topic. A specialist is sent to training courses and then that person is responsible for sharing this knowledge with his or her colleagues through cascade training.

Performance evaluation- Performance evaluation depends on the areas in which the employees work. For instance, field personnel are evaluated for following the established protocols. While customer service employees are evaluated by means of a customer satisfaction survey on the way in which they solve customers' queries. Customers evaluate them on their knowledge, not their attitude. In this way, customers will be more objective. People become more aware of the relevance of training; training becomes a work tool.

CURRENT CHALLENGES FACING THE ORGANIZATIONS

There is no doubt that both knowledge and human capital are increasingly regarded as key levers of competitive advantage in today's global, dynamic and complex business environment. More importantly, people and knowledge are inextricably linked. Individual human beings are the ultimate knowledge creators and bearers. Organizations do not think by themselves, although they may have "knowledge enabling" contexts and memory systems. In fact, both people and knowledge should be regarded as having special potential as scarce and idiosyncratic resources, consistent with the premises of the resource-based approach to strategic management. Indeed, a significant amount of rigorous, resource-based scholarship exists that also deals with KM or HRM.

Table 1 summarizes the results of our research obtained after the cross-case analysis was concluded. In the first column, the knowledge generation process is strongly supported by employees' motivation to innovate and learn lessons that allow them to obtain new and better knowledge (Nonaka et al., 2000). According to our cases, *Systems* and *Satellital* employees describe their firms as organizations that make it possible to continually learn. The role that the employee plays as the bearer of initiatives, suggestions, voluntary effort and greater commitment is very relevant. Thus, organizational structures should be less hierarchically structured and management style more participatory so that the necessary conditions are created, which allow the individual's flexibility to be preserved.

In terms of knowledge transfer, column two exposes how firms operating in dynamic environments face three main challenges pertaining to the reproduction of their: 1) internal competencies, 2) abilities and 3) collective learning developed and its *storage* within the firm over time. *Systems*' employees use different communication channels to interact and share their knowledge. One of these is a knowledge base, which makes the response time to customers (internal and external) more efficient, specifically in the area of maintenance. Employees working on a customer report first need to access the knowledge data base, in order to find out how this same problem has been solved in the past. The knowledge data base works as a decision tree. All maintenance reports are entered into this data base.

Another current communication method involves distribution lists that enable employees to interact with all members of the organization. This communication channel is frequently used to share experiences or to seek assistance from another member of the firm in order to solve a problem; this is the case of *Satellital*. *Systems*' employees have other means of sharing their knowledge, such as the telephone, messenger, face-to-face chats and formal and informal meetings. The use of formal means of communication, which facilitates interaction among employees, in addition to the informal mechanisms of physically present communication, such as face-to-face and hallway chats, can also improve the knowledge transfer process.

Effective knowledge transfer requires an organizational culture that motivates the members of the organization to search for new ways of doing things. Flexibility in the employees' performance and interaction is required, in order to facilitate this search (Zapata, Rialp & Rialp, 2009). This culture should be open so as to encourage employees to share their knowledge and communication among members of the firm, even when time is at a premium. The three cases under study expose how this is possible and is vital in the knowledge transfer process. For example, the *Systems* case shows how culture is a relevant factor in the knowledge transfer process, especially in creating a sharing atmosphere. The employees are committed to sharing their *know-how* among their co-workers. At the same time, the top management has committed itself to acquiring the media that makes knowledge transfer possible.

As regards the contextual factors affecting knowledge transfer, another important aspect to consider is that people are often afraid to share their knowledge. They believe that they will lose the advantage that their expertise gives them among their peers and within the organization. In Mexican organizational culture, knowledge transfer is only possible if the director or top management asks somebody to share or transfer his or her knowledge to somebody else. In some enterprises, the organizational culture supports knowledge transfer. This can mainly be observed in transnational companies. Mexicans are very possessive about what they know because a lot of effort was required to obtain it. In *Systems*, top management is trying to introduce a commitment letter. An employee, who has been given the opportunity to attend a training course outside the firm, has to disseminate this new knowledge throughout the firm. The firm is clearly actively thinking about how its employees can become trainers. In this way, HRM plays an important role in knowledge transfer.

Finally, column three summarizes how KM initiatives are beginning to be supported by HRM. Although managers are usually keen to recognize that human and social issues underpin the success of KM initiatives, a number of structural, organizational embedded factors hinder the KM efforts, and they are quite difficult to overcome, despite the initial managerial commitment to solving these problems. This has been observed in the three Mexican cases described above, specifically, how HRM is transforming its traditional activities.

In reference to recruitment and selection, firms have the possibility of generating a human capital advantage through recruiting and retaining outstanding people: "capturing" a stock of exceptional human talent, latent with powerful forms of "tacit" knowledge (Svetlik and Stavrou-Costea, 2007) by designing a competitive compensation system. The industrial sectors in which *Systems*, *Snacks* and *Satellital* operate depend on the creation of specific knowledge needed for quality work. Only *Snacks* has *a* specific method for recruiting employees, but it believes that employee retention is related to commitment and trust. *Systems* and *Satellital* are beginning to develop operational programs to select the best employees, regarding salary as a relevant method of doing so.

Training is a relevant HRM activity in our three cases that show how knowledge transfer is critical. As has been observed, these cases reveal

Figure 1. Cross-case results

Firm	Knowledge Generation	Knowledge Transfer	KM & HRM
Systems	▪Organizational culture ▪Less hierarchical organizational structure ▪Opportunity to learn	▪Capacity to assimilate ▪Attitude to share ▪Common language	▪Recruitment and selection ▪Train the trainer ▪Performance evaluation-*customer*
Snacks	▪Top management initiatives ▪Innovation year (2008)	▪Organizational culture – *technological communication media*	▪Position profile ▪Cascade training ▪Compensation: High performance bonus
Satellital	▪Organizational culture ▪Top management support ▪Less hierarchical organizational ▪Opportunity to learn	▪Organizational culture - *knowledge base and distribution lists* ▪Physical distance ▪Afraid to share	▪Recruitment and selection ▪Training ▪Compensations

that training is vital as a mechanism to share new knowledge throughout an organization. Depending on the size of the firm or its geographical presence, this HRM activity is supported by different media communication mechanisms, but face-to-face is still the most useful.

Concerning knowledge workers' careers, Evaggelia (2007:322) points out: *"The old career ladders are gone. The old lifetime employees are gone. If career ladders do not exist within one company, they must exist across different companies if they are to exist at all."* Here is where HRM practices have to change in order to retain or maintain employees for a long time. The most popular retention strategies are related to compensation. A high pay package with performance incentives motivates personal knowledge creation, sharing and use, and also helps to attract and retain employees. In this sense, *Systems* is thinking of developing incentives for employees to share their knowledge among co-workers. Organizations must find incentives both to encourage knowledge transfer and to generate knowledge. These incentives could be both monetary and non-monetary.

SOLUTIONS AND RECOMMEDNATIONS

This chapter shows cultural and social drivers, such as management and personal behaviors, should be changed to create environments of trust, and efforts should be made to find the root cause of problems without assigning blame. New management practices must focus on combining understanding, knowledge, skills and attitudes when assembling work teams; analyzing requirements for performing work and establishing new work settings are a huge component of tacit knowledge, or knowledge that is mainly embedded in people.

Hence, to enable successful knowledge transfer, knowledge should be taught in a more practical and appropriate manner and adapted to each firm's specific processes. The use of mentoring systems can help to minimize knowledge uncertainty by informing the members of the organization about each other's respective competencies. Besides that, the internal transfer of knowledge requires that the receivers be willing and committed to adopt the activity and that they have the capacity to assimilate, understand and adopt the recently acquired knowledge.

A central challenge for firms operating in dynamic environments is how best to manage their employees. This can be achieved by ensuring that HRM strategies facilitate the creation and transfer of knowledge. HRM strategies may therefore include: external talent acquisition, internal talent development, job rotation and action learning, and contractually binding the most desired and talented employees to the organization (Evaggelia, 2007). However, in addition to a traditional contract, a new psychological contract (Dávila and Elvira, 2009) is needed to promote loyalty and retention.

HRM deserves an explicit mention in the people-centered KM debate. In leading organizations and industries, people management is assumed to be one of the top strategic priorities. It is worth mentioning the increase in research dealing with the relationship between HR practices and organizational performance (e.g. Arthur, 1994; Huselid, 1995; Guest, 2002). Although from different perspectives and with notable differences in specific practices, empirical support for the positive impact of HRM on organizational performance can be assumed.

Finally, managers must find and develop an environment for learning, knowledge-sharing and knowledge integration procedures (e.g. Teece, 2007), given the fact that people are afraid to share their knowledge, and in some cases, employees may mistrust the source. Knowledge as a part of intellectual capital is the key to a competitive advantage in the knowledge economy; hence, HRM should seek to become an integral part of corporate strategy, enabling knowledge to be managed in the interests of organizational performance (Thite, 2004).

REFERENCES

Arthur, J. B. (1994). Effects of human resource systems on manufacturing pPerformance and turnover. *Academy of Management Journal*, *37*, 670–684. doi:10.2307/256705

Barney, J. B. (1991). Firm resources and sustained competitive advantage. *Journal of Management*, *17*, 99–120. doi:10.1177/014920639101700108

Barrett, M., Cappleman, S., Shoib, C., & Walsham, G. (2004). Learning in knowledge communities: Managing technology and context. *European Management Journal*, *22*(1), 87–98. doi:10.1016/j.emj.2003.11.019

Beatty, R. W., Huselid, M. A., & Schneier, C. E. (2003). New HR Metrics: Scoring on the Business Scorecard. *Organizational Dynamics*, *32*(2), 107–121. doi:10.1016/S0090-2616(03)00013-5

Brinkley, I. (2006). Defining the knowledge economy. Knowledge Economy Programme Report. *The Work Foundation*, July.

Buckley, P., Minette, K., Joy, D., & Michaels, J. (2004). The use of an automated employment recruitment and screening system for temporary professional employees: A case study. *Human Resource Management*, *43*(2&3), 233–241. doi:10.1002/hrm.20017

Cascio, W. F. (1990). Strategic human resource management in high technology industry. In L. Gomez-Mejia, R. & M. Lawless (Eds) *Organizational Issues in High Tech Magazine, pp.* 179-198. Greenwich, CT: JAI Press.

Cohen, W. M., & Levinthal, D. (1990). Absorptive capacity: a new perspective on learning and innovation. *Administrative Science Quarterly*, *35*, 128–152. doi:10.2307/2393553

Cully, M., Woodland, S., O'Reilly, A., & Dix, G. (1999). *England at Work. As Depicted by the 1998 Workplace Employment Relation Survey*. London: Routledge.

Cummings, J. N. (2004). Work groups, structural diversity, and knowledge sharing in a global organization. *Management Science*, *50*(3), 352–264. doi:10.1287/mnsc.1030.0134

Davenport, T. (2003). *A measurable proposal.* CIO Magazine, June.

Davila, A., & Elvira, M. M. (2009a). *Best Human Resource Management Practices in Latin America*. Oxford, UK: Routledge.

Drucker, P. F. (1969). *The age of discontinuity: guidelines to our changing society*. London: Heinemann.

Efimova, L. (2004). *Discovering the iceberg of knowledge work: A weblog case. Submitted to OKLC 2004* [Available at https://doc.telin.nl/dsweb/Get/Document-34786/]. Retrieved November 14th, 2008.

Evaggelia, F. (2007). Human resource management policies and knowledge workers. *Proceedings of the 8th European Conference of Knowledge Management:* (p 319). Barcelona, España.

Gooderham, P. N. (2007). Enhancing knowledge transfer in multinational corporations: a dynamic capabilities driven model. *Knowledge Management Research & Practice*, *5*, 34–43. doi:10.1057/palgrave.kmrp.8500119

Grant, R. M. (1996). Toward a knowledge-based theory of the firm. *Strategic Management Journal*, *17*, 109–122.

Grant, R. M. (2000). Shifts in the world economy: the drivers of knowledge management. In Despres, C., & Chauvel, D. (Eds.), *Knowledge Horizonts: the present and the promise of knowledge management* (pp. 27–53). Boston: Butterworth-Heinemann.

Guest, D. E. (2002). Human resource management, corporate performance and employee wellbeing: Building the worker into HRM. *The Journal of Industrial Relations*, *44*(3), 335–358. doi:10.1111/1472-9296.00053

Hislop, D. (2003). Linking human resource management and knowledge management via commitment, a review and research agenda. *Employee Relations*, *25*(1/2), 182–202. doi:10.1108/01425450310456479

Huo, Y. P., Huang, H. J., & Naiper, N. K. (2002). Divergence or convergence: A cross-national comparison of personnel selection practices. *Human Resource Management*, *41*(1), 31–44. doi:10.1002/hrm.10018

Huselid, M. (1995). The impact of human resource management practices on turnover, productivity, and corporate financial performance. *Academy of Management Journal*, *38*, 635–670. doi:10.2307/256741

Lepak, D., P. & Snell, S., A. (1999). The Human Resource Architecture: Toward a Theory of Human Capital Allocation and Development. *Academy of Management Review*, *24*(1), 31–48. doi:10.2307/259035

Minbaeva, D. B. (2005). HRM practices and MNC knowledge transfer. *Personnel Review*, *34*(1), 125–144. doi:10.1108/00483480510571914

Minbaeva, D. B., Pedersen, T., Björkman, I., Fey, C. F., & Park, H. J. (2003). MNC knowledge transfer, subisidiary absorptive capacity, and HRM. *Journal of International Business Studies*, *34*, 586–599. doi:10.1057/palgrave.jibs.8400056

Nabeth, T., Angehrn, A., & Roda, C. (2002). Towards personalised, socially aware and active Knowledge Management System; *Proceedings: E-2002 e-Business and e-Work Annual Conference, Prague; Prague, Czech Republic 2002.* Retrieved January 10th 2008.

Nadler, L. (1994). *Every Manager's Guide to Human Resource Development.* San Francisco: Jossey-Bass.

Nonaka, I., & Reinmoeller, P. (2000). Dynamic business systems for knowledge creation and utilization. In C. Despres y D. Chauvel(eds.), *Knowledge Horizonts: the present and the promise of knowledge management*, 89-112. Boston: Butterworth-Heinemann.

Nonaka, I., & Takeuchi, N. (1995). *The Knowledge Creating Company.* New York: Oxford University Press.

OECD. (1996). *The knowledge based economy.* Paris.

OECD (2004). *The significance of knowledge management in the business sector,* July.

Polanyi, M. (1962). *Personal knowledge: Towards a post-critical philosophy.* University of Chicago: Chicago Press.

Ramirez, J., & Fornerino, M. (2007). Introducing the impact of technology: a 'neo-contingency' HRM Anglo-French comparison. *International Journal of Human Resource Management*, *18*(5), 924–949.

Rüdiger, K., & McVerry, A. (2007). *Exploiting Europe's knowledge potential: 'Good Work' or Could do better - knowledge work and knowledge workers in Europe.* The Work Foundation.

Sparkes, J., R., & Miyake, M. (2000). Knowledge transfer and human resource development practices: Japanese firms in Brazil and Mexico. *International Business Review*, *9*, 599–612. doi:10.1016/S0969-5931(00)00021-4

Spira, J. B. (2005). In praise of knowledge workers. *KM World*, *12*(2), 25–27.

Svetlik, I., & Stavrou-Costea, E. (2007). Connecting human resources management and knowledge management. *International Journal of Manpower*, *28*(3-4), 197–206. doi:10.1108/01437720710755209

Szulanski, G. (1996). Exploring internal stickiness: Impediments to the transfer of best practices within the firm. *Strategic Management Journal*, *17*, 27–43.

Tayeb, M. H. (2005). *International Human Resource Management: A Multinational Company Perspective*. Oxford, UK: Oxford University Press.

Teece, D. (2000). Strategies for managing knowledge assets: The role of firm structure and industrial context. *Long Range Planning*, *33*(1), 35–54. doi:10.1016/S0024-6301(99)00117-X

Teece, D. (2007). Explication dynamic capabilities: The nature and microfoundations of (sustainable) enterprise performance. *Strategic Management Journal*, *28*, 1319–1350. doi:10.1002/smj.640

Thite, M. (2004). Strategic positioning of HRM of knowledge-based organizations. *The Learning Organizations: An International Journal*, *11*(1), 28–44. doi:10.1108/09696470410515715

Wiig, K. M. (2000). Knowledge Management: an Emerging Discipline Rooted in a Long History. In Despres, C. & Chauvel, D. (eds). *Knowledge Horizons: the present and the promise of knowledge management,pp.* 3-26. Boston: Butterworth-Heinemann.

Winter, S. G. (1995). Four Rs of profitability: Rents, resources, routines, and replication. In Montgomery, C. (Ed.), *Resources-based and Evolutionary Theories of the Firm*. Boston: Kluwer Academic Publishers.

Yin, R. K. (2009). *Case study research: design and methods* (4th Ed.). Thousand, Oaks, California: Sage Publications.

Zágarra, C., & García-Falcón, J. M. (2003). Factors favoring knowledge management in work teams. *Journal of Knowledge Management*, *7*(2), 81–93. doi:10.1108/13673270310477306

Zapata, L. (2004). *Los determinantes de la generación y la transferencia del conocimiento en pequeñas empresas de tecnologías de la información en Barcelona*, Doctoral Thesis, Universidad Autónoma de Barcelona, July.

Zapata, L., Rialp, J., & Rialp, A. (2009). Generation and transfer of knowledge in IT-related SMEs. *Journal of Knowledge Management*, *13*, 246–256.

KEY TERMS AND DEFINITIONS

Tacit Knowledge: An intangible resource and difficult to transfer from one firm to another or for competitors to imitate it; and it is not only connected to individuals but also to the way these individuals interact. Barney (1991) has defined as a valuable, generally essential, rare and inimitable resource.

Knowledge Workers: Participants in the knowledge-based economy, those persons who work in the top three standard occupational classifications (managers, professionals, associate professionals), with high levels skills, indicated by degree or equivalent qualifications (professional diplomas and/or higher education), and perform tasks that require expert thinking and complex communication skills with the assistance of computers.

Knowledge Generation: Institutionalized activity, so each organization must be able to establish its own creative routines, processes and human intervention to make possible to have new knowledge.

Knowledge Transfer: Process that involves sharing of knowledge from a source and accumulation or assimilation of this knowledge in the receiver unit. This process could be useful value if the recipient does see the potential of the new knowledge and hence, does utilize it in his or her own activities.

Informal Communication Mechanism: Are methods utilized to transfer knowledge without being scheduling or mandatory by top managers. These methods such as face-to-face interaction, corridor chats and IT systems, facilitate interaction among colleagues and improve the knowledge transfer process. Knowledge workers prefer to use *personal* tools that they can control and customize to their own needs such as informal communication mechanism.

Knowledge Management: Process that involves planning, coordination and (KM) control of knowledge generation, knowledge transfer and knowledge utilization.

Human Resource Management: Management of practices and activities related with the wealth of organizations' most value asset: human capital.

ENDNOTE

The capacity to assimilate refers to the ability of the firm to recognize the value of new information, assimilate it and apply it to further its commercial goals.

Chapter 18
Perceptions of Factors Influencing Knowledge-Based Technology Management in Conflict Areas

Ian Hipkin
École Supérieure de Commerce de Pau, France

EXECUTIVE SUMMARY

A challenge for developing countries is to enter the global economy. This is impeded by weak infrastructure, poor educational and other formidable barriers as these countries embark on essential projects to improve living standards and ultimately create some sort of technological base. This chapter describes perceptions of knowledge-based technology management in Afghanistan, one of the poorest and least developed countries in the world. Afghanistan's politics and history have always confronted managers with unique demands. The chapter examines the perceptions of Afghan managers regarding technology management by studying the relationship between the importance of different factors in managing new technology, and the extent to which a manager can control them. An importance-control grid framework is used to isolate individual parameters and to assess these in relation to a manager's environment. The research highlights imbalances between importance and control, and suggests reasons therefore. Some broader implications for managers are also discussed.

ORGANIZATION BACKGROUND

Formidable challenges confront developing countries in their attempts to install new technology. Countries afflicted by conflict are further hindered by fear on the part of technology suppliers. Even once technology provision has been agreed, the transfer of technology is inhibited by a lack of knowledge, physical infrastructure, managerial abilities, technical skills, as well as the dangers of operating in such environments. Introducing technology requires consideration of both core technology (equipment, instruments, software) and the context in which the technology is embodied (relating particularly to knowledge and the human dimension).

DOI: 10.4018/978-1-61692-886-5.ch018

Countries subject to endemic conflict are generally excluded from the global operations of multinational firms. These countries are usually in the news because of war, famine or poor governance. Western firms in particular are reluctant to conduct business in turbulent environments such as Afghanistan. Yet, recent reports such as the announcement that the largest copper mine in the world is to open in Afghanistan are likely to entice some international firms into business dealings. At best Afghanistan will be exploited for its raw materials, and is not viewed as a centre for manufacturing. Yet there are firms prepared to endure the risk of operating in a country which may yield attractive long term returns. While operational activities are severely constrained by the current turbulence, some progress is being made in improving infrastructure, electricity generation and water provision.

The subject of this chapter derives from a study of perceptions of factors influencing knowledge-based technology management in Afghanistan. Interviews were conducted in Kabul with managers in a number of utilities. The research is a preliminary study of knowledge management as a factor in introducing basic technology in utilities.

SETTING THE STAGE

This chapter discusses a number of perspectives of knowledge-based technology management in a country where 'normal' rules and practices do not generally apply. There is scant literature in this area, and analyses of such settings invariably only consider the macroeconomic situation and the role of governments, the military and international aid agencies.

While theorists claim to arrive at answers through logical reasoning, the assumptions and boundary conditions imposed by theoretical models frequently do not take into account the operating context, social phenomena and human factors (Hipkin and Bennett, 2003). In uncertain environments new initiatives should be encouraged since it is difficult to predict the formal learning and experimentation required to utilise and adapt knowledge (Govindarajan and Trimble, 2004). Firms basing their strategy on knowledge and new systems are reliant on employee competence and an exchange of know-how to ensure compliance with design, quality, delivery and cost structures (Lynskey, 1999). Poor educational standards in developing countries retard the pursuit of production competence (Voss and Blackmon, 1996), and there is no speedy solution to this problem since mastery of 'implicit' elements in new operational approaches needs a long period of learning (Lall, 1993).

The turbulent economic and political environment prevailing in a country such as Afghanistan calls for different management approaches from those in stable, developed countries (Prochno and Corrêa, 1995). Japanese management techniques, with their apparent simplicity and lower capital investment requirements, are initially appealing to countries in any stage of development. However, Grant and Gregory (1997) identify obstacles in their adoption in developing countries: difficulties lie in areas such as essential competencies, infrastructure, education, unionisation, culture and employee attitudes. Shannon (2009) believes that filling the gap in knowledge might enable organizations to deal better with the constrained environments they are likely to encounter in places such as Afghanistan: this is a call for more research in the country and others afflicted by conflict.

Developing countries can only operate competently and meet global competition through introducing new technology and expanding their infrastructure, training, cost reduction efforts, and productivity improvements. It may be difficult to acquire high technology appropriate for a developing country, since undue extravagance may introduce levels of complexity that ultimately render the process unworkable and worthless as intricate operational procedures cannot be followed or breakdowns become irreparable. A technology policy which claims that the purchase of a new machine will simply give the recipient 'a new technology' is misguided, as this makes unrealistic assumptions of "self-sufficiency, flexibility, and endless reliability on equipment design" (Leonard-Barton, 1995, p.225). The challenge is to select appropriate technology that can be absorbed in, or adapted to, local situations (Blumentritt and Johnson, 1999). Technology implementation requires modification to suit the operating context (Platt and Wilson, 1999).

Figure 1 which shows the percentage of firms in Afghanistan experiencing difficulties with the factors identified on the horizontal axis (results relate to a survey conducted by the World Bank in 2006).

CASE DESCRIPTION

The case context for this research is a number of utility organizations in Afghanistan. The perceptions of some 21 Afghan managers are examined regarding the way in which technology is perceived, transferred and managed. The managers are involved with development projects to improve service provision in the electricity and water sectors in Kabul. In the initial stage of the study managers were invited to discuss the issues which they believed were important in managing and transferring technology in their operational environments. Once definitions had been clarified and overlapping factors eliminated, a set of 'importance' factors was selected. One of the most significant was knowledge management. The managers were then asked to score these factors on a Likert scale of 1 to 5 with respect to the following criteria:

1. How important is this factor now, in so far as it affects your working environment?
2. How much control can you exercise over this factor now?
3. How important will this factor be in three years time?
4. How much control will you be able to exercise over this factor in three years time?

The research follows the methodology of Hipkin (2004) and Naude et al (1990) in studying the

Figure 1. Percentage of firms in Afghanistan citing significant constraints

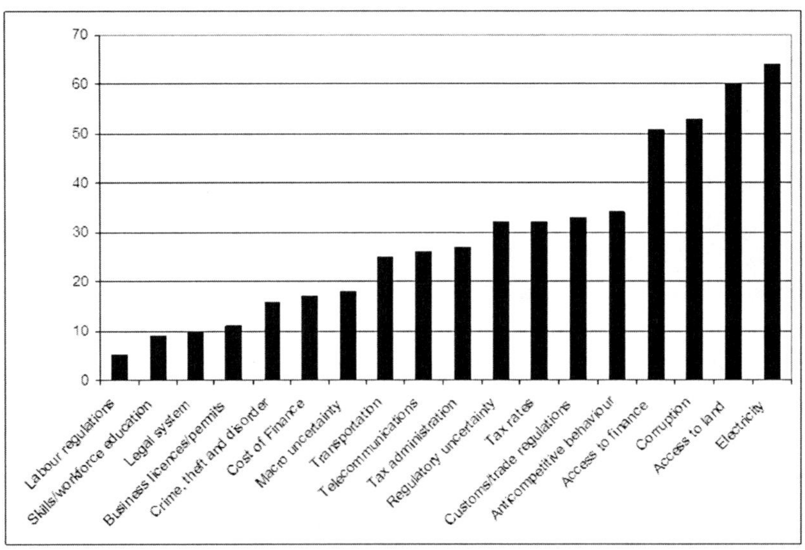

relationship between the importance of different factors in a manager's operational environment, and the extent to which a manager can control them. The framework enables the researcher to isolate individual parameters and to study these in relation to the complexity of a manager's environment. By plotting the scores on a grid (shown in Figure 1, along with empirical data), the following distinct areas may be identified:

- *core issues* which managers see as the most important and over which they can exercise the most control; these issues require the greatest management time, effort and strategic planning
- *complex issues* which are perceived as being important but over which managers can exercise limited control
- *simple issues* which are of lesser importance and which are easily controlled by management
- *peripheral issues* which are generally of limited importance and over which little control can be exercised.

The importance-control grid enables the researcher to establish imbalances between importance and control, which may result in frustration or inappropriate managerial intervention. Concentrating on simple issues represents poor utilisation of managerial resources. Managers wish to control complex factors, but frequently find themselves unable to do so. The grid provides a useful methodology for identifying such problems, and its use can be extended to suggest action which will enhance the use of technology. Both the methodology and the findings will prove useful to organizations contemplating technology investments in conflict areas.

The findings from the importance and control scoring by managers are summarised in Table 1 and illustrated in Figure 2. Table 1 shows individual item scores for the present, and perceptions of three years hence, as well as average factor scores for technology, knowledge, human resources, and so on.

A number of issues emerge from this information. It will be noted that the current scores for individual items are all to be found in quadrant 4

Figure 2. The importance-control grid: current and future perceptions (factor and item scores)

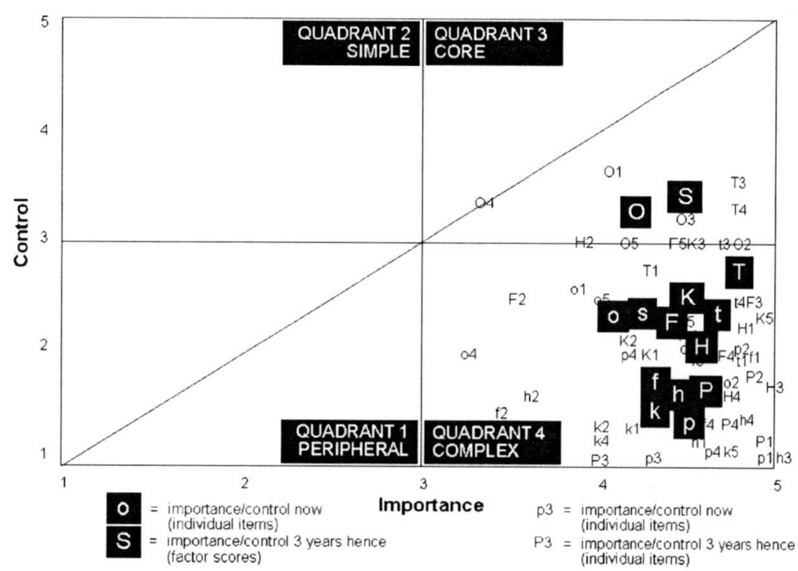

Table 1. Factors influencing the management of technology

FACTORS AND ITEMS	Code	Import-ance now	Control now	Import-ance 3 years	Control 3 years
TECHNOLOGY AND TECHNOLOGY MANAGEMENT INTER-FACE	T	4.7	2.3	4.8	2.8
Ease of use of technology	T1	4.8	2.0	4.8	2.4
Adaptability of technology to local conditions	T2	4.6	1.5	4.6	2.0
Greater output/performance through technology	T3	4.7	3.0	4.8	3.5
Availability and reliability of equipment	T4	4.8	2.5	4.8	3.2
POLITICAL AND ECONOMIC ISSUES	P	4.6	1.4	4.6	1.5
Political instability and corruption	P1	5.0	1.0	5.0	1.2
Security of plant, workers, installations, workshops, stores	P2	4.8	2.0	4.9	1.8
Government regulations and bureaucracy (planning permission, work permits)	P3	4.4	1.0	4.0	1.1
Overall level of economic development and infrastructure	P4	4.6	1.1	4.7	1.3
Reluctance of foreign experts to work Afghanistan	P5	4.2	2.0	4.5	2.1
KNOWLEDGE	K	4.3	1.5	4.5	2.3
Lack of understanding of technology (hardware and software)	K1	4.2	1.3	4.4	2.0
Collection and analysis of failure data	K2	4.0	1.4	4.2	2.1
Understanding of how technology works and how it fails	K3	4.4	2.4	4.6	3.0
Effects of failures on output	K4	4.0	1.2	4.4	2.3
Appropriate maintenance action (time/condition based)	K5	4.7	1.1	4.9	2.3
HUMAN RESOURCES	H	4.5	1.3	4.6	2.1
Lack of labour commitment and productivity/inability to take responsibility	H1	4.6	1.2	4.8	2.2
Difficult to accept new (different) ways of working (learning culture)	H2	3.7	1.5	3.9	3.0
Shortage of skilled personnel for high-technology equipment	H3	5.0	1.0	5.0	1.7
Low educational levels of workforce	H4	4.8	1.3	4.7	1.6
ORGANIZATION FOR TECHNOLOGY	O	4.1	2.2	4.2	3.2
Promoters/champions of the technology	O1	3.8	2.6	4.1	3.7
Belief in need for and commitment to technology	O2	4.7	1.6	4.8	3.0
Internal infrastructure to integrate technology throughout organization	O3	4.5	2.1	4.5	3.2
Create feedback mechanisms to report failures, operating circumstances	O4	3.3	2.0	3.3	3.3
Organization and reporting structures	O5	4.0	2.5	4.2	3.0
SUPPLY CHAIN AND PARTNERSHIPS	S	4.3	2.3	4.5	3.3
Assistance from technology partners	S1	4.6	2.8	4.7	4.1
Compatibility between supplier and user	S2	4.0	1.5	4.5	3.0
Lead time to acquire technology/spares	S3	4.4	1.3	4.5	2.8
Establishing supplier networks and accessing local infrastructure	S4	3.8	2.5	4.1	3.2
Contractual arrangements	S5	4.5	3.3	4.5	3.5
FINANCIAL	F	4.4	1.6	4.5	2.3
Cost of technology acquisition	F1	4.8	1.9	4.7	2.1

continued on following page

Table 1. Continued

Short term returns required from technology	F2	3.5	1.2	3.6	2.5
Hidden costs of technology (including TT, HR development, environmental)	F3	4.6	2.0	4.8	2.5
Justification of technology on a cost/benefit basis (lower production costs)	F4	4.7	1.4	4.7	2.0
Cost of training and developing local workforce	F5	4.5	2.3	4.5	3.0

(complex factors), where Afghan managers believe the items are important, but over which they have no or little control. The fact that these items are referred to as 'complex' points to the difficulties managers experience in managing significant and essential factors, when in reality they have little ability to influence how these items are managed. Managers cannot use 'conventional' means that would rely on position and legitimate power. They may be able to use expert power bases in isolated instances, but will inevitably have to resort to other bases of power. Managers acknowledged that referent power was probably their most powerful mechanism for achieving their goals.

The lowest current control scores relate to political and economic items: managers recognise the detrimental effects of political instability and corruption (p1), the importance of economic development and infrastructure (p4), but are frustrated by their inability to control these. Human resource items, such as shortages of skilled personnel (h3), low educational levels (h4) and lack of labour commitment (h1) are likewise seen as essential for successful delivery of water and electricity, but managers do not envisage they can do much about these in the next 3 years. Current averaged factor scores under each heading (technology, political issues, knowledge, etc) are therefore firmly in quadrant 4.

Afghan managers' predictions of events in 3 years suggest broadly similar levels of importance, and limited increased control. The result is that some items have moved to quadrant 3 (the 'core' quadrant). This represents a more satisfactory situation in that managers have some control over what they deem to be important. This is particularly the case for items pertaining to supply chain and partnerships (S), and organizational issues (O). Managers foresee increased assistance from technology partners (S1), better control over contractual arrangements (S5), promoters/champions of technology (O1), and improvements in internal infrastructure (O3). They visualise a slight increase in control of knowledge elements, such as how technology works (K3), but the overall averaged knowledge factor remains in quadrant 4.

CURRENT CHALLENGES FACING THE ORGANIZATIONS

Afghanistan suffers a lack of laws that integrate asset management across municipalities, provinces and regions. Attempts are being made to free utilities from direct government control. The telecommunications sector has been successfully privatised. A policy decision by the Government of Afghanistan to corporatise the water industry was taken several years ago, an arduous process which is currently underway. While individual ministries have been instructed to release some utilities from direct control, others, including the all-powerful Ministry of Finance, still effectively control the way the utilities operate, and their ability to acquire new technology. One example relates to the water and electricity sectors: the water industry is installing meters in households in Kabul and other larger centres, but it is not permitted to disconnect water supplies for non-payment (among the worst offenders are government offices). The result is that the water industry operates at a massive financial loss, and is unable to pay the electricity utility for power for pumping water. The electricity utility therefore also shows a financial deficit. The Ministry of Finance is then called upon to keep both utilities afloat.

Technology and Technology Management Interface

Respondents acknowledged that the most apparent reason for poor performance arose from technological issues and poor maintenance practice.

The knowledge base referred to in this study is not complex, and the 'technology' referred to is basic. Machines and equipment typically include pumps, electric motors, diesel generators, switchgear, manually controlled switching, water reticulation, and electricity distribution. In the Afghan context ensuring reliable functioning of such equipment would represent a great step forward. The replacement of equipment is a priority for managers, as some plant is over 50 years old. Financial constraints mean utilities are unable to meet all the demands of stakeholders. Many plans for new water wells and associated pumps, or transmission lines, have been considered, but the technology would be the most basic.

The scores in Table 1 reflect a situation where managers accept the importance of simple technology, adapted to local conditions, and the recognition that technology can significantly increase output (for example, electricity generated and water distributed). At the same time they appreciate the lack of understanding by many workers of the most basic concepts. Failure data is not collected, the consequences of failure in one area on another are not recognised. Maintenance is invariably on a breakdown basis. While managers understand the importance of these issues, they have little power to control them.

Political and Economic Issues

At this stage in the challenging environment in Afghanistan most foreign investment comes in the form government- and donor-funded reconstruction projects. New decisions about investments in technology depend on five basic factors (World Bank, 2006): political and economic stability and security, clear unambiguous regulations; reasonable tax rates that are equitably enforced; access to finance and infrastructure; and an appropriately skilled workforce. Clearly, these conditions are lacking in Afghanistan.

Lister and Wilder (2005) explore how administration both shapes and is shaped by complex political dynamics and should be seen as part of broader political processes. They contend that interventions to strengthen administration should focus not only on mechanisms and structures but should take into account the current distribution of power (formal and informal, *de facto* and *de jure*) in Afghanistan. One of the difficulties in creating any form of management structure arises from the lack of an overarching political strategy of state-building and organization-building. An issue frequently raised by managers was the low level of trust in state institutions (as emphasised by Graham and Chattopadhyay, 2009).

In the context of the politics of Afghanistan, Shannon (2009: 23) comments "The lack of the rule of law is driven by a poorly functioning and corrupt government with an ever increasing credibility gap". All businesses are severely restricted by bureaucracy, poor infrastructure, corruption, financial constraints and access to land. Estimates are that firms pay 8% of sales as bribes, which is more than four times that paid in Pakistan. To mitigate security concerns and seek protection, firms have been obliged to form close ties with powerful elements in society, including warlords and government officials. Of particular concern to all businesses is the lack of electricity, water and transport. Even in Kabul, electricity is at best only available for a few hours a day, and prolonged outages are common, since less than half of the country's installed capacity of some 500MW is operational at any one time. Private generators are therefore essential for the furtherance of any meaningful business activity.

Knowledge

Resource-based paradigms assist in explaining the performance of equipment and the contribution of employees. This requires an understanding of technological skills, complementary assets and routines. Capabilities are clusters of tangible and intangible abilities that span individuals and groups (Coates and McDermott, 2002).

The low level of education and technical knowledge may be illustrated by pointing our that in some pumping stations the role of security guard and operators is combined. The individuals concerned are illiterate, so are unable to differentiate between 'off' and 'on' switches. Flowmeters and ammeters are also of little benefit to them. Training is therefore often at the most basic level: colours indicating the status of a motor.

Human Resources

Managers spend much of their time on human resource issues, day-to-day operations and maintenance, and in meetings with government officials. The acute shortage of spares and lack of skilled staff are the main causes for very low equipment availability and reliability, although performance and output are often not measured.

When considering the knowledge base of the country, it is estimated that only one third of the population of Afghanistan is literate. The most skilled Afghans escaped from the country during the various phases of serious upheaval. While some managers and qualified workers have returned, there is still a critical shortage of educated and skilled workers. The education policies of the Taliban were attuned to their madrasa foundations, and certainly not conducive to creating a skills based economy that will yield management and technical knowledge, and of course, their exclusion of the female 50% of the population from education exacerbated the situation. There are still very few females in positions of authority. The World Bank Report concludes: "Afghanistan is caught in a trap. The lack of opportunities discourages workers from improving their skills and education. But the lack of employment opportunities is partly caused by investors not wanting to expand because of the lack of skilled labour. Only a small fraction – 5% of Afghan firms – offer training. Approximately one-third of the workforce in Afghan enterprises has a secondary education or higher. 62% of Afghan managers have secondary or higher education compared with 96 and 98 percent in Pakistan and India respectively. The shortage of skilled labour helps explain the low productivity of Afghan firms" (p.96).

Training has been identified as the area where support and intervention are essential, so training courses have been developed and presented in a number of areas: general and supervisory management, general principles of operations and maintenance, control and instrumentation, quality management, intensive training on individual items of equipment, stores management, performance measurement and recording (financial and technical), basic finance, accounting principles. However, it was discovered that the training has generally not been well received, nor has it proved useful. The following are examples of some of the issues identified:

- The low motivation of employees means that technicians can only be motivated when paid to attend training (in addition to their normal salary), and providing their transport costs to the training venue are paid. (Salaries of technicians are very low: they are in the same pay grade as a cashier or administrative assistant).
- Technical training tends not to acknowledge the reality of the operating context. Employees are trained to operate and maintain fully functional equipment. The reality is that spares and materials are usu-

ally not available, but training does not acknowledge the extent to which employees have to improvise.
- There is a lack of clear lines of reporting responsibility and transparent decision making. Instructions are frequently overruled, and employees do not understand the reporting processes. The result is that all queries were referred to a manager, who in turn is unable to delegate.

Managers in the utilities do not have adequate capability, time and resources to undertake detailed investigations and make rational choices.

Organization for Technology

Afghanistan has embarked on a road of political, economic, and social transformation, but no one can be assured this will be successful. Mills (2007) paints a bleak picture of the Afghan situation in that after more than two decades of conflict and a brief period of delicate respite, Afghanistan is again under threat from extremists, and many of the early gains have disappeared. Throughout the country capacity remains weak; education is poor; skills are at an extremely low level; the central government holds little sway outside the capital. The majority of Afghans live in dire poverty without access to the basics of human existence: clean drinking water, electricity, adequate medicine or the promise of improving their lives. Mills (2007) describes economic activity in Afghanistan as "Ramshackle roadside 'businesses' making things from scrap, selling vegetables, battering steel, heaving wood, bartering, trading, haggling, and just generally surviving" (p. 66). The utilities investigated suffer from a host of difficulties that are almost inconceivable in a developed country environment. In particular, this relates to security. In terms of Max Weber's definition, Afghanistan is a 'failed state' in that the central government does not control the means of organised violence.

Organizational structures exist in principle, but some managers do not actually understand them. Usually, only a manager can give the instruction for a task to be undertaken. The notion of a job description is foreign, and concepts such as key performance indicators are completely beyond most managers' comprehension. The effective organization structure is therefore flat, with managers having to contend with a wide span of control, frequently in excessive of 50 people.

Supply Chain and Partnerships

The lack of political and economic stability in Afghanistan, and its appalling violence and difficult internal problems do not represent an attractive investment destination for technology suppliers. The lead time to supply equipment and spares is long, and supply routes are insecure. Technology suppliers from many countries are reluctant to conduct business in Afghanistan because of security fears and concern about corruption.

Financial

Knowledge of technology is essential for addressing the poor service delivery of services in Afghanistan. As part of this overall requirement, improvements will demand a massive improvement in managerial abilities and substantial development of technical skills. The country does not have the funds to achieve this on its own so will remain dependent on foreign aid. However, it finds itself in a vicious circle where foreign direct investment will not enter the country until investors see some potential return, but returns will not be forthcoming unless the country can rise from its lawlessness. Only then can technology be called upon to contribute to the economic transformation that has occurred in some other emerging economies.

It is noteworthy that managers do not mention the important of a technology strategy in their responses. They are thus clearly not thinking in strategic terms, which is necessary for technology transfer to be successful. Managers are too occupied with operational matters at the lower levels. The outlook is therefore one where items of equipment may be imported into the countries, but the transfer and sound management of technology remain a distant hope. Knowledge plays little role in enhancing performance.

SOLUTIONS AND RECOMMENDATIONS

The findings show that managers in utilities in Afghanistan are obliged to deal with operational issues in most trying circumstances. On the basis of the underlying premise of the importance-control framework, the implications for managers are that positive action is required for 'complex' items found in quadrant 4, and those that are some distance from the diagonal in Figure 2. A number of recommendations can be made for each of the factors presented in Table 1 and Figure 2.

Technology and Technology Management Interface

Low levels of operating and maintenance skills mean that technology should be as simple as possible. The selection of equipment requires a delicate balance: the choice between complex equipment that is largely automatic and does not call for high levels of operator skills, and simple technology that demands more operator intervention, but which is easier to maintain. Managers should be placed in a position where they have a meaningful say in the type of equipment selected (control), with an emphasis on simplicity and adaptability to local conditions (such as correct electric motor cooling fins and adequate filters to cope with dusty conditions). While most design work (such as reticulation design work, for example) is performed by external consultants, the purchasing process should recognise the importance of an operational manager's input to the technology selection process. Technology interfacing management is an operational issue, which should be firmly in the domain of operational managers (meaning that authority and responsibility for performance should pass to an operations function – in terms for Figure 2, to the core quadrant).

Political and Economic Issues

The political situation and security concerns are beyond the control of managers. Utility workers are frequently viewed as working for the Government of Afghanistan, so they and their families are considered legitimate targets for insurgent attacks. There is a need to provide managers with access to funds to address such issues, but this factor will remain intractable in Afghanistan.

It is clear that lack of good governance is a major impediment to the operation of any kind of business, but especially those controlled directly by the government. Afghanistan suffers from a total lack of vision for government controlled organizations. The result is few sound policies emanating from the highest levels of government. It is necessary to bring policymakers and operations staff together, but any form of planning at national level that would bring this about is weak. So many of the dealings and rulings by senior staff in government are only about power and what comes to those in powerful positions.

These issues will remain intractable in the current Afghan context. Politics affect all aspects of life in Afghanistan, and will continue to complicate the control functions of line managers. It is likely that these matters will remain in the 'complex' quadrant for many years.

Knowledge

Items under this heading are closely associated with human resource issues. The notion of 'control' under the knowledge heading means the power and ability that managers have to increase operators' and maintainers' understanding of the technology (machines, processes, etc). Control implies that the manager can take the decision as to how this may best be done. Collection and analysis of failure data require proactive and persistent intervention and supervision of adequately educated staff. Managers need to assume responsibility to change maintenance policies as appropriate. Training and development are very much part of a knowledge agenda, and with additional funds the longer term hope is that these factors can emerge in the core quadrant.

Human Resources

The legacy of the country's past will remain for decades, with education, in particular, not having being transformed for years. Shortages of technical and managerial skills will continue to constrain technological development. Practical and technical education is fundamental to the use of knowledge-based technology. It is essential that managers are empowered to provide such technical training. Resources are extremely limited in Afghanistan, so it is not possible to promise great monetary rewards to aspirant technicians and engineers. If they speak English, they will earn far higher salaries working for international aid agencies. However, offering a career path in one of the utilities would be attractive. One manager commented: "Democracy is all very well, but we have taken it too far". By this he meant the frequent strikes by apprentices and students who express dissatisfaction on a regular basis, and who make direct appeals to a minister or parliamentarian. This occurs because managers are deemed to be powerless, and recourse to higher authority is seen as the only way to achieve anything. There is of course a 'political' component to the ability to take responsibility: decision making in Afghanistan is complex and relational. Decisions are invariably deferred to senior authority because of a reluctance to take action without higher approval. This results in inefficiencies and frustration. Ministers and senior officials must be prepared to delegate decision making to operational managers.

Human resource issues will remain in the complex quadrant for as long as managers are powerless to manage their subordinates. This is frequently the fault of politicians and others in authority in that they permit and even encourage undue pressures to be exerted on managers (through external means).

Organization for Technology

This factor is in many ways intangible. It also relates to knowledge-based decision making. Many instances were recounted where equipment is purchased, but no provision is made for installation. There is little infrastructure for technical interface management: pumps are purchased, but no provision made for switchgear and cabling to the electric motor, so months pass before the pumps can be commissioned. Operating staff are afraid of reporting failures for fear of retribution. Maintenance technicians frequently do not report the true cause of failure for the same reason. This raises a cultural challenge to managers: people should not be punished for making mistakes, and only truthful reporting will assist meaningful decision making. Managers must likewise be reassured by their superiors. The low present control scores in Table 1 for organizational issues clearly reflect anxiety on the part of managers to assume control. Future control scores show cautious optimism on the part of managers that some of these issues will be addressed favourably.

Supply Chain and Partnerships

Managers envisage more control in the future of supply chain and partnership matters. They clearly see working with technology suppliers as advantageous since this will improve compatibility issues. In the past contractual matters have been a cause of great concern and animosity because of fundamental differences in understanding what is meant by a contract. One manager who had worked in Germany commented that few Afghan managers understand an international contract: their way of doing business is to reach an equitable agreement to ensure completion of a project. For example, local managers do not understand the notion of having to bear the risk of raw material price fluctuations and construe insistence by a technology supplier as unreasonable. International equipment suppliers frequently encounter difficulties when dealing with Afghan subcontractors (such as a civil or a cabling contractor). Inevitably it is the utility manager that is called upon to arbitrate and work out a solution. Training such managers in elements of contractual arrangements and project management is one way to enhance such managers' control of a situation for which they are held responsible.

The issue of corruption was also raised by managers under this heading. As one manager noted: "What you in the West call corruption, we consider to be giving gifts." The supply chains that businesses have created to obtain good from Pakistan and other countries are complex and intricate. They may be not familiar to a western perspective of a supply chain, but that is because the context is so radically different to what western technology suppliers would expect. Items under this heading will remain in the complex quadrant until a more conventional approach to supply chain management can be adopted.

Financial

New technology and operating processes are simply not adequately investigated because of limited funds. Even if funds are forthcoming managers invariably do not have the skills to perform a competent evaluation of difference technologies. Further, supporting departments fail to understand the functional requirements for a particular technology. One example provided by a manager was the purchase of office computers. When the computers arrived it was discovered that no provision had been made for tables, cabling and the necessary: the budget was insufficient. The computers were stored for months awaiting delivery of associated items, but slowly disappeared (theft, used elsewhere). The cause of this problem was that the purchasing and finance functions had not consulted with the operational manager on what was required. While lack of funding is certainly a constraint, difficulties that are ostensibly 'financial' are in reality organizational. Managers need to be provided with the tools and authority to deal with these.

In using the framework to analyse managerial perceptions, Naudé et al (1990: 524) speak of "an underlying dynamic nature to strategic issues implying a migration around the grid" (p. 524). Peripheral issues representing a balance between importance and control are expected to be relatively stable. Factors in other quadrants may respond to long-term trends or unexpected events, which result in repositioning, particularly in the complex quadrant, where high importance is not associated with commensurate control. The resulting misalignment will require conscious management efforts to increase control over these. Once balance has been achieved, importance recedes, and managers need merely to maintain these issues, rather than concentrate on their control. Movement is towards the simple issue quadrant. The challenge is then to address those issues that, over time, move from the diagonal into the complex quadrant.

In this study in Afghanistan the differences between current and future scores show a small move towards the diagonal where importance is associated with commensurate control, but there is a long way to go before Afghan managers are in a position to control the variables which they deem significant.

Technology is as important for developing countries like Afghanistan as it is elsewhere. The linkages between knowledge and technology include non-technical and operational issues.

Developing countries suffer from many disadvantages preventing them from benefiting fully from a technology. This study suggests that knowledge has a vital role to play in implementing technology, but in the Afghan utilities studied, the issues impeding this lie beyond the field of knowledge management: political, cultural, organizational and other factors are an essential ingredient in the management process. Technology management includes the control of a broad selection of technical and non-technical factors that enhance technology through a network of knowledge-related capabilities.

REFERENCES

Blumentritt, R., & Johnston, R. (1999). Towards a strategy for knowledge management. *Technology Analysis and Strategic Management, 11*(3), 287–300. doi:10.1080/095373299107366

Coates, T. T., & McDermott, C. M. (2002). An exploratory analysis of new competencies: a resource based view perspective. *Journal of Operations Management, 20*, 435–450. doi:10.1016/S0272-6963(02)00023-2

Govindarajan, V., & Trimble, C. (2004). Strategic innovation and the science of learning. *MIT Sloan Management Review, 45*(2), 67–75.

Graham, C., & Chattopadhyay, S. (2009). Well-being and Public Attitudes in Afghanistan. *World Economy, 10*(3), 105–146.

Grant, E. B., & Gregory, M. J. (1999). Adapting manufacturing processes for international transfer. *International Journal of Operations & Production Management, 17*(10), 994–1005. doi:10.1108/01443579710176997

Hipkin, I. B. (2004). Determining technology strategy in developing countries. *Omega International Journal of Management Science, 32*(3), 245–260. doi:10.1016/j.omega.2003.11.004

Hipkin, I. B., & Bennett, D. J. (2003). Managerial perceptions of factors influencing technology management in South Africa. *Technovation, 23*, 719–735. doi:10.1016/S0166-4972(02)00031-7

Lall, S. (1993). Promoting technology development: the role of technology transfer and indigenous effort. *Third World Quarterly, 14*(1), 95–109. doi:10.1080/01436599308420315

Leonard-Barton, D. (1995). *Wellsprings of knowledge*. Boston: Harvard University Press.

Lister, S., & Wilder, A. (2005). Strengthening subnational administration in Afghanistan: technical reform or state-building? *Public Administration and Development, 25*, 39–48. doi:10.1002/pad.346

Lynskey, M. J. (1999). The transfer of resources and competencies for developing technological capabilities - the case of Fujitsu-ICL. *Technology Analysis and Strategic Management, 11*(3), 317–336. doi:10.1080/095373299107384

Mills, G. (2007). *From African to Afghanistan*. Johannesburg, South Africa: Wits University Press.

Naudé, P., Human, P., & Malan, L. (1990). Managerial perceptions of the future in a volatile society: the South African case. *Omega, 18*(5), 521–528. doi:10.1016/0305-0483(90)90008-W

Platt, L., & Wilson, G. (1999). Technology development and the poor/marginalised: context, intervention and participation. *Technovation, 19*, 393–401. doi:10.1016/S0166-4972(99)00030-9

Prochno, P. J. L. C., & Corrêa, H. L. (1995). The development of manufacturing strategy in a turbulent environment. *International Journal of Operations & Production Management, 15*(11), 20–36. doi:10.1108/01443579510102873

Shannon, R. (2009). Playing with principles in an era of securitized aid: negotiating humanitarian space in post-9/11 Afghanistan. *Progress in Development Studies, 9*(1), 15–36. doi:10.1177/146499340800900103

Voss, C., & Blackmon, K. (1996). The impact of national and parent company origin on world-class manufacturing. *International Journal of Operations & Production Management, 16*(11), 98–115. doi:10.1108/01443579610131492

World Bank Report, (2006). *Doing Business in Afghanistan Report.* Washington: World Bank.

KEY TERMS AND DEFINITIONS

Technology: may be construed as differentiated knowledge about specific applications: typically the assets that form the basis of a firm's competitiveness.

Technology Transfer: is the process by which science and technology are diffused throughout human activity. Wherever systematic rational knowledge developed by one group or institution is embodied in a way of doing things by other institutions or groups, we have TT. This can either be transfer from more basic scientific knowledge into technology or adaptation of an existing technology to a new use. TT differs from ordinary scientific information transfer in that to be really transferred, it must be embodied in an actual operation of some kind.

Developing Countries: are those countries that generally lack a high degree of industrialization, infrastructure, and other capital investment, sophisticated technology, widespread literacy, and advanced living standards among their populations as a whole (AGOA.info).

Knowledge-Based: implies the use of an organization's knowledge and intellectual assets to create value in terms of better internal management, grantmaking decisions, etc. For example, creating a guide of lessons learned or best practices (http://www.pfc.ca/cms_en/page1112.cfm).

Chapter 19
Knowledge Management to Promote Organizational Change in India

Juha Kettunen
Turku University of Applied Sciences, Finland

Manodip Ray Chaudhuri
Future Business School, India

EXECUTIVE SUMMARY

This chapter contributes to the literature of knowledge management by providing a conceptual framework to promote organizational change. The chapter demonstrates that knowledge management can be used as a general framework which integrates the approaches of strategic and change management. A business company is an organization that must continually respond to environmental change and adjust to fluctuations to gain competitive advantage. Business intelligence produces tacit and explicit information about the markets that are used in the strategy process. The tools of change management provided in this chapter can be used in different kinds of organisations to increase competitiveness for the future. In addition, this chapter presents cases of successful change management. This chapter is useful for those who want to enhance change to increase competitive advantage of companies and other organisations.

INTRODUCTION

The purpose of this paper is to present a general conceptual framework of knowledge management to promote organisational change and competitive advantage. The framework integrates strategic and change management and presents useful tools of change management. The study helps managers to use different management approaches to develop organization and identify the landmarks of change management that are essential in the planning and implementation of change. The study also presents practical cases of how change management has been applied in Indian companies to promote competitiveness and a better future.

During the last few years, many change projects were not successful (Carter, 2008; Holden et al., 2008; Zink, 2008). There are different causes for these failures. The company may not have necessary resources available to implement

DOI: 10.4018/978-1-61692-886-5.ch019

the long-term strategy (Westphal & Fredrickson, 2001). There may be an adequate participation of the individuals and groups affected by the changes (Lines, 2004). The company's or plant's culture may not support the change (Jones et al., 2005). Knowledge-intensive organizations need to harness knowledge and to analyse knowledge sharing and change mechanisms in the whole organization. It is well recognized that knowledge-sharing mechanisms are highly complex processes (Bechina, 2006). Zack (2009) found that knowledge management practices are directly related to organizational performance. This study intends to make sense from the comprehensive change management tools in order to sustain the competitive advantage.

Change management falls within the broader theoretical approach of social change and has been a perennially popular topic in the organizational effectiveness and management literature (Asgary & Walle, 2002; Shaurhofer & Peschi, 2005). Identifying the need for organization-wide change and the process of leading organizations through that change is widely recognized as one of the most critical and challenging responsibilities of organizational leadership. These responsibilities have become even more important in recent years, because organizations are today proceeding through an age of change, innovation, and creativity at every stage of their operations. Business contexts are changing enormously and forcefully around the globe. There are strong contemporary issues such as globalization, the influence of new technologies, the emergence of e-business, growing e-connectivity among far-flung financial markets and evolving consumer choices and preferences to be taken into account. All these factors are accelerating the pace of commerce and changing the style and mode of doing business throughout the world in virtually every industry.

The paper is organized as follows. The next section provides a conceptual framework of knowledge management to support organizational change. The following section enhances the role of an organization in relation to its environment. Then the approach of strategic management is presented to produce competitive advantage through the change process. Thereafter, the concepts of change management are presented, including the context of a macro organization, the dimensions of intentional organizational change, resentment to change, and coping strategies. Based on the theoretical background, the study then offers cases of how the competitiveness of companies has been increased through these change initiatives. Finally, the results of the study are briefly discussed in the concluding section.

KNOWLEDGE MANAGEMENT AS AN OVERALL FRAMEWORK

Successful and widespread management approaches can be transmitted to other organizations via repetition. The success of a management approach is also enhanced if it sounds reasonable, if its definition is malleable enough so that its adoption creates little cognitive dissonance (Birnbaum, 2000). A prominent feature in the management literature and consulting business is that new management approaches emerge and meet in the organisations. Once dissonance exists, the new management approaches challenge the old approaches. A cognitive and practical challenge is to combine the old and new management approaches in working life and use them flexible in different situations.

A challenge of this study is to align knowledge, strategic and change management especially in knowledge-intensive organizations. Knowledge is a context-dependent collective phenomenon arising from social interactions and effective dialogue (Freyens and Martin, 2007). Efficient knowledge acquisition is crucial for the success of an organization. Strategic plans should be designed to take the knowledge creation and sharing into account. When strategic objectives have been defined, the resources, knowledge, processes and structures should be developed to achieve the strategic objectives. Efficient knowledge sharing improves the renewal and competitive advantage of the organization to meet the needs of their environments.

Even though the cycle of knowledge creation introduced by Nonaka and Takeuchi (1995) is widely understood (Boateng, 2006), it provides a useful framework to integrate the different management approaches in an innovative manner. Especially it offers a framework which allows consistently use the valuable management tools of different management approaches. Once problems occur, an experienced manager is able to reflect on its past experiences and explore solutions using different management approaches. Consequently knowledge management can be seen as an overall framework that embraces other specific approaches.

Figure 1 describes the integration of environmental analysis, strategic and change management and change implementation using the spiral of knowledge conversion adapted from Takeuchi and Nonaka (2004). Environment including many changes such as social and technological change is constantly analyzed in a market economy. An organisation responds to the environmental changes in its strategy process, which produces strategic plans including vision statements, and strategic themes and objectives. Change management combines the organisations financial and human resources and develops its processes and structures to achieve the strategic objectives. Learning by doing characterizes this phase. Personal motivation and the opportunity to learn from other employees are critical to generate knowledge (Zapata Cantú et al., 2009). Change is implemented to meet the customer needs and the demand for the products and services of the organisation.

In the socialization phase of the spiral of knowledge conversion, the tacit knowledge about the environment is disseminated to the tacit knowledge of other people. Typically market situation and social change have incomplete and unclear information. In the externalization phase, the tacit knowledge is transformed into explicit knowledge in the strategy process. The strategy process must produce statements and documents which are understandable to wide audience including personnel, customers, partners, owners and other stakeholders. In the combination phase, the change management combines and disseminates explicit knowledge to other interest groups using all the necessary tools of change management. Finally, in the internalization phase, explicit knowledge is transferred to tacit knowledge to meet the customer needs. The spiral of knowledge conversion is a never-ending process, which collects information about the markets, defines strategies and implements necessary changes to meet the customer needs.

ORGANIZATION IN ITS ENVIRONMENT

This section of the study describes the socialization phase of knowledge management, which analyses the organization in its environment. An organization is likely to know more about its market and have more success in its product selection when it collects and analyzes environmental knowledge. Market intelligence is the process of acquiring and analyzing information in order to understand the market including existing and potential customers, to determine the current and future needs and preferences, attitudes and behavior of the market and to assess changes in the business environment that may affect the size and nature of the market in the future (Cornish, 1997a,b).

Figure 2 presents the identification of an organization in relation to its environment. Studies of organizational change remain quite shapeless without considering the implicit and explicit assumptions about that organization and the relationship of that organization to its environment. Primarily, an organization is understood as an organism that continually responds to its environment and fluctuations in that environment (Kettunen, 2004a, b, 2006). It adjusts and evolves in response to fluctuations or contingencies in the environment. The mechanistic form of an organization may be appropriate in situations of stable and certain markets, but an organic organization is strongly recommended in unstable and uncertain market situations (Dessler, 1976; Khadwall, 1995).

An organization, in relation to its business environment, is further identified in terms of its resources and internal capabilities. On the face of both planned and unplanned variations in business processes and systems, an organization needs to adapt to its environment through optimum utilization of its resources and capabilities. Companies with superior structures and practices are more able to survive environmental variations while those with inferior methods and processes often die out in resource-constrained competition that pulls them down within their own structures and systems (Rao, 1991).

An organization is also identified as a subsystem within the environment where it operates. Environment is functionally known as context. An organization influences its context at every phase of operation and is also constantly influenced in turn by the various constituent elements in its context, namely, how the organization behaves in a crisis situation, what makes it move toward success, in what way it measures its pool of talent, and how it orients itself towards its goals. Answers to these and many similar questions depend on the context of the organization. Context, however, remains the determining factor of the ability of an organization and its agility in executing its various functions.

Information processing and decision-making are other criteria when identifying an organization in relation to its environment. An organization is defined as efficient or inefficient depending on its information base and the particular nature of information processing within its operations. An efficient organization is understood to have

Figure 1. Knowledge management responds to market needs

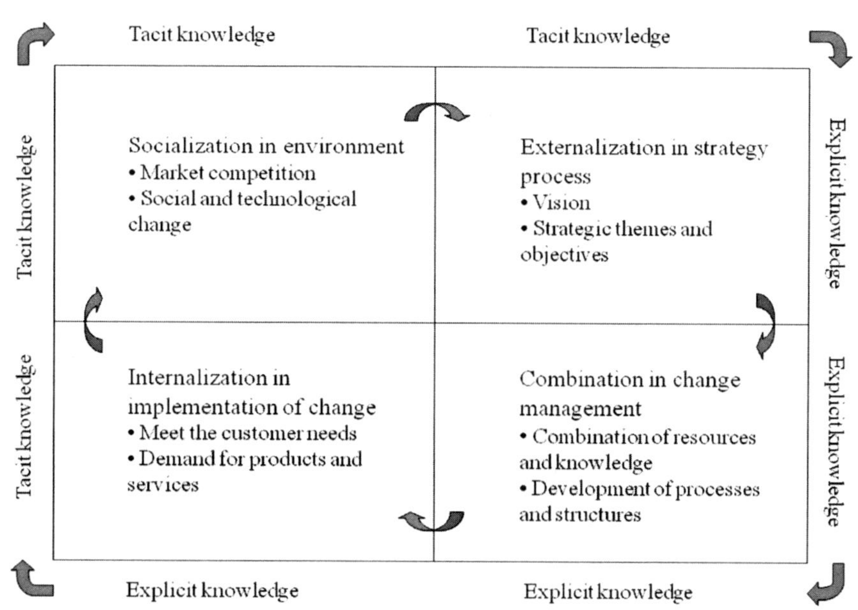

a strong information base and a quick, efficient information processing system in all its functions. Such an organization is known to be operationally efficient in terms of information utilization and decision-making at all levels of operations. Faster information processing and decision-making can lead an organization toward operational efficiency in a fast growing business environment (Dessler, 1976).

The primary perception of change management thus hovers around the notion of an organic pattern of organization in the context of sustaining competition in a changing business environment. An organization that emphasizes an organic system appreciates the notion of environment reactions and its associated responses, teamwork, the participative style of operation, the elements of motivation, employee empowerment and delegation, the dynamic aspects of change, and adaptation and learning. This orientation is expressed in Kurt Lewin's popular model of change where the fact of a changing business environment is understood in association with a related response to such change and fluctuations within work systems. It incorporates the consideration that the basic tenet of human nature is to resist change and answers the question regarding what should be the correct approach of individuals in order to move toward change and accept change in their strides at work (French & Bell, 1989; Rao, 1991). The resistance to organizational change has been analyzed, and in some circumstances, it can be an obstacle to achieving successful organizational change (Bell & Gomez, 2008; Furst & Cable, 2008; Laframboise et al., 2003).

STRATEGIC MANAGEMENT TO PROMOTE COMPETITIVE ADVANTAGE

This section of the study describes the externalization phase of knowledge management, which describes the strategy process of the organization. The most elementary factor to consider how to gain competitive advantage by applying effective

Figure 2. Identification of an organization in relation to its environment

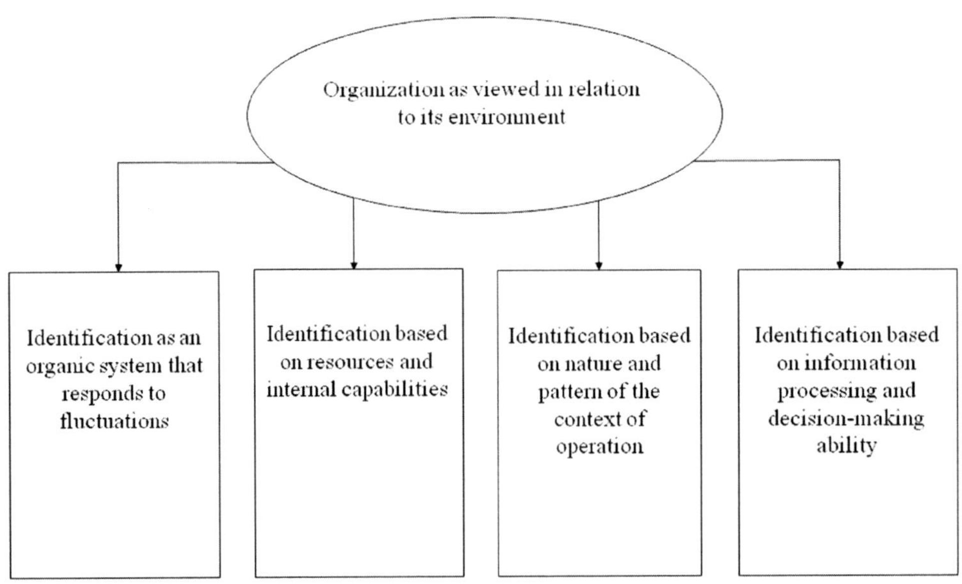

change management is to redesign the tools and techniques of change management in light of the changing definition of certain essential strategies that affect the process of change in the workplace. There are a number of ways to approach change, but interestingly, no single strategy works for every passing eventuality. It is important to remember, therefore, that choosing a specific strategy to effect change must identify the most appropriate strategy to implement a particular change in the context of specific circumstances of a situation. One needs a clear understanding about the business operations of an organization before a particular strategy is adopted and produces effective implementation of change in that organization (Kettunen, 2007).

Figure 3 presents the strategies to manage change for competitive advantage. The effective change process using a strategic management approach may take the form of command or decree strategies, specialist or professional strategies, consulting or bargaining strategies, educative or learning strategies, and involvement or membership strategies (Robbins, 1990).

Command or decree strategies refer to the use of authority and directions for the task of change management. There is generally little or no involvement of individuals. Although such a strategy is less time consuming, it typically does not consider the opinions or views and emotions of individual employees. This strategy may not be particularly helpful in situations where there is strong resistance to change by the employee groups at large.

Specialist or professional strategies resort to professional experts in handling issues pertaining to change management at work. The expert may be a specialist team or a group made up of senior management or may be specifically a change manager. This approach guarantees the quality of solutions and ensures clarity in problem-solving within the change management process. One major point of concern, however, is the extent to which the viewpoints of those affected are able to coincide with the viewpoints of the professional experts. In other words, the question remains as to whether the expert opinion is forcibly imposed on individuals who will be affected by the change.

Consulting or bargaining strategies refer to the willingness of management to negotiate the change management exercise with employee groups at large. This process acknowledges the right of the individuals affected by change and brings forward their views and opinion regarding the proposed change. The expectations of employees are precisely understood and identified before launching any change elements at work, a strategy that in turn often results in smoother acceptance of the change in the workplace.

Educative or learning strategies aim at securing the full commitment of employee groups to accept the planned change to be implemented at work. Thus, management, before actually implementing change in the organization, considers a combination of activities that relate to persuasion, teaching, educating, instructing, coaching, and training employees regarding the change elements to be introduced at work and also the specific need for such education. The concern for this approach related to the time and costs consumed to complete the learning process.

Lastly, involvement or membership strategies take participative approach that involves both management and employee groups in deciding on the change management strategies and policies that will be implemented in the workplace. An essence of learning is embedded in this style of change management since employees are given the chance of active participation in strategic planning and implementation. This approach addresses and ensures the significance of workers in participating in management and applies a more modern management approach that is now being applied in many business companies.

As a reflection of the strategic orientation of the change process, practicing managers will outline typical strategic tools for managing change in organizations. The choice of tools depends on the organization's nature, resources, and problems and must be tailored to the specific context and situational demands of the change. Business analysts have identified typical techniques in this regard that strengthen the strategic orientation of change management. They include backward mapping, balanced scorecards, benchmarking, business process reengineering, continuous improvement in quality, culture change, employee involvement, international standardization of products and services, knowledge management, a learning environment, management by objectives, organization design, job analysis, outcome-based evaluation, and total quality management (Birnbaum, 2000).

The ultimate consideration of a successful change management strategy will emphasize factors that affect the choice of change strategies in the real-life situation of each business. The choice of the correct strategy depends on the urgency of the need for change, the degree and extent of any resentment to change, the power of the change context that will be implemented, the reliability and validity of the change process, the extent of the prevailing information about the change, the degree of commitment toward acceptance of the change, and the clarity of the change process that needs to be launched.

CHANGE MANAGEMENT

Change Management in the Context of a Macro Organization

This section of the study describes the combination and internalization phases of knowledge management, which describe the change management of the organization. Fundamentally, the idea of business change rests on certain precise core elements. Change at work corresponds to inculcating a change orientation in employees,

Figure 3. Strategies for managing change for competitive advantage

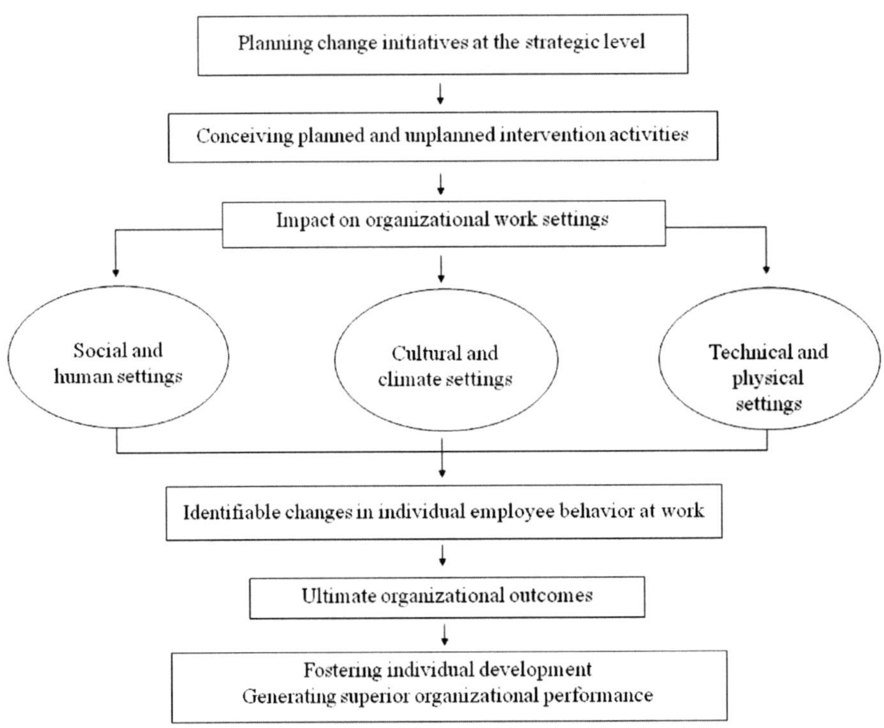

or more basically individuals, in relation to their skills, values, attitudes, and behavioral norms and practices with an eye for operational organizational change. Change further emphasizes a reorientation of organizational structures and systems in terms of training systems, performance and reward systems, communication patterns and human relationships, work structures, and work designs. Most importantly, the notion of change and change management points to an investigation of the organizational climate, interpersonal work styles, and the work culture. Managing culture change remains the imperative and most vital phenomenon in any milieu of change management in a macro organization (French & Bell, 1989; Schein, 1992).

Change management also identifies change as an element that is embedded within the technology and physical work settings of an organization. Such aspects of change begin with identifiable, planned change interventions in the prevailing work settings of the organization. This reasoning in turn leads to the need for individual employees to bring change into their organizational behavior and their approaches to organizational processes and constituent elements within the work system. Finally, changes in the human approach to work and work settings give rise to related behavioral changes that will impact individual development and organizational performance that are both key organizational outcomes (Nilson, 1995).

Figure 4 describes the maneuvering of planned and unplanned change interventions. Understanding the notion of change management and the elements of change at work, offers certain elemental landmarks in the context of macro organization. These landmarks include planning change initiatives at the strategic level, conceiving planned and unplanned intervention activities, impacting organizational work settings that include the related impacts on social and human settings, cultural and climate settings, and technical and physical settings. These separate impacts can produce identifiable changes in individual employee behavior at work and ultimate organizational outcomes. Eventually, the various phases of landmarks lead to the fostering of individual development and thus superior organizational performance.

An attempt to maneuver change in organizations can begin by planning change initiatives at the strategic level. Nevertheless the heart of these activities lies in the actual implementation of the various planned and unplanned interventions at work. Indeed, whether change management has been successful or not depends on the extent to which change interventions can impact organizational work settings, organizational culture, and individual behavior. The ultimate outcome of change management rests on the degree to which individual employee behavior can be molded in favor of the change as well as the acceptance of the change being implemented within the organization.

Dimensions of Intentional Organizational Change

One elemental debate exists in terms of answering the crucial question regarding whether change in individual attitudes does foster change at work or whether change in individual attitudes is instead an offshoot of initiating the process of change. This question leads to identifying change and change management in business organizations as a dynamic and intentional process with certain inherent critical dimensions (Nilson, 1995).

The following elements are essential in change management:

- the reasons for initiating organizational change and the goals of that change
- the nature of leadership required for the successful initiation of organizational change
- correct gauging of the focus of the organizational change
- outlining of the process of implementation of any change initiation strategies

- a reward system associated with the success of organizational change initiation
- use of consultancy services in the context of organizational change management

Table 1 describes the dimensions of change when generating economic value and capacity. The dimensions of intentional organizational change bear relevance in organizations and relates to both the economic value and the capacity of organizations and their processes. Economic value refers to the gross profitability of an organization as a system with constituent subsystems. Capacity is synonymous with capability, which refers to skills, knowledge, and the resource base of an organization. Each dimension of intentional change management will have clear identification in the organization in terms of its economic value and the capacity of the business. Each dimension must be identified in terms of what way it is perceived in order to generate positive economic value and capacity for the organization and produce effective change management (Nilson, 1995; Rao, 1991).

Resentment to Change and Coping Strategies

Although the notion of change management reflects a building up of a certain level of competitive advantage, so an organization can survive in a globally competitive business environment, the implementation of change is not often either as smooth or simple as it sounds at the start. It can be quite difficult for change managers and line managers to implement change at work in the practical sense of that term because of prevailing and ongoing opposition to change. These kinds of opposition to change appear to be expressions of antipathy toward the change. The elemental issues that form the roadblocks to accepting change in organizations are primarily behaviors that pinpoint individual attitudes toward change in business companies. These may be explored in terms of more specific human feelings, such as the loss of control, face and identity, threat to competence, the high degree of change uncertainty, surprise

Figure 4. Maneuvering of planned and unplanned change interventions

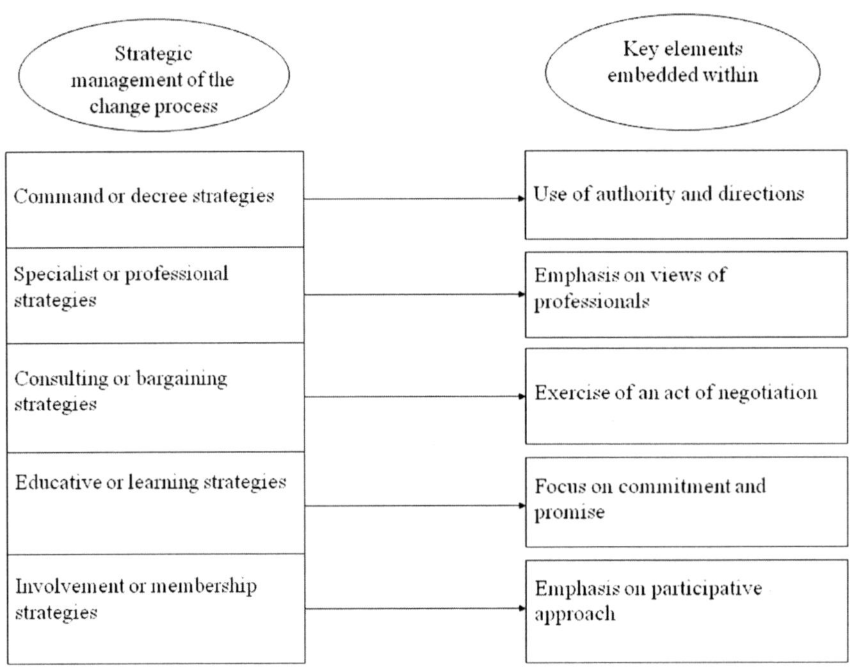

and shock, and attitudes toward the nature of the change content (Nilson, 1995).

To reduce such resentment and opposition to change, change managers, organizational behavioral experts, and change agents need to look into avoiding any implementation of change that is unnecessary. If top management considers a change necessary and commences the change process, it is important to communicate the strategic objectives and need for the change to all levels of the organization effectively, listen to those people who are likely to be affected by the planned change as to why they resent the change, and make sure that everyone participates and is involved in the actual change process. Negotiations and bargaining should be preferred over commands and directions when undertaking change management. The change managers should emphasize the identification of the major drivers of change within the organization and the facilitation of any change by involving external consultants, if required. In addition, it is important to monitor the change to see that the objectives of the change process are achieved.

Modern business companies increasingly use strategic planning in change management to enhance global competitiveness in their operations. Care should be taken to address all the necessary aspects of an organization when the change is implemented at work. Skilled management, total quality commitment, and the need to apply an all-pervasive change drive are the major identifications needed for successful change management projects in modern organizations.

Figure 5 describes the strategic orientation needed to cope with resentment and any antipathy to change. The primary focus of management should revolve around the theme of creating suitable structures and space for change in the organization or unit where a certain kind of planned change is initiated. Any implementation of change should be preceded by a clear understanding about the visible capabilities, values, and processes of the organization that will in turn determine the organization's ability to implement a particular change strategy effectively in a given context or situation. It is well to note too that planned change or transformation can be successful only when the drive for change is fundamentally vision-led and not based on temporary problem-solving (Khadwall, 1995).

Executive sponsorship, participation, and leadership are also very critical to the success of change initiatives. Studies on organizational behavior have identified leadership and participation by top management as the most significant contributors to success of any change management agenda. The essence of effective leadership binds an organization together and holds it strongly focused with a view toward overcoming resistance and inertia within the organization. The major requirements of leadership during planned orga-

Table 1. The dimensions of change in generating economic value and capacity

The dimensions of intentional organizational change	*Notion of economic value*	*Notion of capacity*
Outlining goals of change	• Maximize monetary value	• Build capacity
Defining leadership styles	• Emphasize essence of authority	• Delegation and participation
Gauging the focus of change	• Build systems and structure	• Build culture and climate
Implementing the process of change	• Generate value for money	• Gain competitive edge
Defining rewards for successful change initiation	• Motivate through monetary rewards	• Motivate through fostering of a socio-technical environment
Consultancy of outsourcing in change initiatives	• Depend on consultants and change management experts for greater profit	• Assist management in better problem-solving

Figure 5. Strategic orientation coping with resentment and antipathy to change

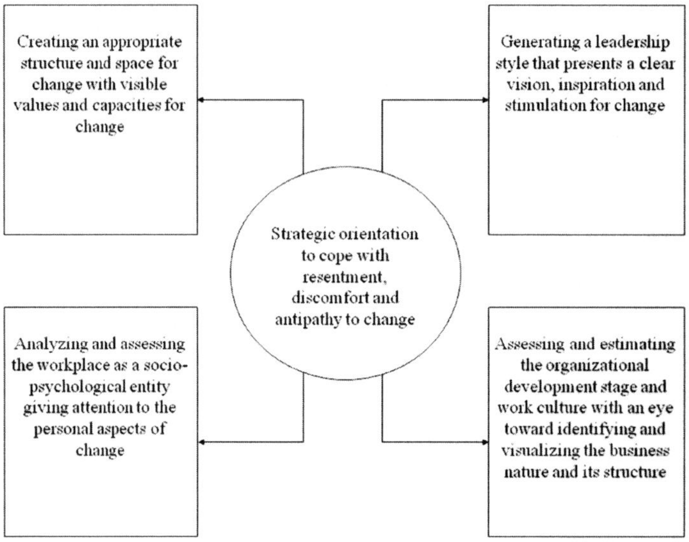

nizational changes are the courage to confront disinterested forces or resistors and deal proactively with conflict, the ability to provide vision and coherence to the entire organization in times of change and change initiation, and the ability to motivate and overcome resistance to conquer crises.

The right leader and the correct leadership style can make or break an organization in times of change implementation. To be a change leader, a considerable emphasis needs to be placed on providing vision, inspiration, encouragement, stimulation and motivation, together with confidence, conviction, and the assurance to demonstrate integrity, give meaning, generate trust, and communicate values that relate positively to the change initiative planned by the organization. It remains the leader's task to know how to pace and sequence change efforts to avoid change weariness and pessimism (Rao, 1991).

The core strategic tool for managing change lies in how management recognizes and gives attention to the personnel aspects of change within its organization. This aspect focuses on managing the workplace as a socio-psychological entity. Correct identification of how managers and employees are responding to change, providing advice on how to handle stress, conflicts, and emotional issues that accompany change, and learning how to gain support for and participation in the change effort can make an organization-wide change less upsetting, less hurtful, and non-traumatic. A study of the human aspects of change, therefore, remains a crucial aspect to enable the elements of change to glide smoothly and efficiently at work (Balaji et al., 1998).

Finally, the assessment of an organization, its current styles of functioning, and its developmental stage are further considered to be effective strategic tools for change management in organizations. Such assessment also signifies the importance of understanding organizational culture, making an effort to unfreeze the current cultural orientation, and creating the motivation to change as imperative aspects of managing change effectively at work. Organizational analysis and assessment for effective change management means creating a shared purpose in the organization, enhancing internal and external communication, cultivating diversity, specialization, differentiation and integration, creating the integrative principles of action, demarcating behavioral boundaries between individuals and units, creating psychological safety and emotional comfort embedded within the spirit of change, chalking out effective risk-management techniques, making realistic, pragmatic and practical transition plans, and having the change target perceived as both concrete and clear. Such an orientation, in terms of identifying and assessing the nature and structure of actual business processes, generally goes a long way toward having successful implementation of change at work (Balaji et al., 1998).

Indian Cases

Knowledge management can be seen as an overall framework which embraces strategic and change management. The integration of different management approaches enables managers to observe different views at the same time and allow them to consecutively accomplish the tasks of knowledge management that earlier have been interpreted using isolated and distinct management approaches. The integration of different management approaches offers plenty of R&D opportunities.

The management of business companies needs effective management tools and clear concepts. This necessitates outlining the task of change management in terms of an amalgamation of a visionary, decision-making, political, business, and transactional social system and manipulating skills that managers need to control management of change for success in a globally competitive business environment that is in the operational limelight. The practical implication of such an understanding about change management can be traced in the success stories of many modern organizations of recent times. Some of these are highlighted below.

Electrosteel Castings Ltd., incorporated in 1955 as a public limited company, is a water infrastructure company that provides techno-economic solutions for water supply and sewage systems. The company is one of India's largest manufacturers of ductile iron spun pipes and ductile iron fittings. It also produces cast iron spun pipes and hard coke. In the face of globalization, the mantra of the company for future growth and survival was "change" in its management practices at all levels (http://www.electrosteel.com/).

The company has gained considerably from the change initiatives in terms of gaining an integrated vision, developing a clear customer focus, generating respect for people, learning a paradigm shift of change as the only permanent feature, focusing on teamwork and small group activities, developing trust in people and empowerment, building a safe and clean work place, emphasizing people involvement, looking into waste reduction, and building on productivity improvement and community development. Today, the strength of the organization rests on its culture, leadership style, and the quality of its people, products and technology.

Customer expectations change continuously. The company has taken the initiative to meet these challenges and regularly upgraded its quality management systems and facilities. Some major initiatives include ISO 9000 and ISO 14000 – Quality and Environment Policy; TQM – Vision, Empowerment, Kaizen and Quality Culture; HRD – Vibrant Organization and People Development; Business Process Improvement – Material Process and Order Management Process and IT – Smarter and Faster Decisions.

Electrosteel Castings Ltd. provides an environment that contributes to the well-being, satisfaction and motivation of all employees. The company believes in "Hiring for attitude and training for skills". The employees are the real assets of the company and it believes in preserving and nurturing these valuable assets through various employment engagement activities, ongoing education, training and development programs. The company believes in appreciating, recognizing and rewarding the talent within the organization. The employees of the company are a motivated lot who is always willing to take up higher professional challenges. It has been emphasized that knowledge assets lead to a sustainable competitive advantage (Moustaghfir, 2009; Werr et al., 2009).

Emami Group is a diversified professional business group with an interest in fast-moving consumer goods, paper, writing instruments, healthcare, information technology, retail departmental stores, and real estate. Emami Group is not only a major success story in Eastern India, but also one of the prominent upcoming business companies in all of India. The company's major change drivers in the recent past have involved a focus on personnel skills and building on a dedicated workforce of experienced employees backed by research and creation of an efficient work culture that establishes a strong psychological contract between the individual employee and the organization. A positive outlook toward having an effective change management agenda has made the organization flourish in terms of having a strong marketing network of distributors, sub-distributors and retail outlets, being identified as a "one star export house", being the only major newsprint in Eastern India, making a significant mark in the healthcare field, and being a part of many social work projects in the country (http://www.emamiltd.in/).

In view of the fact that innovation represented the company's principal driver, the company continued to strengthen its R&D infrastructure. The R&D department, located in Kolkata and Mumbai, follows an extensive process, introducing products that are safe and effective. Market study integrated with ideas from the internal database is the starting point of the process followed by the series of trials, internal specifications at each level and external approval. The coordination between production, quality control and R&D team is important. Finally toxicity, pharmacology and clinical trials are made.

The company's research has carefully focused on its priorities. The successful completion of a series of tests required to develop safe and effective products. The extensive study of a database of age-old tomes (granths) had led to the identification of effective ayurvedic formulations. Innovative introduction of differentiated or improved products has offered a superior value-for-money proposition. The company adhere to specifications (raw material use, formulation development, flavour development, shelf-life testing, finished goods specification, validation protocols) apart from external certifications from the Drug Control Board. Updated equipment in the laboratory certified by National Accreditation Board for Testing and Calibration Laboratories (NABL). Laboratory practices comply with current Good Manufacturing Practice (cGMP). The company conduct clinical trials in association with renowned ayurvedic hospitals and colleges.

Hindustan National Glass & Industries Ltd. aims to create a world-class glass manufacturing plant that pursues quality, cost reduction, and productivity improvement in a truly holistic manner that leads to customer, shareholder, employee and supplier total quality satisfaction. The vision of the company is to become an industry benchmark and a role model for systems, processes, and results. Established in 1952, the company's success story has been built on major strategic and intentional change initiatives that include technology absorption and development, improvement in productivity and quality, focus on cost reduction, increased customer satisfaction, interactive work culture, increased usage of information science, and a high degree of social commitment. The company's compelling vision is to achieve a world-class structure, methods, and outcomes (http://www.hngindia.com/).

The company has successfully developed wide and diversified products to meet the customer needs. The company offers premium glass containers for every kind of application. The company is an acknowledged market leader of today and producing 6 million bottles per day ranging from 5 ml to 3200 ml containers for multifarious industries like pharmaceuticals, beverages, processed foods, cosmetics and liquors. The clientele includes leading companies like Pepsi, Coca-Cola, Cadbury's, Nestle, Raun Pollack, Dabur, Bayer, Lakme, Glaxo, Pfizer, Reckitt & Coleman, Shaw Wallace and UB group.

Quality control and R&D sections are well equipped with sophisticated instruments. With a strong quality control system, the company has been able to achieve quick identification and rectification defects, substantial reduction in job changeover and job upbringing time and shift from break down maintenance to preventive maintenance philosophy. The design cell of the company is equipped with CAD-CAM facilities for designing and manufacturing moulds and mould accessories and is ready to offer containers of unique shapes and sizes in various colors. The quality checks include among others the inspection of the incoming raw materials and packing materials, tight glass density control as a routine tool for glass properties monitoring and computerized, camera based and light-based automatic inspection machines followed by the visual inspection and use of statistical quality control.

Globalisation and the economic transformation take place in the emerging economies and bring new opportunities and challenges for the Indian companies. Falling trade barriers and technological advancements are changing the domestic and international trade patterns of India. The change of the business environment and the development of new technology needs foresight and adaptability on the part of management. The companies in the emerging economies must continually respond to the environmental change and adjust their management tools for gaining competitive advantage.

CONCLUSION

Knowledge management aligned with strategic and change management is the most successful tool for building business competitiveness that evolves at a different level of the organizational life cycle. These management approaches reflect a sense of moving from a regular reactive style of operations toward a stage that produces more effective, stable, and planned development. Change should always reflect a strategy or a set of strategies that focuses on accomplishing the overall goals of an organization and the process should be an all-pervasive exercise that touches all levels of the organization and not done for the sake of change alone.

An organization must always be identified in terms of its relation to its environment. It must respond to changes in that environment and adjust its activities to create success for a better future within that environment. An organization must also identify its optimal use of resources and its internal capabilities to plan positive actions of its internal processes. The business cases presented herein support the argument that companies do pursue different forms of change management. The general pattern seems to be that companies that apply the adequate methods of change management are more likely to be successful and survive ongoing environmental variations.

Change management is a very demanding area of management, but if wisely used, it can foster individual development and generate superior organizational performance. Building organizational competitiveness through change management identifies change as a macro concern. Successful change initiatives focus on visualizing the very element of change as a tool for macro organizational development that builds on the platform of a philosophy of organizational transformation most commonly referred to as organization-wide change.

REFERENCES

Asgary, N., & Walle, A. H. (2002). The cultural impact of globalisation: Economic activity and social change. *Cross Cultural Management: An International Journal*, *9*(3), 58–75. doi:10.1108/13527600210797433

Balaji, C., Chandrashektar, S., & Dutta, R. (1998). *Leading Change Through Human Resources: Towards a Globally Competitive India*. New Delhi: McGraw-Hill.

Bechina, A. (2006). Knowledge sharing practices: Analysis of a global Scandinavian consulting company. *Electronic Journal of Knowledge Management*, *4*(2), 109–116.

Bell, M. C., & Gomez, B. E. (2008). Effect of unsignaled delays between stimuli in a chain schedule on responding and resistance to change. *Behavioural Processes*, *77*(3), 343–350. doi:10.1016/j.beproc.2007.08.004

Birnbaum, R. (2000). *Management Fads in Higher Education: Where They Come From, What They Do, Why They Fail*. San Francisco: Jossey-Bass.

Boateng, W. (2006). Knowledge management working tool for agricultural extension practice: the case of Chana. *Knowledge Management for Development Journal*, *2*(3), 19–29.

Carter, E. (2008). Successful change requires more than change management. *Journal for Quality and Participation*, *31*(1), 20–23.

Cornish, S. L. (1997a). Product innovation and the spatial dynamics of market intelligence: Does proximity to markets matter? *Economic Geography*, *73*(2), 143–165. doi:10.2307/144445

Cornish, S. L. (1997b). Strategies for the acquisition of market intelligence and implications for the transferability of information inputs. *Annals of the Association of American Geographers. Association of American Geographers*, *87*(3), 451–470. doi:10.1111/1467-8306.00063

Dessler, G. (1976). *Organization and Management*. Upper Saddle River, NJ: Prentice Hall.

French, W. H., & Bell, C. H. Jr. (1989). *Organization Development*. New Delhi, India: Prentice Hall.

Freyens, B., & Martin, M. (2007). Multidisciplinary knowledge transfer in training multimedia projects. *Journal of European Industrial Training*, *31*(9), 680–705. doi:10.1108/03090590710846666

Furst, S. A., & Cable, D. M. (2008). Employee resistance to organizational change: Managerial influence tactics and leader-member exchange. *The Journal of Applied Psychology*, *93*(2), 453–462. doi:10.1037/0021-9010.93.2.453

Holden, R. J., Or, C. K. J., Alper, S. J., Rivera, A. J., & Karsh, B. T. (2008). A change management framework for macroergonomic field research. *Applied Ergonomics*, *39*(4), 459–474. doi:10.1016/j.apergo.2008.02.016

Jones, R., Jimmieson, N., & Griffiths, A. (2005). The impact of organizational culture and reshaping capabilities on change implementation success: the mediating role of readiness for change. *Journal of Management Studies*, *42*(2), 361–386. doi:10.1111/j.1467-6486.2005.00500.x

Kettunen, J. (2004a). Bridge building to the future of Finnish polytechnics. *Journal of Higher Education Outreach and Engagement*, *9*(2), 43–57.

Kettunen, J. (2004b). The strategic evaluation of regional development in higher education. *Assessment & Evaluation in Higher Education*, *29*(3), 357–368. doi:10.1080/0260293042000197591

Kettunen, J. (2006). Strategic planning of regional development in higher education. *Baltic Journal of Management*, *1*(3), 259–269. doi:10.1108/17465260610690917

Kettunen, J. (2007). Innovativeness in higher education management. *Bhavishya. Journal of Future Business School*, *1*(2), 65–74.

Khadwall, P. N. (1995). *Management Styles*. New Delhi: McGraw-Hill.

Laframboise, D., Nelson, R. L., & Schmaltz, J. (2003). Managing resistance to change in workplace accommodation. *Journal of Facilities Management*, *1*(4), 306–321. doi:10.1108/14725960310808024

Lines, R. (2004). Influence of participation in strategic change: resistance, organizational commitment and change goal achievement. *Journal of Change Management*, *4*(3), 193–215. doi:10.1080/1469701042000221696

Moustaghfir, K. (2009). How knowledge assets lead to a sustainable competitive advantage: are organizational capabilities a missing link? *Knowledge Management Research & Practice*, *7*(4), 339–356. doi:10.1057/kmrp.2009.26

Nilson, C. (1995). *Games that Drive Change*. New York: McGraw-Hill.

Nonaka, I., & Takeuchi, H. (1995). *The Knowledge-creating Company: How Japanese Companies Create The Dynamics of Innovation*. New York: Oxford University Press.

Rao, T. V. (1991). *Readings in Human Resource Development*. New Delhi, India: Oxford and IBH Publishing.

Robbins, S. R. (1990). *Organization Theory: Structure Design and Application*. New Delhi: Prentice Hall.

Schaurhofer, M., & Peschi, M. F. (2005). Autonomy: Starting point and goal of personal and social change: A constructivist perspective on knowledge management in empowerment processes. *Kybernetics*, *34*(1/2), 261–277. doi:10.1108/03684920510575834

Schein, E. H. (1992). *Organization Culture and Leadership*. San Francisco: Jossey-Bass.

Takeuchi, H., & Nonaka, I. (2004). *Hitotsubashi on Knowledge Management*. Singapore: John Wiley & Sons.

Werr, A., Blomberg, J., & Löwstedt, J. (2009). Gaining external knowledge – boundaries in manager's knowledge relations. *Journal of Knowledge Management*, *13*(9), 448–463. doi:10.1108/13673270910997114

Westphal, J., & Fredrickson, J. (2001). Who directs strategic change? Director experience, the selection of new CEOs, and change in corporate strategy. *Strategic Management Journal*, *22*(12), 1113–1137. doi:10.1002/smj.205

Zack, M., McKeen, J., & Satyendra, S. (2009). Knowledge management and organizational performance: an exploratory analysis. *Journal of Knowledge Management*, *13*(6), 392–409. doi:10.1108/13673270910997088

Zapata Cantú, L., Rialp Criado, J., & Rialp Criado, A. (2009). Generation and transfer of knowledge in IT-related SMEs. *Journal of Knowledge Management*, *13*(5), 243–256. doi:10.1108/13673270910988088

Zink, K. J. (2008). Comprehensive change management concept: Development of a participatory approach. *Applied Ergonomics*, *39*(4), 527–538. doi:10.1016/j.apergo.2008.02.015

KEY TERMS AND DEFINITIONS

Change Management: Change management emphasizes a reorientation of organizational structures and systems. Change at work corresponds to inculcating a change orientation in employees in relation to their skills, values, attitudes, and behavioral norms and practices.

Combination Phase: The new knowledge is combined with the other explicit knowledge of the organization in the combination phase of knowledge creation.

Explicit Knowledge: Explicit knowledge is easy to communicate. It can be described, for example, in written documents, tapes and databases.

Externalization Phase: The externalization phase transforms the tacit knowledge into explicit knowledge so that it can be communicated.

Internalization Phase: The explicit knowledge created in an organization is internalized in this phase. Learning by doing characterizes the emergence of tacit knowledge in this phase.

Socialization Phase: Socialization is a process of sharing experiences and creating tacit knowledge as shared mental models and technical skills.

Strategic Planning: Strategic planning involves taking a view of the long-term future planning with clearly articulated values, mission, vision and strategic objectives.

Tacit Knowledge: Tacit knowledge consists of the culture of an organization and in the skills, habits and informal decisions of individuals.

Chapter 20
Knowledge Management and Entrepreneurship Cases in India

Neeta Baporikar
College of Applied Sciences Salalah, Sultanate of Oman

EXECUTIVE SUMMARY

Entrepreneurs in general, do contribute to the development of national economies. We need their innovation and risk-taking to create wealth, generate employment and fuel the economy. Hence, knowledge management – both soft and hard is crucial. The benefits of great entrepreneurial venture using the most advanced technology can be nullified by poor knowledge management practices. It is knowledge and management that is the link between success and growth. The key objective of this chapter is to explain different perspectives on what is meant by the Knowledge Management (KM) in relation to entrepreneurship through two cases from India. It is essential that research and policy development fully take into account the differing perspectives of entrepreneurship and knowledge as no single definition fully captures the concepts, nor their underlying assumptions. The chapter focuses on understanding KM as a tool for developing competitive edge and identifies areas of KM application for entrepreneurship success.

INTRODUCTION

As commercial organizations face up to modern pressures to downsize and outsource they have lost knowledge as people leave and take with them what they know. This knowledge is increasingly being recognized as an important resource and organizations are now taking steps to manage it. In addition, as the pressures for globalization increase, collaboration and co-operation are becoming more distributed and international. Knowledge sharing in a distributed international environment is becoming an essential part of Knowledge Management (KM). In this chapter I make a distinction between hard and soft knowledge within an organization and argue that much of what is called KM deals with hard knowledge and emphasizes capture-codify-store. This is a

DOI: 10.4018/978-1-61692-886-5.ch020

major weakness of the current approach to KM. Therefore, new systems are required that not only locate, capture, store, share and leverage data and information, but also knowledge that too soft knowledge.

The chapter focuses on understanding KM as a tool for developing competitive edge and tries to identify areas of KM application for entrepreneurship success. Knowledge that supports the decision making process is an obvious vital resource; however, knowledge has often suffered from under management in the past. It is only in recent years that knowledge has been taken more seriously.

The structure of the chapter is as follows: the first part gives the theoretical backdrop like what is knowledge, relationship of data, information and knowledge, soft and hard knowledge, emergence of entrepreneurial economy and strategy for KM. The second part of the chapter would comprise of understanding the unique challenges faced by the entrepreneurs during the adoption of KM, the switch over, the impact on the working of enterprises. For this part the methodology adopted is as below:

BACKGROUND

What is Knowledge?

According to Webster's Dictionary, knowledge is "the fact or condition of knowing something with familiarity gained through experience or association". In practice, though, there are many possible, equally plausible definitions of knowledge. A frequently used definition of knowledge is "the ideas or understandings which an entity possesses that are used to take effective action to achieve the entity's goal(s). This knowledge is specific to the entity which created it."

An understanding of knowledge requires some grasp of its relationship to information. In everyday language, it has long been the practice to distinguish between information—data arranged in meaningful patterns—and knowledge—which has historically been regarded as something that is believed, that is true (for pragmatic knowledge, that works) and that is reliable.

In recent times, theoretical objections to the concept of truth (e.g. by post-modernists) or to that of reliability (e.g. by positivists) have led to some blurring of the distinction. The interchangeable use of information and knowledge can be confusing if it is not made clear that knowledge is being used in a new and unusual sense, and can seem unscrupulous insofar as the intent is to attach the prestige of (true) knowledge to mere information. It also tends to obscure the fact that while it can be extremely easy and quick to transfer information from one place to another, knowledge is sticky: it is often very difficult and slow to transfer knowledge from person to another. (C.f. the World Bank's 1998 World Development Report on Knowledge for Development which begins with the false assertion that knowledge travels at the speed of light.) In assessing attempts to define knowledge it can be helpful to remember that the human mind has often been seen as capable of two kinds of knowledge—the rational and the intuitive.

In the West, intuitive knowledge has often been devalued in favor of rational scientific knowledge, and the rise of science has even led to claims that intuitive knowledge is not really knowledge at all. However, recognition of the difficulties inherent in transferring knowledge from one person to another has tended to highlight the importance of tacit knowledge e.g. notably in the writings of Polanyi (1975), and Nonaka and Takeuchi (1995). In an effort to distinguish knowledge from mere information, some Western analysts (Karl Erik Sveiby) have tried defining "knowledge" as wholly tacit (i.e. as capacity in action), thus consigning what others have considered as explicit knowledge to mere information

In the East, the tradition has been to celebrate the importance of the intuitive, in comparison with the rational. The Upanishads for instance speak about a higher and a lower knowledge, and associate lower knowledge with the various sciences.

Chinese philosophy has emphasized the complementary nature of the intuitive and the rational and has represented them by the archetypal pair yin and yang.

Debates about the meaning of knowledge have continued for thousands of years, and seem likely to continue for some time to come

Most definitions and explanations of knowledge seem to cover the same vocabulary, concepts and words. Rather than provide a standard definition, the paper addresses the general themes and fundamentals that have become evident in recent years.

- Knowledge goes through a process of sharing tacit with tacit knowledge, tacit to explicit, explicit leverage, and explicit back to tacit.
- Knowledge can be created and tested.
- Knowledge can be distinguished from data and information.
- Explicit knowledge is usually filtered, stored, retrieved and dispersed across the organization.
- A culture that does not foster and reward the sharing of knowledge cannot expect technology to solve its problems (Srinivas 2000).

Relationship of Data, Information and Knowledge

Knowledge is built up from data and information as well as prior knowledge. Data have no meaning or significance in themselves. Examples include individual mental symptoms shown by a particular patient or items in a computer spreadsheet. Information is data which have meaning because of a relational connection. In other words, information is data which have been processed to be useful. Information aims to provide answers to the questions 'Who?', 'What?', 'Where?' and 'When?' It is worth noting that although information is intended to be useful, it is not necessarily so. Merely aggregating data and identifying relationships between variables do not guarantee utility. Knowledge is information to which a process has been applied, which may eventually become expertise (Liebowitz, 2000). It is 'the collation of information for a particular purpose, intended to be useful' (Bellinger *et al*, 1999).

Knowledge aims to answer the question 'How?' Developing new knowledge from that which already exists to answer the question 'Why?' may be defined as understanding. However, while knowledge is a necessary prerequisite for understanding, the availability of appropriate knowledge does not guarantee understanding.

The relationships between data, information and knowledge are summarized in Figure 1. However, this is not necessarily unidirectional, but is often circular or iterative. For example, individual items in a clinical data-set constitute data. When aggregated, they can yield information, which in turn can be appraised and interpreted to give knowledge. However, designing an entrepreneurial data-set includes making a decision about which items are needed and how they should be coded. These decisions are informed by existing knowledge, but they may need revision in the light of new knowledge developed from the use of the data-set. In the same way, reviewing the formulation of an individual entrepreneur's presenting problems (knowledge) may lead to a reappraisal of individual symptoms (data).

Figure 1 also aims to distinguish knowledge management from data management, information management and information technologies. Information technologies, as the term is used in everyday language, are tools which support the management of data, information and knowledge. They include computer hardware and software, storage, indexing and retrieval systems, and so on. Data management and the technologies available to support this are relatively simple. An example would be a simple spreadsheet. Aggregating and analyzing data from such a spreadsheet using statistical software, or bringing data together from different sources using a relational database, illustrate information management. In practice, these technologies clearly overlap. However, this overlap has been exaggerated by conceptual confusion. For example, the main purpose of the KM and Entrepreneurial Strategies appears to be to optimize knowledge management, even though most of the time the focus is on information management and does not make a clear distinction between these two areas of activity. The distinction is important because it helps to highlight the central role in entrepreneurship development methodologies and technologies supporting knowledge management.

Enterprises and entrepreneurs realize that access to quality information and knowledge will help them remain competitive. However, with the advent of rapidly changing business environments,

Figure 1. Data-information-knowledge relationships

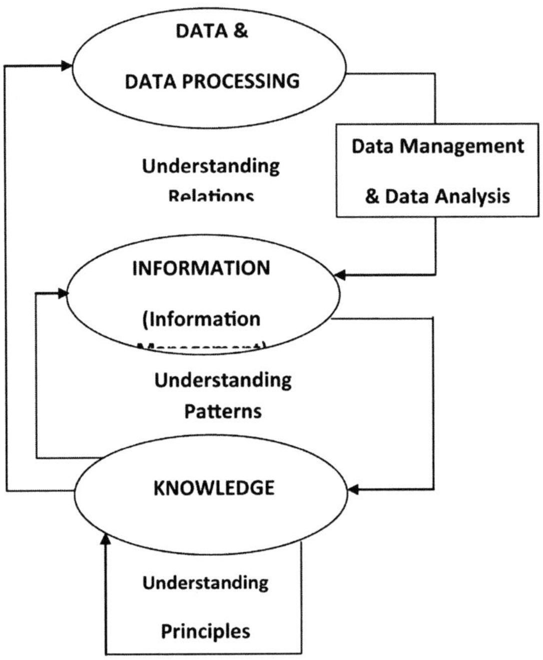

entrepreneurs are now realizing they need to develop an effective knowledge strategy and provide enterprise and employees with best available knowledge to support the decision making process. Data warehousing initiatives, utilizing various data mining techniques, have found common place in many business infrastructures for supporting the decision making process. However, as the vast majority of knowledge exists in the minds, the quality of support these provide, especially for intensive queries, is somewhat uncertain (Nemati, Steiger et al. 2002).

Knowledge management has recently become a fashionable concept, although many entrepreneurs and enterprises are still unable to explain what knowledge is. More importantly, they are unable to develop and leverage knowledge to improve performance. This is due to organizations becoming increasingly more complex in structure, resulting in knowledge that is fragmented, hard to locate, leverage, share and difficult to reuse (Zack 1999). Tacit knowledge is subconsciously understood and applied, difficult to articulate and usually developed from immersing oneself in an activity for an extended period. Explicit knowledge can be easily communicated to others through a system of language, symbols, rules, equations and objects. It consists of quantifiable data, written procedures, mathematical models etc (Nemati, Steiger et al. 2002). An established discipline since 1991 (Nonaka 1991), KM comprises a range of practices used in an organization to identify, create, represent, distribute and enable adoption of insights and experiences. Such insights and experiences comprise knowledge, either embodied in individuals or embedded in organizational processes or practice. Bradley (2007) points out a recent development in the world of business. There has been the emergence of new departments in corporations which carry the word "Insight".

A general definition of knowledge management, adapted from Macintosh (1997), is that it 'comprises the identification and analysis of available and required knowledge, and the subsequent planning and control of actions to develop knowledge assets so as to fulfill individual and/or organizational objectives'. The processes and outcomes of knowledge management are summarized in Figure 2. Within entrepreneurship, knowledge management can be defined as the creation, acquisition, analysis, maintenance and dissemination of knowledge for the benefit of entrepreneurship care and development.

Managing Knowledge: The Challenge

Three major issues facing organizations are globalization, downsizing and outsourcing and all three have implications for knowledge sharing and management. Downsizing and outsourcing (Davenport and Prusak, 1998) mean a reduction in personnel. As people leave, organizations have come to realize that they take with them valuable knowledge. Globalization is a separate issue which affects most organizations in some form (Castells, 1996). Many organizations are now undergoing some form of structural change to cope with the increased internationalization of business. For example, Castells (1996) has observed the emergence of what he calls the Network Enterprise, made up of several organizations of different sizes working together. These changes mean that information and knowledge have to be shared between individuals and companies who perhaps never expected to work together. As globalization impacts upon organizations, they are finding they have to turn to international teams to maintain an essential flexibility (Manheim, 1992). These teams may find themselves operating in different locations, which mean that groups need to share knowledge asynchronously between different locations. There

Figure 2. KM elements

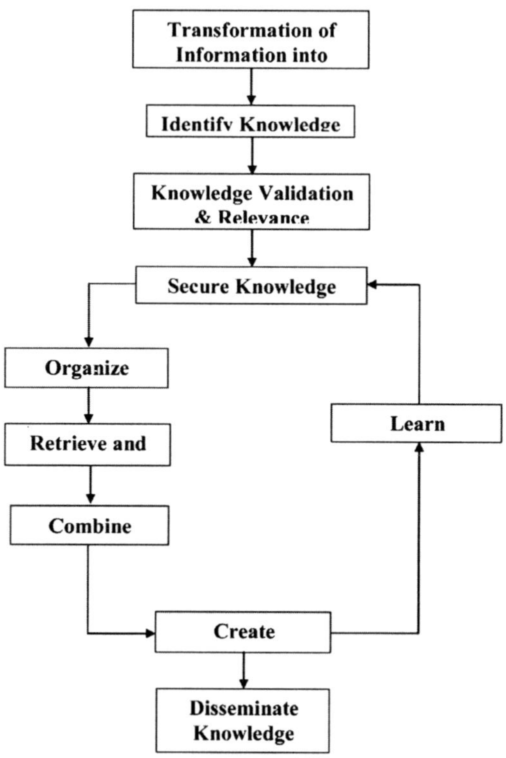

is clearly a need to manage such knowledge and Knowledge Management (KM) can address this. But there is a need to understand the state of KM and distinguish between 'hard' and 'soft' knowledge. Soft knowledge is an aspect of KM that is currently under-explored. Understanding and examining the role of KM for entrepreneurship in international environment is also crucial.

Managing Hard Knowledge

Much of the KM literature has a common view of knowledge that continues to concentrate on the capture-codify-store cycle of management. In this sense, KM does not seem to have moved on from what was previously termed Information Management. For example, the view of knowledge as being 'hard', that is codifiable, has led to attempts to extract knowledge from one group of 'experts' so that it can be used by another, less skilled, group. However, the results of such expert systems have been disappointing (Roschelle, 1996; Davenport and Prusak, 1998).

Another 'hard' knowledge approach aims to support, as opposed to replace, the knowledge worker. With this approach, knowledge is codified into operating procedures or other forms of instruction for action. Orr (1990) reported a study of copier repairers who had manuals containing the procedures to be followed when repairing a copier. These were laid down by the designers and catered only for the problems foreseen by them. However, there were occasions when problems occurred that were not covered by the procedures. The repairers tackled such problems by creating 'workarounds'. Workarounds in this context constituted an example of what we refer to as soft knowledge.

Despite its evident problems, the management of 'hard' knowledge is now well established and there are many tools and frameworks available for this form of KM. The soft knowledge embedded in the day-to-day working practices of groups is much less amenable to a capture-codify-store approach. Some researchers have begun to recognize the challenges raised by soft knowledge (e.g. Macintosh (1998); Buckingham Shum, 1997) but there is a need to understand more fully the nature of soft knowledge and the means by which it might be managed

Managing Soft Knowledge

There is a wide body of literature that suggests that there are 'softer' types of knowledge (Nonaka, 1991; Kogut and Zander, 1992). This knowledge is less quantifiable (Kidd, 1994; Skyrme, 1998) and cannot be captured, codified and stored so easily. Examples of such knowledge might include tacit knowledge that cannot be articulated, internalized experience and automated skills, and internalized domain knowledge and cultural knowledge, and embedded in practice.

Soft knowledge is acquired through the praxis of work and consequently when an organization loses staff, the soft knowledge that is lost cannot easily be replaced. As companies have cut out layers of middle management they find that they have lost the people who knew whom to approach for specific problems; how to deal with different people and who best to use for different tasks. In short, people who knew how to make things happen. The loss of such personnel creates a problem for organizations as they move to cheaper, less knowledge-rich, workers.

As a first step towards the management of such knowledge we need to understand the social processes that govern its construction and its sustenance in an organization. Lave and Wenger (1991) suggest that soft knowledge is created, sustained and shared through communities of practice by a process called legitimate peripheral participation (LPP). They describe how groups are regenerated by newcomers joining and eventually, replacing existing members. The newcomers learn from "old-timers" through co-practice that is graduated, permitting them to undertake more central and critical tasks. In so doing, they not only learn the domain skills associated with the practice but they also learn the language of the community, its values and its attitudes. Through this kind of participation newcomers move from peripheral positions to more central ones and in so-doing are transformed into old-timers. Membership is legitimated though participation and participation is legitimated through membership.

Seely Brown and Duguid (1991), have extended Lave and Wenger's community of practice model and applied it to technological communities. Their example is based on Orr's description of the work of copier repairers given above (Orr 1990, 1997). When a problem could not be solved by adherence to the manual, or when newcomers had difficulties, they would enlist the help of colleagues. By applying their shared experience, they would arrive at a solution to the problem. But such solutions were not then forgotten, the new knowledge was shared with other members by what Orr describes as the telling of 'war stories'. War stories not only represent the soft knowledge of the community, but their telling also serves to legitimate a newcomer as they move from peripheral to fuller participation. Soft knowledge is embedded in the practices of, and relationships within, the group. Secondly, the source of the legitimacy of the knowledge differs from hard knowledge. 'Hard knowledge' is accepted as legitimate by virtue of the formal authority of the designer of the system or the author of the procedure. Soft knowledge becomes accepted by virtue of informal authority and consensus within the group. Although newcomers might have a degree of hard domain knowledge, their soft knowledge only develops as they move from being newcomers to fully-fledged members of the community.

Thus, KM efforts typically focus on organizational objectives such as improved performance, competitive advantage, innovation, the sharing of lessons learned, and continuous improvement of the organization. KM efforts overlap with Organizational Learning, and may be distinguished from that by a greater focus on the management of knowledge as a strategic asset and a focus on encouraging the sharing of knowledge.

KM efforts can help individuals and groups to share valuable organizational insights, to reduce redundant work, to avoid reinventing the wheel per se, to reduce training time for new employees, to retain intellectual capital as employees turnover in an organization, and to adapt to changing environments and markets (McAdam & McCreedy 2000)(Thompson & Walsham 2004).

Duality of Knowledge

If we view knowledge as a duality then by implication, all knowledge is to some degree both hard and soft: it is simply that the balance between the two varies as shown in Figure 3. This approach need not exclude some of the categorizations discussed earlier: the softer aspects of knowledge are those that cannot be externalized and the harder aspects are those that can be articulated, captured and stored. However, viewing knowledge as a duality means that both perspectives are needed and both must be taken into account in any attempt to manage knowledge.

Nonaka contends that tacit knowledge is hard to formalize and therefore difficult to communicate: it is (in Polanyi's terms) knowledge that is not at the forefront of consciousness. However, that which is tacit varies from one situation to another. For example when reading a text, words and linguistic rules function as tacit subsidiary knowledge while the attention of the reader is focused on the explicit meaning of the text. In another situation, those same linguistic rules might become explicit knowledge.

Although under certain circumstances tacit knowledge can become explicit, it is perhaps more accurate to say that even then only part of what is tacit is made explicit. This is because even what we normally think of as explicit knowledge has a whole history of culture, conventions of language and cross-referencing of thought that are never

Figure 3. Duality of knowledge balance

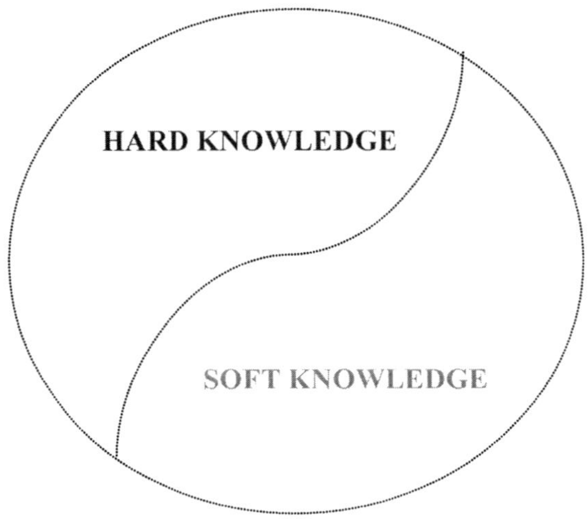

made explicit. There is always more that is embedded, implied, assumed and presupposed than can ever be externalized and made explicit. In practice, the tacit and explicit dimensions to knowledge are inexorably and inextricably interwoven.

Viewing knowledge as a duality can help to explain the failure of some KM initiatives. When the harder aspects are abstracted in isolation the representation is incomplete: the softer aspects of knowledge must also be taken into account. As they put it, 'the important knowledge is all in people's heads', that is the solutions on the server only represents the harder aspects of the knowledge. For a complete picture, the softer aspects are also necessary. Similarly, the expert systems of the 1980s can be seen as failing because they concentrated solely on the harder aspects of knowledge. Ignoring the softer aspects meant the picture was incomplete and the system could not be moved from the environment in which it was developed.

However, even knowledge that is 'in people's heads' is not sufficient - the interactive aspect of 'knowing' must also be taken into account. This is one of the key aspects to the management of the softer side to knowledge. The notion of knowledge as a duality is clearly useful, but this in turn leads to new problems. Seeing KM as including the management of both hard and soft knowledge poses a different set of challenges for entrepreneurs dedicated to developing workable KM solutions in their enterprises.

Finerty (1997) points out that technology has a role to play, but that the emphasis needs to move from trying to package knowledge as an object to using technology as a way of sharing experience. This view is supported by Davenport and Prusak who emphasize the potential of technology as a means to create links between people:*...the more rich and tacit knowledge is, the more technology should be used to enable people to share that knowledge directly. It's not a good idea to try and contain or represent the knowledge itself using technology.* Information Technology alone is not a satisfactory solution and many authors (e.g. Leonard & Sensiper, 1998; Kimble, *et al.*, 2000) emphasize the continuing need for physical contact. However, even with physical contact, there remain difficulties in the management of softer knowledge, for example, the importance of language - not just national language but also professional or technical language.

Clearly there needs to be a shift from simply capturing and leveraging knowledge to supporting learning and the sharing of knowledge. Capturing and storing the harder aspects of knowledge in procedures, books or reports and then disseminating them is now well established in the field of information resource management and is part of the solution. But how can it be applied for entrepreneurship? How can it be managed if it cannot be captured and stored? What happens when there is no opportunity for situated learning, and what happens when entrepreneurs are not co-located? Viewing knowledge as a soft-hard duality begs the question - how can this duality be managed when entrepreneurs as a group are where the softer aspects of knowledge are created, nurtured and sustained.

The Strategy of Knowledge Management

The first and perhaps most difficult of launching a knowledge management program is to put in place a strategy for sharing knowledge. It entails a collective visioning as to how sharing knowledge can enhance organizational performance, and the reaching of a consensus among the senior management of the organization that the course of action involved in sharing knowledge will in fact be pursued. Implicit in such a process is a set of decisions about the particular variety of knowledge management that the organization intends to pursue, including:

What knowledge to share? Knowledge-sharing programs may aim at making available various types of content. The program will be very different depending on whether the intent is to share know-how, best or good practices, or knowledge of clients or customers, or competitive intelligence, or knowledge of processes. The knowledge-sharing program will differ considerably depending on the type of knowledge being shared. Comprehensive, organization-wide programs for sharing knowledge typically emerge when the organization's know-how is perceived as critical to its mission, where the value of the organization's knowledge is high, and where the enterprise is geographically dispersed. Some examples: the knowledge management strategies and programs at Ernst & Young, Arthur Andersen, and the World Bank. In other cases, knowledge sharing programs are limited to a specific function, such as sales and marketing, or research, or a specific area of expertise such as engineering.

The question of "what to share" includes not only the type of knowledge, but also its quality. In organizing knowledge-sharing programs, it is common to put processes in place to ensure that the content that is shared reaches a certain minimal threshold of value and reliability. Some programs make no explicit distinction between different levels of reliability of the material offered, once the initial threshold has been met, thus allowing users to reach their own conclusions as to its ultimate value. Other programs, particularly those that offer external knowledge sharing, provide explicit guidance on whether the material has been authenticated, so that users can make inferences about its reliability. Most knowledge-sharing systems also allow in varying degrees the inclusion of new and promising ideas that have not yet been authenticated and in this sense is not yet knowledge.

Choices about what knowledge to share must go beyond generic prescriptions. The local context is often unpredictable; hence knowledge of the local context and local know-how become very important. Often in complex environments, problems will appear as wicked problems, in which the issue is not so much finding the answer to a defined problem, as it is one of defining the problem to which one is trying to solve. Once the problem is successfully defined, the answer may be obvious.

In entrepreneurial cases one of the major decisions concerns is intended beneficiaries of the knowledge-sharing system. Knowledge sharing programs may aim at sharing with either an internal or an external audience. Internal knowledge sharing programs typically aim at making the existing business work better, faster or cheaper, by arming the front-line staff of an organization with higher quality, more up-to-date and easily accessible tools and inputs to do their jobs, and so add value for clients or save costs. External knowledge sharing poses greater risks than internal sharing programs — raising complex issues of confidentiality, copyright, and in the case of the private sector, the protection of proprietary assets — but it may also offer greater potential benefits.

There needs to be a consensus within the entrepreneurs class as to the principal channels by which knowledge will be shared, whether face-to-face, or by way of help desks, by telephone, fax, email, collaborative tools or the web, or some combination of the above. It is important not to ignore face-to-face communications, since this is still the best and highest quality to transfer knowledge between individuals.

Why will knowledge be shared? Knowledge management is not something that is undertaken for its own sake, but rather something that supports the business of the organization. Reaching explicit agreement as to why knowledge is being shared, and its likely contribution to the enterprise performance, is crucial to sustaining support over the medium term. These motivations may relate to:

- increasing speed,
- lowering costs of operation,
- accelerating innovation, or
- widening the client base.

Since pursuing all of these worthy objectives simultaneously may result in a failure to achieve any of them, it will be useful to make an explicit choice about objectives from the outset.

New firms are usually small and pose no oligopolistic threat in national or international markets. In the model of the entrepreneurial economy, no external costs – in the form of higher prices – are imposed on consumers in the national economy as is the case in the model of the managed economy. Fostering local economies imposes no cost on consumers in the national economy.

Entrepreneurs differ in the degree and type of novelty that they introduce to the economy. But why should entrepreneurs share knowledge or information? This part examines why entrepreneurs share when others can steal their ideas. I present an information sharing framework which builds on the assumption that ideas are abundant and of predominantly poor quality. Rather than hold ideas as stable constructs, the framework examines the case where profiting from innovation is related to the complementarities of specific assets, facing the "lemons" selection problem under conditions of novelty.

A fundamental concern to entrepreneurs, and academics alike, is that someone will steal their idea. This concern is so pervasive it is represented across a broad range of theoretical disciplines, from utilitarianism theories of social welfare and labor economics, to philosophical theories of justice, personhood, and individual identity (Fisher, 2000). In information economics, the concern for the theft of ideas is named the fundamental paradox of information (Arrow, 1971). Arrow's fundamental paradox argues that there is the problem of opportunism in the sharing information as information must be shared to be valued, but after it is shared it has no additional value (Williamson, 1985).

The problem of Arrow's fundamental paradox has been recognized in some form for hundreds of years (Granstrand, 2000), and is reflected in a broad range "intellectual property rights" theories and practices. It argued that when intellectual property rights are available and strong, the problem of information sharing can be ameliorated and the entrepreneur can benefit from a "market for ideas", rather than from product markets (Gans & Stern, 2003). Yet, in most industries, formal intellectual property rights play a small role in the appropriation of innovation (Cohen, Nelson, & Walsh, 2000; Levin, Klevorick, Nelson, &

Winter, 1987), and many firms "freely reveal" much about their innovations to rivals (Harhoff, Henkel, & von Hippel, 2003). Similarly, information sharing from cold calling is pervasive, from entrepreneurs (Birley, 1985) through to venture capital firms (Shane & Cable, 2002).

While histories of entrepreneurship are replete with stories of stolen ideas, they tend to focus on the hazards of information sharing rather than the motivation for sharing. Implicitly, these histories have adopted two simplifying assumptions. First, history typically treats ideas as stable constructs. Implicit in this notion is that entrepreneurs did not update with new information, even under early conditions of uncertainty. Second, the literature treats ideas as valuable. Implicit in this notion is that ideas were scarce and were of relatively low cost to assess.

SETTING THE STAGE

While business schools thrive training young people for jobs in large scale operations, these same schools house researchers establishing a revival of small-scale operations. In the late 1980s and early 1990s fascinating data material is published: the share of smallness varies in modern economies, but increases everywhere. In the United States the average real GDP per firm increased by nearly two-thirds between 1947 and 1989 – from $150,000 to $245,000 – reflecting a trend towards larger enterprises and a decreasing importance of small firms. However, within the subsequent seven years it had fallen by about 14 percent to $210,000, reflecting a sharp reversal of this trend and the re-emergence of small business (Brock and Evans, 1989). Similarly, small firms accounted for one-fifth of manufacturing sales in the United States in 1976, but by 1986 the sales share of small firms had risen to over one-quarter (Acs and Audretsch, 1993).

Such a U-shaped relation between number of firms and time, or inverse U-shaped relation between average firm size and time, seems to be ubiquitous. There is much debate about its meaning, but two things seem evident: the trough, or the summit, is not determined by the calendar year but by the level of economic development of a country. It is as if the trough, or the summit, marks a regime switch. The first can be best illustrated using the material of the Global Entrepreneurship Monitor (GEM). The second is documented by Audretsch and Thurik (2001a and 2004) distinguishing between the managed and the entrepreneurial economy.

CASE DESCRIPTION

The case studies were carried out for entrepreneurs located in India. Secondary sources used are from published records either from government or registered associations which are focusing on development of entrepreneurship like Chamber of Commerce, Entrepreneurship Development Cell, and Confederation of Indian Industry etc. Information and data collected to make assessment has been suitably analyzed.

The respondents were from Pune, Mumbai and Hyderabad. These places being highly developed regions of India in terms of industrial development, education and financial services with excellent climatic conditions and has seen a rise in entrepreneurial activities in the past few years especially in knowledge based industries.

The case studies have been undertaken from India where various entrepreneurs especially in knowledge based industries are setting their shops. The study has been conducted in two parts.

The first part was a survey collecting factual information. A questionnaire was issued to 50 units and 19 were returned (a response rate of 38%). The questionnaire looked for respondents who:

- are in regular contact with entrepreneurs/peers doing similar jobs;
- talk with colleagues to solve problems;

- share projects with others;
- swap anecdotes/experiences;
- learn from discussions..

These five metrics were used as indicators of membership in communities of practice. Using the above it was possible to identify potential examples of soft knowledge being shared in a community of practice. The second stage consisted of semi-structured interviews with 6 entrepreneurs. The aim of the interviews was to obtain richer data regarding the types of KM practices in vogue, issues and challenges for entrepreneurs in KM and strategies adopted for betterment of entrepreneurship.

Case 1. Ideaforge Birth: Formal Approaches to KM

As a kid, he dreamt of the impossible. When he saw Superman, he wanted to be like him and do whatever he did. At school, his teacher told him his ideas were good, but there is still a lot more to learn. He started thinking about the mechanics of engines very early in life, wondering how they worked and how processes could be changed to make things work better.

His father used to have long chats with him everyday on the achievements of great inventors like Albert Einstein and Thomas Edison. This made a profound impact on him and fuelled his imagination further. 'Why can't I be an Einstein?' he wondered. The ability to question and probe everything was his motivation to keep moving forward.

This deep interest in engineering brought Ankit Mehta to the Indian Institute of Technology. Ankit, a top ranked student at IIT Bombay, did not attend a single placement interview when his friends got the greatest offers from the best of companies. For him, it was a pursuit of ideas, a journey to achieve what he had always dreamt of. "The journey was tough, it took us an year to actually give shape to the start-up company. The lack of experience, opposition from parents, and financial problems made it tougher to get started," Ankit says.

But the strong belief in the idea and support from friends, Rahul Singh and Ashish Bhat, and a grant from the government led to the birth of Ideaforge in 2007 when Ankit Mehta was just 24 years old. Ideaforge got started with its pioneering work at the Technology Business Incubator, SINE at IIT Bombay. "I have been very fortunate to get like-minded people like Ashish, who has been innovative and designing electronic gadgets ever since he was a school kid, and Rahul who gave up a plum post to be a part of this start-up," Ankit says.

Today, Ideaforge develops human powered cell phone chargers and fully automated Micro Air Vehicles (MAVs). With these chargers, you can charge your phone while in a car, bike, etc. Ankit lives with an idea bank and that's why he calls his company Ideaforge. "The joy of creating a product gives the satisfaction which no high-paying job can," he says happily. It gives him immense pleasure to work even if it means working till the wee hours. For him, every day is a chance to strive harder and make his dreams come true. He has no regrets about giving up the luxuries, which many of his classmates are used to. "A start-up is a great way to be selfish," he adds.

Ankit Mehta, managing director, Ideaforge, reveals more about the exciting world of entrepreneurship and also gives hints regarding formal approached to KM.

With an initial investment of Rs 1.5 lakh (Rs 150,000) and a government grant. The company was registered in February 2007 and it got incubated at IIT.

The day starts at 10-11 and our work goes on till 1-2 a.m. With a team of 10 members, the core team working late the work goes on. People see it as a great opportunity to work in a start-up. It gives them a sense of fulfillment, which no other job can give. There are parallel roles, people love the challenge. The risk and rewards too are equally exciting.

Education at premier institution like IIT plays a huge role in overall development. It boosts confidence, to become better than the best. The environment is very competitive, yet friendly. The atmosphere opens up the brain so much. Education plays a vital role. Education gives tremendous confidence and knowledge is essential for making products.

The effort is to popularize clean technology products and lead the way to an environment-friendly world. Their mission is to make energy sources available everywhere.... be mobile.

The amount they invested less than Rs 1 crore (Rs 10 million) so far. They have developed a mechanical energy storage and release device and applied for a patent. It will take about 18 months to get the patent for this. They have about seven different types of charging devices, which include a mechanical charger, a USB charger, a bike charger, a car charger and even a wall charger. These chargers can be used with any handset. They have also developed fully automated Micro Air Vehicles (MAVs) like a small helicopter that can fly, take photos and videos. While on the on GPS-hold mode, it acts like a virtual pole in the sky streaming high quality live video to the ground station with all camera controls in user's hand. These can be trained to even go inside buildings and take photos. In the age of increased surveillance, these products come in handy.

The major milestones achieved are of establishing a good distribution network. Their national distributor is Brightpoint India Pvt Ltd. Today, products are available through major retail chains like eZone, Hot Spot, RPG Cellucon, The Mobile Store and Exide's retail outlets and sites like eBay, Indiatimes, homeshop18, et cetera. They have also started a pilot project with the Tatas so products will be available in Tata stores as well and will soon tie up with Airtel. They have been able to reach out to the rural areas with a tie-up with ITC's e-choupal.

Innovation is vital for any entrepreneurship to succeed and it is absolutely essential to create value with new ideas and offer a product that the market requires. An idea pays and it is the stepping stone to success. The product should have a good market value. These are the issues of utmost importance for increasing the rate of entrepreneurial growth. The most important things that entrepreneurs must have to succeed are patience to go on, perseverance of the goal. Impatience also as it helps to get things done faster. A great team is essential. One must be able think differently. One has to believe in the idea and it should, of course, have a logical end. Make the right use of opportunities. There are billion ways of doing the same thing. One must make the right choice. Make sure to get finance and see to it that there is someone to bail you out in the event of a crisis. Ability to listen to others' ideas, and be receptive to new ideas is the key to succeed and for these ingredients there is a need to adopt the formal approach to knowledge management

Case 2. A Nutritious Business: Benchmarking for KM Adoption

At 17, when most girls of her age were set to go to college for higher studies, Sarala Bastian, now a successful entrepreneur got married.

But Sarala Bastian did not wish to just be confined to her house. She wanted to do something on her own. She wanted to be independent and carve out identify for herself. "My father gave me the initial capital of Rs 15,000 to start a business in 2004. I started a mushroom farm in my backyard," says Sarala. There has been no looking back since.....today; she is a successful entrepreneur in a field not explored by many. It has been an incredible journey for the 32-year-old Sarala. To add to her glory, she has been short listed for the prestigious Youth Business International (YBI) Entrepreneur of the Year award for 2009. YBI is a UK-based network established by Prince Charles to encourage young entrepreneurs. Ghada Ba-ageel of Saudi Arabia and Linet Indiazi of Kenya are the other two nominees in the female category for this year's award. The award, to be announced on November 16, will comprise a cash prize of $1,000, a certificate and a medal.

Here's how Sarala transformed herself from a simple housewife into a successful entrepreneur with minimum resources at her disposal and has been facilitating benchmarking for KM adoption to small entrepreneurs.

The Beginning Years

She spent her childhood in Coimbatore where her father worked as an auditor in the state co-operative society. In school, she learnt typewriting and shorthand. She was married to Sebastian after passing class 12 exams and so came to Chennai with him. She has two daughters, elder one is 14-years old and seven year old younger daughter is mentally challenged since as a baby, she had brain fever and life changed for the worst because of this.

Becoming an Entrepreneur

There never was any financial difficulty and she was happy with what husband earned. But it was a dream to start something on their own, be independent and offer jobs to other needy women. That is why 95 per cent of employees are women. The dream is to make all women financially independent. The idea to cultivate mushrooms came in mind as she felt it was a very 'nutritious' business and had not been attempted by many.

Funds to Start the Business

She got a lot of support from my father who gave the initial capital of Rs 15,000 to start the business in 2004. She started a mushroom farm in backyard. Before starting mushroom farming, she attended training classes at eleven different places, including the agricultural college in Coimbatore.

The Art of Cultivating Mushrooms

It was very difficult to get seeds in those days, bought one seed for Rs 60 and got started. The seeds are kept in a perforated plastic cylinder between layers of cleansed hay. Within 17 days, mushrooms start coming out of the holes in the cylinder. From one cylinder, they get about 2 kg of mushrooms.

Initial Struggles

There was a time when she used to go to the supermarkets and beg them to keep their mushrooms in their shops. They used to reluctantly agree after a lot of persuasion. Today, early in the morning, they call and ask, 'Madam, how many packets would you send today?' It takes time for any business to flourish, but ultimately you succeed. Within three months of starting business, they made a profit of Rs 65,000. Whatever profit have made so far, they have put back into the business. More than making money, the aim is to expand the business. Besides fresh milky and oyster mushrooms, they also sell processed foods like pickles, soup powder, jam, et cetera. The turnover is around Rs 250,000 every month these days. Now, the aim is to inspire more women. So, she meets many self-help groups and asks them to do something worthwhile.

Working with Jail Inmates

Some college students from Oorambakkam (near Chennai) wanted her to teach them mushroom cultivation for a project. Later, she found that the students had adopted a village and they wanted her to teach the villagers too. She decided to teach them free of cost. When these college students were planning to train jail inmates, she went to meet the female inmates of Puzhal prison in Chennai. Initially, though a bit scared she was able to convince them to work hard on their own, after they come out and only after they promised, she agreed to teach them how to cultivate mushrooms.

The process is really simple, anyone can do it. She plans to give them regular work, and the process for the same is on.

International Award

It was the Bharatiya Yuva Sakthi Trust that had been helping with lot with business advise from the time she started my enterprise who recommended for her name to the Youth Business International (YBI) Entrepreneur of the Year award for 2009. About 50 names from India had been sent and she is the only Indian short listed from the thousands of names they got from all over the world.

CURRENT CHALLENGES FACING THE ORGANIZATIONS

The challenges faced by the organizations are mostly in the area of developing innovative approaches and application of knowledge management techniques. They also have limitations in terms of adapting to technological changes which are at a faster rate than their own ability to cope up due to being small in size of operations.

Entrepreneurial knowledge is an important construct because it shifts attention from "hoarding information" theories (Aldrich, 1979; Campbell, 1960) about appropriability to theories which address specific asset configurations in addition to novel information. There are a number of reasons why this is useful. First, a "more novel information is better" approach is subject to increasing misinformation costs (Ackoff, 1967) and search costs (Fiet, 2007). Second, entrepreneurial knowledge anchors novelty in a place where it is more tangible (Schumpeter, 1934) and therefore easier for both researchers and entrepreneurs to assess (Carlile, 2002). Finally, entrepreneurial knowledge shifts attention from the *ex post* moral hazard problem of information disclosure (Arrow, 1971), typically associated with agency and monitoring (Jensen & Meckling, 1996), to the *ex ante* selection problems associated with entrepreneurial markets (Akerlof, 1970) and development (Schumpeter, 2005).

This of course is subject to number of limitations. First, the historical record regarding the Kearns' windshield wiper is complex and continues to be contested. Second, the construct of entrepreneurial knowledge has not been directly observed as much as it has been inferred from anomalies in specific cases. Third, the cases chosen are not representative of the population in general, and may not even be representative of many entrepreneurial processes. The cases are chosen to be illustrative and to ground the theory in more tangible phenomena. However, the role of entrepreneurial knowledge is important because it provides a pragmatic and theoretical rationale for the sharing of information under hazards of opportunism.

KM Facilitates Entrepreneurial Innovativeness

Entrepreneurial innovativeness depends both on individual factors *and* on the environment in which the individual acts. In particular, high educational attainment, unemployment and a high degree of self-confidence are significantly associated with entrepreneurial innovativeness at the individual level. Furthermore, the distribution of innovative and imitative entrepreneurship varies across countries. Entrepreneurs in highly developed countries are significantly more likely to engage in innovative rather than purely imitative activities.

Why are some entrepreneurs more innovative than others? This question surely belongs to the core of entrepreneurship research (Shane and Venkataraman 2000). Surprisingly, our knowledge about what exactly leads to innovative rather than purely imitative business ventures is still very limited. To some extent, this may be due to a lack of suitable empirical data to investigate this question. Studying different types and degrees of entrepreneurial innovativeness obviously poses some challenges to identify a relevant population and to define, disentangle, operationalize and empirically measure the concepts of entrepreneurship and innovation in a precise way. In addition, it constitutes a theoretical challenge because it requires researchers to think about and take a stand on the nature and origins of entrepreneurial opportunities and the question why some rather than other individuals exploit these opportunities. Until today, no comprehensive theory is available that answers these questions. Arguably, a more problematic issue is that we still lack a common understanding of what entrepreneurship, innovation, and opportunity actually mean (Davidsson 2005, Koppl 2007, McMullen et al. 2007).

SOLUTIONS AND RECOMMENDATIONS

Many histories of technology entrepreneurs have demonstrated that the firms who are the first to commercialize new product or process innovations often profit less than competitors and imitators (Schumpeter, 2001; Teece, 1986). Gans and Stern (2003) extend Teece's classical analysis to propose that the principle challenge that technology entrepreneurs face is that the owners of complementary assets have the greatest incentives to imitate the innovator. They argue that when innovation value is reinforced by existing complementary assets, innovators will use a "marketplace for ideas" (p.341). When the technology is excludable, innovators will share information when they can increase their bargaining power by "credibly threatening to pursue an idea with a third party" (p. 338). And when the technology is not-excludable, innovators will choose to cooperate when "incumbents foster a reputation for ensuring mutual advantage" (p. 334).

In the knowledge management literature, the novelty of information is represented as information associated with novel action (Nonaka, 1998). Some knowledge, such as explicit knowledge can be readily transferred to others, whereas other knowledge is tacit and difficult to imitate. The implication of this view is that individuals possessing more tacit knowledge may accrue advantages from novel knowledge that is difficult to codify. Both the information economics and knowledge management literature objectify novelty as a construct associated with an individual actor. However, knowledge is more than a fad; it is now at the centre of an organization's strategic thinking. The essence of any knowledge management strategy can be summed up by the following quote, from Drucker (Drucker 1993) "*A company's key to success resides not so much in its work and capital as in the capacity to treat knowledge, corporate knowledge, be it explicit or tacit.*"

Despite all the inherent difficulties of such research work and without claiming to resolve them, the aim here is to contribute to the entrepreneurship literature by providing some theoretical considerations and empirical evidence that helps us better understand the emergence of different types and degrees of entrepreneurial innovativeness. The conceptual approach is based on a judgment and decision making framework that analyses factors influencing individual decision making combined with additional insights that help us understand where different decision alternatives come from. Knowledge as an input into economic activity is inherently different from land, labour, and capital. It is characterized by high uncertainty, high asymmetries across people, and high transaction costs. An economy where knowledge is the main source of competitive advantage is more consistent with the model of the entrepreneurial economy. Apart from these the entrepreneurs who adopt KM will certainly derive the following benefits:

- Avoidance of costly mistakes
- Sharing of best practices
- Faster problem solving
- Faster development time

- Better customer solutions
- Opportunities for new business
- Improved customer service

A broad range of thoughts on the KM discipline exists with no unanimous agreement. Approaches vary by author and school. As the discipline matures, though the academic debates have increased regarding both the theory and practices of KM, the following perspectives need further research:

- **Entrepreneurial and organizational** - with a focus on how an enterprise design at business plan level itself to facilitate knowledge processes best.
- **Holistic issues** - ecological, environmental, ethical and sustainable aspects with a focus on the interaction of people, identity, cross culture and knowledge.

A Personal Reflection

The motivation of this paper is to draw attention to important issues of KM in entrepreneurship. It reflects the need to store not just different forms of knowledge, but different types of knowledge. However, it should be remembered that an over-emphasis on technological aspects of KM might force an enterprise to concentrate on knowledge storage, rather than knowledge flow. New insights and opportunities are available to entrepreneurs if they are able to integrate knowledge across shared and different contexts. The Internet has enabled the creation of virtual communities, networked through technologies only available just a few years ago. As the Internet is becoming the standard form of collaboration, the trend of the e-knowledge network looks set to continue. While technology can greatly enhance an enterprise's knowledge management strategy, it does not necessarily ensure an enterprise is managing its resources and capabilities in the right way. However, technology is vital to enable the capturing, indexing, storing and distribution of knowledge across and with other organizations. Knowledge can be viewed in a number of other contexts, it is vital each is addressed if an organization is to improve performance.

- Successful knowledge strategies depend on whether organizations can link their business strategy to their knowledge requirements. This articulation is vital to allocating resources and capabilities for explicating and leveraging knowledge.
- The competitive value of knowledge must be addressed to assess areas of weakness. Strategic efforts should be made to close these knowledge gaps to ensure the organization remains competitive. The strategic value of knowledge should be addressed, focusing on the uniqueness of knowledge.
- Finally, an organization should address the social aspects affecting knowledge initiatives, namely cultural, political and reward systems. Beyond the entrepreneurial and management roles proposed herein, the environment should promote co-operation, innovation and learning for those partaking in knowledge based roles.
- Preparing this paper has itself been a useful personal experience of knowledge management and entrepreneurship. Much of the literature I found was from unfamiliar sources and it was striking that, despite the clear importance of knowledge management to entrepreneurship, most of the literature referred to business settings. In appraising this literature, I found it very difficult to apply the standard methods used. Two main factors influenced the decisions about what to include in the paper. First, some concepts and/or authors were cited in multiple sources, suggesting that they had been appraised as useful and relevant. Second, and even more important, it was possible to identify examples from enterprises, venture setting up practice, indicating that they at least had face value.

This process has certainly given me new knowledge.

REFERENCES

Acs, Zoltan J. (ed.) *Are Small Firms Important?* Boston. Kluwer Academic Publishers, 1999.

Aldrich, H. (1979). *Organizations and environments*. Upper Saddle River, NJ: Prentice-Hall.

Audretsch, D. B. (2007a). *The Entrepreneurial Society*. Oxford, UK: Oxford University Press. doi:10.1093/acprof:oso/9780195183504.001.0001

Audretsch, D. B., Grilo, I., & Thurik, A. R. (2007). *The Handbook of Research on Entrepreneurship Policy*. Cheltenham, UK and Northampton, MA, US: Edward Elgar Publishing Limited.

Baumol, William. *The Free Market Innovation Machine: Analyzing the Growth Miracle of Capitalism*. Princeton: Princeton University Press, 2002.

Bhide, Amar. *The Origin and Evolution of New Businesses*. New York: Oxford University Press, 2000.

Birch, D. (1987). *Job Creation in America*. New York: Free Press.

Birley, S. (1985). *The Role of Networks in the Entrepreneurial Process*. Cranfield School of Management.

Brush, Condida *et al.*, *The Diana Project: Women Business Owners and Equity Capital: The Myths Dispelled*. Kansas City. Kauffman Center for Entrepreneurial Leadership, 2001.

Davenport, T., & Prusak, L. (1998). *Working knowledge. How organizations manage what they know*. Cambridge, MA: Harvard Business School Press.

Drucker, P. (1992). The new society of organizations. *Harvard Business Review*, *70*(Sept-Oct), 95–104.

Florida, Richard. *The Rise of the Creative Class*. New York: Basic Books, 2002.Nonaka, I. (1991). The knowledge creating company. *Harvard Business Review*, *69*(Nov-Dec), 96–104.

Nonaka, I., & Takeuchi, H. (1995). *The Knowledge-Creating Company: How Japanese Companies Create the Dynamics of Innovation*. New York: Oxford University Press.

Petzinger, Thomas. *The New Pioneers*. New York: Simon and Schuster, 1999.Polanyi, M. (1967). *The tacit dimension*. London: Routledge and Kegan Paul.

Smilor, Ray. *Daring Visionaries: How Entrepreneurs Build Companies, Inspire Allegiance and Create Wealth*. Holbrook, MA: Adams Media, 2001.

Wilson, T. D. (2002) "The nonsense of 'knowledge management'" *Information Research*, 8(1), paper no. 144 [Available at http://InformationR.net/ir/8-1/paper144.html]

World Bank. (1999). *World Development Report: Knowledge for Development. Washington, D.C.Piore, M.J. & C.F. Sabel, (1984), The Second Industrial Divide: Possibilities for Prosperity*. New York: Basic Books.

ADDITIONAL READING SECTION

Acs, Z. J., & Audretsch, D. B. (1993). Conclusion. In Acs, Z. J., & Audretsch, D. B. (Eds.), *Small Firms and Entrepreneurship; an East-West Perspective*. Cambridge, UK: Cambridge University Press.

Allee, V. (1997). *The knowledge evolution. Expanding organizational intelligence*. Boston, MA: Butterworth-Heinemann.

Argyris, C. (1998). Teaching smart people how to learn. In *Harvard Business Review on Knowledge Management* (pp. 81–108). Boston, MA: Harvard Business School.

Audretsch, D. B. (2007b). Entrepreneurship capital and economic growth. *Oxford Review of Economic Policy*, *23*, 63–78. doi:10.1093/oxrep/grm001

Audretsch, D. B., & Thurik, A. R. (2001b). Globalization and the strategic management of regions. In Audretsch, D. B., & Bonser, C. F. (Eds.), *Globalization and Regionalization: Challenges for Public Policy* (pp. 49–70). Boston, Dordrecht: Kluwer Academic Publishers.

Bellinger, G., Castro, D., & Mills, A. (1999) Data, information, knowledge and wisdom. http://www.outsights.com/systems/dikw/dikw.htm.

Bosma, N., Jones, K., Autio, E., & Levie, J. (2008). *Global Entrepreneurship Monitor 2007, Babson Park, MA, US/ London*. UK: Babson College/ London Business School.

Brock, W. A., & Evans, D. S. (1989). Small business economics. *Small Business Economics*, *1*, 7–20. doi:10.1007/BF00389913

Chandler, A. D. (1977). *The Visible Hand: The Managerial Revolution in American Business*. Cambridge, MA: Harvard University Press.

D'Andrade, R. (1995). *The Development of Cognitive Anthropology*. Cambridge: Cambridge University Press.

Davenport, T., & Prusak, L. (1998). *Working Knowledge*. How Organizations Manage What They Know. Harvard Business School Press.

Davenport, T. H. (1998) Ten principles of knowledge management. http://www.bus.utexas.edu/kman/kmprin.htm.

Davies, H. T. O., & Nutley, S. M. (2000). Developing learning organizations in the new NHS. *BMJ (Clinical Research Ed.)*, *320*, 998–1001. doi:10.1136/bmj.320.7240.998

Flanagan, J. C. (1954). The critical incident technique. *Psychological Bulletin*, *5*, 327–358. doi:10.1037/h0061470

Garvin, D. A. (1998). Building a learning organization. In *Harvard Business Review on Knowledge Management* (pp. 47–80). Boston, MA: Harvard Business School.

Liebowitz, J. (2000). *Building Organizational Intelligence: A Knowledge Management Primer*. Boca Raton, CA: CRC Press.

Macintosh, A. (1997) Position paper on knowledge asset management. http://www.aiai.ed.ac.uk/~alm/kam.html.

Naudé, W. (2007). *Entrepreneurship in economic development*. Helsinki: World Institute for Development Economics Research, United Nations University.

Thurik, A. R. (2008forthcoming). Entreprenomics: entrepreneurship, economic growth and policy. In Acs, Z. J., Audretsch, D. B., & Strom, R. (Eds.), *Entrepreneurship, Growth and Public Policy*. Cambridge, UK: Cambridge University Press.

WEF. 2007, *Global Competitiveness Report 2007-2008*, Geneva: World Economic Forum.

Wennekers, S., van Stel, A., Thurik, R., & Reynolds, P. (2005). Nascent entrepreneurship and the level of economic development'. *Small Business Economics*, *24*, 293–309. doi:10.1007/s11187-005-1994-8

KEY TERMS AND DEFINITIONS

Knowledge Management: The systematic process of finding, selecting, organizing, distilling and presenting information that improve the comprehension in a specific area of interest.

Individual Competence: refers to the capacity of individuals/employees to act in a wide variety of situations. It's their education, skills, experience, energy and their attitudes that will make or break the relationships with the customers and the products or services that are provided. Sveiby uses it instead of Knowledge.

Development of Knowledge: The development of knowledge includes not only processes of external knowledge procurement (i.e. through cooperative efforts, consultants, new contacts, etc.) or the creation of specific knowledge resources like research and development departments. The formation of personal and technical knowledge networks is also part of the development of knowledge.

Knowledge Management and the Individual: The individual as the initial point of knowledge management have been neglected, especially as knowledge management has become a topic important in the business world. Most companies at first relied on technology-based knowledge management, which has mostly led to the implementation of databases.

Startup: A new business venture in its earliest stage of development.

Entrepreneur: an individual engaged in the process of starting and growing one's own business or idea.

The Entrepreneurial Process: Creativity and its links to enterprise, the three stage process of entrepreneurship; Opportunity screening; Innovation and competitive advantage; Acquiring resources – financing ventures and the role of family, friends and fools.

Innovation: Introduction of a new idea into the marketplace in the form of a new product or service, or an improvement in organization or process.

Patent: A property right granted to an inventor to exclude others from making, using, offering for sale, or selling the invention for a limited time in exchange for public disclosure of the invention when the patent is granted.

APPENDIX 1

Online Resources

www.ncoe.org

www.entreworld.org

www.ruraleship.org

www.entre-ed.org

www.cfed.org

www.nfte.com

www.sba.gov

APPENDIX 2

Questions and Answers

1. What is knowledge management in entrepreneurship?

Knowledge management refers to the systematic process of finding, selecting, organizing, distilling and presenting information to improve the comprehension in a specific area of interest in promoting and developing entrepreneurs.

2. Is entrepreneurship growth related to better knowledge management practices?

Entrepreneurship growth in general is related to better resource management,, never the less the most important resource especially in the globalized world is human resource and since more economies are becoming knowledge based economies the growth in entrepreneurship is definitely going to be related to adoption of better practices.

3. What stage of entrepreneurial process should the design of knowledge management is incorporated?

Ideally speaking the design of knowledge management should be incorporated right from the preparation of business plan, but we all know that enterprises and entrepreneurs come into existence in both planned and unplanned manner. Research in entrepreneurship reflect that all the reasons of how entrepreneurs are made are not known – however wherever possible it is better to incorporate knowledge management right from the business plan stage.

4. What are difficulties in implementing knowledge management techniques?

The difficulties in implementation of knowledge management techniques are both on the front of technology and human resource. Highly enable technology driven KM tools and techniques tend to be costly and human interface based tools and techniques tend to be messy and vague.

5. Which are the ideal knowledge management practices for small entrepreneurs?

Some of the simple knowledge management practices which are ideal for small entrepreneurs are: creation of knowledge repositories with respect to new enterprises and entrepreneurs, innovation models and practices, lines of business and venture capital and funding practices. Apart from this entrepreneurs' experience sharing and developing case studies would be helpful.

6. In the cases given what are the underlying principles of knowledge management?

The underlying principles of knowledge management in the cases discussed are:

1. Knowledge based entrepreneurship will help in establishing market position.
2. Knowledge benchmarking needs to be done even if the enterprise is small or non-technical.
3. Facilitating development of entrepreneurs.
4. The institutions of learning can be creators and knowledge facilitators for entrepreneurs.

Lessons learned

1. Knowledge Management is crucial for development of entrepreneurial class.
2. More research needs to be undertaken for developing effective tools and techniques for management of knowledge.
3. Knowledge management has cultural aspects, bias and implications for entrepreneurs and society as a whole.

Compilation of References

Aadne, J. H., Von Krogh, G., & Roos, J. (1996). Representationism: The Traditional Approach to Cooperative Strategies. In Von Krogh, G., & Roos, J. (Eds.), *Managing Knowledge Perspectives on Cooperation and Competition* (pp. 9–31). London: Sage.

Abraham, R. (1999). Emotional intelligence in organizations: A conceptualization. *Genetic, Social, and General Psychology Monographs, 125*(2), 209–225.

Accenture (2007). *The Rise of the Multi-Polar World.* Retrieved February 10, 2010, from www.accenture.com.

Adams, K. C. (2001). The Web as a database: new extraction technologies & content management. *Online, 25*(2), 27–32.

Aharoni, Y. (1993). In search for the unique: can firm-specific advantages be evaluated? *Journal of Management Studies, 31*(1), 31–49. doi:10.1111/j.1467-6486.1993.tb00294.x

Ahmad, H., Sharom, N., & Abdullah, C. S. (2006). *Knowledge Sharing Behaviour in the public sector: the business process management perspectives.* Retrieved from http://www.kmice.edu.my/files/kmice06.PDF.

Ahmed, K. A. (2002). *Learning Through Knowledge Management.* Oxford, UK: Butterworth-Heinemann.

Ahmed, A., & Hegazy, K. (2006). Knowledge management perception in the Middle Eastern region: an empirical investigation within Egypt context. *International Journal of Management Practice, 2*(2), 109–126.

Al-Athari, A., & Zairi, M. (2001). Building benchmarking competence through knowledge management capability: An empirical study of the Kuwaiti context. *Benchmarking: An International Journal, 8*(1), 70–80. doi:10.1108/14635770110383489

Alavi, M., & Tiwana, A. (2002). Knowledge integration in virtual teams: the potential role of KMS. *Journal of the American Society for Information Science and Technology, 53*(12), 1029–1037. doi:10.1002/asi.10107

Alavi, M., & Leidner, D. (2001). Review: Knowledge management and knowledge management systems: Conceptual foundations and research issues. *Management Information Systems Quarterly, 25*(1), 107–136. doi:10.2307/3250961

Alavi, M., & Leidner, D. (2001). Review: knowledge management and knowledge management systems: conceptual foundations and research issues. *Management Information Systems Quarterly, 25*(1), 107–136. doi:10.2307/3250961

Alavi, M., & Leidner, D. E. (2001). Review: Knowledge Management and Knowledge Management Systems: Conceptual Foundations and Research Issues. *Management Information Systems Quarterly, 25*(1), 107–136. doi:10.2307/3250961

Alavi, M. (2000). Managing organizational knowledge. In Zmud, R. (Ed.), *Framing the domains of IT management* (pp. 29–50). Cincinnati, OH: Pinnaflex.

Al-Busaidi, K. A., & Olfman, L. (2005). An investigation of the determinants of knowledge management systems success in Omani organizations. *Journal of Global Information Technology Management*, *8*(3), 6–27.

Al-Busaidi, K. A., Olfman, L., Ryan, T., & Leroy, G. (2007). Revealing the antecedents and benefits of KMS use: An exploratory study in petroleum company in Oman, *The Electronic Proceeding of the 9th International Conference on Decision Support Systems*, Kolkata, India.

Aldrich, H. (1979). *Organizations and environments*. Upper Saddle River, NJ: Prentice-Hall.

Al-Hawamdeh, S. (2003). *Knowledge management: Cultivating knowledge professionals*. Oxford, UK: Chandos Publishing.

Ali, A. J. (1990). Management theory in a transitional society: The Arab's experience. *International Studies of Management and Organization*, *20*(3), 7–35.

Ali, J. (2006). *Mengurus sumber manusia*. Dewan Bahasa dan Pustaka, Kuala Lumpur.

Alves, J., Marques, M. J., Saur, I., & Marques, P. (2007). Creativity and innovation through multidisciplinary and multisectoral cooperation. *Creativity and Innovation Management*, *16*(1), 27–34. doi:10.1111/j.1467-8691.2007.00417.x

Alwis, S.-R., & Hartmann, E. (2008). The use of tacit knowledge within innovative companies: Knowledge management in innovative enterprises. *Journal of Knowledge Management*, *12*(1), 133–147. doi:10.1108/13673270810852449

Amabile, T. M. (1996). *Creativity in context*. Boulder, CO: Westview Press.

Amabile, T. M., & Gryskiewicz, S. S. (1987). *Creativity in the R&D laboratory*. Greensboro, NC: Center for Creative Leadership.

Ambrosini, V., & Bowman, C. (2001). Tacit Knowledge: Some Suggestions for Operationalization. *Journal of Management Studies*, *38*, 811–829. doi:10.1111/1467-6486.00260

Anderson, E., & Weitz, B. (1992). The use of pledges to build and sustain commitment in distribution channels. *JMR, Journal of Marketing Research*, *29*(1), 18–34. doi:10.2307/3172490

Anderson, J. C., & Narus, J. A. (1990). A model of distributor firm and manufacturer firm working partnerships. *Journal of Marketing*, *54*(1), 42–58. doi:10.2307/1252172

Andersson, K. (2000). *Knowledgetechnology applications for knowledge management*. Sweden: Tesis Dr. Falsafah. Uppsala University.

Appan, P., Sundaram, H., & Birchfield, D. (1999). Communicating Everyday Experiences. *Proceedings of the 1st ACM workshop on Story representation, mechanism and context,* New York. 17-24.

Arent, J., Nørbjerg, J., & Pederson, M. (2000). Creating organizational knowledge in software process improvement. In K. Althoff, W. Müller (Eds.), *Proceedings of the second International Workshop on Learning Software Organizations*. (pp. 81-92) Finland.

Argote, L., & Ingram, P. (2000). Knowledge transfer: A basis for competitive advantage in firms. *Organizational Behavior and Human Decision Processes*, *82*(1), 150–169. doi:10.1006/obhd.2000.2893

Argote, L., & Ingram, P. (2000). Knowledge Transfer: A Basis for Competitive Advantage in Firms. *Organizational Behavior and Human Decision Processes*, *82*(1), 150–169. doi:10.1006/obhd.2000.2893

Argote, L. (1999). *Organizational Learning: Creating, Retaining and Transferring Knowledge*. Boston: Kluwer Academic, Boston

Aron, R., Clemons, E.K. & Reddi, S. (2005), Just Right Outsourcing: Understanding and Managing Risk. *Journal of Management Information Systems* 22(2),.37-55.

Arora, R. (2002). Implementing KM - a balances score card approach. *Journal of Knowledge Management*, *6*(3), 240–249. doi:10.1108/13673270210434340

Arthur, J. B. (1994). Effects of human resource systems on manufacturing pPerformance and turnover. *Academy of Management Journal*, *37*, 670–684. doi:10.2307/256705

Artz, K. W., & Brush, T. H. (2000). Asset specificity, uncertainty and relational norms: An examination of coordination costs in collaborative strategic alliances. *Journal of Economic Behavior & Organization*, *41*, 337–362. doi:10.1016/S0167-2681(99)00080-3

Asgary, N., & Walle, A. H. (2002). The cultural impact of globalisation: Economic activity and social change. *Cross Cultural Management: An International Journal*, *9*(3), 58–75. doi:10.1108/13527600210797433

Asibong, C., & Barlow, J. (1997, April). *Barriers to innovation and change in the housebuilding industry*. Paper presented at the Housing Studies Association Conference, York, England.

Audretsch, D. B. (2007a). *The Entrepreneurial Society*. Oxford, UK: Oxford University Press. doi:10.1093/acprof:oso/9780195183504.001.0001

Audretsch, D. B., Grilo, I., & Thurik, A. R. (2007). *The Handbook of Research on Entrepreneurship Policy*. Cheltenham, UK and Northampton, MA, US: Edward Elgar Publishing Limited.

Autio, E. (2000). Effects of age at entry, knowledge intensity, and imitability on international growth. *Academy of Management Journal*, (Oct): 2000.

Awad, E. M., & Ghaziri, H. M. (2004). *Knowledge management*. Upper Saddle River, NJ: Pearson Education Inc.

Baba, M. L., Gluesing, J., Ratner, H., & Wagner, K. (2004). The contexts of knowing: Natural history of a globally distributed team. *Journal of Organizational Behavior*, *25*(5), 547–587. doi:10.1002/job.259

Baer, M., & Oldham, G. R. (2006). The curvilinear relation between experienced creative time preassure and creativity: Moderating effects of openness to experience and support for creativity. *The Journal of Applied Psychology*, *91*(4), 963–970. doi:10.1037/0021-9010.91.4.963

Baeza-Yaes, R., Ciatamita, M., Mika, P., & Zaragoza, H. (2008). Towards Semantic Search. *Natural Language and Information Systems*, 4-11.

Bailey, J., & Clark, N. (2008). Supplier problems led to new delay of Boeing 787. Retrieved February 10, 2010, from www.nytimes.com.

Bailey, T. (2005). *The story of JPL stories.* Retrieved September 12, 2009, from http://askmagazine.nasa.gov/issues/11/special/index.html

Balaji, C., Chandrashektar, S., & Dutta, R. (1998). *Leading Change Through Human Resources: Towards a Globally Competitive India*. New Delhi: McGraw-Hill.

Balakrishnan, K., Mohan, U., & Seshadri, S. (2008). Outsourcing of front-end business processes: Quality information and customer contact. *Journal of Operations Management*, *26*(2), 288–302. doi:10.1016/j.jom.2007.08.001

Barley, S. R., & Orr, J. E. (1997). *Between craft and science: Technical settings in U.S.Settings*. London: ILR Press.

Barlow, J. (1999). From craft production to mass customization. Innovation requirements for the UK Housebuilding Industry. *Housing Studies*, *14*(1), 23–42. doi:10.1080/02673039982984

Barney, J. (1991). Firm resources and sustained competitive advantage. *Journal of Management*, *17*(1), 99–120. doi:10.1177/014920639101700108

Barney, J. B. (1991). Firm resources and sustained competitive advantage. *Journal of Management*, *17*, 99–120. doi:10.1177/014920639101700108

Barratt, M., & Oke, A. (2007). Antecedents of supply chain visibility in retail supply chains: A resource-based theory perspective. *Journal of Operations Management*, *25*(6), 1217–1233. doi:10.1016/j.jom.2007.01.003

Barrett, M., Cappleman, S., Shoib, C., & Walsham, G. (2004). Learning in knowledge communities: Managing technology and context. *European Management Journal*, *22*(1), 87–98. doi:10.1016/j.emj.2003.11.019

Bartol, K. M., & Srivastava, A. (2002). Encouraging knowledge sharing: The role of organizational reward systems. *Journal of Leadership & Organizational Studies*, *9*(1), 64–76. doi:10.1177/107179190200900105

Baumard, P. (1999). *Tacit knowledge in organization*. London: Sage.

Beatty, R. W., Huselid, M. A., & Schneier, C. E. (2003). New HR Metrics: Scoring on the Business Scorecard. *Organizational Dynamics*, *32*(2), 107–121. doi:10.1016/S0090-2616(03)00013-5

Becerra-Fernandez, I., Gonzalez, A., & Sabherwal, R. (2004). *Knowledge management: Challenges, solutions and technologies*. New Jersey: Pearson Education Inc.

Bechina, A. (2006). Knowledge sharing practices: Analysis of a global Scandinavian consulting company. *Electronic Journal of Knowledge Management*, *4*(2), 109–116.

Bechina, A. A., & Bommen, T. (2006). Knowledge sharing practices: Analysis of a global scandinavian consulting company. *The Electronic Journal of Knowledge Management,* 4(2), 10 November 2009.

Becker, M. C., & Zirpoli, F. (2003). Organizing new product development: Knowledge hollowing-out and knowledge integration – the FIAT Auto case. *International Journal of Operations & Production Management*, *23*(9), 1033–1061. doi:10.1108/01443570310491765

Beckman, T. J. (1999). The current state of knowledge management. In *Knowledge management handbook*. New York: CRC Press.

Beckman, T. J. (1999). The Current State of Knowledge Management. In Liebowitz, J. (Ed.), *Knowledge Management Handbook*. Boca Raton, FL: CRC Press.

Beckman, T. J. (1999). The Current State of Knowledge Management. In J. Liebowitz (Ed.), *Knowledge Management Handbook* (pp. 1-1 to 1-22). New York: CRC Press.

Bell, M. C., & Gomez, B. E. (2008). Effect of unsignaled delays between stimuli in a chain schedule on responding and resistance to change. *Behavioural Processes*, *77*(3), 343–350. doi:10.1016/j.beproc.2007.08.004

Bennet, A., & Bennet, D. (2003). The Partnership between Organizational Learning and Knowledge Management. In Holsapple, C. W. (Ed.), *Handbook on Knowledge Management 1: Knowledge Matters* (pp. 439–455). Berlin: Springer-Verlag.

Bergh, B. V., & Stuhlfaut, M. (2006). Is advertising creativity primarily an individual or a social process? *Mass Communication & Society*, *9*(4), 373–397. doi:10.1207/s15327825mcs0904_1

Best, J. W., & Kahn, J. V. (1998). *Research in Education*. Boston: Library of Congress Cataloguing-in-Publication data.

Bhardwaj, M., & Monin, J. (2006). Tacit to Explicit: An Interplay Shaping Organization Knowledge. *Journal of Knowledge Management*, *10*(3), 72–85. doi:10.1108/13673270610670867

Bhatt, G. D. (2002). Management Strategies for Individual Knowledge and Organizational Knowledge. *Journal of Knowledge Management*, *6*(1), 31–39. doi:10.1108/13673270210417673

Binney, D. (2001). The knowledge management spectrum - understanding the KM landscape. *Journal of Knowledge Management*, *5*(1), 33–42. doi:10.1108/13673270110384383

Birch, D. (1987). *Job Creation in America*. New York: Free Press.

Birley, S. (1985). *The Role of Networks in the Entrepreneurial Process*. Cranfield School of Management.

Birnbaum, R. (2000). *Management Fads in Higher Education: Where They Come From, What They Do, Why They Fail*. San Francisco: Jossey-Bass.

Bjørnson, F., & Dingsøyr, T. (2008). Knowledge management in software engineering: a systematic review of studied concepts, findings and research methods. *Information and Software Technology*, *50*(11), 1055–1068. doi:10.1016/j.infsof.2008.03.006

Blackler, F. (1995). Knowledge, Knowledge Work and Organization: An Overview and Interpretation. *Organization Studies*, *16*, 1021–1046. doi:10.1177/017084069501600605

Blumentritt, R., & Johnston, R. (1999). Towards a strategy for knowledge management. *Technology Analysis and Strategic Management*, *11*(3), 287–300. doi:10.1080/095373299107366

Boateng, W. (2006). Knowledge management working tool for agricultural extension practice: the case of Chana. *Knowledge Management for Development Journal*, *2*(3), 19–29.

Bohn, R. E. (1994). Measuring and managing technological knowledge. *Sloan Management Review*, *36*(1), 61–73.

Bosch-Sijtsema, P. M., Ruohomaki, V., & Vartiainen, M. (2009). Knowledge work productivity in distributed teams. *Journal of Knowledge Management*, *13*(6), 533–546. doi:10.1108/13673270910997178

Bradner, E., Kellogg, W. A., & Erickson, T. (1999). The adoption and use of BABBLE: a field study of chat in the workplace. *Proceedings of the Sixth European Conference on Computer Supported Cooperative Work*, Copenhagen, Denmark. 139-158.

Braunscheidel, M. J., & Suresh, N. C. (2009). The organizational antecedents of a firm's supply chain agility for risk mitigation and response. *Journal of Operations Management*, *27*(2), 119–140. doi:10.1016/j.jom.2008.09.006

Brézillon, P. (1999). Context in problem solving: A survey. *The Knowledge Engineering Review*, *14*(1), 1–34. doi:10.1017/S0269888999141018

Brief, A. P., & Motowidlo, S. J. (1986). Prosocial organizational behaviors. *Academy of Management Review*, *11*, 710–725. doi:10.2307/258391

Brinkley, I. (2006). Defining the knowledge economy. Knowledge Economy Programme Report. *The Work Foundation*, July.

Brockman, E. N., & Anthony, W. P. (2002). Tacit Knowledge and Strategic Decision Making. *Group & Organization Management*, *27*(4), 436–455. doi:10.1177/1059601102238356

Brown, J. S., & Duguid, P. (1998). Organizing Knowledge. *California Management Review*, *40*(3), 90–111.

Brown, J.S. & Duguid, P. (1991).Organization Learning and Communities of Practice: Toward a unified view of working, learning and innovation. *Organization Science* 2(2), 1991, pp. 40-57.

Buckley, P., Minette, K., Joy, D., & Michaels, J. (2004). The use of an automated employment recruitment and screening system for temporary professional employees: A case study. *Human Resource Management*, *43*(2&3), 233–241. doi:10.1002/hrm.20017

Bukowitz, W. R., & Williams, R. L. (1999). *The Knowledge Management Fieldbook*. London: Prentice Hall.

Bunkley, N., & Maynard, M. (2010). With recall expanding, Toyota gives an apology. Retrieved February 10, 2010, from www.nytimes.com.

Bureš, V., & Čech, P. (2008). Knowledge Management: the Central European Experience. *Effective Executive*, *6*(XI), 47–53.

Bureš, V. (2003). Cultural Barriers in Knowledge Sharing. *E+M Economics and Management (6)*special issue, 57-62.

Bureš, V. (2005). Methodology of Knowledge Management Implementation. *3rd International Conference on Multimedia and Information & Communication Technologies*. Cáceres, Spain, Vol. 3, pp. 963-968.

Bureš, V. (2009). Conceptual Perspective of Knowledge Management. *E+M Economics and Management (12)*2, 84-96.

Burgess, D. (2005). *What* Motivates Employees to Transfer Knowledge outside their work unit? *Journal of Business Communications* 42(4), 2005, pp. 324-348.

Burrows, P. (2005). *Commentary: Apple's Blueprint for Genius*. Retrieved on February 10, 2010, from www.businessweek.com.

Buyya, R., Yeo, C. S., & Venugopal, S. (2008). Market-Oriented Cloud Computing: Vision, Hype, and Reality for Delivering IT Services as Computing Utilities. *Proc. of the 10th IEEE International Conference on High Performance Computing and Communications*, China, 2008.

Cadman, D., & Austin-Crowe, L. (1983). *Property Development* (2nd ed.). London: Spon.

Cai, S., & Yang, Z. (2008). Development of cooperative norms in the buyer-supplier relationship: The Chinese Experience. *Journal of Supply Chain Management, 44*(1), 55–70. doi:10.1111/j.1745-493X.2008.00045.x

Calof, J. L., & Viviers, W. (2001). Adding competitive intelligence to South Africa's knowledge management mix. *Africa Insight, 31*(2), 61–67.

Cannon, J. P., Achrol, R. S., & Gundlach, G. T. (2000). Contracts, norms, and plural form governance. *Academy of Marketing Science, 28*(2), 180–194. doi:10.1177/0092070300282001

Cannon, J. P., & Perreault, W. D. (1999). Buyer-seller relationships in business markets. *JMR, Journal of Marketing Research, 36*(4), 439–460. doi:10.2307/3151999

Cantu, L. Z., Criado, J. R., & Criado, A. R. (2009). Generation and transfer of knowledge in IT-related SMEs. *Journal of Knowledge Management, 13*(5), 243–256. doi:10.1108/13673270910988088

Cantú, F. J., Bustani, A., Molina, A., & Moreira, H. (2009). A knowledge-based development model: the research chair strategy. *Journal of Knowledge Management, 13*(1), 154–170. doi:10.1108/13673270910931233

Carlile, P. R. (2004). Transferring, translating, and transforming: An integrative framework for managing knowledge across boundaries. *Organization Science, 15*(5), 555–568. doi:10.1287/orsc.1040.0094

Carlile, P. R., & Rebentisch, E. S. (2003). Into the black box: The knowledge transformation cycle. *Management Science, 49*(9), 1180–1195. doi:10.1287/mnsc.49.9.1180.16564

Carmel, E., & Agarwal, R. (2002). The maturation of Offshore Sourcing of Information Technology Work. *MIS Quarterly Executive, 1*(2), 65–77.

Carter, E. (2008). Successful change requires more than change management. *Journal for Quality and Participation, 31*(1), 20–23.

Cascio, W. F. (1990). Strategic human resource management in high technology industry. In L. Gomez-Mejia, R. & M. Lawless (Eds) *Organizational Issues in High Tech Magazine, pp.*179-198. Greenwich, CT: JAI Press.

Casselman, R. M., & Samson, D. (2005). Moving Beyond Tacit and Explicit: Four Dimensions of Knowledge. *Proceedings of the 38th Hawaii International Conference on System Sciences.* 3 – 6 January. Hawaii: IEEE Computer Society Press.

Cavusgil, S. T., Calantone, R. J., & Zhao, Y. (2003). Tacit Knowledge Transfer and Firm Innovation Capability. *Journal of Business and Industrial Marketing, 18*(1), 6–21. doi:10.1108/08858620310458615

Čech, P. (2005). *The Potential of Knowledge in Organizations*. Ph.D. Thesis, Hradec Králové: University of Hradec Králové.

Čech, P., & Olševičová, K. (2009). Proposal of Intelligent Assistant for Graduation Thesis Preparation. *Intelligent environments: workshops proceedings of the 5th international conference*, Barcelona, Spain, pp. 321-327.

Chang, S., & Lee, M. (2007). The effects of organizational culture and knowledge management mechanisms on organizational innovation: An empirical study in Taiwan. *The Business Review-Cambridge, 7*(1), 295–301.

Chang, B., & Phi, J. (2007). The IDS story: Reinventing distribution through supply chain logistics. In Lee, H. L., & Lee, C.-Y. (Eds.), *Building Supply Chain Excellence in Emerging Economies.* New York: Springer.

Chau, K. W., Wong, S. K., & Yiu, C. Y. (2007). Housing quality in the forward contracts market. *The Journal of Real Estate Finance and Economics, 34*, 313–325. doi:10.1007/s11146-007-9018-x

Chaudhry, A. S. (2005). *Knowledge sharing practices in Asian institutions: a multi perspective from Singapore.* Retrieved from http://archive.ifla.org/IV/ifla71/papers/066e-Chaudhry.pdf.

Chen, H. (2003). Introduction to the JASIST special topic section on Web retrieval and mining: A Machine Learning Perspective. *Journal of the American Society for Information Science and Technology, 54*(7), 621–624. doi:10.1002/asi.10254

Chen, D. H. C., & Dahlman, C. J. (2005). The Knowledge Economy, the KAM Methodology and World Bank Operations. Washington DC: World Bank. Retrieved, March 4, 2009, from http://siteresources.worldbank.org.

Chen, H., Chau, M., & Zeng, D. (2002). *CI Spider: a tool for competitive intelligence on the Web.* [Electronic] Available: http://ai.bpa.arizona.edu/~mchau/papers/CISpider.pdf

Cheung, M.-S., Myers, M. B., & Mentzer, J. T. (in press). Does relationship learning lead to relationship value? A cross-national supply chain investigation. *Journal of Operations Management.*

Chin, W. W. (1998). The partial least square approach to structural equation modeling. In G.A Marcoulides (Ed), *Modern methods for business research*(pp. 295-336). Mahawah, London, UK: Lawrence Erlbaum Associates.

Chiu, C. M., Hsu, M. H., & Wang, E. T. G. (2006). Understanding knowledge sharing in virtual communities: An integration of social capital and social cognitive theories. *Decision Support Systems, 42*(3), 1872–1888. doi:10.1016/j.dss.2006.04.001

Chong, S. C. (2006). KM critical success factors: A comparison of perceived importance versus implementation in Malaysian ICT companies. *The Learning Organization, 13*(3), 230–256. doi:10.1108/09696470610661108

Chong, C. W., Holden, T., Wilhelmij, P., & Schmidt, R. A. (2000). Where does knowledge management add value? *Journal of Intellectual Capital, 1*(4), 366–380. doi:10.1108/14691930010359261

Choo, C. W. (2006). *The Knowing Organization How Organizations Use Information to Construct Meaning, Create Knowledge and Make Decisions* (2nd ed.). New York: Oxford University Press.

Choo, W. C. (2002). Sensemaking, Knowledge Creation, and Decision Making. In Choo, C. W., & Bontis, N. (Eds.), *The Strategic Management of Intellectual Capital and Organizational Knowledge* (pp. 79–88). New York: Oxford University Press.

Clark, S. (2005). Refinery expertise preservation. *Control Engineering, 52*(8), IE5.

Coates, T. T., & McDermott, C. M. (2002). An exploratory analysis of new competencies: a resource based view perspective. *Journal of Operations Management, 20*, 435–450. doi:10.1016/S0272-6963(02)00023-2

Coates, T. T., & McDermott, C. M. (2002). An exploratory analysis of new competencies: a resource based view perspective. *Journal of Operations Management, 20*, 435–450. doi:10.1016/S0272-6963(02)00023-2

Cohen, S. G., & Bailey, D. E. (1997). What makes teams work: group effectiveness research from the shop floor to the executive suite. *Journal of Management, 23*(3), 239–290. doi:10.1177/014920639702300303

Cohen, W. M., & Levinthal, D. A. (1990). Absorptive capacity: a new perspective on learning and innovation. *Management Science, 5*(1), 227–251.

Cohen, W. M., & Levinthal, D. A. (1990). Absorptive Capacity: A new perspective on learning and innovation. *Administrative Science Quarterly, 35*(1), 128–152. doi:10.2307/2393553

Cohen, W. M., & Levinthal, D. (1990). Absorptive capacity: a new perspective on learning and innovation. *Administrative Science Quarterly, 35*, 128–152. doi:10.2307/2393553

Cohn, M. (2009). *Suceeding with Agile: Software development using Scrum.* Reading, MA: Addison-Wesley Professional.

Collins, H. M. (2001). Tacit Knowledge, Trust, and the Q of Sapphire. *Social Studies of Science*, *31*(1), 71–85. doi:10.1177/030631201031001004

Cong, X. dan Pandya, K.V. (2003). Issues of knowledge management in the public sector. *Electronic Journal of Knowledge Management*, *1*(2), 25–33.

Connell, N. A., Klein, J. H., & Powell, P. L. (2003). It's Tacit Knowledge But Not As We Know It: Redirecting the Search for Knowledge. *The Journal of the Operational Research Society*, *54*, 140–152. doi:10.1057/palgrave.jors.2601444

Conner, K. R., & Prahalad, C. K. (1996). A resource-based theory of the firm: knowledge versus opportunism. *Organization Science*, *7*(5), 477–501. doi:10.1287/orsc.7.5.477

Cook, S. D., & Brown, J. S. (1999). Bridging Epistemologies: The Generative Dance between Organization Knowledge and Organizational Knowing. *Organization Studies*, *10*(4), 381–400. doi:10.1287/orsc.10.4.381

Cook, D. P., Goh, C.-H., & Chung, C. H. (1999). Service typologies: A state of the art survey. *Production and Operations Management*, *8*(3), 318–339. doi:10.1111/j.1937-5956.1999.tb00311.x

Corbin, R., Dunbar, C., & Zhu, Q. (2007). A three-tier knowledge management scheme for software engineering support and innovation. *Journal of Systems and Software*, *80*(9), 1494–1505. doi:10.1016/j.jss.2007.01.013

Cornish, S. L. (1997a). Product innovation and the spatial dynamics of market intelligence: Does proximity to markets matter? *Economic Geography*, *73*(2), 143–165. doi:10.2307/144445

Cornish, S. L. (1997b). Strategies for the acquisition of market intelligence and implications for the transferability of information inputs. *Annals of the Association of American Geographers. Association of American Geographers*, *87*(3), 451–470. doi:10.1111/1467-8306.00063

Costa, P. T., & McCrae, R. R. (1995). Domains and facets: Hierarchical personality assessment using the Revised NEO Personality Inventory. *Journal of Personality Assessment*, *64*, 21–50. doi:10.1207/s15327752jpa6401_2

Côté, S., & Miners, C. T. H. (2006). Emotional intelligence, cognitive intelligence, and job performance. *Administrative Science Quarterly*, *51*(1), 1–28.

Coviello, A. (2001). *Standardised KM Implementation Approach. IST Project No 2000-26393, Deliverable D 3.1*. European KM Forum.

Craighead, C. W., Hult, G. T. M., & Ketchen, D. J. (2009). The effects of innovation–cost strategy, knowledge, and action in the supply chain on firm performance. *Journal of Operations Management*, *27*(5), 405–421. doi:10.1016/j.jom.2009.01.002

Creswell, J. W. (2003). *Research Design: Qualitative, Quantitative, and Mixed Methods Approaches*. London: Sage.

Cristofolini, I., Filippi, S., & Bandera, C. (2009). The role of product feature relations in a knowledge based methodology to manage design modifications for product measurability. *International Journal of Production Research*, *47*(9), 2373–2389. doi:10.1080/00207540701666246

Crossan, M. M., & Berdrow, I. (2003). Organizational learning and strategic renewal. *Strategic Management Journal*, *24*, 1087–1105. doi:10.1002/smj.342

Cruz, N. M., Perez, V. M., & Cantero, C. T. (2009). The influence of employee motivation on knowledge transfer. *Journal of Knowledge Management*, *13*(6), 478–490. doi:10.1108/13673270910997132

Cudahy, G., Mulani, N., & Cases, C. (2008). Mastering global operations in a multipolar world. *Supply Chain Management Review*, *12*(2), 22–29.

Cudanov, M., Jasko, O. & Jevtic M. (2009). Influence of Information and Communication Technologies on Decentralization of Organizational Structure. *Computer science and information systems – COMSIS Journal*, 6(1), 93-109.

Cully, M., Woodland, S., O'Reilly, A., & Dix, G. (1999). *England at Work. As Depicted by the 1998 Workplace Employment Relation Survey*. London: Routledge.

Cummings, J. N. (2004). Work groups, structural diversity, and knowledge sharing in a global organization. *Management Science*, *50*(3), 352–264. doi:10.1287/mnsc.1030.0134

D'Eredita, M. A., & Barreto, C. (2006). How Does Tacit Knowledge Proliferate? An Episode-Based Perspective. *Organization Studies*, *27*(12), 1821–1841. doi:10.1177/0170840606067666

Dahane, M., Clementz, C., & Rezg, N. (2008). Analysis of joint maintenance and production policies under a subcontracting constraint. *International Journal of Production Research*, *46*(19), 5393–5416. doi:10.1080/00207540802273686

Davenport, T., & Prusak, L. (1998). *Working Knowledge: How Organizations Manage What They Know*. Boston: Harvard Business School Press.

Davenport, T. H., & Smith, D. E. (2000). Managing Knowledge in Professional Service Firms. In *The Knowledge Management Yearbook 2000-2001*. Oxford, UK: Butterworth-Heinemann.

Davenport, T. H., De Long, D. D., & Beers, M. C. (1999). Successful Knowledge Management Projects. In Cortada, J. W., & Woods, J. A. (Eds.), *The Knowledge Management Yearbook 1999-2000* (pp. 89–107). Washington, DC: Butterworth-Heinemann Ltd.

Davenport, T. (2003). *A measurable proposal.* CIO Magazine, June.

Davenport, T. H., & Marchand, D. A. (2000). Is KM Just Good Information Management? In Marchand D.A., DAvenport T.H., & Dickson T. (eds.) *Mastering Information Management*, London, UK: Prentice Hall, 165-169.

Davila, A., & Elvira, M. M. (2009a). *Best Human Resource Management Practices in Latin America*. Oxford, UK: Routledge.

Day, A. L., & Carroll, S. A. (2004). Using an ability-based measure of emotional intelligence to predict individual performance, group performance, and group citizenship behaviors. *Personality and Individual Differences*, *36*(6), 1443–1458. doi:10.1016/S0191-8869(03)00240-X

De Long, D. W., & Fahey, L. (2000). Diagnosing cultural barriers to knowledge management. *The Academy of Management Executive*, *14*(4), 113–127.

Dean, J. (2007). The Forbidden City of Terry Gou. *Wall Street Journal, 11 Aug 2007, A1*.

Dearstyne, B. W. (2007). Blogs, Mashups, & Wikis Oh, My!. *The information management journal, 41* (4), 24-33.

Debowski, S. (2006). *Knowledge management*. Australia: Wiley.

Debowski, S. (2006). *Knowledge Management*. Milton, QLD: John Wiley & Sons.

Dekkers, R. (2009). Distributed Manufacturing as co-evolutionary system. *International Journal of Production Research*, *47*(8), 2031–2054. doi:10.1080/00207540802350740

Deloitte Touche Tohmatsu. (2008). *Innovation in emerging markets: Managing product sourcing risks in emerging markets*. Retrieved February 10, 2010, from www.deloitte.com.

DeLone, W., & McLean, E. R. (2003). The DeLone and McLean model of information systems success: A ten-year update. *Journal of Management Information Systems*, *19*(4), 9–30.

DeLone, W. H., & McLean, E. R. (2003). The DeLone and McLean model of information systems success: a ten year update. *Journal of Management Information Systems*, *19*(4), 9–30.

Denison, D. R., & Mishra, A. K. (1995). Toward a theory of organisational culture and effectiveness. *Organization Science*, *6*(2), 204–223. doi:10.1287/orsc.6.2.204

Dennis, A. R. (1996). Information exchange and use in group decision making: You can lead a group to information, but you can't make it think. *Management Information Systems Quarterly*, *20*(4), 433–457. doi:10.2307/249563

Denzin, N. K., & Lincoln, Y. S. (1998). *The Landscape of Qualitative Research*. London: Sage.

Desouza, K. (2003). Facilitating tacit knowledge exchange. *Communications of the ACM, 46*(6), 85–88. doi:10.1145/777313.777317

Dessler, G. (1976). *Organization and Management*. Upper Saddle River, NJ: Prentice Hall.

Dierickx, I., & Cool, K. (1989). Asset stock accumulation and sustainability of competitive advantage. *Management Science, 35*(12), 1504–1511. doi:10.1287/mnsc.35.12.1504

Dimnik, T. P., & Johnston, D. A. (1993). Manufacturing managers and the adoption of advanced manufacturing technologies. *Omega, 21*(2), 155–162. doi:10.1016/0305-0483(93)90048-P

Dingsøyr, T. (2005). Postmortem reviews: purpose and approaches in software engineering. *Information and Software Technology, 47*(5), 293–303. doi:10.1016/j.infsof.2004.08.008

Disterer, G. (2001). *Individual and social barriers to knowledge transfer.* Retrieved from http://www.hicss.hawaii.edu/HICSS_34/PDFs/OSTTA01.pdf.

Dixon, N. M. (2000). *Common Knowledge: How Companies Thrive by Sharing What They Know*. Boston: Harvard Business School Press.

Dixon, N. M. (2002). The neglected receiver of knowledge sharing. *Ivey Business Journal* (March), 35-40.

Dong, S. (2009). The Tacit Knowledge Transfer of Transnational Corporation. International. *Journal of Business and Management, 3*(6), 148–150.

Dreher, G., & Ash, R. (1990). A comparative study of mentoring among men and women in managerial, professional, and technical positions. *The Journal of Applied Psychology, 75*(5), 539–546. doi:10.1037/0021-9010.75.5.539

Drucker, P. F. (1993). *The post capitalist society*. Oxford, UK: Butterworth–Heinemann.

Drucker, P. (2007). *Innovation and Entrepreneurship - Practice and Principles. The classic Drucker Collection*. Oxford, UK: Butterworth-Heinemann.

Drucker, P. F. (1993). *The post capitalist society*. New York: HarperBusiness.

Drucker, P. F. (1969). *The age of discontinuity: guidelines to our changing society*. London: Heinemann.

Drucker, P. (1992). The new society of organizations. *Harvard Business Review, 70*(Sept-Oct), 95–104.

Druskat, V. U., & Wolff, S. B. (2001). Building the emotional intelligence of groups. *Harvard Business Review, 79*(3), 80–90.

Du Plesis, M. (2007). Knowledge management: what makes complex implementations successful? *Journal of Knowledge Management, 11*(2), 91–101. doi:10.1108/13673270710738942

Du Toit, A. S. (2003). Competitive intelligence in the knowledge economy: what is in it for South African manufacturing enterprises? *International Journal of Information Management, 23*(2), 111–120. doi:10.1016/S0268-4012(02)00103-2

Dyer, J. H., & Hatch, N. W. (2006). Relation-specific capabilities and barriers to knowledge transfers: Creating advantage through network relationships. *Strategic Management Journal, 27*(8), 701–719. doi:10.1002/smj.543

Dyer, J. H., & Nobeoka, K. (2000). Creating and managing a high-performance knowledge-sharing network: The Toyota case. *Strategic Management Journal, 21*(3), 345–367. doi:10.1002/(SICI)1097-0266(200003)21:3<345::AID-SMJ96>3.0.CO;2-N

Edwards, J. S., Collier, P. M., & Shaw, D. (2005). Knowledge Management and Its Impact on the Management Accountant [Electronic Version]. *The Chartered Institute of Management Accountants (CIMA), London, U.K.*, 1 - 33. Retrieved June 20, 2006 from www.cimaglobal.com.

Efimova, L. (2004). *Discovering the iceberg of knowledge work: A weblog case. Submitted to OKLC 2004* [Available at https://doc.telin.nl/dsweb/Get/Document-34786/]. Retrieved November 14th, 2008.

Eisendiel, V. (1997, May). *Towards a Sustainable Housing Strategy*. Paper presented at the National Housing Convention. Kuala Lumpur, Malaysia.

Ellram, L. (1995). Partnering pitfalls and success factors. *International Journal of Purchasing and Materials Management, 31*(2), 36–44.

Ellram, L., & Carr, A. (1994). Strategic purchasing: A history and review of the literature. *International Journal of Purchasing and Materials Management, 30*(2), 10–18.

Endres, M. L., Endres, S. P., Chowdhury, S. K., & Alam, I. (2007). Tacit Knowledge Sharing, Self-Efficacy Theory, and Application to the Open Source Community. *Journal of Knowledge Management, 11*(3), 92–103. doi:10.1108/13673270710752135

Engstrom, T. E. J. (2003). Sharing knowledge through mentoring. *Performance Improvement, 42*(8), 36–42. doi:10.1002/pfi.4930420811

Ensor, J., Cottam, A., & Band, C. (2001). Fostering knowledge management through the creative work environment: A portable model from the advertising industry. *Journal of Information Science, 27*(3), 147–155. doi:10.1177/016555150102700304

Eraut, M. (2000). Non-formal Learning and Tacit Knowledge in Professional Work. *The British Journal of Educational Psychology, 70*, 113–136. doi:10.1348/000709900158001

Ester, M., Kriegel, H., & Schubert, M. (2002). Web site mining: a new way to spot competitors, customers and suppliers in the world wide Web.*Proceedings of SIGKDD*, Edmonton, 249-258.

European Council. (2000). *Presidency Conclusions - Lisbon European Council*. Lisbon. Retrieved December 20, 2008, from http://europa.eu.int.

Evaggelia, F. (2007). Human resource management policies and knowledge workers. *Proceedings of the 8th European Conference of Knowledge Management:* (p 319). Barcelona, España.

Evans, K., Kersh, N., & Kontiainen, S. (2004). Recognition of Tacit Skills: Sustaining Learning Outcomes in Adult Learning and Work Re-entry. *International Journal of Training and Development, 8*(1), 54–72. doi:10.1111/j.1360-3736.2004.00196.x

Falbo, R., Borges, L., & Valente, F. (2004). Using knowledge management to improve software process performance in a CMM level 3 organization. In *Proceedings of the fourth International Conference on Quality Software*(pp. 162-169). Washington, DC: IEEE Computer Society.

Fan, W., Wallace, L., Rich, S., & Zhang, Z. (2005). Tapping into the power of text mining. *Communications of the ACM, 49*(2), 76–82.

Feldman, R., Aumann, Y., Liberzon, Y., Ankori, K., Schler, J., & Rosenfeld, B. (2001). Information retrieval and text mining: a domain independent environment for creating information extraction modules.*Proceedings of the CIKM'01*, Atlanta, 586-588.

Fengjie, A., Fei, Q., & Xin, C. (2004). Knowledge Sharing and Web-Based Knowledge-Sharing Platform. *CEC-EAST '04: Proceedings of the E-Commerce Technology for Dynamic E-Business, IEEE International Conference,* 278-281.

Fenstermacher, K. D. (2005). The Tyranny of Tacit Knowledge: What Artificial Intelligence Tells Us About Knowledge Representation. *Proceedings of the 38th Hawaii International Conference on System Sciences.* 3 – 6 January. Hawaii: IEEE Computer Society Press.

Ferdows, K., & De Meyer, A. (1990). Lasting improvements in manufacturing performance: in search of a new theory. *Journal of Operations Management, 9*, 168–184. doi:10.1016/0272-6963(90)90094-T

Fernandes, K., Raja, V., & Austin, S. (2005). Portals as a knowledge repository and transfer tool – VIZCon case study. *Technovation, 25*(11), 1281–1289. doi:10.1016/j.technovation.2004.01.005

Fernie, S., Green, S. D., Weller, S. J., & Newcombe, R. (2003). Knowledge Sharing: Context, Confusion and Controversy. *International Journal of Project Management*, *21*(3), 177–186. doi:10.1016/S0263-7863(02)00092-3

Feyerherm, A. E., & Rice, C. L. (2002). Emotional intelligence and team performance: The good, the bad and the ugly. *The International Journal of Organizational Analysis*, *10*(4), 343–362. doi:10.1108/eb028957

Figallo, C., & Rhine, N. (2002). *Building the knowledge management network: best practices, tools and techniques for putting conversation to work*. New York: John Wiley & Sons.

Fih.foxconn.com (2007). *FIH: Vertical Integration*. Retreived February 10, 2010, from fih.foxconn.com.

Fine, C. (1998). *Clockspeed: Winning Industry Control in the Age of Temporary Advantage*. Cambridge: Perseus Publishing.

Firestone, J. (1999). The Artificial Knowledge Manager Standard: A "Strawman". *Working Paper No. One*. KM ANSI/ISO Standards Committee Meeting. Executive Information Systems, Inc. Retrieved March 12, 2009, from http://www.dkms.com/papers/akmsstrawbak.pdf.

Fisher, G., & Ostwald, J. (2001). Knowledge management: problems, promises, realities and challenges. *IEEE Intelligent Systems*, *1*(16), 60–72. doi:10.1109/5254.912386

Flesca, S., Manco, G., Masciari, E., Rende, E., & Tagarelli, A. (2004). Web wrapper induction: a brief survey. *AI Communications*, *17*(2), 57–61.

Forrest, D., & Simmons, R. (2003). Sport and gambling. *Oxford Review of Economic Policy*, *19*(419), 598–611. doi:10.1093/oxrep/19.4.598

Frappaolo, C. (2006). *Knowledge Management*. London: Capstone Publishing.

French, W. H., & Bell, C. H. Jr. (1989). *Organization Development*. New Delhi, India: Prentice Hall.

Freyens, B., & Martin, M. (2007). Multidisciplinary knowledge transfer in training multimedia projects. *Journal of European Industrial Training*, *31*(9), 680–705. doi:10.1108/03090590710846666

Fugate, B. S., Stank, T. P., & Mentzer, J. T. (2008). Linking improved knowledge management to operational and organizational performance. *Journal of Operations Management*, *27*(3), 247–264. doi:10.1016/j.jom.2008.09.003

Furquim, T., Oliveira, M., & Amaral, S. (2007). Enabling knowledge creation and sharing in software organizations: the case of the Brazilian federal software agency. In C. Stary, F. Barachini, & Hawamdeh, S. (Eds.), *Knowledge management: innovation, technology and cultures: Proceedings of the fourth International Conference on Knowledge Management* (pp. 101-109). New Jersey: World Scientific.

Furst, S. A., & Cable, D. M. (2008). Employee resistance to organizational change: Managerial influence tactics and leader-member exchange. *The Journal of Applied Psychology*, *93*(2), 453–462. doi:10.1037/0021-9010.93.2.453

Gabriel, Y. (2000). *Storytelling in organizations: Facts, fictions, and fantasies*. London: Oxford University Press.

Galbreath, J. (2005). Which resources matter the most to firm success? An exploratory study of resource-based theory. *Technovation*, *25*(9), 979–987. doi:10.1016/j.technovation.2004.02.008

Ganesan, S. (1994). Determinants of long-term orientation in buyer–supplier relationships. *Journal of Marketing*, *58*(2), 1–19. doi:10.2307/1252265

Ganesh, U., Miree, C. E., & Prescott, J. (2003). Competitive intelligence field research: moving the field forward by setting a research agenda. *Journal of Competitive Intelligence and Management*, *1*(1), 1–12.

Garnsey, E. (1998). A Theory of the Early Growth of the Firm. *Industrial and Corporate Change*, *7*(3), 523–556.

Gassmann, O. (2001). Multicultural teams: Increasing creativity and innovation by diversity. *Creativity and Innovation Management*, *10*(2), 88–95. doi:10.1111/1467-8691.00206

Gavalec, M., & Mls, K. (2005). Individual cognitive maps and evaluation of their dynamic changes. *International conference Kognice 2005,* Hradec Králové, Czech Republic, pp.24-29.

Germain, R., Droge, C., & Christensen, W. (2001). The mediating role of operations knowledge in the relationship of context with performance. *Journal of Operations Management, 19*, 453–469. doi:10.1016/S0272-6963(00)00067-X

Gibb, K. (1999). Regional Differentiation and the Scottish private housebuilding sector. *Housing Studies, 14*(1), 43–56. doi:10.1080/02673039982993

Gibbert, M., & Krause, H. (2002). Practice exchange in a best practice market place. In Davenport, T. H., & Gilbert, J. B. (Eds.), *Knowledge management case book: Siemen Best Practices*. Erlangen, Germany: Publicis Corporate Publishing.

Gilmour, D. (2003). How to fix knowledge management. *Harvard Business Review, 81*(10), 16–17.

Gilson, L. L., & Shalley, C. E. (2004). A little creativity goes a long way: An examination of team's engagement in creative processes. *Journal of Management, 30*(4), 453–470. doi:10.1016/j.jm.2003.07.001

Glaser, B. G., & Strauss, A. L. (1967). *The Discovery of Grounded Theory: Strategies for Qualitative Research.* Mill Valley, CA: Sociology Press.

Gloet, M., & Terziovski, M. (2004). Exploring the relationship between knowledge management practices and innovation performance. *Journal of Manufacturing Technology Management, 15*(5), 402–409. doi:10.1108/17410380410540390

Goddard, A. (1998). Facing Up to Market Forces. *Times Higher Education Supplement, 13*(November), 6–7.

Goh, S. (2002). Managing effective knowledge transfer: An integrative framework and some practice implications. *Journal of Knowledge Management, 1*(6), 23–30. doi:10.1108/13673270210417664

Gold, A. H., Malhotra, A., & Segars, A. H. (2001). Knowledge management: An organizational capabilities perspective. *Journal of Management Information Systems, 18*(1), 185–214.

Gold, A., Malhotra, A., & Segars, A. (2001). Knowledge management: an organizational capabilities perspective. *Journal of Management Information Systems, 18*(1), 185–214.

Goman, C. K. (2007). The hoarding/sharing instinct. *Knowledge Management Review, 9*(6), 6–6.

Gooderham, P. N. (2007). Enhancing knowledge transfer in multinational corporations: a dynamic capabilities driven model. *Knowledge Management Research & Practice, 5*, 34–43. doi:10.1057/palgrave.kmrp.8500119

Gore, C., & Gore, E. (1999). Knowledge Management: The Way Forward. *Total Quality Management, 10*(4-5), 554–560.

Gourlay, S. (2006). Conceptualizing Knowledge Creation: A Critique of Nonaka's Theory. *Journal of Management Studies, 43*(7), 1415–1436. doi:10.1111/j.1467-6486.2006.00637.x

Gourlay, S. (2006). Towards Conceptual Clarity for 'Tacit Knowledge': A Review of Empirical Studies. *Knowledge Management Research and Practice., 4*(1), 60–69. doi:10.1057/palgrave.kmrp.8500082

Gourlay, S. (2004). Tacit Knowledge: The Variety of Meaning in Empirical Research. *Proceedings of the Fifth European Conference on Organizational Knowledge, Learning and Capabilities.* 2-3 April. Innsbruck, Austria.

Govindarajan, V., & Trimble, C. (2004). Strategic innovation and the science of learning. *MIT Sloan Management Review, 45*(2), 67–75.

Graham, C., & Chattopadhyay, S. (2009). Well-being and Public Attitudes in Afghanistan. *World Economy, 10*(3), 105–146.

Grant, R. (1996). Prospering in Dynamically-competitive Environments: Organizational Capability as Knowledge Integration. *Organization Science*, *7*(4), 375–387. doi:10.1287/orsc.7.4.375

Grant, R. M. (1996a). Prospering in dynamically-competitive environments: Organizational capability as knowledge integration. *Organization Science*, *7*(4), 375–387. doi:10.1287/orsc.7.4.375

Grant, R. M., & Baden-Fuller, C. (2004). A knowledge accessing theory of strategic alliances. *Journal of Management Studies*, *41*(1), 61–84. doi:10.1111/j.1467-6486.2004.00421.x

Grant, R. M. (1996). Toward a knowledge-based theory of the firm. *Strategic Management Journal*, *17*, 109–122.

Grant, E. B., & Gregory, M. J. (1999). Adapting manufacturing processes for international transfer. *International Journal of Operations & Production Management*, *17*(10), 994–1005. doi:10.1108/01443579710176997

Grant, R. M. (2000). Shifts in the world economy: the drivers of knowledge management. In Despres, C., & Chauvel, D. (Eds.), *Knowledge Horizonts: the present and the promise of knowledge management* (pp. 27–53). Boston: Butterworth-Heinemann.

Grant, R. M. (1996b). Towards a knowledge-based theory of the firm. *Strategic Management Journal,* 17(Winter Special Issue), 109-122.

Grant, R.M. (1996). Toward a Knowledge-based theory of the firm. *Strategic Management Journal,* 17, Winter Special Issue, 109-122.

Greiner, L. (1972). Evolution and revolution as organizations grow. *Harvard Business Review*, *50*(4), 37–46.

Grimes, S. (2004). Consumer and enterprise search: not an exact match. *Intelligent Enterprise*, *7*(9), 22–31.

Groce, R. (2004). An experiential study of elementary teachers with the storytelling process: Interdisciplinary benefits associated with teacher training and classroom integration. *Reading Improvement*, 41.

Groff, T., & Jones, T. (2003). *Introduction to knowledge management: KM in business*. Oxford: Butterworth-Heinemann.

Grover, V., & Malhotra, M. K. (2003). Transaction cost framework in operations and supply chain management research: theory and measurement. *Journal of Operations Management*, *21*, 457–473. doi:10.1016/S0272-6963(03)00040-8

Grover, V., & Davenport, T. (2001). General perspectives on knowledge management: fostering a research agenda. *Journal of Management Information Systems*, *18*(1), 5–21.

Gruhl, D., Chavet, L., Gibson, D., Meyer, J., Pattanayak, P., Tomkins, A., & Zien, J. (2004). How to build a WebFountain: an architecture for very large-scale text analytics. *IBM Systems Journal*, *43*(1), 64–77. doi:10.1147/sj.431.0064

Guba, E., & Lincoln, Y. (1994). Competing paradigms in qualitative research. In Denzin, N., & Lincoln, Y. (Eds.), *Handbook of qualitative research* (pp. 105–177). Thousand Oaks, CA: Sage.

Guest, D. E. (2002). Human resource management, corporate performance and employee wellbeing: Building the worker into HRM. *The Journal of Industrial Relations*, *44*(3), 335–358. doi:10.1111/1472-9296.00053

Gulati, R., & Sytch, M. (2007). Dependence asymmetry and joint dependence in interorganizational relationships: Effects of embeddedness on a manufacturer's performance in procurement relationships. *Administrative Science Quarterly*, *52*, 32–69.

Gulev, R. (2005). Economic Culture Impacts on Knowledge Flows and Knowledge Sharing within Selected EU Countries. *Journal of Knowledge Management Practice*. Retrieved September 20, 2009, from http://www.tlainc.com/articl102.htm.

Gupta, B., Iyer, L. S., & Aronson, J. E. (2000). Knowledge Management: Practices and Challenges. *Industrial Management & Data Systems*, *100*(1), 17–21. doi:10.1108/02635570010273018

Gupta, A., & Govindarajan, V. (2000). Knowledge management's social dimension: lessons from Nucor Steel. *MIT Sloan Management Review*, *42*(1), 71–81.

Guptara, P. (1999). Why knowledge management fails. *Knowledge Management Review*, *9*(1), 26–29.

Gurteen, D. (1998). Knowledge, creativity and innovation. *Journal of Knowledge Management*, *2*(1), 5–13. doi:10.1108/13673279810800744

Guzman, G. A., & Wilson, J. (2005). The "Soft" Dimension of Organizational Knowledge Transfer. *Journal of Knowledge Management*, *9*(2), 59–74. doi:10.1108/13673270510590227

Hackbarth, G., & Grover, V. (1999). The Knowledge Repository: Organizational Memory Information *Systems. IS Management*, *16*(3), 21–30.

Haesli, A., & Boxall, P. (2005). When knowledge management meets HR strategy: an exploration of personalization-retention and codification-recruitment configurations. *International Journal of Human Resource Management*, *16*(11), 1955–1975. doi:10.1080/09585190500314680

Halbwachs, M. (1990). *A memória colectiva*. São Paulo: L.L. Shaffter.

Haldin-Herrgard, T. (2000). Difficulties in Diffusing of Tacit Knowledge in Organization. *Journal of Intellectual Capital*, *1*(4), 357–363. doi:10.1108/14691930010359252

Haldin-Herrgard, T. (2000). Difficulties in diffusion of tacit knowledge in organizations. *Journal of Intellectual Capital*, *1*(4), 357–365. doi:10.1108/14691930010359252

Haldin-Herrgard, T. (2004). Diving Under the Surface of Tacit Knowledge. *Proceedings of Fifth European Conference on Organizational Knowledge, Learning and Capabilities*. 2-3 April. Innsbruck, Austria.

Handfield, R. B., & Bechtel, C. (2002). The role of trust and relationship structure in improving supply chain responsiveness. *Industrial Marketing Management*, *31*, 367–382. doi:10.1016/S0019-8501(01)00169-9

Handy, C. (1995). Trust and the virtual organization. *Harvard Business Review*, *73*(3), 40–50.

Handy, C. (1996). *Gods of Management* (p. 14). New York: Oxford University Press USA.

Hansen, M. T., Nohria, N., & Tierney, T. (1999). What's Your Strategy for Managing Knowledge? *Harvard Business Review*, (March-April): 106–116.

Hansen, B. H., & Kautz, K. (2005). Grounded Theory Applied- Studying Information Systems Development Methodologies in Practice. *Proceedings of the 38th Hawaii International Conference on System Sciences*. 3 – 6 January. Hawaii: IEEE Computer Society Press.

Hargadon, A. B., & Bechky, B. A. (2006). When collections of creatives become creative collectives: A field study of problem solving at work. *Organization Science*, *17*(4), 484–500. doi:10.1287/orsc.1060.0200

Harlow, H. (2008). The effect of tacit knowledge on firm performance. *Journal of Knowledge Management*, *12*(1), 148–163. doi:10.1108/13673270810852458

Harrison, J. R., & Carroll, G. R. (1991). Keeping the Faith: A Model of Cultural Transmission in Formal Organizations. *Administrative Science Quarterly*, *36*, 552–582. doi:10.2307/2393274

Harvey, P., & Butcher, D. (1998). Those who make a difference: Developing business through developing individuals. *Industrial and Commercial Training*, *30*(1), 12–15. doi:10.1108/00197859810197690

Hassanuddin, T. (2009). *House Buyer's Awareness: the case of Penang, Malaysia*. Unpublished M. Sc. Project Management dissertation, Universiti Sains Malaysia, Malaysia.

Hawamdeh, S. (2003). *Knowledge management cultivating knowledge professionals*. Oxford, UK: Chandos Publishing.

Haythornthwaite, C. (2002). Strong, weak and latent ties and the impact of new media. *The Information Society*, *18*(5), 385–401. doi:10.1080/01972240290108195

Hearst, M. A. (1999). Untangling text data mining. *Proceedings of ACL '99: 37th annual meeting of the association for computational linguistics*, University of Maryland, June 20-26, pp. 1-8.

Hedesstrom, T., & Whitley, E. A. (2000). What is meant by tacit knowledge? Towards a better understanding of the shape of actions. *Proceedings of the 8th European Conference on Information Systems*. 3-5 July. Vienna, Austria: ECIS 2000, 46-51.

Heide, J. B., & John, G. (1992). Do norms matter in marketing relationships? *Journal of Marketing*, *56*(2), 32–44. doi:10.2307/1252040

Heide, J. B., & Miner, A. S. (1992). The shadow of the future: Effects of anticipated interaction and frequency of contact on buyer-seller cooperation. *Academy of Management Journal*, *35*, 265–291. doi:10.2307/256374

Heier, H., & Borgman, H. P. (2002). Knowledge Management Systems Spanning Cultures: The Case of Deutsche Bank's HRBASE. *Symposium conducted at the meeting of ECIS 2002 in Gdansk*, Poland.

Helfat, C. E., & Raubitschek, R. S. (2000). Product sequencing: Co-evolution of knowledge, capabilities and products. *Strategic Management Journal*, *21*(10-11), 961–979. doi:10.1002/1097-0266(200010/11)21:10/11<961::AID-SMJ132>3.0.CO;2-E

Hellstrom, T., & Husted, K. (2004). Mapping Knowledge and Intellectual Capital in Academic Environments. *Journal of Intellectual Capital*, *5*(1), 165–180. doi:10.1108/14691930410512987

Hemenway, K., & Calishain, T. (2004). *Spidering Hacks: 100 Industrial-Strength Tips & Tools*. Sebastopol, CA: O'Reilly.

Henderson, R. M., & Clark, K. B. (1990). Architecturai innovation: The reconfiguration of existing product technologies and the failure of established firms. *Administrative Science Quarterly*, *35*, 9–30. doi:10.2307/2393549

Henderson, R. M., & Cockburn, I. (1994). Measuring competence? Exploring firm effects in pharmaceutical research. *Strategic Management Journal*, *15*(Winter), 63–84. doi:10.1002/smj.4250150906

Herkema, S. (2003). A complex adaptive perspective on learning within innovation projects. *The Learning Organization*, *10*(6), 340–346. doi:10.1108/09696470310497177

Hernandez, M. (2000). *The impact of the dimensions of the Learning Organization on the transfer of tacit knowledge process and performance improvement within private manufacturing firms in Colombia.* Doctoral Dissertation. University of Georgia Athens, Georgia. ProQuest Digital Dissertation, UMI No:9994101, Retrieved on May 20, 2007 from http://wwwlib.umi.com/dissertations/fullcit9994101.

Hijazi, S., & Kelly, L. (2003). Knowledge Creation in Higher Education Institutions: A Conceptual Model. *Proceedings of Association of Small Computer Users in Education (ASCUE) Conference.* 8-12 June. South Carolina, United States of America.

Hill, C. W. L. (1990). Cooperation, opportunism, and the invisible hand: Implications for transaction cost theory. *Academy of Management Review*, *15*(3), 500–513. doi:10.2307/258020

Hipkin, I. B. (2004). Determining technology strategy in developing countries. *Omega International Journal of Management Science*, *32*(3), 245–260. doi:10.1016/j.omega.2003.11.004

Hipkin, I. B., & Bennett, D. J. (2003). Managerial perceptions of factors influencing technology management in South Africa. *Technovation*, *23*, 719–735. doi:10.1016/S0166-4972(02)00031-7

Hislop, D. (2002). Mission Impossible? Communicating and Management Knowledge via Information technology. *Journal of Information Technology*, (17): 165–177. doi:10.1080/02683960210161230

Hislop, D. (2003). Linking human resource management and knowledge management via commitment, a review and research agenda. *Employee Relations*, *25*(1/2), 182–202. doi:10.1108/01425450310456479

Hitchins, D. (2007). *Systems Engineering: A 21st Century Systems Methodology*. Chichester, UK: John Wiley and Sons.

Hoetker, G. (2005). How much you know versus how well I know you: Selecting a supplier for a technically innovative component. *Strategic Management Journal*, *26*(1), 75–96. doi:10.1002/smj.453

Hofstede, G. (1980). *Culture's Consequences: International Differences in Work-Related Values*. London: Sage Publications.

Hofstede, G. (2001). *Culture's Consequences* (2nd ed.). London: Sage Publications.

Hofstede, G., Neuijen, B., Ohayv, D. D., & Sanders, G. (1990). Measuring Organizaitonal Cultures: A Qualitative and Quantitative Study Across Twenty Cases. *Administrative Science Quarterly*, (35): 286–316. doi:10.2307/2393392

Hofstede, G. (1991). *Cultures and organizations: Software of the mind*. London: McGraw-Hill.

Hofstede, G. (2008) Geert Hofstede Cultural Dimensions. Retrieved from http://www.geert-hofstede.com, Accessed May 15, 2008.

Holcomb, T. R., & Hitt, M. A. (2007). Toward a model of strategic outsourcing. *Journal of Operations Management*, *25*(2), 464–481. doi:10.1016/j.jom.2006.05.003

Holden, R. J., Or, C. K. J., Alper, S. J., Rivera, A. J., & Karsh, B. T. (2008). A change management framework for macroergonomic field research. *Applied Ergonomics*, *39*(4), 459–474. doi:10.1016/j.apergo.2008.02.016

Holsapple, C. W., & Joshi, K. D. (2001). Organizational Knowledge Resources. *Decision Support Systems*, *1*(31), 39–54. doi:10.1016/S0167-9236(00)00118-4

Holsapple, C. W., & Singh, M. (2003). The Knowledge Chain Model: Activities for Competitiveness. In Holsapple, C. W. (Ed.), *Handbook on Knowledge Management 2: Knowledge Directions* (*Vol. 215-251*). Heidelberg, Germany: Springer-Verag Berlin.

Holsapple, C. W., & Joshi, K. D. (2003). A Knowledge Management Ontology. In Holsapple, C. W. (Ed.), *Knowledge matters* (pp. 89–124). Berlin: Springer.

Hoog, R. (1999). Investigating a Theoretical Framework for Knowledge Management: A Gaming Approach. In Liebowitz, J. (Ed.), *Knowledge Management Handbook*. Boca Raton, FL: CRC Press.

Horvath, J. A., Forsythe, G. B., Bullis, R. C., Sweeny, P. J., Williams, W. M., & McNally, J. A. (1999). Experience, Knowledge, and Military Leadership. In Sternberg, R. J., & Horvath, J. A. (Eds.), *Tacit Knowledge in Professional Practice* (pp. 39–58). Mahwah, NJ and London: Lawrence Erlbaum Associates.

Horvath, J. A., Williams, W. M., Forsythe, G. B., Sweeney, P. J., Sternberg, R. J., McNally, J. A., et al. (1994). *Tacit knowledge in military leadership: A review of the literature* (Technical Report 1017). Alexandria, VA: United States Army Research Institute for the Behavioral and Social Sciences.

House Buyers Association. (2003), Adopt build-then sell variant. *NST-PROPERTY*. Retrieved January 11, 2007 from http://www.hba.org.my/articles/buyer_watch/2003/adopt_build.htm

House Buyers Association. (2005, June). *Sell then Build or Build then Sell or 10:90 Variant Models for the Housing Delivery System*. Paper presented at a Housing Workshop, Kajian Dasar Perumahan Negara, Penang, Malaysia.

Howells, J. (1996). Tacit Knowledge, Innovation and Technology Transfer. *Technology Analysis and Strategic Management*, *8*, 91–106. doi:10.1080/09537329608524237

Hsu, I.-C. (2008). Knowledge sharing practices as a facilitating factor for improving organizational performance through human capital: A preliminary test. *Expert Systems with Applications*, *35*(3), 1316–1326. doi:10.1016/j.eswa.2007.08.012

Huber, G. P. (1991). Organizational learning: The contributing processes and the literatures. *Organization Science*, *2*(1), 88–115. doi:10.1287/orsc.2.1.88

Huber, G. (1991). Organizational learning: The contributing processes and the literatures. *Organization Science*, *2*(1), 88–115. doi:10.1287/orsc.2.1.88

Hughes, J., & Howcroft, D. A. (2000). Grounded Theory: Never Knowingly Understood. *Information Systems Research*, *4*(1), 181–1974.

Hughes, N. (2009). *Foxconn reportedly tapped to manufacture next-gen iPhone.* Retrieved February 10, 2010, from www.appleinsider.com.

Hult, G. T., Ketchen, D. J., Cavusgil, S. T., & Calantone, R. J. (2006). Knowledge as a strategic resource in supply chains. *Journal of Operations Management*, *24*, 458–475. doi:10.1016/j.jom.2005.11.009

Hult, G. T. M., & Ketchen, J. D. J., S. T. & Arrfelt, M. (2007). Strategic supply chain management: Improving performance through a culture of competitiveness and knowledge development. *Strategic Management Journal*, *28*, 1035–1052. doi:10.1002/smj.627

Hult, G. T. M., Ketchen, J. D. J., Cavusgil, S. T., & Calantone, R. J. (2006). Knowledge as a strategic resource in supply chains. *Journal of Operations Management*, *24*(5), 458–475. doi:10.1016/j.jom.2005.11.009

Hunt, D. P. (2003). The Concept of Knowledge and How to Measure It. *Journal of Intellectual Capital*, *4*(1), 100–112. doi:10.1108/14691930310455414

Huo, Y. P., Huang, H. J., & Naiper, N. K. (2002). Divergence or convergence: A cross-national comparison of personnel selection practices. *Human Resource Management*, *41*(1), 31–44. doi:10.1002/hrm.10018

Huselid, M. (1995). The impact of human resource management practices on turnover, productivity, and corporate financial performance. *Academy of Management Journal*, *38*, 635–670. doi:10.2307/256741

Husted, K., & Michailova, S. (2002). Knowledge sharing in Russian companies with Western participation. *Management International*, *6*(2), 17.

Husted, K., & Michailova, S. (2002). Diagnosing and fighting knowledge-sharing hostility. *Organizational Dynamics*, *31*(1), 60–73. doi:10.1016/S0090-2616(02)00072-4

Hutchings, K., & Michailova, S. (2004). Facilitating knowledge sharing in Russian and Chinese subsidiaries: The role of personal networks and group membership. *Journal of Knowledge Management*, *8*(2), 84–94. doi:10.1108/13673270410529136

Huxley, A. (2004). *The Doors of Perception and Heaven and Hell* (p. 13). New York: Harper Perennial Modern Classics.

IBM. (2005). *IBM Blogging policy and guidelines*.Retrieved from http://www.snellspace.com/IBM_Blogging_Policy_and_Guidelines.pdf

Idsgroup.com (2009). *Profile*. Retreived February 10, 2010, from idsgroup.com.

Igbaria, M. (1990). End-user computing effectiveness: A structural equation model. *Omega*, *18*(6), 637. doi:10.1016/0305-0483(90)90055-E

Inkpen, A.C. & Tsang, E.W.K. (2005).Social Capital, Networks and Knowledge Transfer. *Academy of Management Review* 30(1), 2005, pp. 146-165.

Ipe, M. (2003). Knowledge sharing in organizations: A conceptual framework. *Human Resource Development Review*, *2*(4), 337–359. doi:10.1177/1534484303257985

Irwin, H., More, E., & McGrath, M. (1998). Relationship management for innovation: the central role of communication in Australia's participation in two high-tech industries. *Technology Analysis and Strategic Management*, *10*(4), 467–481. doi:10.1080/09537329808524328

Ives, W., Torrey, B., & Gordon, C. (2000). Knowledge sharing is human behavior. In Morey, D., Maybury, M., & Thuraisingham, B. (Eds.), *Knowledge Management: Classic and Contemporary Works*. Massachusetts: Massachussetts Institute of Technology Press.

Jahn, M., et al. 2005. *Strategy of economic growth of Czech Republic*. Prague: Czech Republic. Retrieved May 12, 2009, from http://www.hospodarskastrategie.org.

Jap, S. D. (1999). Pie-expansion efforts: Collaboration processes in buyer-supplier relationships. *JMR, Journal of Marketing Research*, *36*(November), 461–475. doi:10.2307/3152000

Jap, S. D., & Anderson, E. (2003). Safeguarding interorganizational performance and continuity under ex post opportunism. *Management Science*, *49*(12), 1684–1701. doi:10.1287/mnsc.49.12.1684.25112

Jayaratna, N. (1994). *Understanding and Evaluating Methodologies – NIMSAD A Systemic Framework*. London: McGraw-Hill.

Jennex, M., & Olfman, L. (2006). A model of knowledge management success. *International Journal of Knowledge Management*, *2*(3), 51–68.

Jennex, M., & Olfman, L. (2005). Assessing knowledge management success. *International Journal of Knowledge Management*, *1*(2), 33–49.

Jennex, M., & Olfman, L. (2002). Organizational memory/knowledge effects on productivity, A longitudinal study. *Proceedings of the 35th Hawaii International Conference on System Sciences(HICSS'02), Vol.6*(pp. 1316 – 1316). Hawaii: HICSS.

Jiang, X., & Lia, Y. (2008). An empirical investigation of knowledge management and innovative performance: The case of alliances. *Research Policy*, *38*(2), 358–368. doi:10.1016/j.respol.2008.11.002

Johannessen, J. A., Olaisen, J., & Olsen, B. (2001). Mismanagement of Tacit Knowledge: The Importance of Tacit Knowledge, The Danger of Information Technology, and What To Do About It. *International Journal of Information Management*, *21*, 3–20. doi:10.1016/S0268-4012(00)00047-5

John, O. P., & Srivastava, S. (1999). The Big Five trait taxonomy: History, measurement and theoretical perspectives. In Pervin, L. A., & John, O. P. (Eds.), *Handbook of personality: Theory and research* (pp. 102–138). New York: Guilford Press.

Johnson, J. M. (2002). In depth interviewing. In Gubrium, J. F., & Holsteins, J. A. (Eds.), *Handbook of Interview Research: Context and Methods* (pp. 103–120). Thousand Oaks, CA: Sage.

Jones, O., & Craven, M. (2001). Expanding capabilities in a mature manufacturing firm: absorptive capacity and the TCS. *International Small Business Journal*, *19*(3), 39–55. doi:10.1177/0266242601193003

Jones, G. R., & George, J. M. (1998). The experience and evolution of trust: Implications for cooperation and teamwork. *Academy of Management Review*, *23*(3), 531–546. doi:10.2307/259293

Jones, R., Jimmieson, N., & Griffiths, A. (2005). The impact of organizational culture and reshaping capabilities on change implementation success: the mediating role of readiness for change. *Journal of Management Studies*, *42*(2), 361–386. doi:10.1111/j.1467-6486.2005.00500.x

Jordan, P. J., Ashkanasy, N. M., Hartel, C. E. J., & Hooper, G. S. (2002). Workgroup emotional intelligence development and relationship to team process effectiveness and goal focus. *Human Resource Management Review*, *12*(2), 195–214. doi:10.1016/S1053-4822(02)00046-3

Jordan, P. J., & Troth, A. C. (2002). Emotional intelligence and conflict resolution: Implications for human resource development. *Advances in Developing Human Resources*, *4*(1), 62–79. doi:10.1177/1523422302004001005

Jordan, P. J., & Troth, A. C. (2004). Managing emotions during team problem solving. *Human Performance*, *17*(2), 195–218. doi:10.1207/s15327043hup1702_4

Jordan, P.J., & Lawrence, S.A., Troth, & A.J., Jordan. (2006). The impact of negative mood on team performance. *Journal of Management & Organization*, *12*, 131–145.

Jordan, P. J., & Troth, A. J. (2002). Emotional Intelligence and Conflict Resolution: Implications for Human Resource Development. *Advances in Developing Human Resources*, *4*, 62–79. doi:10.1177/1523422302004001005

Josephson, P. E., & Hammarlund, Y. (1999). The causes and costs of defects in construction: A study of seven building projects. *Automation in Construction*, *8*(6), 681–687. doi:10.1016/S0926-5805(98)00114-9

Joshi, K. D., Sarker, S., & Sarker, S. (2007). Knowledge transfer within information systems development teams: Examining the role of knowledge source attributes. *Decision Support Systems*, *43*(2), 322–335. doi:10.1016/j.dss.2006.10.003

Kakabadse, N. K., Kakabadse, A., & Kouzmin, A. (2003). Reviewing the Knowledge Management Literature: Towards a Taxonomy. *Journal of Knowledge Management*, *7*(4), 75–91. doi:10.1108/13673270310492967

Kanhanhali, A., Tan, B. C. Y., & Wei, K. K. (2005). Contributing Knowledge to Electronic Knowledge Repositories: An Empirical Investigation. *Management Information Systems Quarterly*, *29*(1), 113–143.

Kankanhalli, A., & Tan, B. (2004). A review of metrics for knowledge management systems and knowledge management initiatives. *Proceedings of the 37th Annual Hawaii International Conference on System Sciences (HICSS'04),* vol. 8.(pg.8). Hawaii: HICSS.

Kao, J. (1996). *Jamming*. London: Haper Collins Business.

Kaplan, R. S., & Norton, D. P. (1996). *Balanced Scorecard: Translating Strategy into Action*. Boston: Harvard Business School Press.

Kasi, P. (1992, April). *Pros and Cons towards Build Then Sell concept*. Paper presented at the Seminar of the Build then Sell Housing Development Approach Towards 2020, Kuala Lumpur, Malaysia.

Katz, R., Rebentisch, E. S., & Allen, T. J. (1996). A study of technology transfer in a multinational cooperative joint venture. *IEEE Transactions on Engineering Management*, *43*(1), 97–105. doi:10.1109/17.491273

Katz, R., & Allen, T. (1982). Investigating the not invented here (NIH) syndrome: A look at the performance, tenure and communication patterns of 50 R&D projects groups. *R & D Management*, *12*(1), 7–19. doi:10.1111/j.1467-9310.1982.tb00478.x

Katzeff, C., & Ware, V. (2006). Video storytelling as mediation of organizational learning. *NordiCHI '06: Proceedings of the 4th Nordic conference on Human-computer interaction,* Oslo, Norway. 311-320.

Kazanjian, R. K. (1988). Relation of Dominant Problems to Stages of Growth in Technology-Based New Ventures. *Academy of Management Journal*, *31*(2), 257–279. doi:10.2307/256548

KDNuggets. (2009). Top Data Mining Tools Used. KDNuggets, 2009(10), item 1. [Electronic] Available: http://www.kdnuggets.com/news/2009/n10/1i.html.

Kearns, G. S., & Lederer, A. L. (2004). The impact of industry contextual factors on IT focus and the use of IT for competitive advantage. *Information & Management*, *41*(7), 899–919. doi:10.1016/j.im.2003.08.018

Kennedy, S. D. (1998). Competitive intelligence on the Web 101. *Information Today*, *15*(9), 40–41.

Kermally, S. (1998). *New Economy Energy: Unleasing Knowledge for Competitive Advantage*. Sussex, England: John Wiley and Sons.

Kettinger, W. J., & Lee, C. C. (1994). Perceived service quality and user satisfaction with the Information services function. *Decision Sciences*, *25*(5/6), 737–765. doi:10.1111/j.1540-5915.1994.tb01868.x

Kettunen, J. (2004a). Bridge building to the future of Finnish polytechnics. *Journal of Higher Education Outreach and Engagement*, *9*(2), 43–57.

Kettunen, J. (2004b). The strategic evaluation of regional development in higher education. *Assessment & Evaluation in Higher Education*, *29*(3), 357–368. doi:10.1080/0260293042000197591

Kettunen, J. (2006). Strategic planning of regional development in higher education. *Baltic Journal of Management*, *1*(3), 259–269. doi:10.1108/17465260610690917

Kettunen, J. (2007). Innovativeness in higher education management. *Bhavishya. Journal of Future Business School*, *1*(2), 65–74.

Khadwall, P. N. (1995). *Management Styles*. New Delhi: McGraw-Hill.

Khalfan, A., & Alshawaf, A. (2004). Adoption and implementation problems of e-banking: A study of the managerial perspective of the banking industry in Oman. *Journal of Global Information Technology Management*, *7*(1), 47–64.

Khamseh, H. M., & Jolly, D. R. (2008). Knowledge transfer in alliances: Determinant factors. *Journal of Knowledge Management*, *12*(1), 37–50. doi:10.1108/13673270810852377

Khurana, R. (2007). *From Higher Aims to Hired Hands: The Social Transformation of American Business Schools and the Unfulfilled Promise of Management as a Profession*. Princeton: Princeton University Press.

Kim, Y. G., Yu, S. H., & Lee, J. H. (2003). Knowledge strategy planning: methodology and case. *Expert Systems with Applications*, *3*(24), 295–307. doi:10.1016/S0957-4174(02)00158-6

Kim, H., & Gong, Y. (2009). The roles of tacit knowledge and OCB in the relationship between group-based pay and firm performance. *Human Resource Management Journal*, *19*(2), 120–139. doi:10.1111/j.1748-8583.2009.00093.x

King, J. (2001). Shell strikes knowledge gold. *Computerworld*, *1*(3), 34.

Klein, H. K., & Myers, M. D. (1999). A Set of Principles for Conducting and Evaluating Interpretive Field Studies in Information Systems. *Management Information Systems Quarterly*, *23*(1), 67–93. doi:10.2307/249410

Kneuper, R. (2002). Supporting software processes using knowledge management. In Chang, S. K. (Ed.), *Handbook of software engineering and knowledge engineering* (*Vol. II*, pp. 579–608). New Jersey: World Scientific Publishing.

Ko, D. G., Kirsch, L. J., & King, W. R. (2005). Antecedents of Knowledge Transfer from Consultants to Clients in Enterprise System Implementations. *Management Information Systems Quarterly*, *29*(1), 59–85.

Kobayashi, M., & Takeda, K. (2000). Information retrieval on the Web. *ACM Computing Surveys*, *32*(2), 144–173. doi:10.1145/358923.358934

Koenig, M. (2002). The Third Stage of KM Emerges. *KMWorld.*, *11*(3), 20–21.

Kogut, B., & Zander, U. (1992). Knowledge of the firm, combinative capabilities, and the replication of technology. *Organization Science*, *3*(3), 383–397. doi:10.1287/orsc.3.3.383

Kogut, B., & Zander, U. (1996). What firms do? Coordination, identity, and learning. *Organization Science*, *7*(5), 502–518. doi:10.1287/orsc.7.5.502

Kosala, R., & Blockheel, H. (2000). Web mining research: a survey. *Proceedings of SIGKDD*, *2*(1), 1–15. doi:10.1145/360402.360406

Koskinen, K. U. (2003). Evaluation of Tacit Knowledge Utilization in Work Units. *Journal of Knowledge Management*, *7*(5), 67–80. doi:10.1108/13673270310505395

Koudal, P., & Engel, D. A. (2007). Globalization and emerging markets: The challenge of continuous global network optimization. In Lee, H. L., & Lee, C.-Y. (Eds.), *Building Supply Chain Excellence in Emerging Economies*. New York: Springer.

Kouvelis, P., & Niederhoff, J. (2007). On the globalization of operations and supply chain strategies: A conceptual framework and its application. In Lee, H. L., & Lee, C.-Y. (Eds.), *Building Supply Chain Excellence in Emerging Economies*. New York: Springer.

Kowch, E. G. (2005). Do we plan the journey or read the compass? An argument for preparing educational technologists to lead organizational change. *Research Journal of Educational Technology*, *36*(6), 1067–1070. doi:10.1111/j.1467-8535.2005.00577.x

Krause, D. R., Handfield, R. B., & Tyler, B. B. (2007). The relationships between supplier development, commitment, social capital accumulation and performance improvement. *Journal of Operations Management*, *25*(2), 528–545. doi:10.1016/j.jom.2006.05.007

Kristensson, P., & Norlander, T. (2003). The creative product and the creative processes in virtual environments. *Creativity and Innovation Management*, *12*(1), 32–40. doi:10.1111/1467-8691.00266

Krogh, G. (1998). Care in knowledge creation. *California Management Review*, *40*(3), 133–153.

Kukko, M., Helander, N., & Virtanen, P. (2008). Knowledge management in renewing software development processes. In: *Proceedings of the 41st Annual Hawaii International Conference on System Sciences* (pp. 332-342). Washington, DC: IEEE Computer Society.

Kumar, K., Von Glinow, M., & Van Fenema, P. (2009). Offshoring and global distribution of work: Implications for task interdependence theory and practice. *Journal of International Business Studies*, *40*(4), 642–667. doi:10.1057/jibs.2008.77

Kurtzberg, T. R., & Amabile, T. A. (2001). From Guilford to creative synergy: Opening the black box of team-level creativity. *Creativity Research Journal*, *13*(3), 285–294. doi:10.1207/S15326934CRJ1334_06

Kylen, S., & Shani, A. B. R. (2002). Triggering creativity in teams: an exploratory investigation. *Creativity and Innovation Management*, *11*(11), 17–30. doi:10.1111/1467-8691.00233

Lacity, M. C., & Willcocks, L. P. (1998). An empirical Investigation of Information Technology Sourcing Practices: Lessons from Experience. *Management Information Systems Quarterly*, *22*(3), 363–408. doi:10.2307/249670

Laframboise, D., Nelson, R. L., & Schmaltz, J. (2003). Managing resistance to change in workplace accommodation. *Journal of Facilities Management*, *1*(4), 306–321. doi:10.1108/14725960310808024

Lai, H., & Chu, T. (2000). Knowledge management: a review of theoretical frameworks and industrial cases. In R. Sprague Jr (Ed.) *Proceedings of the 33rd Hawaii International Conference on System Sciences* (pp. 1-10) Washington, DC: IEEE Computer Society.

Lall, S. (1993). Promoting technology development: the role of technology transfer and indigenous effort. *Third World Quarterly*, *14*(1), 95–109. doi:10.1080/01436599308420315

Lam, A. (2000). Tacit Knowledge, Organizational Learning and Societal Institutions: An Integrated Framework. *Organization Studies*, *21*(3), 487–513. doi:10.1177/0170840600213001

Larson, A. (1992). Network dyads in entrepreneurial settings: A study of the governance of exchange relationships. *Administrative Science Quarterly*, *37*, 76–104. doi:10.2307/2393534

Larsson, K. J., & Ohlin, L. (2002). *Implementing a knowledge sharing initiative – principles for success*. Sweden: Tesis Sarjana. Linkoping Institute of Technology.

LeBlanc, S. M., & Hogg, J. (2006). *Storytelling in knowledge management: An effective tool for uncovering tacit knowledge*. Retrieved January 23, 2007, from http://www.stcatlanta.org/currents06/proceedings/leblanc.pdf

Lee, H. L. (2004). The triple-A supply chain. *Harvard Business Review*, *82*(10), 102–112.

Lee, H. L., & Lee, C.-Y. (Eds.). (2007). *Building Supply Chain Excellence in Emerging Economies*. New York: Springer.

Lee, C. K., & Al-Hawamdeh, S. (2002). Factors impacting knowledge sharing. *Journal of Information and Knowledge Management*, *1*(1), 49–56. doi:10.1142/S0219649202000169

Lee, A. S. (1989). A scientific methodology for MIS studies. *Management Information Systems Quarterly*, *13*(1), 33–50. doi:10.2307/248698

Lee, H., & Choi, B. (2003). Knowledge management enablers, processes and organizational performance: an integrative view and empirical examination. *Journal of Management Information Systems*, *20*(1), 179–228.

Leenders, R. T. A. J., Van Engelen, J. M. L., & Kratzer, J. (2003). Virtuality, communication, and new product team creativity: A social network perspective. *Journal of Engineering and Technology Management*, *20*, 69–92. doi:10.1016/S0923-4748(03)00005-5

Lehaney, B., Clarke S., Coakes, E, and Jack Gillian. (2004). *Beyond knowledge management*. Hershey, PA,: Idea Group Publishing.

Leonard, D., & Sensiper, S. (1998). The Role of Tacit Knowledge in Group Innovation. *California Management Review*, *40*(3), 112–132.

Leonard-Barton, D. (1990). A dual methodology for case studies: synergistic use of a longitudinal single site with replicated multiple sites. *Organization Science*, *1*(1), 248–266. doi:10.1287/orsc.1.3.248

Lepak, D., P. & Snell, S., A. (1999). The Human Resource Architecture: Toward a Theory of Human Capital Allocation and Development. *Academy of Management Review*, *24*(1), 31–48. doi:10.2307/259035

Levina, N. (2005). Collaborating on multiparty information systems development projects: A collective reflection-in-action view. *Information Systems Research*, *16*(2), 109–130. doi:10.1287/isre.1050.0055

Lewin, K. (2008). *A Dynamic Theory Of Personality: Selected Papers*. Lewin Press.

Li, M., & Gao, F. (2003). Why Nonaka Highlights Tacit Knowledge: A Critical Review. *Journal of Knowledge Management*, *7*(4), 6–14. doi:10.1108/13673270310492903

Li, H., & Atuahene-Gima, K. (2001). Product innovation strategy and the performance of new technology ventures in China. *Academy of Management Journal*, *44*(6), 1123–1134. doi:10.2307/3069392

Li, Y., Liu, Y., Li, M., & Wu, H. (2008). Transformational offshore outsourcing: Empirical evidence from alliances in China. *Journal of Operations Management*, *26*, 257–274. doi:10.1016/j.jom.2007.02.011

Liebeskind, J. P. (1996). Knowledge, strategy, and the theory of the firm. *Strategic Management Journal*, 17(Winter Special Issue), 93-107.

Liebowitz, J., & Chen, Y. (2003). Knowledge Sharing Proficiencies: The Key to Knowledge Management. In Holsapple, C. W. (Ed.), *Handbook on Knowledge Management 1: Knowledge Matters* (*Vol. 409-424*). Berlin, Germany: Springer-Verlag, Berlin, Germany.

Liebowitz, J. (2006) *Strategic Intelligence: Business Intelligence, Competitive Intelligence, and Knowledge Management*. Boca Raton FL: Auerbach Publications (Taylor & Francis)

Liker, J. K., & Choi, T. Y. (2004). Building deep supplier relationships. *Harvard Business Review*, *82*(12), 104–113.

Lim, C. Y. (2003). *Factors contributing to E-readiness of SMEs*. Unpublished MBA thesis, Universiti Sains Malaysia, Malaysia.

Lin, C.-P. (2007). To share or not to share: Modeling tacit knowledge sharing, its mediators and antecedents. *Journal of Business Ethics*, *70*, 411–428. doi:10.1007/s10551-006-9119-0

Linde, C. (2001). Narrative and social tacit knowledge. *Journal of Knowledge Management*, *5*(2), 160–170. doi:10.1108/13673270110393202

Linderman, K. W., Schroeder, R. G., Zaheer, S., Liedtke, C., & Choo, A. S. (2004). Integrating quality management practices with knowledge creation processes. *Journal of Operations Management*, *22*, 589–607. doi:10.1016/j.jom.2004.07.001

Lines, R. (2004). Influence of participation in strategic change: resistance, organizational commitment and change goal achievement. *Journal of Change Management*, *4*(3), 193–215. doi:10.1080/1469701042000221696

Lippman, S. A., & Rumelt, R. P. (1982). Uncertain imitability: An analysis of interfirm differences in efficiency under competition. *The Bell Journal of Economics*, *13*(2), 418–438. doi:10.2307/3003464

Lister, S., & Wilder, A. (2005). Strengthening subnational administration in Afghanistan: technical reform or state-building? *Public Administration and Development*, *25*, 39–48. doi:10.1002/pad.346

Liu, Y., Luo, Y., & Liu, T. (2008). Governing buyer–supplier relationships through transactional and relational mechanisms: Evidence from China. *Journal of Operations Management*, *27*(4), 294–309. doi:10.1016/j.jom.2008.09.004

Liu, P., & Tsai, C. (2007). Effect of knowledge management systems on operating performance: An empirical study of hi-tech companies sing the balanced scorecard approach. *International Journal of Management*, *24*(4), 734–744.

Liu, B., Chin, C. W., & Ng, H. T. (2003). Mining topic-specific concepts and definitions on the Web. *Proceedings of WWW 2003*, May 20-24, Budapest, 251-260.

Liu, S. C. (2003). *A study of factors that facilitate use of knowledge management systems and the impact of use on individual learning. Unpublished* doctoral dissertation, Claremont Graduate University, Claremont, CA.

Lusch, R. F., & Brown, J. R. (1996). Interdependency, contracting, and relational behaviour in marketing channels. *Journal of Marketing*, *60*, 19–38. doi:10.2307/1251899

Lyles, M. A., & Salk, J. E. (1996). 'Knowledge Acquisition from Foreign Parents in International Joint Ventures: An Empirical Examination in the Hungarian Context. *Journal of International Business Studies*, *27*(5), 877–903. doi:10.1057/palgrave.jibs.8490155

Lynskey, M. J. (1999). The transfer of resources and competencies for developing technological capabilities - the case of Fujitsu-ICL. *Technology Analysis and Strategic Management*, *11*(3), 317–336. doi:10.1080/095373299107384

Macher, J. T. (2006). Technological Development and the Boundaries of the Firm: A Knowledge-Based Examination in Semiconductor Manufacturing. *Management Science*, *52*(6), 826–843. doi:10.1287/mnsc.1060.0511

MacLeod, M., & Davidson, E. (2007). Organizational storytelling and technology innovation. *Hawaii International Conference on System Sciences,* 0, 248.

Madhok, A., & Tallman, S. B. (1998). Resources, transactions and rents: Managing value through interfirm collaborative relationships. *Organization Science*, *9*(3). doi:10.1287/orsc.9.3.326

Magnusson, M. G. (2004). Managing the Knowledge Landscape of an MNC: Knowledge Networking at Ericsson. *Knowledge and Process Management*, *11*(4), 262–272. doi:10.1002/kpm.210

Maier, R. (2002). *Knowledge Management Systems: Information and communication technologies for knowledge management*. Berlin: Springer.

Malhotra, A., Majchrzak, A., Carman, R., & Lott, V. (2001). Radical innovation without collocation: A case study at Boeing-Rocketdyne. *MIS Quartely*, *25*(2), 229–249. doi:10.2307/3250930

Maqsood, T., Finegan, A. D., & Walker, D. H. (2004). Biases and Heuristics in Judgment and Decision Making: The Dark Side of Tacit Knowledge. *Issues in Informing Science and Information Technology.*, *1*, 295–301.

Marmier, F., F., Varnier, C. & Zerhouni, N. (2009). Proactive, dynamic and multi-criteria scheduling of maintenance activities. *International Journal of Production Research*, *47*(8), 2185–2201. doi:10.1080/00207540802311106

Mathiassen, L., & Pourkomeylian, P. (2003). Managing knowledge in a software organization. *Journal of Knowledge Management*, *7*(2), 63–80. doi:10.1108/13673270310477298

Mayer, J. D., & Beltz, C. M. (1998). Socialization, society's "emotional contract," and emotional intelligence. *Psychological Inquiry*, *9*(4), 300–304. doi:10.1207/s15327965pli0904_12

Mayer, J. D., & Salovey, P. (1997). What is emotional intelligence? In Salovey, P., & Sluyter, D. (Eds.), *Emotional development and emotional intelligence: Education implication*. New York: Basic Books.

McAdam, R., & Reid, R. (2000). A comparison of public and private sector perception and use of knowledge management. *Journal of European Industrial Training*, *24*(6), 317–329. doi:10.1108/03090590010346424

McCallum, A. (2005). Information Extraction: Distilling Structured Data from unstructured Text. *ACM Queue; Tomorrow's Computing Today*, *3*(9), 48–57. doi:10.1145/1105664.1105679

McElroy, M. (2004). How to Operationalize KM, KM World. Santa Clara, California. Retrieved May 12, 2009, from http://www.macroinnovation.com.

McEvily, S. K., Das, S., & McCabe, K. (2000). Avoiding competence substitution through knowledge sharing. *Academy of Management Review*, *25*(2), 294–311. doi:10.2307/259015

McInerney, C. (2002). Knowledge Management and the Dynamic Nature of Knowledge. *Journal of the American Society for Information Science and Technology*, *53*(12), 1009–1018. doi:10.1002/asi.10109

McKinney, V., Yoon, K., & Zahedi, F. M. (2002). The measurement of web-customer satisfaction: An expectation and disconfirmation approach. *Information Systems Research*, *11*(3), 296–315. doi:10.1287/isre.13.3.296.76

McKone, K. E., Schroeder, R. G., & Cua, K. O. (1999). Total productive maintenance: a contextual view. *Journal of Operations Management*, *17*, 123–144. doi:10.1016/S0272-6963(98)00039-4

Mens, T., & Demeyer, S. (2001). Future trends in software evolution metrics. *Proceedings of the 4th International Workshop on Principles of Software Evolution,* Vienna, Austria. 83-86. New York, NY, USA: ACM Publishing

Meredith, J. R. (1987). Automating the factory. *International Journal of Production Research*, *25*(10), 1493–1510.

Metaxiotis, K., & Psarras, J. (2003). Applying Knowledge Management in Higher Education: The Creation of a Learning Organisation. *Journal of Information and Knowledge Management.*, *2*(4), 1–7.

Meyer, J. P., & Allen, N. J. (1997). *Commitment in the workplace: Theory, research and application*. Thousand Oaks, CA: Sage.

Mikulecký, P. (2001). *Information Filtration in Knowledge Management Support* (pp. 190–193). Bratislava, Slovak Republic: Quantitative Methods in Economy and Business.

Mikulecký, P., Ponce, D., & Hynek, J. (1997) Knowledge-Based Projects in Industrial Applications, *Proceedings of 4th International Seminar on Mathematical Simulation in the Glass Melting*, Horní Bečva, Czech Republic, pp. 191-201.

Miles, M. B., & Huberman, A. M. (1994). *Qualitative data analysis* (2nd ed.). Newbury Park, CA: Sage Publication.

Miller, N. E., & Dollard, J. (1950). *Personality and Psychotherapy*. New York: McGraw-Hill.

Miller, J. G. (1978). *Living Systems*. New York: McGraw-Hill.

Mills, A., Love, P. E. D., & Williams, P. (2009). Defect costs in residential construction. *Journal of Construction Engineering and Management*, (January): 12–16. doi:10.1061/(ASCE)0733-9364(2009)135:1(12)

Mills, G. (2007). *From African to Afghanistan*. Johannesburg, South Africa: Wits University Press.

Minbaeva, D. B. (2005). HRM practices and MNC knowledge transfer. *Personnel Review*, *34*(1), 125–144. doi:10.1108/00483480510571914

Minbaeva, D. B., Pedersen, T., Björkman, I., Fey, C. F., & Park, H. J. (2003). MNC knowledge transfer, subisidiary absorptive capacity, and HRM. *Journal of International Business Studies*, *34*, 586–599. doi:10.1057/palgrave.jibs.8400056

Ministry of Housing & Local Government. (1999). *Housing in the New Millenium. A Malaysian Perspective.* Article 6. Retrieved March 5, 2007 from http://www.kpkt.gov.my/jpn/artikel/arti04.htm

Ministry of Housing & Local Government. (2003). *Number of Projects, Houses, House Buyers and the Estimated Value of Abandoned Housing Projects in Malaysia*. Retrieved March 5, 2007, from http://www.kpkt.gov.my/statistik/ perangkaan2002/kandungan.htm.

Mitchell, H. J. (2006). Knowledge Sharing- The value of story telling. *International Journal of Organisational Behaviour*, *9*(5), 632–641.

Moberg, C. R., Cutler, B. D., Gross, A., & Speh, T. W. (2002). Identifying antecedents of information exchange within supply chains. *International Journal of Physical Distribution and Logistics Management, 32*(9/10), 755–770.

Modi, S. B., & Mabert, V. A. (2007). Supplier development: Improving supplier performance through knowledge transfer. *Journal of Operations Management, 25*(1), 42–64. doi:10.1016/j.jom.2006.02.001

Mohayidin, M. G., Man, N., Kamaruddin, N., & Margono, M. I. (2007). The Application of Knowledge Management in Enhancing the Performance of Malaysian Universities. *Electronic Journal of Knowledge Management, 5*(3), 301–312.

Mohd Ridzuan Nordin. (2001). The Challenge for Higher Education in Malaysia in an Era of a Knowledge-Based Economy. In Abdulai, D. N. (Ed.) *Malaysia and the K-Economy: Challenges, Solutions and the Road Ahead.* (pp. 1-30). Subang Jaya: Pelanduk Publications (M) Sdn. Bhd.

Mohd. Hanipah, B., Abd Rahman, M. Y., & Ishak, Z. (2005, September). *The Role of Financial Support and Strength in Ensuring the Completion of Housing Development Projects in Malaysia.* Paper presented at Inaugural Meeting & Seminar of the Real Estate Educators & Researchers Association Malaysia, Universiti Teknologi Mara City Campus, Kuala Lumpur, Malaysia.

Molina, L. M., Llorens-Montes, J., & Ruiz-Moreno, A. (2007). Relationship between quality management practices and knowledge transfer. *Journal of Operations Management, 25*, 682–701. doi:10.1016/j.jom.2006.04.007

Moore, G., & Benbasat, I. (1991). Development of an instrument to measure the perceptions of adopting an information technology innovation. *Information Systems Research, 2*, 192–222. doi:10.1287/isre.2.3.192

Morgan, R. M., & Hunt, S. D. (1994). The commitment-trust theory of relationship marketing. *Journal of Marketing, 58*, 20–38. doi:10.2307/1252308

Morgan, R. M., & Hunt, S. D. (1994). The commitment-trust theory of relationship marketing. *Journal of Marketing, 58*, 20–38. doi:10.2307/1252308

Mostert, N. M. (2007). Diversity of the mind as the key to successful creativity at Unilever. *Creativity and Innovation Management, 16*(1), 93–100. doi:10.1111/j.1467-8691.2007.00422.x

Motavalli, J. (2010). *Accelerator Pedal Supplier in Toyota's Recall Has Many Customers.* Retreived on February 10, 2010, from www.nytimes.com.

Moustaghfir, K. (2009). How knowledge assets lead to a sustainable competitive advantage: are organizational capabilities a missing link? *Knowledge Management Research & Practice, 7*(4), 339–356. doi:10.1057/kmrp.2009.26

Mowery, D. C., & Oxley, J. E. (1995). Inward technology transfer and competitiveness: the role of national innovation systems. *Cambridge Journal of Economics, 19*, 67–93.

Nabeth, T., Angehrn, A., & Roda, C. (2002). Towards personalised, socially aware and active Knowledge Management System; *Proceedings: E-2002 e-Business and e-Work Annual Conference, Prague; Prague, Czech Republic 2002.* Retrieved January 10th 2008.

Nadler, L. (1994). *Every Manager's Guide to Human Resource Development.* San Francisco: Jossey-Bass.

Nahapiet, J. & Ghoshal, S. (1998) Social Capital, Intellectual Capital and the Organizational Advantage. *Academy of Management Review* 23(2), 1998, 242-268.

National Storytelling Association. (1997). *What is storytelling*? 2008, from http://www.eldrbarry.net/roos/st_defn.htm

Naudé, P., Human, P., & Malan, L. (1990). Managerial perceptions of the future in a volatile society: the South African case. *Omega, 18*(5), 521–528. doi:10.1016/0305-0483(90)90008-W

Ndlela, L. T., & Toit, A. S. A. (2001). Establishing a knowledge management programme for competitive advantage in an enterprise. *International Journal of Information Management*, *21*(2), 151–165. doi:10.1016/S0268-4012(01)00007-X

Nelson, R. R., & Winter, S. G. (1982). *An Evolutionary Theory of Economic Change*. Cambridge, MA: Harvard University Press.

Nemiro, J. E. (2004). *Creativity in virtual teams: key components for success. The Collaborative Work Systems Series. Center for the study of work teams*. San Francisco: Pfeiffer.

Nielsen, L., & Madsen, S. (2006). Storytelling as Method for Sharing Knowledge across IT Projects. *Proceedings of the 39th Annual Hawaii International Conference on System Sciences (HICSS'06)* Kauai, Hawaii., 8 191a.

Nilson, C. (1995). *Games that Drive Change*. New York: McGraw-Hill.

Ninth Malaysia Plan 2006-2010. (2006). *Retrieved from*http://www.pmo.gov.my/ RancanganWeb/Rancangan1.nsf/fec3c825c96ab6be482571a20024010b/520459cd0c6d953848257 1a90028cf68?OpenDocument [15 November 2006].

Nonaka, I., & Takeuchi, H. (1995). *The Knowledge Creating Company*. New York: Oxford University Press.

Nonaka, I. (1994). A dynamic theory of organizational knowledge creation. *Organization Science*, *5*(1), 14–37. doi:10.1287/orsc.5.1.14

Nonaka, I., & Takeuchi, H. (1995). *The Knowledge-creating company: how Japanese companies create the dynamics of innovation*. Oxford, UK: Oxford University Press.

Nonaka, I., & Takeuchi, H. (1997). *Criação de conhecimento nas empresas*. Rio de Janeiro: Campus.

Nonaka, I. (1991). The knowledge creating company. *Harvard Business Review*, *69*(Nov-Dec), 96–104.

Nonaka, I., & Reinmoeller, P. (2000). Dynamic business systems for knowledge creation and utilization. In C. Despres y D. Chauvel(eds.), *Knowledge Horizonts: the present and the promise of knowledge management*, 89-112. Boston: Butterworth-Heinemann.

Noordewier, T. G., John, G., & Nevin, J. R. (1990). Performance outcomes of purchasing arrangements in industrial buyer-vendor relationships. *Journal of Marketing*, *54*, 80–93. doi:10.2307/1251761

O'Dell, C., & Grayson, C. J. J. (1999). Knowledge transfer: Discover your value proposition. *Strategy and Leadership*, 10–15. doi:10.1108/eb054630

O'Dell, C., & Grayson, C. J. (1998). If only we knew what we know: Identification and transfer of internal best practices. *California Management Review*, *40*(3), 14–37.

O'Dell, C., & Grayson, C. J. Jr. (1999). Knowledge tranfer: Discover your value proposition. *Strategy and Leadership*, *27*(2), 10–16. doi:10.1108/eb054630

OECD (2004). *The significance of knowledge management in the business sector,* July.

OECD, (2003).*Conclusions from the result of the survey of Knowledge Management practices for Ministries/Departments/agencies of Central Government in OECD Member Countries, February 3-4, 2003.* Retrieved on December 20, 2005 from GOV/PUMA/HRM (2003)2

OECD. (1996). *The Knowledge-Based Economy*. Paris. Retrieved December 27, 2008, from http://www.oecd.org.

Olševičová, K., Ponce, D., & Haviger, J. (2006). Procedural knowledge Representation in Learning Management System. *5th European Conference on e-Learning*, (pp.262-272) Reading, UK.

Ong, C., & Lai, J. (2007). Measuring user satisfaction with knowledge management systems: scale development, purification, and initial test. *Computers in Human Behavior*, *23*(3), 1329–1346. doi:10.1016/j.chb.2004.12.012

Ong, S. E. (1997). Building defects, warranties and project financing from pre-completion marketing. *Journal of Property Finance*, *8*(1), 35–51. doi:10.1108/09588689710160507

Organ, D. W. (1988). *Organizational citizenship behaviors: The good soldier syndrome*. Lexington, MA: Lexington Books.

Orlikowski, W. J. (2002). Knowing in Practice: Enacting a Collective Capability in Distributed Organizing. *Organization Science*, *13*(3), 249–273. doi:10.1287/orsc.13.3.249.2776

Orlikowski, W. J., & Baroudi, J. J. (1991). Studying Information Technology in Organizations: Research Approaches and Assumptions. *Information Systems Research*, *2*, 1–28. doi:10.1287/isre.2.1.1

Orlikowski, W. J. (1992). The duality of technology: Rethinking the concept of technology in organizations. *Organization Science*, *3*(3), 398–427. doi:10.1287/orsc.3.3.398

Orr, J. E. (1997). *Talking about machines: An ethnography of a modern job*. Ithaca, NY: Cornell University Press.

Othman, A. K. (2010). *The influence of emotional intelligence on job performance in service businesses*. Unpublished PhD thesis, Universiti Teknologi MARA, Shah Alam.

Paiva, E. L., Roth, A. V., & Fensterseifer, J. E. (2008). Organizational knowledge and the manufacturing strategy process: A resource-based view analysis. *Journal of Operations Management*, *26*(1), 115–132. doi:10.1016/j.jom.2007.05.003

Paliouras, G., Karkaletsis, V., Papatheodorou, C., & Spyropoulos, C. D. (1999). Exploiting Learning Techniques for the Acquisition of User Stereotypes and Communities. *Proceedings of the Seventh International Conference of User Modeling*, 169-178.

Pan, S. L., & Scarbrough, H. (1999). Knowledge management in practice: An exploratory case study. *Technology Analysis and Strategic Management*, *11*(3), 359–374. doi:10.1080/095373299107401

Pandey, A. C. (2010). *Y2010 & Ahead– value chain trends in emerging economy.* Retrieved February 15, 2010, from www.infosysblogs.com.

Parasuraman, A., Zeithaml, A. V., & Berry, L. L. (1985). A conceptual model of service quality and its implications for future research. *Journal of Marketing*, *49*(4), 41–50. doi:10.2307/1251430

Paulraj, A., Lado, A. A., & Chen, I. J. (2008). Interorganizational communication as a relational competency: Antecedents and performance outcomes in collaborative buyer-supplier relationships. *Journal of Operations Management*, *26*(1), 45–64. doi:10.1016/j.jom.2007.04.001

Paulus, P. B. (2000). Groups, teams and creativity: the creative potential of idea-generating groups. *Applied psychology. International Review (Steubenville, Ohio)*, *49*(2), 237–262.

Paulus, P. B., Larey, T. S., & Dzindolet, M. T. (2001). Creativity in groups and teams. In Turner, M. E. (Ed.), *Groups at work: Theory and research* (pp. 319–338). Mahwah, NJ: Lawrence Erlbaum Associates Publishers.

Peleg-Gillai, B. (2007). Esquel Group: Going beyond the traditional approach in the apparel industry. In Lee, H. L., & Lee, C.-Y. (Eds.), *Building Supply Chain Excellence in Emerging Economies*. New York: Springer.

Penrose, E. (1959). *The theory of the growth of the firm*. New York: Wiley.

Penrose, E. (1995). *The Theory of the Growth of the Firm*. Oxford, UK: Oxford University Pres. doi:10.1093/0198289774.001.0001

Perlini, A. H., & Halverson, T. R. (2006). Emotional intelligence in the National Hockey League. *Canadian Journal of Behavioural Science*, *38*(2), 109–120. doi:10.1037/cjbs2006001

Petersen, K. J., Handfield, R. B., & Ragatz, G. L. (2005). Supplier integration into new product development: coordinating product, process and supply chain design. *Journal of Operations Management*, *23*(3-4), 371–388. doi:10.1016/j.jom.2004.07.009

Piccoli, G., Powell, A., & Ives, B. (2004). Virtual teams: team control structure, work processes, and team effectiveness. *Information Technology & People*, *17*(4), 359–379. doi:10.1108/09593840410570258

Pinilla-Cisneros, S. I. (2004). *Condiciones de éxito de los emprendedores emergentes de Lima en el contexto de globalización. Magister disertation*. Perú: Universidad de San Marcos.

Pintelon, L. M., & van Wassenhove, L. N. (1990). A maintenance management tool. *Omega*, *18*, 59–70. doi:10.1016/0305-0483(90)90018-5

Piore, M., & Sabel, Ch. (1990). *La segunda ruptura industrial*. Madrid: Alianza.

Piore, M., & Ruiz, C. (1998). Innovation cases into the mexican maquila. In Kagami, M., Humphrey, J., & Piore, M. (Eds.), *Learning, Liberalization and Economic Adjustment*. Tokyo: Institute of Developing Economies.

Platt, L., & Wilson, G. (1999). Technology development and the poor/marginalised: context, intervention and participation. *Technovation*, *19*, 393–401. doi:10.1016/S0166-4972(99)00030-9

Podsakoff, P. M., MacKenzie, S. B., Moorman, R. H., & Fetter, R. (1990). Transformational leader behaviors and their effects on followers' trust in leader, satisfaction, and organizational citizenship behaviors. *The Leadership Quarterly*, *1*, 107–142. doi:10.1016/1048-9843(90)90009-7

Polanyi, M. (1966). *Personal Knowledge: Towards a Post-Critical Philosophy*. Chicago: The University of Chicago Press.

Polanyi, M. (1966). *The Tacit Dimension*. New York: Doubleday.

Polanyi, M. (1967). *The tacit dimension*. London: Routledge and Kegan Paul.

Polanyi, M. (1962). *Personal knowledge: Towards a post-critical philosophy*. University of Chicago: Chicago Press.

Ponce, D. (2007). Model of slow process as ontological design pattern. *International conference Znalosti 2007*, (pp.348-351). Ostrava, Czech Republic.

Porter, M. E. (1985). *Competitive advantage: Creating and sustaining superior performance*. New York: Free Press.

Porter, M. E. (1996). What is strategy? *Harvard Business Review*, *74*(6), 61–78.

Post, T. (2002). The impact of storytelling on NASA and EDUTECH. *Knowledge Management Review*, *5*(1), 26–29.

Pothukuchi, V., Damanpour, F., Choi, J., Chen, C. C., & Park, S. H. (2002). National and Organizational Culture Differences and International Joint Venture Performance. *Journal of International Business Studies*, *33*(2), 243–265. doi:10.1057/palgrave.jibs.8491015

Pozzebon, M., & Pinsonneault, A. (2005). Challenges in conducting empirical work using structuration theory: Learning from IT research. *Organization Studies*, *26*(9), 1353–1376. doi:10.1177/0170840605054621

Prahalad, C., & Hamel, G. (1990). The core competencies of the corporation. *Harvard Business Review*, *68*(3), 79–91.

Prasad, K., & Akhilesh, K. B. (2002). Global virtual teams: What impacts their design and performance? *Team Performance Management: An International Journal*, *8*(5/6), 102–112. doi:10.1108/13527590210442212

Prochno, P. J. L. C., & Corrêa, H. L. (1995). The development of manufacturing strategy in a turbulent environment. *International Journal of Operations & Production Management*, *15*(11), 20–36. doi:10.1108/01443579510102873

PROEXPANSION. (2006). *Estudio sobre cluster y asociatividad*. Document prepared for PROMPyme. Lima. Retrieved February 02, 2010, from http://www.programa-area.org/img/documentos/309-estudio-sobre-clusters-y-asociatividad-peru-pdf.pdf

Quin, T. Y., Yusof, M., & Hamdan, A. R. (2005). *Knowledge management readiness in organization: a case of public sector in Malaysia.* Retrieved from http://www.ickm.edu.my.

Quinn, J. B., Anderson, P., & Finkelstein, S. (1996). Managing professional intellect: Making the most of the best. *Harvard Business Review*, (March - April): 71–80.

Rabellotti, R., (1993). Industrial districts in Mexico: The case of the footwear industry. *Small Enterprise Development*, 4(3).

Raman, M. (1997). Housing and the Concerns of Consumers. In Berhad, C. (Ed.), *Housing The Nation: A Definitive Study* (pp. 709–741). Kuala Lumpur: ISIS.

Ramirez, J., & Fornerino, M. (2007). Introducing the impact of technology: a 'neo-contingency' HRM Anglo-French comparison. *International Journal of Human Resource Management*, *18*(5), 924–949.

Rao, T. V. (1991). *Readings in Human Resource Development.* New Delhi, India: Oxford and IBH Publishing.

Ravichandran, T., & Rai, A. (2003). Structural analysis of knowledge creation and knowledge embedding on software process capability. *IEEE Transactions on Engineering Management*, *50*(3), 270–284. doi:10.1109/TEM.2003.817278

Reagans, R., Argote, L., & Brooks, D. (2005). Individual experience and experience working together: predicting learning rates from knowing who knows what and knowing how to work together. *Management Science*, *51*(6), 869–881. doi:10.1287/mnsc.1050.0366

Rebernik, M., & Sirec, K. (2007). Fostering innovation by unlearning tacit knowledge. *Kybernetes*, *36*(3/4), 406–419. doi:10.1108/03684920710747039

Reis, R. S., & Chevalier, F. (2009, August). *Unfolding creative process in development of new products within globally distributed teams*. Paper presented at the meeting of Academy of Management, Chicago, IL.

Reuber, A. R., Dyke, L. S., & Fischer, E. M. (1990). Using Tacit Knowledge Methodology to Define Expertise. *Proceedings of ACM SIGBDP Conference on Trends and Directions in Expert Systems.* November 1990. Orlando, Florida. ACM, (pp. 262-274).

Revilla, E., & Cury, T. (2008) Antecedents and Consequences of Knowledge Integration in Product Development: An Empirical Evidence. Available at SSRN: http://ssrn.com/abstract=1134491

Rhee, S.-Y. (2005). *How shared emotions among group members influence group effectiveness? The role of broadening-and-building interactions.* Unpublished PhD dissertation, The University of Michigan.

Ribiere, V. M., & Sitar, A. S. (2003). Critical role of leadership in nurturing a knowledge - supporting culture. *KnowledgeManagement Research & Practice*, *1*, 39–48. doi:10.1057/palgrave.kmrp.8500004

Rice, C. L. (1999). *A quantitative study of emotional intelligence and its impact on team performance.* Unpublished master's thesis, Pepperdine University, Malibu, CA.

Richardson, J. (1993). Parallel sourcing and supplier performance in the Japanese automobile industry. *Strategic Management Journal*, *14*(5), 339–350. doi:10.1002/smj.4250140503

Rico, Z. (2003). Fast Growth in Control. *Accountability, 203-221*

Riege, A. (2005). Three-dozen knowledge sharing barriers managers must consider. *Journal of Knowledge Management*, *9*, 18–35. doi:10.1108/13673270510602746

Ring, P. S., & van de Ven, A. H. (1994). Developmental processes of cooperative interorganizational relationships. *Academy of Management Review*, *19*(1), 90–118. doi:10.2307/258836

Robb, D. (2004). Taming text. *Computerworld*, *38*(25), 40–41.

Robbins, S. R. (1990). *Organization Theory: Structure Design and Application.* New Delhi: Prentice Hall.

Rosenkopf, L., & Nerkar, A. (2001). Beyond local search: boundary spanning, exploration, and impact in the optical disk industry. *Strategic Management Journal, 22*, 287–306. doi:10.1002/smj.160

Rottman, J.W. & Lacity, M.C. (2004). *Twenty Practices for Offshore Sourcing. MIS Quarterly Executive* 3(3), 2004, 117-130.

Rouse, M. J., & Daellenbach, U. S. (2002). More thinking on research methods for the resource-based perspective. *Strategic Management Journal, 23*, 963–967. doi:10.1002/smj.256

Rüdiger, K., & McVerry, A. (2007). *Exploiting Europe's knowledge potential: 'Good Work' or Could do better - knowledge work and knowledge workers in Europe.* The Work Foundation.

Ruggles, R. (1998). The State of the Notion: Knowledge Management in Practice. *California Management Review, 40*(3), 80–90.

Rus, I., Lindvall, M., & Sinha, S. (2001). *Knowledge management in software engineering: a state-of-the-art report.* Research Report, Fraunhofer Center for Experimental Software Engineering Maryland and the University of Maryland.

Ryu, S., Hee, H. S., & Han, I. (2003). Knowledge sharing behavior of physicians in hospitals. *Expert Systems With Application, 25*, 113–122. doi:10.1016/S0957-4174(03)00011-3

Sackmann, S. A., & Friesl, M. (2007). Exploring cultural impacts on knowledge sharing behavior in project teams: results from a simulation study. *Journal of Knowledge Management, 11*(6), 142. doi:10.1108/13673270710832226

Sahay, S., & Robey, D. (1996). Organizational Context, Social Interpretation, and The Implementation and Consequences of Geographic Information Systems.Accounting. *Management and Information Technology., 6*(4), 255–282. doi:10.1016/S0959-8022(96)90016-8

Salleh, K. (2008). *Knowledge Management and Public Sector Accountants: Knowledge Sharing Process in the Accountant General's Department of Malaysia.* Doctoral Dissertation, Universiti Teknologi MARA, Shah Alam, Malaysia.

Salleh, K. dan Syed Noh, S.A. (2005). *Knowledge management in the local authorities – a suitable platform for e-government.* Retrieved from http://www.kmice.edu.my.

Salleh, K., Ahmad, S. N. S., & Syed-Ikhsan, S. O. S. (2008). *Knowledge Management in a Malaysia Public Sector Accounting Organization: An Integrated KM Framework.* Paper presented at the 9th European Conference in Knowledge Management (ECKM 2008), Southampton Solent University, Southampton, United Kingdom,Publishing Limited, Reading, UK, pp.763 – 774

Salovey, P., Kokkonen, M., Lopes, P. N., & Mayer, J. D. (2004). Emotional Intelligence: What do we know? In Manstead, A. S. R., Frijda, N., & Fischer, A. (Eds.), *Feelings and Emotions: The Amsterdam symposium* (pp. 321–340). Cambridge, UK: Cambridge University Press.

Sanders, N. R. (2008). Pattern of information technology use: The impact on buyer–suppler coordination and performance. *Journal of Operations Management, 26*, 349–367. doi:10.1016/j.jom.2007.07.003

Santoro, F. M., & Brezillon, P. (2005). Group Storytelling Approach to Collect Contextualized Shared Knowledge. *Proceedings. Sixteenth International Workshop on Database and Expert Systems Applications, 2005*, Copenhagen, Denmark. 388-392.

Scarbrough, H. (2003). Knowledge Management, HRM and the Innovation Process. *International Journal of Manpower, 24*(5), 501–516. doi:10.1108/01437720310491053

Scarbrough, H., & Swan, J. (2001). Knowledgecommunities and innovation. In Husyman, M., & Van Baalen, P. (Eds.), *Trends in communication cpecial issues on communities of practice.* Amsterdam: Boom.

Schaurhofer, M., & Peschi, M. F. (2005). Autonomy: Starting point and goal of personal and social change: A constructivist perspective on knowledge management in empowerment processes. *Kybernetics*, *34*(1/2), 261–277. doi:10.1108/03684920510575834

Schein, E. (1985). *Organizational culture and leadership*. San Francisco, CA: Jossey-Bass.

Schein, E. H. (1992). *Organization Culture and Leadership*. San Francisco: Jossey-Bass.

Schmid, A. A. (2000). Affinity as social capital: its role in development. *Journal of Socio-Economics*, (29): 159. doi:10.1016/S1053-5357(00)00059-7

Schmidt, F. L.& Hunter, J. E. (1993). Tacit Knowledge, Practical Intelligence, General Mental Ability, and Job Knowledge. *Current Directions in Psychological Science: A Journal of American Psychological Society. 2*(1), 8-9.

Schmitz, H. (1995). Small shoemakers and fordist giants: tale of a supercluster. *World Development*, *23*(1), 9–28. doi:10.1016/0305-750X(94)00110-K

Scholl, W., Konig, C., & Meyer, B., & Heisig, p. (2004). The future of knowledge management: An international Delphi study. *Journal of Knowledge Management*, *8*(2), 19–35. doi:10.1108/13673270410529082

Schönström, M., & Carlsson, S. (2003). Methods as knowledge enablers in software development organizations. In C. Ciborra *et al.* (Eds.) *Proceedings of the 11th European Conference on Information Systems*. Naples: CUEN.

Schultze, U., & Leidner, D. E. (2002). Knowledge management in IS research. *Management Information Systems Quarterly*, *26*(3), 213–242. doi:10.2307/4132331

Schumpeter, J. A. (1942). *Capitalism, Socialism, and Democracy*. New York: Harper.

Schutte, N. S., Malouff, J. M., Bobik, C., Coston, T. D., Greeson, C., & Jedlicka, C. (2001). Emotional Intelligence and Interpersonal Relations. *The Journal of Social Psychology*, *141*(4). doi:10.1080/00224540109600569

Selnes, F., & Sallis, J. (2003). Promoting relationship learning. *Journal of Marketing*, *67*(July), 80–95. doi:10.1509/jmkg.67.3.80.18656

Senge, P. (1990). *The fifth discipline: the art and practice of the learning organization* (1st ed.). Currency Doubleday. [s.l.]

Shannon, R. (2009). Playing with principles in an era of securitized aid: negotiating humanitarian space in post-9/11 Afghanistan. *Progress in Development Studies*, *9*(1), 15–36. doi:10.1177/146499340800900103

Sharratt, M., & Usoro, A. (2003). Understanding knowledge-sharing in online communities of practice. *Electronic Journal of Knowledge Management*, *1*(2), 187–196.

Shields, R., Holden, T., & Schmidth, R. A. (2000). *A critical analysis of knowledge management initiatives in the Canadian public service: the impact of a knowledge-based economy on work in the public service, the virtual of expertise and knowledge-based economy on work in the public service, the virtual of expertise and knowledge.* Retrieved from http://www.carleton.ca/innovation/ km-fed.pdf.

Siemieniuch, C. E., & Sinclair, M. A. (2004). A framework for organisational readiness for knowledge management. *International Journal of Operations & Production Management*, *24*(1), 79–98. doi:10.1108/01443570410511004

Siemsen, E., Roth, A. V., & Balasubramanian, S. (2008). How motivation, opportunity, and ability drive knowledge sharing: The constraining-factor model. *Journal of Operations Management*, *26*, 426–445. doi:10.1016/j.jom.2007.09.001

Silvestri, F., Perego, R., & Orlando, S. (2004). Assigning document identifiers to enhance compressibility of Web search engines indexes. *Proceedings of SAC '04*, Nicosia, 600-605.

Silvestro, R., Fitzgerald, L., Johnson, R., & Voss, C. (1992). Towards a classification of service processes. *International Journal of Service Industry Management*, *3*(3), 62–76. doi:10.1108/09564239210015175

Simon, H. A. (1991). Bounded rationality and organizational learning. *Organization Science*, *2*(1), 125–134. doi:10.1287/orsc.2.1.125

Simonin, B. (1999). Transfer of marketing know-how in international strategic alliances: An empirical investigation of the role and antecedents of knowledge ambiguity. *Journal of International Business Studies*, *30*(3), 463–490. doi:10.1057/palgrave.jibs.8490079

Skyrme, D. J. (1997). *From information to knowledge management: Are you prepared?* Retrieved from http://www.skyrme.com/pubs/on97full.htm.

Small, C. T., & Sage, A. P. (2006). Knowledge management and knowledge sharing: A review. *Information Knowledge Systems Management*, *5*, 153–169.

Socitm Insight. (2003). *Rediscovering knowledge: an overview of knowledge sharing in the public sector*. retrieved from http://www.socitm.gov.uk.

Sole, D. (2002). *Sharing knowledge through storytelling*. Retrieved September 14, 2007, from http://www.providersedge.com/docs/km_articles/Sharing_Knowledge_Through_Storytelling.pdf

Song, M., Song, I.-Y., & Hu, X. (2003). KPSpotter: a flexible information gain-based keyphrase extraction system.*Proceedings of WIDM'03*, New Orleans, pp. 50-53.

Sorenson, O., Rivkin, J. W., & Fleming, L. (2006). Complexity, networks and knowledge flow. *Research Policy*, *35*(7), 994–1017. doi:10.1016/j.respol.2006.05.002

Sparkes, J., R., & Miyake, M. (2000). Knowledge transfer and human resource development practices: Japanese firms in Brazil and Mexico. *International Business Review*, *9*, 599–612. doi:10.1016/S0969-5931(00)00021-4

Spector, P. E., & Fox, S. (2002). An emotional-centered model of voluntary work behavior: Some parallels between counterproductive work behavior and organizational citizenship behavior. *Human Resource Management Review*, *12*, 269–292. doi:10.1016/S1053-4822(02)00049-9

Spender, J. C. (1996). Making Knowledge the Basis of a Dynamic Theory of the Firm. *Strategic Management Journal*, *17*, 45–62.

Spender, J. C. (1996). Competitive Advantage from Tacit Knowledge? Unpacking The Concept and Its Strategic Implication. In Moingeon, B., & Edondson, A. (Eds.), *Organizational Learning and Competitive Advantage* (pp. 56–73). London: Sage.

Spender, J.-C. (1996). Making knowledge the basis of a dynamic theory of the firm. *Strategic Management Journal,* 17(Winter Special Issue), 45-62.

Spira, J. B. (2005). In praise of knowledge workers. *KM World*, *12*(2), 25–27.

Stahl, B. C. (2008). *Information Systems: Critical Perspectives*. Routledge.

Stankosky, M. A. (2005). Advances in Knowledge Management: University Research Toward an Academic Discipline. In M. Stankosky (ed.), *Creating The Discipline of Knowledge Management - The Latest in University Research.(pp.1-14)*. Washington, DC: Elsevier-Butterworth_Heninemann.

Star (2008). Slow response from Developers. 4 June 2008, pg 5.

Steinberg, D. (2002). Share knowledge (ignorance costs millions) Shell international exploration and production. *Ziff Davis Smart Business*, *15*(4), 45.

Steinfeld, E. S. (2004). China's shallow integration: Networked production and the new challenges for late industrialization. *World Development*, *32*(11), 1971–1987. doi:10.1016/j.worlddev.2004.04.003

Stenmark, D. (2000). Turning Tacit Knowledge Tangible. *Proceedings of the 33rd Hawaii International Conference on System Sciences.* 4-7 January. Maui, Hawaii: IEEE Computer Society Press.

Stenmark, D. (2001). Leveraging Tacit Organisational Knowledge. *Journal of Management Information Systems.* 17(3), (Winter 2000-2001), 9-24.

Sternberg, R. J., Forsythe, G. B., Hedlund, J., Horvath, J. A., Wagner, R. K., & Williams, W. M. (2000). *Practical Intelligence in Everyday Life*. New York: Cambridge University Press.

Sternberg, R. J., & Horvath, J. A. (Eds.). (1999). *Tacit Knowledge in Professional Practise: Researcher and Practitioner Perspectives*. Mahwah, NJ: Lawrence Erlbaum Associated Inc.

Sternberg, R. J. (1985). *Beyond IQ: A triarchic theory of human intelligence*. Cambridge, UK: Cambridge University Press.

Stough, C., & De Guara, D. (2003). Examining the relationship between emotional intelligence and job performance. *Australian Journal of Psychology*, *55*, 145.

Supar, N., Ibrahim, A. A., Mohamed, Z. A., Yahya, M., & Abdul, M. (2005). *Factors affecting knowledge sharing and its effects on performance: a study of three selected higher academic institutions*. Retrieved from http://www.kmice.edu.my.

Sure, Y., Staab, S., & Studer, R. (2004). On-To-Knowledge Methodology. In Staab, S., & Studer, R. (Eds.), *Handbook on Ontologies*. Berlin: Springer-Verlag.

Sussman, S. W., & Siegal, W. S. (2003). Informational influence in organizations: An integrated approach to knowledge adoption. *Information Systems Research*, *14*(1), 47–65. doi:10.1287/isre.14.1.47.14767

Svetlik, I., & Stavrou-Costea, E. (2007). Connecting human resources management and knowledge management. *International Journal of Manpower*, *28*(3-4), 197–206. doi:10.1108/01437720710755209

Swap, W., Leonard, D., Shields, M., & Abrams, L. (2001). Using mentoring and storytelling to transfer knowledge in the workplace. *Journal of Management Information Systems*, *18*(1), 95–114.

Swart, J., & Pye, A. (2002). Conceptualizing Organization Knowledge as Collective Tacit Knowledge: A Model of Redescription. *Proceedings of the 3rd European Conference on Organization Knowledge, Learning and Capabilities*. 5 – 6 April. ALBA, Athens, Greece.

Syed Omar, S. S. I., & Rowland, F. (2004). Benchmarking knowledge management in a public organisation in Malaysia. *Benchmarking: An International Journal*, *11*(3), 238–266. doi:10.1108/14635770410538745

Syed Omar, S. S. I. (2005). *Knowledge Management in a public organisation: A Study on the performance of knowledge transfer in the Ministry of Entrepreneur Development of Malaysia*. unpublished PhD thesis University of Lougborough United Kingdom.

Syed-Ikhsan, S. O. S., & Rowland, F. (2004a). Benchmarking Knowledge Management in a Public Organization in Malaysia. *Benchmarking: An International Journal*, *11*(3), 238–266. doi:10.1108/14635770410538745

Syed-Ikhsan, S. O. S., & Rowland, F. (2004b). Knowledge Management in a public organization: A study on the relationship between organizational elements and the performance of knowledge transfer. *Journal of Knowledge Management*, *8*(2), 95–111. doi:10.1108/13673270410529145

Syed-Ikhsan, S. O. S. (2005). *Knowledge Management in a public organization: A study on the relationship between organizational elements and the performance of knowledge transfer.* Doctoral Dissertation, Loughborough University, United Kingdom, Loughborough, UK.

Szulanski, G. (2003). *Sticky Knowledge: Barriers to Knowing in the Firm*. London: Sage.

Szulanski, G. (1996). Exploring internal stickiness: Impediments to the transfer of best practice within the firm. *Strategic Management Journal*, *17*, 27–43.

Szulanski, G. (1996). Exploring internal stickiness: Impediments to the transfer of best practices within the firm. *Strategic Management Journal*, *17*, 27–43.

Takeuchi, H., & Nonaka, I. (2004). *Hitotsubashi on Knowledge Management*. Singapore: John Wiley & Sons.

Tashakkori, A., & Teddlie, C. (1998). *Mixed Methodology: Combining Qualitative and Quantitative Approaches*. London: Sage.

Tayeb, M. H. (2005). *International Human Resource Management: A Multinational Company Perspective*. Oxford, UK: Oxford University Press.

Taylor, W. A., & Wright, G. H. (2004). Organizational Readiness for Successful Knowledge Sharing: Challenges for Public Sector Managers. *Information Resources Management Journal*, *17*(2), 22–37.

Taylor, D. W. (2004). Knowledge Activism by CPAs in Public Sector Organizations. *Malaysian Accounting Review*, *3*, 171–192.

Taylor, W. A., & Wright, G. H. (2004). Organizational readiness for Successful Knowledge Sharing: Challenges for Public Sector Managers. *Information Resources Management Journal*, *17*(2), 22–37.

Taylor, P. (1999). Thinking about new learning environments (report no. TAY99147), *Australian Association for Research in Education. RetrievedSeptember18, 2009*http://www.aare.edu.au/99pap/tay99147.htm

Teece, D. J., Pisano, G., & Shuen, A. (1997). Dynamic capabilities and strategic management. *Strategic Management Journal*, *18*(7), 509–533. doi:10.1002/(SICI)1097-0266(199708)18:7<509::AID-SMJ882>3.0.CO;2-Z

Teece, D. (2000). Strategies for managing knowledge assets: The role of firm structure and industrial context. *Long Range Planning*, *33*(1), 35–54. doi:10.1016/S0024-6301(99)00117-X

Teece, D. (2007). Explication dynamic capabilities: The nature and microfoundations of (sustainable) enterprise performance. *Strategic Management Journal*, *28*, 1319–1350. doi:10.1002/smj.640

The_Supply_Chain_Council. (2008). Retrieved 11/1/2008, from http://www.supply-chain.org

Thite, M. (2004). Strategic positioning of HRM of knowledge-based organizations. *The Learning Organizations: An International Journal*, *11*(1), 28–44. doi:10.1108/09696470410515715

Thomas, J. C., Kellogg, W. A., & Erickson, T. (2001). The knowledge management puzzle: human and social factors in knowledge management. *IBM Systems Journal*, *4*(40), 863–884. doi:10.1147/sj.404.0863

Thomas, E. J. (2005). *Knowledge management in the public and private sector: a synthetic analysis the contemporary literature*. unpublished Master Thesis, School of Information Sciences and Technology, The Pennsylvania State University.

Thompson, J. D. (1967). *Organizations in Action*. New York: McGraw-Hill.

Thuraisingham, B. M. (2003). *Web Data Mining and Applications in Business Intelligence and Counter-Terrorism*. Boca Raton, FL: CRC Press. doi:10.1201/9780203499511

Tiwana, A. (2000). *The Knowledge Management Toolkit, Practical Techniques for Building a Knowledge Management System*. Upper Saddle River, NJ: Prentice Hall.

Triandis, H. C. (2006). Cultural intelligence in organizations. *Group & Organization Management*, *31*(1), 20–26. doi:10.1177/1059601105275253

Tschannen-Moran, M., & Nestor-Baker, N. (2004). Tacit knowledge of productive scholars in education. *Teachers College Record*, *106*(7), 1484–1511. doi:10.1111/j.1467-9620.2004.00387.x

Tsoukas, H. (2003). Do We Really Understand Tacit Knowledge? In Easterby-Smith, M., & Lyles, M. A. (Eds.), *The Blackwell Handbook of Organizational Learning and Knowledge Management* (pp. 410–427). Oxford, UK: Blackwell Publishing Ltd.

Tu, Q., Vonderembse, M. A., Ragu-Nathan, T. S., & Sharkey, T. W. (2006). Absorptive capacity: enhancing the assimilation of time-based manufacturing practices. *Journal of Operations Management*, *24*, 692–710. doi:10.1016/j.jom.2005.05.004

Tučník, P., & Mohelská, H. (2009). Multicriterial Decision-Making Can Drive Strategy, International conference Hradec Economic Days 2009, Hradec Králové, Czech Republic, pp. 80-83.

Turban, E., McLean, E., & Whetherbe, J. (2001). *Information Technology for Management: Making connections for strategic advantage*. New York: John Wiley and Sons Inc.

Turnipseed, D. (1996). Organization citizenship behaviour: An examination of the influence of the workplace. *Leadership and Organization Development Journal*, *17*(2), 42–47. doi:10.1108/01437739610111222

Usoro, A. (2005). Cultural Dimensions of Knowledge-Sharing in Global Virtual teams: An Exploratory Study. *Computing and Information Systems Journal*, *3*(9), 85–91.

Uzzi, B., & Lancaster, R. (2003). Relational embeddedness and learning: The case of bank loan managers and their clients. *Management Science*, *49*, 383–399. doi:10.1287/mnsc.49.4.383.14427

Van den Brink, P. (2003). *Social, organizational and technological conditions that enable knowledge sharing*. unpublished PhD Thesis, Faculty of Technology, Policy and Management, Delft University of Technology, Holland.

Van den Hooff, B., Elving, W. J. L., Meeuwsen, J. M., & Dumoulin, C. M. (2003). Knowledge Sharing in Knowledge Communities. In Huysman, M. H., Wulf, V., & Wenger, E. (Eds.), *Communities and Technologies*. Deventer: Kluwer Academic Publishers.

Velder, R. G., Vaneck, M. T., Guynes, C. S., & Cappel, J. J. (1999). CEO and CIO perspectives on competitive intelligence. *Communications of the ACM*, *42*(8), 109–116.

Villaran, F. (1998). *Riqueza Popular, pasión y gloria de la pequeña empresa*. Peru/Lima. Congreso del Perú.

Viviers, W., Saayman, A., Muller, M.-L., & Calof, J. (2002). Competitive intelligence practices: A South African study. *South African Journal of Business Management*, *33*(3), 27–37.

von Krogh, G., Ichijo, K., & Nonaka, I. (2001). *Facilitando a criação de conhecimento: reinventando a empresa com o poder da inovação contínua*. Rio de Janeiro: Campus.

Von Krogh, G., & Roos, J. (Eds.). (1996). *Managing Knowledge Perspectives on Cooperation and Competition*. London: Sage.

Voss, C., & Blackmon, K. (1996). The impact of national and parent company origin on world-class manufacturing. *International Journal of Operations & Production Management*, *16*(11), 98–115. doi:10.1108/01443579610131492

Wagner, R. K., & Sternberg, R. J. (1985). Practical Intelligence in Real-World Pursuits: The Role of Tacit Knowledge. *Journal of Personality and Social Psychology*, *49*, 436–458. doi:10.1037/0022-3514.49.2.436

Wagner, R. K., & Sternberg, R. J. (1985). Practical intelligence in real-world pursuits: The role of tacit knowledge. *Journal of Personality and Social Psychology*, *49*(2), 436–458. doi:10.1037/0022-3514.49.2.436

Wallace, D. P. (2007). *Knowledge Management: Historical and Cross-Disciplinary Themes*. Westport, CT: Libraries Unlimited.

Walsh, J., & Ungson, G. R. (1991). Organizational memory. *Academy of Management Review*, *16*(1), 57–91. doi:10.2307/258607

Walsham, G. (1993). *Interpreting Information Systems in Organisations*. Chichester, UK: Wiley.

Walsham, G. (2001). *IT in a Global Context*. Chichester, UK: Wiley.

Walsham, G. (2005). Knowledge Management Systems: Representation and Communication in Context. *An International Journal on Communication. Information Technology and Work.*, *1*(1), 6–18.

Ward, P. T., McCreery, J. K., Ritzman, L. P., & Sharma, D. (1998). Competitive priorities in operations management. *Decision Sciences*, *29*(4), 1035–1046. doi:10.1111/j.1540-5915.1998.tb00886.x

Ward, J., & Aurum, A. (2004). Knowledge management in software engineering: describing the process. In *Proceedings of the 18th Australian Software Engineering Conference* (pp. 137-146), Washington, DC: IEEE Computer Society.

Wasko, M. M., & Faraj, S. (2000). It is what one does: why people participate and help others in electronic communities of practice. *The Journal of Strategic Information Systems*, *9*(2-3), 155–173. doi:10.1016/S0963-8687(00)00045-7

Wasko, M. M., & Faraj, S. (2005). Why Should I Share? Examining Social Capital and Knowledge Contribution in Electronic Networks of Practice. *Management Information Systems Quarterly*, *29*(1), 35–57.

Weber, M. (1904). Die Objektivität sozial-wissenschaftlicher und sozialpolitischer Erkenntnis. In *Gesammelte Aufsatze zur Wissenschaftslehre*. Tubingen: Mohr.

Weber, M. (1906). Kritische Studien auf dem Gebiet der kultur-wissenschaftlichen Logik. In *Gesammelte Aufsatze zur Wissenschaftslehre*. Tubingen: Mohr.

Weber, M. (1922). *Wirtschaft und Gesellschaft*. Tubingen: Mohr.

Weber, M. (2003). *Ciência e política: duas vocações*. Brazil, São Paulo: Martin Claret.

Weiss, L. (1999). Collection and connection: The anatomy of knowledge sharing in professional service firms. *Organization Development Journal*, *17*(4), 61–77.

Welbourne, T. M., Johnson, D. E., & Erez, A. (1998). The role-based performance scale: Validity analysis of a theory-based measure. *Academy of Management Journal*, *41*(5), 540–555. doi:10.2307/256941

Wenger, E., & Synder, W. (2000). Communities of practice: the organizational frontier. *Harvard Business Review*, *78*(1), 139–145.

Wernerfelt, B. (1984). A resource based view of the firm. *Strategic Management Journal*, *5*(5), 171–180. doi:10.1002/smj.4250050207

Werr, A., Blomberg, J., & Löwstedt, J. (2009). Gaining external knowledge – boundaries in manager's knowledge relations. *Journal of Knowledge Management*, *13*(9), 448–463. doi:10.1108/13673270910997114

Westphal, C. (2008). *Data Mining for Intelligence, Fraud and Criminal Detection: Advanced Analytics and Information Sharing Technologies*. Boca Raton, FL: CRC Press. doi:10.1201/9781420067248

Westphal, J., & Fredrickson, J. (2001). Who directs strategic change? Director experience, the selection of new CEOs, and change in corporate strategy. *Strategic Management Journal*, *22*(12), 1113–1137. doi:10.1002/smj.205

Wetlauffer-Adcock, C. (2004). *The Storyteller as a Knowledge Transfer Agent: Facilitating Transformational Learning in the Adult Classroom*. PhD, Walden University.

Wiig, K. (1999). Introducing Knowledge Management into the Enterprise. In *J. Liebowitz (1999). Knowledge Management Handbook*. Boca Raton, FL: CRC Press.

Wiig, K., de Hoog, R., & van der Spek, R. (1997). Supporting Knowledge Management: A Selection of Methods and Techniques. *Expert Systems with Applications*, *1*(13), 15–27. doi:10.1016/S0957-4174(97)00019-5

Wiig, K. M. (2002). Knowledge management in public administration. *Journal of Knowledge Management*, *6*(3), 224–239. doi:10.1108/13673270210434331

Wiig, K. M. (1999). Introducing knowledge management into the enterprise. In Liebowitz, J. (Ed.), *Knowledge management handbook*. New York: CRC Press.

Wiig, K. M. (2000). Knowledge Management: an Emerging Discipline Rooted in a Long History. In Despres, C. & Chauvel, D. (eds). *Knowledge Horizons: the present and the promise of knowledge management,pp.* 3-26. Boston: Butterworth-Heinemann.

Wilkesmann, U., Fischer, H., & Wilkesmann, M. (2009). Cultural characteristics of knowledge transfer. *Journal of Knowledge Management*, *13*(6), 464–477. doi:10.1108/13673270910997123

Willcocks, L. P., Hindle, J., & Feeny, D. Lacity, M. 2004. IT and Business Process Outsourcing: The Knowledge Potential. *Information Systems Management*, Summer, pp. 7-15.

Williamson, O. E. (1981). The economics of organization: The transaction cost approach. *American Journal of Sociology*, *87*(3), 548–577. doi:10.1086/227496

Williamson, O. E. (2008). Outsourcing: Transaction cost economics and supply chain management. *Journal of Supply Chain Management*, *44*(2), 5–16. doi:10.1111/j.1745-493X.2008.00051.x

Wilson, T. D. (2002) "The nonsense of 'knowledge management'" *Information Research*, 8(1), paper no. 144 [Available at http://InformationR.net/ir/8-1/paper144.html]

Winter, S. G. (1995). Four Rs of profitability: Rents, resources, routines, and replication. In Montgomery, C. (Ed.), *Resources-based and Evolutionary Theories of the Firm*. Boston: Kluwer Academic Publishers.

Witten, I. H., & Frank, E. (2005). *Data Mining: Practical Machine Learning Tools and Techniques*. San Francisco: Elsevier.

Wong, C.-S., & Law, K. S. (2002). The effects of leader and follower emotional intelligence on performance and attitude: An exploratory study. *The Leadership Quarterly*, *13*(3), 243–274. doi:10.1016/S1048-9843(02)00099-1

World Bank. (1999). *World Development Report: Knowledge for Development. Washington, D.C.Piore, M.J. & C.F. Sabel, (1984), The Second Industrial Divide: Possibilities for Prosperity*. New York: Basic Books.

World Bank Report, (2006). *Doing Business in Afghanistan Report*. Washington: World Bank.

World Bank. (1998). *World development report 1998/1999: Knowledge for development*. Retrieved May 20, 2005, from http://www.worldbank.org/wdr/wdr98/contents.htm

World Bank. (2003). *Technical cooperation program brief on GCC*. Retrieved May 20, 2005, from http://web.worldbank.org/WBSITE/EXTERNAL/COUNTRIES/MENAEXT/BAHRAINEXTN/0,menuPK:312668~pagePK:141132~piPK:141107~theSitePK:312658,00.html

Yahya, S., & Goh, W. K. (2002). Managing human resources toward achieving knowledge. *Journal of Knowledge Management*, *6*(5), 457–468. doi:10.1108/13673270210450414

Yang, J. T. (2004). Job-related knowledge sharing: comparative case studies. *Journal of Knowledge Management*, *8*(3), 118–126. doi:10.1108/13673270410541088

Yang, S., Lee, H., & Kurnia, S. (2009). Social Capital in Information and Communications Technology research: Past, Present and Future. *Communications of the AIS*, *25*(23), 183–220.

Yao, L. J., Kam, T. H. Y., & Chan, S. H. (2007). Knowledge sharing in Asian public administration sector: The case of Hong Kong. *Journal of Enterprise Information Management*, *20*(1), 51–69. doi:10.1108/17410390710717138

Yeh, Y., Lai, S., & Ho, C. (2006). Knowledge management enablers: a case study. *Industrial Management & Data Systems*, *106*(6), 793–810. doi:10.1108/02635570610671489

Yin, R. K. (2009). *Case study research: design and methods* (4th Ed.). Thousand, Oaks,California: Sage Publications.

Young, G., Sapienza, H., & Baumer, D. (2003). The influence of flexibility in buyer–seller relationships on the productivity of knowledge. *Journal of Business Research*, *56*, 443–451. doi:10.1016/S0148-2963(01)00243-0

Yu, T.-K., Lu, L.-C., & Liu, T.-F. (2010). Exploring factors that influence knowledge sharing behavior via weblogs. *Computers in Human Behavior*, *26*(1), 32–41. doi:10.1016/j.chb.2009.08.002

Yusof, N., Mohd Shafiei, M. W., & Yahya, S. (2009). *Assessment of the Readiness of Private Housing Developers and Financial Institutions to the BTS concept*. Unpublished Final Report. NAPREC Research Grant.

Yusof, N., Mohd Shafiei, M. W., Yahya, S., & Ridzuan, M. (2010). Strategies to implement the "build then sell" housing delivery system in Malaysia. *Habitat International*, *34*, 53–58. doi:10.1016/j.habitatint.2009.06.001

Zack, M. (1999). Managing Codified Knowledge. *Sloan Management Review*, *40*(4), 45–58.

Zack, M., McKeen, J., & Satyendra, S. (2009). Knowledge management and organizational performance: an exploratory analysis. *Journal of Knowledge Management, 13*(6), 392–409. doi:10.1108/13673270910997088

Zágarra, C., & García-Falcón, J. M. (2003). Factors favoring knowledge management in work teams. *Journal of Knowledge Management*, *7*(2), 81–93. doi:10.1108/13673270310477306

Zahra, S. A., & George, G. (2002). Absorptive capacity: a review, reconceptualization, and extension. *Academy of Management Review*, *2*(2), 185–203. doi:10.2307/4134351

Zand, D. E. (1972). Trust and managerial problem solving. *Administrative Science Quarterly*, *17*(2), 229–239. doi:10.2307/2393957

Zapata, L., Rialp, J., & Rialp, A. (2009). Generation and transfer of knowledge in IT-related SMEs. *Journal of Knowledge Management, 13*, 246–256.

Zapata Cantú, L., Rialp Criado, J., & Rialp Criado, A. (2009). Generation and transfer of knowledge in IT-related SMEs. *Journal of Knowledge Management, 13*(5), 243–256. doi:10.1108/13673270910988088

Zapata, L. (2004). *Los determinantes de la generación y la transferencia del conocimiento en pequeñas empresas de tecnologías de la información en Barcelona*, Doctoral Thesis, Universidad Autónoma de Barcelona, July.

Zapf, D. (2002). Emotion work and psychological well-being: A review of the literature and some conceptual considerations. *Human Resource Management Review*, *12*, 237–268. doi:10.1016/S1053-4822(02)00048-7

Zarifian, P. (2001). *Objetivo competência por uma nova lógica*. Tradução [Trylinski. Brazil/São Paulo. Atlas. Key Terms and Definitions]. *Maria Helena, C*, V.

Zhang, L., Li, J., & Shi, Y. (2005). Study on improving efficiency of knowledge sharing in knowledge-intensive organizations. Retrieved from http://www.springerlink.com/content/f4nh 617100065813/fulltext.pdf (accessed 7 March 2007).

Zheng, W., Yang, B., & McLean, G. N. (2009). Linking organizational culture, structure, strategy, and organizational effectiveness: Mediating role of knowledge management. *Journal of Business Research*, article in press - http://dx.doi.org/10.1016/j.jbusres.2009.06.005

Zhihong, L., & Tao, Z. (2009). Empirical study on the influence of social capital to informal knowledge transfer among individuals. *Journal of Software*, *4*(4), 291–298.

Zhong, N., Liu, J., & Yao, Y. (2003). *Web intelligence*. Berlin: Springer.

Zink, K. J. (2008). Comprehensive change management concept: Development of a participatory approach. *Applied Ergonomics*, *39*(4), 527–538. doi:10.1016/j.apergo.2008.02.015

Zucker, L. G. (1986). Production of Trust: Institutional sources of economic structures, 1840-1920. In Shaw, B., & Cummings, L. (Eds.), *Research in Organizational Behavior* (*Vol. 8*, pp. 53–111). Greenwich, CT: JAI Press.

About the Contributors

Minwir Al-Shammari is Professor of Management and Director of Graduate Studies at the University of Bahrain, College of Business Administration. He holds a PhD in Business Administration (Industrial Management) from University of Glasgow (UK, 1990) and M.S in Industrial Management from Central Missouri State University (USA, 1986). He has been involved for about 20 years in teaching, research, training, and/or consultancy in the areas of operations management, knowledge management, supply chain management, management information systems, business process re-engineering, organization theory, organizational change and development, project management, spreadsheet modeling, management science, and research methodology. He is Editor-in-Chief of Journal of Supply Chain and Customer Relationship Management, IBIMA Publishing. He is the author of the premier reference source Customer Knowledge Management: People, Processes, and Technology, IGI-Global Publishing. He has published more than 30 research papers that have appeared in international refereed journals such as International Journal of Knowledge Management, Logistics Information Management, International Journal of Information Management, European Journal of Operational Research, Expert Systems with Applications, Journal of Computer Information Systems, International Journal of Information Management, International Journal of Information Communication Technologies and Human Development, International Journal of Operations and Production Management, Production and Inventory Management Journal, Business Process Management Journal, International Journal of Commerce and Management, International Journal of Computer Applications in Technology, Cross-Cultural Management, International Journal of Management, Leadership and Organization Development Journal, and Creativity and Innovation Management.

* * *

Syed Noh Syed Ahmad is a Professor of Accounting with the Faculty of Accounting, Universiti Teknologi MARA. He holds PhD (Accounting) and MSc (Accounting and Finance) from University of Stirling, Scotland, United Kingdom. He has held professorial positions at Ohio University, United States of America and University Utara Malaysia. He is active as a consultant and has conducted seminars and workshops on Forensic Accounting techniques for various agencies including the Malaysian Anti-Corruption Agency, Income Tax Department, Accountant General's Department, Bank Negara Malaysia and the Royal Malaysian Police (PDRM). He is also being appointed by the government to be a member of the Operations Review Panel of the Malaysian Anti-Corruption Commission. His teaching areas include financial and managerial accounting, corporate governance, professional development and business ethics and forensic accounting. His main research interest includes corporate governance, fraud, financial statement reporting and analysis and knowledge management.

Kamla Ali Al-Busaidi is an assistant professor of information systems at Sultan Qaboos University in Oman. She received her PhD in management information systems from Claremont Graduate University in California in 2005, MSc in information systems management from Duquesne University in Pennsylvania in 1999, and BSc in information systems from Sultan Qaboos University in Oman in 1997. Her research interests include knowledge management systems, decision support systems, learning management systems, e-government and the deployment of information and communication technologies in Arab countries. She has published several conference proceedings and journal papers, and served as a reviewer for several conference proceedings.

Rose Alinda Alias. With over 20 years of experience in the academic field of Information Systems, Professor Dr Rose Alinda Alias is a strong proponent of a multiple perspectives approach in Information Systems research and teaching in Malaysia. The multiple perspectives approach guides her research, teaching and supervision, that focus on the topics of IS Quality, IS Strategy, and Knowledge Management. In 2005 she was appointed by Universiti Teknologi Malaysia as a full Professor for Information Systems Management. She is currently the Dean of Graduate Studies at the university, a post that she has held since 2007. She recently completed a contract research on Adoption, Appropriation and Impact of Mobile Phones in Malaysian Society, funded by the Malaysian Communications and Multimedia Commission. Her previous research projects funded by Malaysia's Ministry of Science, Technology and Innovation (MOSTI) earned a Silver Award at National Science & Technology Expo 2003 for SQSP: A Toolkit for IS Strategic Planning and Service Quality Assessment and a Special Award from PIKOM for A Methodology and Toolkit for IS Planning during IPTA R&D Exposition 2003. She has graduated seven PhD students and currrently supervising 15 doctoral candidates at UTM.

Sueli Angélica do Amaral has a PhD in information science and she works as an associated professor at the Information Science Department of the University of Brasília. From 2000 to 2004 she coordinated the graduate course on information science at University of Brasília. She is the leader of a research group on information marketing in the broad area of information and knowledge management. She authored and co-authored books about marketing in libraries as well as several papers on information and knowledge management in journals and conferences. She was member of the standing committee of the Section on Latin America and the Caribbean of International Federation of Library Associations and Institutions (IFLA). She also worked as head librarian of the Administration Section at the Brazilian Library of Senate and as specialized librarian at other libraries in Rio de Janeiro, Brazil.

Ryan Atkins is currently a PhD student at Queen's University Belfast in Northern Ireland. His research focus is supply chain management, with an emphasis on knowledge in supply chain management. He holds a Bachelor's degree in Business Logistics and International Business and a Master's of Manufacturing Management from Penn State University. Ryan has several years of industry experience in various aspects of supply chain management.

Neeta Baporikar is a Professor in the area of Strategic Management and Entrepreneurship. Currently with College of Applied Sciences, Salalah, she holds PhD in Management, MBA (Distinction) and Law (Hons.) degrees. With more than a decade of experience in industry, consultancy and training, she made a lateral switch to research and academics in 1995. Apart from this, she is an Accredited Management

Teacher, Qualified Trainer, Doctoral Guide and Board Member of Academics and Selection Committee in accredited B-Schools. At University of Pune, she supervises a number of doctoral research students and currently collaborates with colleagues and outside on multidisciplinary research projects. Reviewer for international journals, she has to her credit more than 40 research papers and authored books in the area of Entrepreneurship, Strategy and Management Education.

Jean-Paul Van Belle is associate professor and Head of the Department of Information System at the University of Cape Town. In the last 8 years, he has authored or co-authored about 15 books/chapters, 15 journal articles and more than 60 peer-reviewed published conference papers. His key research area is the social and organisational adoption of emerging information technologies in a developing world context, including M-commerce, e-government and open source software.

Richard Brunet-Thornton, FIM, FBA, MIM, MBA, FRSA, is a doctoral candidate and a management practitioner. While in Montréal, he is a partner at one of the world's largest IS/IT providers where he is responsible for an international business practice that caters to a niche industry. In the Czech Republic, besides collaborating in research with a number of academics, he is also a visiting lecturer at the University of Economics in the Czech capital, Prague. He is also a member of the Academy of International Business (Michigan State University) and the European Institute for Advanced Studies in Management. Richard is a Fellow of the prestigious Royal Society for the encouragement of Arts, Manufactures, and Commerce.

Vladimír Bureš, is associate professor that specialises in areas such as Knowledge Management, Ambient Intelligence, or Systems Theory. Alongside the Department of Information Technologies at the faculty of Informatics and Management of the University of Hradec Kralove Vladimír cooperates with Faculty of Military Health Sciences, University of Defence. He has participated and holds experience on a couple of European research projects (e.g. 6th and 7th Framework Program of the EU - Enhanced Learning Unlimited (ELU) and VitalMind, or projects from Socrates/Minerva, and Leonardo da Vinci programmes) and the Czech Science Foundation projects (406/04/2140 KNOMEDIAS, 402/06/1325 AMIMADES, or 403/10/1310 SMEW). He is also a member of international professional organisations (ACM SIGCSE, IADIS) and conference programme committees (Hradec economic days, International Congress on Pervasive Computing and Management, IADIS Web-Based Communities and Ambient Intelligence Forum).

Manodip Ray Chaudhuri is Professor and Dean at the Future Business School, Future Institute of Engineering and Management, Kolkata, India. He holds a M.Sc. (Economics) from the University of Calcutta, a MBA in Human Resources from Visva-Bharati, and a MA in International Development from the University of Guelph, Ontario, Canada. His interest areas cover a broad variety of management approaches in emerging economies. They include among others knowledge management, strategic management, human resources management, organizational development, change management and employee training. He is an active editor of the Bhavishya, Journal of Future Business School.

Mladen Čudanov got his magister degree at Faculty of organizational sciences in 2006 and is finishing his PhD thesis. Currently he is in assistant position at Faculty of Organizational Sciences, University of Belgrade. He has been visiting for one semester as an assistant professor in China. He (co)authored

3 books and more than 40 journal and conference articles. His major research interests are influence of ICT on organizational design, restructuring of business systems and organizational change.

Ylenia Curzi is Post-doc fellow at the Department of Business Administration of the Faculty of Economics of the University of Modena and Reggio Emilia (Italy). Visiting scholar at Cardiff Business School (Wales, UK), in 2009. Her research interests encompass: organization theory, organizational change and Information Communication Technologies, the methodology of the social sciences, with a particular interest in the social research methodology put forward by Max Weber. Her current projects concern an original approach to the action research aiming at improving organizational efficiency and wellbeing. She's also teaching assistant to the courses of Organization Theory and Human Resources Management delivered within first and second level degree courses at the Faculty of Economics of the University of Modena and Reggio Emilia, and within the international PhD programme in Labour Relations of the Marco Biagi Foundation.

Tatiana de Almeida Furquim has a PhD degree in Information Science. She works at the Brazilian Federal Data Processing Agency (SERPRO) since 1997 and she has more than 15 years of experience in the software industry. At SERPRO she worked as a software engineer, project manager, and consultant as well as a member of the corporative Software Engineering Process Group (SEPG). She also collaborated with the corporative Software Quality Assurance (SQA) group and participated on SERPRO's Project Management Committee. She coordinated one of SERPRO's business units' Project Office and she recently joined the company's newly created P&D unit. She authored and co-authored papers in conferences and journals.

Biswadip Ghosh is Assistant Professor of Computer Information Systems at Metropolitan State College of Denver, USA. He received his Ph.D. in Information Systems from the University of Colorado in 2007. His research interests are in Healthcare information systems, knowledge management and decision support systems. Prior to joining Metro State in 2007, he worked on global process/technology management at AT&T Bell Laboratories/Lucent/Avaya on the research, design, development and implementation of telecommunications systems and expert systems for global service delivery. He has also consulted with the Veterans Health Administration, USA for their National Surgical Quality Improvement Program. He has published over 40 papers in ACM, AIS and IEEE sponsored conferences and Information Systems journals such as MIS Quarterly executive, IEEE Transactions, IEEE Networks, Information Systems Management, Information Systems Education, and International Journal of Information Technology Management. He has received numerous awards including the best paper award at the 12th AMCIS conference (AIS).

Haryani Haron is an associate professor in the Center for Systems Sciences at the Faculty of Computers and Matematical Sciences, Universiti Teknologi MARA (UiTM) Malaysia. Dr. Haryani has over 20 years of experience in teaching computer science and information technology. Her doctoral dissertation is on tacit knowledge exploration in Malaysian academia. She has been activety performing research with grants from the university in information technology and knowledge management disipline for the past 15 years. Her research and development interest include knowledge managent in academia, knowledge managent technologies, organization intelligence and e-learning. She has authored and co-authored over 30 publications nationally and internationally.

Ian Hipkin teaches strategic and operations management. Recent institutions include École Supérieure de Commerce de Pau, France and the University of Exeter, UK. His first degree was in electrical engineering. He obtained his PhD at Aston University, UK. He has wide working experience in the power utility industry and has worked as a management consultant for many large companies in the Europe, US, Canada, the Far East, the Middle East, Russia and Africa. He teaches in a number of top business schools as a visiting professor. He has published in a wide range of international journals on topics relating to strategy, technology and knowledge management, and technology transfer. His current research interests include management in conflict environments, the strategic management of technology, and the use of knowledge in strategic decision-making, technology transfer and strategic alliances.

Syed Omar Sharifuddin Syed Ikhsan is the Head of Leadership Program for Leadership Center in National Institute of Public Administration of Malaysia. He holds PhD in Knowledge Management from Loughborough University, United Kingdom and Master of Business Administration (MIS) from University of Dallas, Irving, Texas, U.S.A. His field of specialization is on knowledge management in public sector. He is actively involved as a training or workshop facilitator to government officers who attended leadership programs organized by National Institute of Public Administration of Malaysia. He had published a few papers for international and local refereed journal and book chapters on KM in public sector organizations.

Mohd Bakhari Ismail is currently a Phd Candidate at the Department of Information Science, Faculty of Information Science and Technology, Universiti Kebangsaan Malaysia. He obtained his M.Sc in Strategic Information Management from Luton Business School, University of Luton, United Kingdom. His research interests are in Information Management, Knowledge Management, and IT & Society.

Khairul Shafee Kalid is doing his PhD in Information systems at Universiti Teknologi PETRONAS. His thesis is on the use of stories to support knowledge transfer in organization. The research covers issues on the practices of knowledge-embedded storytelling and the construction of such stories. He received his Masters degree in Information Technology from Queensland University of Technology, Australia in 2001. His research interest includes knowledge management and information systems.

Juha Kettunen is the Rector of the Turku University of Applied Sciences in Finland and Adjunct Professor at the University of Jyväskylä in Finland. He was previously the Director of the Vantaa Institute for Continuing Education of the University of Helsinki and Director of the Advanced Management Education Centre of the University of Jyväskylä. He holds a Ph.D. from the University of Bristol in the UK, a DSc (Economics and Business Administration) from the University of Jyväskylä and a DSc (Technology) from the University of Oulu, Finland. His interest areas include knowledge management, strategic management, change management and quality assurance in higher education.

Kathrin Kirchner finished her studies in computer science in 2000 and in adult education in 2006. She completed her doctorate in 2006 on spatial decision support systems for the rehabilitation of gas pipeline networks in Jena. Since then, she has held a post-doctoral research and teaching position in the Department of Business Information Systems at Friedrich Schiller University Jena. Her research includes domains such as Data Mining, Business Process Modelling, Decision Support and Knowledge Management. Previous work includes around 30 refereed journal articles, book chapters and conference papers.

José Manuel Cárdenas Medina, M.Sc. is a PhD student at São Paulo University – Department of Production Engineering. He is a graduate in Industrial Engineering from the University of San Agustin. He earned his bachelor's degree by an analytical study of the behavior of small enterprises and how these ones develop emerging strategies. Those small firms were intended into the sector of agricultural services. Later, he earned master's degree at the University of São Paulo for a comparative study on innovation within small firms from Brazil and Peru. Part of this experience was used in the chapter. Nowadays, he is interested on knowledge management aspects that include IT tools and metrics of IT success.

Jacobo Ramírez Núñez holds a Doctor of Business Administration, U. of Newcastle upon Tyne, England in collaboration with Grenoble School of Management, France. He is assistant professor of Human Resource Management at the Tecnológico de Monterrey (Mexico). Dr. Ramirez holds the research Chair in European Studies. Currently, Dr. Ramirez is Assistant Professor at the Copenhagen Business School (CBS), Copenhagen, Denmark, where he is attached to the graduate program on cross-cultural management. His current research focuses on cross-cultural studies of the formulation and impact of human resource strategy in workplaces. Since 2008, he is member of Mexico's National Researchers System.

Abdul Kadir Othman is a senior lecture at the Faculty of Office Management and Technology, Universiti Teknologi MARA, Puncak Alam Campus, Selangor, Malaysia. He has just completed his PhD on the topic of the Influence of Emotional Intelligence on Job Performance in Service Businesses at the Faculty of Administrative Science and Policy Studies, Universiti Teknologi MARA, Shah Alam, Selangor, Malaysia. He has produced numerous articles for local as well as international journals and book chapters on various themes such as Human Resource Management, Quality Management, Knowledge Management and others. He has conducted a number of research projects on the area of employees' attitudes and behaviors, organizational culture, leadership, training effectiveness, and knowledge management practices in private as well as government offices. Currently, he is conducting a research project funded by Malaysian Fundamental Research Grant Scheme (FRGS) on Multiple Intelligences and Career Orientation.

José Luis Pineda is associate professor at the Department of Marketing, within the Business School of the Tecnológico de Monterrey, Monterrey Campus, in Mexico. His experience in mass media includes a term as a journalist and cartoonist in newspapers. In radio and television he has produced and conducted several shows within promotional strategies for non-profit institutions. He has been involved in the management of public relations events for academic institutions. As a full time professor, his teaching activities are related to advertising, marketing communications and business ethics. He was awarded a PhD from the Universitat Autonoma de Barcelona and his research interests have been focused on mass media consumption and domestication of new technologies, and recently on the impact of the academic research in the business practice.

Rosana Silveira Reis is Professor and Consultant in Human Resources. She has 22 years of experience in Human Resources, 15 of them she has been a Manager in large companies. Since 2000 she has been teaching Human Resources Management and Organizational Behavior in graduate and MBA courses. Currently, she works in collaboration with professors in the fields: organization studies at the University of Bologna, Italy; organization behavior at Positivo University, Curitiba, Brazil; and, creativity & innovation at HEC – Paris, France. Her research focuses on Human Resources, Organization Behavior, Management, Creativity and Innovation. Her current projects concern unfolding and understanding several aspects (i.e. trust, knowledge integration, cultural diversity) involved in the creative process of Globally Distributed Teams.

Kalsom binti Salleh is a senior lecturer in the Faculty of Accountancy, Universiti Teknologi MARA, Malaysia. She obtained her PhD (Accountancy) from Universiti Teknologi MARA and Master in Business Administration from Ohio University, USA. She is also a registered Chartered Accountant as well as an appointed council member with Malaysian Institute of Accountants. Besides her teaching focus on auditing and public sector accounting, her research areas of interest include knowledge management, intellectual capital, electronic government and corporate governance. She has published a few papers for journal articles and book chapters on KM in public sector organizations. She was invited to become a co-chairman of a mini track conference for ECKM 2010 and appointed as the committee member and paper reviewer for the ECKM 2009 and ECKM 2010. She was appointed as an editorial board member and paper reviewers for UiTM's Journal of Administrative Science and Social and Management Research Journal.

Mohd Wira Mohd Shafiei is a senior lecturer at the School of Housing, Building and Planning, Universiti Sains Malaysia. He studied quantity surveying formally in Sheffield Hallam University, before obtaining his MSc (Eng), and PhD in Construction Management from the School of Civil Engineering, University of Leeds. He has been deeply involved in teaching and developing new teaching materials in quantity surveying and construction management related subjects. His main research interests are marketing, green buildings, and strategic management in the quantity surveying profession and construction industry. He has also been involved in several research and consultancy projects within the fields of construction marketing and quantity surveying.

Mauro de Mesquita Spinola, Ph.D. and Associated Professor, has obtained his electronic engineering undergraduation at ITA – Instituto Tecnologico da Aeronautica, Computer Systems Master Degree at INPE – Instituto Nacional de Pesquisas Espaciais, Engineering Ph. D. and Associated Title at Escola Politecnica da USP. His main research and educacation focus is IT process management. He is undergraduation and graduation Professor since 1995. He developed relevant IT research and development projects related to traffic control system development and systems engineering process improvement. Recent researches are focused on implementation effectiveness of IT models and knowledge management modeling and measuring.

Lynnda Wagner was an honours student in information systems at the Department of Information Systems. She did her research whilst in South Africa during her stint as operations director for "Affordable Medicines for Africa". Her first degree was a Bachelor of Science in Business Administration majoring in both Human Resources and International Business at the California Polytechnic University, Pomona. Her professional IT experience is in ERP and HR systems.

Nor'Aini Yusof is an associate professor and a deputy dean at the School of Housing, Building and Planning, Universiti Sains Malaysia (USM). She has a Diploma in Valuation and a Bachelor of Surveying (Hons) in Property Management from Universiti Teknologi Malaysia, an M.Sc. in Project

Management and a Ph.D in housing from USM. Her research interests are housing development and management, and sustainable tourism development. She has published in national and international journals and conference proceedings. She is a member of Editorial Board for Journal of Construction in Developing Countries and an Assistant Editor for International Journal of Organizational Innovation. She regularly reviews technical papers for international and local journals, and conferences. At present she supervises 5 Master by research and 5 Ph.D students. Two of her research students had successfully completed their Master by research and Ph.D degrees respectively. She is also a member of expert panel for NAPREC Grant Scheme and a committee member of USM's tourism cluster.

Zawiyah M. Yusof is currently an Associate Professor at Department of Information Science, Faculty of Information Science and Technology, Universiti Kebangsaan Malaysia. She obtained her PhD in Records Management from Department of Information and Library Science, Aberystwyth, United Kingdom. Her research interests are in Records and Information Management; Knowledge Management, Information Policy; and IT and Society. She is the author of Issues in Records Management book published by The Universiti Kebangsaan Malaysia Press.

Laura Zapata-Cantú is associate professor of Organizational Learning and Business Consulting at Tecnológico de Monterrey (Mexico). Dr. Zapata holds the research Chair in European Studies. Among her main research interests are knowledge management in small and medium enterprises, intellectual capital and organizational learning. She has presented her research work at international conferences and has several publications in international journals such as International Journal of Manpower, Estudios de Administración (Universidad de Chile), European Journal of International Management, Journal of Knowledge Management. Since 2006, she is member of Mexico's National Researchers System.

Index

D

E

F

G

H

I

J

K

R

S

Breinigsville, PA USA
12 October 2010
247133BV00003B/6/P

9 781616 928865